T0184347

Logistic
Regression Models

CHAPMAN & HALL/CRC
Texts in Statistical Science Series

Series Editors

Bradley P. Carlin, *University of Minnesota, USA*
Julian J. Faraway, *University of Bath, UK*
Martin Tanner, *Northwestern University, USA*
Jim Zidek, *University of British Columbia, Canada*

Texts in Statistical Science

Logistic Regression Models

Joseph M. Hilbe

Jet Propulsion Laboratory
California Institute of Technology
and
Arizona State University
Tempe, U. S. A.

CRC Press
Taylor & Francis Group
Boca Raton London New York

CRC Press is an imprint of the
Taylor & Francis Group an **informa** business

A CHAPMAN & HALL BOOK

Chapman & Hall/CRC
Taylor & Francis Group
6000 Broken Sound Parkway NW, Suite 300
Boca Raton, FL 33487-2742

First issued in paperback 2017

© 2009 by Taylor & Francis Group, LLC
Chapman & Hall/CRC is an imprint of Taylor & Francis Group, an Informa business

No claim to original U.S. Government works

ISBN 13: 978-1-138-10671-0 (pbk)
ISBN 13: 978-1-4200-7575-5 (hbk)

This book contains information obtained from authentic and highly regarded sources. Reasonable efforts have been made to publish reliable data and information, but the author and publisher cannot assume responsibility for the validity of all materials or the consequences of their use. The authors and publishers have attempted to trace the copyright holders of all material reproduced in this publication and apologize to copyright holders if permission to publish in this form has not been obtained. If any copyright material has not been acknowledged please write and let us know so we may rectify in any future reprint.

Except as permitted under U.S. Copyright Law, no part of this book may be reprinted, reproduced, transmitted, or utilized in any form by any electronic, mechanical, or other means, now known or hereafter invented, including photocopying, microfilming, and recording, or in any information storage or retrieval system, without written permission from the publishers.

For permission to photocopy or use material electronically from this work, please access www.copyright.com (http://www.copyright.com/) or contact the Copyright Clearance Center, Inc. (CCC), 222 Rosewood Drive, Danvers, MA 01923, 978-750-8400. CCC is a not-for-profit organization that provides licenses and registration for a variety of users. For organizations that have been granted a photocopy license by the CCC, a separate system of payment has been arranged.

Trademark Notice: Product or corporate names may be trademarks or registered trademarks, and are used only for identification and explanation without intent to infringe.

Library of Congress Cataloging-in-Publication Data

Hilbe, Joseph.
   Logistic regression models / Joseph M. Hilbe.
      p. cm. -- (Chapman & Hall Texts in statistical science)
   Includes bibliographical references and index.
   ISBN 978-1-4200-7575-5 (alk. paper)
   1. Logistic regression analysis--Data processing. I. Title. II. Series.

QA278.2.H53 2009
519.5'36--dc22
                                             2008042667

Visit the Taylor & Francis Web site at
http://www.taylorandfrancis.com

and the CRC Press Web site at
http://www.crcpress.com

Contents

Preface

Logistic regression has become one of the most used statistical procedures employed by statisticians and researchers for the analysis of binary and proportional response data. All major statistical software applications have a basic binary logistic regression procedure as a primary offering, and most packages provide their users with the capability of modeling a wide variety of more complex logistic routines. This text is specifically devoted to providing students and researchers with an overview of the full range of logistic models, including binary, proportional, ordered, and categorical response regression procedures. It is intended to serve as a handbook that researchers can turn to as a resource for applying this range of models to their data.

Logistic Regression Models examines the theoretical foundation of each type of logistic model discussed and offers the reader a solid description of how each type of model is to be established, interpreted, and evaluated as to its goodness of fit. Before beginning our explication of the various models, however, we provide an overview of the foremost methods used in estimating logistic models—maximum likelihood (ML) estimation and iteratively reweighted least squares (IRLS) estimation. Other estimation methods exist, for example, quadrature, simulation, Monte Carlo, Firth (1993), and exact methods, but we shall delay discussing them until needed for a particular model. For example, exact logistic regression is based on a complex hierarchy of permutation methods and has only been available to the majority of statisticians within the past 10 years. However, exact logistic methods can be used to model data that otherwise cannot be modeled using traditional methods. Because it can play such an important role in modeling, we shall devote an entire chapter to it.

The basic logistic regression model—the one commonly found in general-purpose statistical packages—has a binary response, or dependent variable, and one or more predictors or explanatory independent variables. Predictors may take the form of continuous, categorical, or indicator/binary variables. In fact, like ordinary least squares or linear regression, logistic regression models need not have any predictor. Instead, they can be used to model only the constant. Several important statistical tests employ a constant-only model as a term in their formulae. But more of this later.

Logistic regression models are used to understand data from a wide variety of disciplines. Such models are perhaps best known in the medical and health fields with the notion of odds ratio applied to studies such as smoking, cardiovascular ailments, and a host of other "risk" events. This brings us to another important topic of discussion: How is the odds ratio to be understood? There is considerable controversy among statisticians on the relationship of odds ratio to risk ratio. We shall devote comparably substantial space to untwisting the concepts involved and we will offer suggestions as to how logistic regression models are most appropriately interpreted. We will discover that how we understand the odds ratio will in part depend on how the model is structured as well as the type of logistic model being used.

Aside from the medical and health fields, logistic models are commonly used in all of the social sciences, in economic research, and in the physical sciences. Binary logistic regression, which some consider as the base logistic model, is one of the statistical tools used in Six Sigma quality control analyses, and it plays an important role in the area of data mining. With the exception of multiple linear regression, binary and proportional, or binomial, logistic regression is perhaps used more than any other statistical regression technique for research of every variety. There is little question of its importance in current scientific research.

Stata will be used to display most modeling results, as well as ancillary output that is necessary when developing and evaluating the various models we discuss in this text. The reason for selecting Stata is its scope of capabilities. Nearly every model we discuss can be analyzed using Stata. If a newly developed capability or test appears in the literature, it is easy to write a program for it using Stata's proprietary higher language. Unlike SAS or SPSS, Stata's official statistical procedures are written in both C and as ado files. Ado is understood as an automatic do, or batch, program. It runs upon calling or selecting the command name and appropriate options. Stata's `logit` command, the original procedure for estimating logistic models, is written entirely in C. The `logistic` command, however, is written as an ado file, but calls `logit` for its core estimation. The `glm` command, Stata's generalized linear models procedure, is entirely written as an ado file, with all programming code available for viewing. We shall employ all three of these commands for displaying results of the basic binary response logistic model. The `glm` command, though, also will be used for binomial or proportional response logistic models.

More complex logistic models, that is, ordered and multinomial logistic regression, continuation ratio logistic regression, and so forth, are all modeled using Stata's maximum likelihood capabilities. Stata has developed comprehensive tools for estimating models using maximum likelihood (ML). Several types of ML mechanisms are available and can be employed to create a wide range of statistical procedures. Examples of the various methods used by commercial statistical software for estimating logistic models will be examined in Chapter 3. These methods, in turn, can be implemented by readers to develop other models that may be of interest to them but that are not available in the commercial Stata package. For example, although it is possible to write the command in Stata, alternating logistic regression (ALR), which is used for modeling binary response population averaged longitudinal or clustered data, is not part of the commercial package nor has it been published by a user. It is, however, an option available in SAS's GENMOD command, authored by Gordon Johnston.

Students in my statistics.com logistic regression course recommended adding an appendix providing an overview of basic Stata commands and examples of how to use them for statistical analysis. They also suggested that I provide a list of commonly used commands, in particular, commands that are used for examples in this text. Students who do not use Stata and know nothing about it also thought that this suggestion would be helpful to them for converting examples to their preferred statistical application. By better understanding the commands used in the text and how they generate the example output, they believed that it would be easier for them to duplicate what is displayed in the text. I have, therefore, written Appendix A, which provides such an overview as well as a list of the commands used in this text. Appendix B provides an extensive list of Stata commands for estimating the models discussed in this text. Appendix C lists all of the Greek letters and how they are typically used for the models discussed in the book. Appendix D displays code for a complete binary response logistic regression. The code can be amended to extend estimation to alternative models. I have placed a derivation of the beta binomial distribution from its two constituent distributions in Appendix E, and I give the log-likelihood of the adaptive Gauss–Hermite distribution in Appendix F. These have been provided in this book because they are difficult to locate elsewhere and can be of value to more advanced readers. Appendix G provides a listing of the various data sets and user-created Stata commands used in the text. Finally, Appendix H provides an overview of the meaning and application of marginal effects and discrete change. Both concepts are particularly important in the field of economic research and in the social sciences.

I have placed R commands at the end of all but a few chapters that are associated with the Stata commands used to display example output. The R code is provided to allow R users an understanding of how they may duplicate output displayed in the text. R is a free-ware application that can be accessed at www.r-project.org. Interested readers can obtain information on the software by downloading the R manual without cost. R code is not given for a corresponding Stata command or calculation when none exists in the default libraries that come with R (which are not in the libraries I have found related to those commands) or because none currently exists. No R code is given for examples in Chapters 12, 14, and 15 because the majority of models discussed have no R support. As support becomes available, I shall list the code on the book's Web site, as well as in the *Solutions and Support Manual,* which will be available from the publisher.

I assume that everyone reading this text has a solid background in linear regression—in particular multivariable linear regression. If you wish to get full benefit from the discussion of the estimating equations that produce parameter estimates, standard errors, and so forth, a background in both basic differential and integral calculus as well as probability theory is necessary. An understanding of partial differentiation and matrix calculus is necessary to gain a working mastery of maximum likelihood estimation (MLE) and Fisher Scoring, upon which traditional GLM estimation is based. However, in order to become fluent with logistic model building, interpretation, and goodness-of-fit analysis, a knowledge of linear models and good reasoning ability should suffice.

The majority of examples in the text were developed using Stata statistical software. A few examples are provided in R and SAS. Data sets used for examples in the text can be accessed on the publisher's Web site for the book. Data sets will be available in Stata, R, Excel, SAS, SPSS, and Limdep format.

It is important to remember that this volume is concerned with learning how to use logistic regression and extended logistic models for research. It is not about learning Stata or any particular statistical application. The methods used and examples displayed are intended to explain the concepts and techniques of successful logistic modeling. Many readers prefer to follow along with the text and try modeling the same examples as shown in the discussion. From my perspective, this is not of paramount importance. One can develop a thorough understanding of the methods discussed without duplicating what is in the text. Individual readers may find such duplication to be helpful to their learning, but it is secondary to the purpose of the book.

I wish to thank Peter Bruce, CEO of statistics.com, who first suggested that I expand material I had prepared for my Web course on logistic regression into a book. I have twice used early incomplete draft versions of this text for the course and have been grateful for the feedback obtained from students, many of whom themselves teach courses related to logistic regression or who have used or desire to use logistic models for their research. Several points discussed in this text are not given much attention in other books dealing with logistic regression. This has been a direct result of such feedback. Although many students provided excellent suggestions, several have spent substantial time researching and discussing with me certain topics. I owe them considerable thanks for their efforts.

Three statisticians of note kindly offered their time and expertise in helping this book become more than it would be by my efforts alone. Each contributed in different ways, but they all contributed greatly to its fruition. I wish to thank them for their efforts.

Garry Anderson, University of Melbourne, Australia, read the various chapters of the manuscript as they developed, offering valuable suggestions, insight, and references. He also re-read the manuscript at near completion, ensuring that it reflected what I intended to express. Professor Justine Shults, Center for Epidemiology and Biostatistics, University of Pennsylvania, carefully read through the entire manuscript near the end of its completion, checking for typos, inconsistencies, or where wording might be better expressed. She also offered software she had developed and advice on areas related to GEE. Dr. Robert LaBudde, Least Cost Formulations, not only provided feedback on the earliest stages of the book, but also revisited it at its final stages, offering suggestions for its improvement. He also contributed many of the R commands associated with the Stata commands used for examples throughout the text.

I also thank other statisticians who contributed to the book, including Professor Sudha Purohit, University of Pune, India, who helped with R code I wrote for generalized additive models; Jichun Xie, University of Pennsylvania, who helped with R code related to quasi-least squares regression; Professor T. V. Ramanathan, University of Pune, India, teaching assistant for my logistic regression course with statistics.com; Professor Paul Wileyto, University of Pennsylvania School of Medicine, who provided suggestions regarding the beta binomial and associated models; Kristin MacDonald of Stata Corporation who provided valuable suggestions on creating various simulations for proportional slopes models; Professor J. Scott Long, University of Indiana, for his suggestions on ordered response models and survey data; and Professor Richard Williams, University of Notre Dame, who provided useful discussion and reference materials for areas related to categorical response logistic models. Professor Williams also read Chapters 10 through 12, making certain that formulae were correct and that the concepts were properly expressed. I thank them all for their kind help.

Roger Payne, VSN International, U.K.; Gavin Ross, Rothamsted Experimental Station, U.K., retired; and John Nelder, Emeritus Professor, Imperial College, London; provided me with excellent historical background and, in the case of Professor Nelder, many hours of face-to-face discussion related to the history, theory, and application of generalized and hierarchical generalized linear models. Their personal experiences and pages of recollections regarding the initial software implementations of logistic regression proved to be invaluable for the writing of Section 1.3.

I also wish to express my appreciation to Rob Calver, my editor, who encouraged me to write this book when I had previously sworn that I would not write another. He provided me with texts I needed to research particular areas related to the book. This, too, has proved to be invaluable. I also express sincere appreciation to Pat Branton of Stata Corporation for her truly invaluable assistance and support over the years. Lastly, I'd like to thank the team at Taylor & Francis, especially Joette Lynch, and the team at diacriTech, particularly Nick Maier and Carolyn Albee, for helping make production of this book possible.

I alone am responsible for any mistakes or errors found in the book. I have tried to make certain that no errata exist, but if such is identified, please feel free to contact me. Errata are posted to the book's Web site at: http://www.crcpress.com/e_products/downloads/download.asp?cat_no=C7575.

Finally, I express my love and appreciation to my wife, Cheryl, and sons Michael and Mitchell, for having to live through my writing of another book. Few people realize the sacrifice in time and attention that are taken from family members when writing a book such as this. My grandsons, Austin and Shawn O'Meara, and my daughter, Heather, also lost time I would have otherwise given to them. Finally, I also wish to express my appreciation to Sirr Hilbe, who sat close by my side during nearly all of this book's writing, and who provided me with diversions when needed. Sirr is a white Maltese pup, but that takes nothing away from his unique contribution to this book's creation.

Note: The preferred method of converting Stata data files into R differs from what is shown in the text when using Stata versions 10 and later. Using the **medpar.dta** Stata file, stored in c:/data, as an example, I suggest using the following code:

```
library("Hmisc")
medpar <- stata.get("c://data/medpar.dta")"
```

1

Introduction

1.1 The Normal Model

Regression is a statistical method by which one variable is explained or understood on the basis of one or more other variables. The variable that is being explained is called the **dependent, or response, variable;** the other variables used to explain or predict the response are called **independent variables.** Many times independent variables are simply referred to as predictors. I shall use the terms response and predictors throughout this volume.

Linear regression is the standard or basic regression model in which the mean of the response is predicted or explained on the basis of a single predictor. The basic model is easily extendable such that it becomes a multivariable linear model, that is, a linear regression having more than one predictor.

Since linear regression is used to predict the mean value of the response, there will be error involved. Models do not—or at least should not—perfectly predict the mean response. In the case of linear regression, the error is normally distributed—normal meaning as a bell curve. The response is, likewise (at least theoretically), normal for a given value of the predictor(s). That is why the linear model is many times referred to as **Gaussian regression.**

In short, normal, or Gaussian, regression assumes that the response and error terms are normally distributed. Normal distribution, theoretically, includes both positive and negative numbers with 0 as the theoretically ideal mean. The linear regression model, however, is extremely robust to violations of the strict adherence to these requirements. What is requisite, though, is that the response is continuous and allows the possibility of negative values. It is also required that the variance is constant.

1.2 Foundation of the Binomial Model

What if the model response is no longer continuous, but rather binary, taking the values of 1 (success) or 0 (failure)? First, the error terms will take the values of 1 and 0 as well. Second, the variance becomes nonconstant. And third, the predicted mean values can fall outside the range of 1 through 0. As we shall see a bit later, a primary feature of the logistic model is that the **predicted mean,** or **fitted value,** is a probability ranging from having a value of 1—or perfect prediction—to having a probability of 0. In actuality, predicted probabilities only rarely take the values of 1 or 0, but rather take the values between. Regardless, modeling a binary response using Gaussian or normal regression—commonly referred to as **linear probability regression**—violates the fundamental assumptions upon which both the linear and binomial models are based.

These violations make the Gaussian or linear regression model inappropriate for the analysis of binary responses. The linear probability model was once fairly popular when no other means of modeling binary response data were readily available. But now all major commercial statistical software applications have the capability to satisfactorily model these types of data situations.

Logistic regression is the foremost method used to model binary responses. Other models capable of handling binary responses exist as well; probit, complementary loglog, and loglog are the most noteworthy examples. However, only logistic regression can be used to estimate the odds ratio for the model predictors. This is an important feature that plays a vital role in areas such as medical research statistics. We shall discuss this aspect of the logistic model in more detail as we progress in this text.

Although some binary logistic regression applications accept a binary response of 1 or 2, the internal mechanism of the algorithm converts these values to 0 and 1. Typically, the same order is maintained, so that 1/2 becomes 0/1. The reason for this has to do with the underlying 0/1 **Bernoulli distribution.** The Bernoulli distribution is named after Jakob (or James) Bernoulli (1654–1705), and is, as we shall examine in some detail, a subset of the more general binomial distribution. The distribution is defined by two factors: the comparative number or ratio of 1s to 0s, and the number, n, of observations in the distribution. The 1s are usually considered to be some measure of success or meeting some defined criteria. Values of 0 then represent failure to meet those criteria. What we mean by criteria can take nearly any form and is to be understood in the context of the particular study model. For example, if we are modeling death on smoking, 1 would normally be assigned to someone who died, 0 to cases where the person did not die. Note that there are software applications, such as SAS, where 0 is understood to be the criterion rather than 1, but this is a rarity.

It might be helpful at this point to provide a bit of historical background and context to the binary and binomial logistic models we shall be addressing in this text. Historical reference will be given to each of the extended logistic models as they are presented. Since statistical modeling is so closely tied to the software used to estimate parameters, standard errors, and so forth, I shall also give a brief overview of some of the major software applications that have been and are being used by statisticians to model this form of data.

1.3 Historical and Software Considerations

First, the **logistic model,** which is commonly referred to as the **logit model** in econometrics, was initially introduced by Joseph Berkson in 1944. Berkson derived the logit model from his work with probit regression and bioassay studies, which originated with Chester Bliss 10 years earlier. Rather than employ the sigmoid function used in the probit algorithm, which is the **inverse cumulative distribution function** (CDF) of the normal or Gaussian probability distribution, Berkson argued that the logistic function could also be used to effectively model binomial data. Both models can be used to model probability with little difference in output (Hardin and Hilbe 2007). For his new model, Berkson amended the term probit, substituting an L in place of the initial P to represent the logit model. Both logit and probit, as well as another binomial model called complementary loglog, were estimated using a weighted minimum *Chi2* least squares algorithm.

It was not until 1949 that George Barnard, while at Imperial College in London, developed the term **log-odds** in the context of Berkson's logit model. Given that p is the probability of success, that is, the probability of the response taking a value of 1, the odds of success is defined as $p/(1-p)$. The log-odds is $\ln(p/(1-p))$. It is the linear predictor, $x\beta$, of the binary logistic model and, therefore, plays an essential role in logistic modeling.

Barnard et al. were also the first to use the term "likelihood inference" in 1962, although the concept and methodology were employed by Ronald Fisher as far back as 1922. Fisher is also credited with the development of maximum likelihood estimation as well as a host of other important statistical concepts. Finney (1952) was the first to program a maximum likelihood probit, but it was not until 1972 that Rothamsted's Gavin Ross first created a maximum likelihood program that estimated both probit and logit models. In any case, by the 1960s, statisticians were using both the probit and logit models to estimate binary and binomial response models, albeit mostly using a variety of weighted least squares methods of estimation.

Contemporary maximum likelihood estimation routines, found in most major statistical software applications, are examples of sophisticated software programming. Separately developed maximization procedures exist in SAS, Stata, Limdep, and GAUSS but are wholly integrated into their respective applications. One can use the software's maximum likelihood module to create one's own maximum likelihood-based statistical model. Maximum likelihood algorithms are all modifications of the basic Newton–Raphson algorithm, which will be described in Chapter 3. Until recently, models that were estimated using maximum likelihood could prove tenuous. Starting values were required to be specified at the outset, and many times model data were structured such that convergence would fail. In the days when batch programs defining the data and model were encoded on cards and submitted to a remote mainframe, failure of model convergence or even failure to provide appropriate parameter estimates and standard errors cost money and time.

As a result of the difficulties that were commonly incurred when modeling binary response data using maximum likelihood estimation, many statisticians instead used linear probability regression when modeling binomial data. In fact, the linear probability model was used directly with binary response data; grouped or proportional binomial logistic models were modeled as weighted binary models, with a bit of prefatory data management required. The estimates produced using this method were biased and, as previously noted, could result in fitted probability values outside the 0–1 range.

However, the most common alternative method of modeling binary response data was not a regression model at all—it was multivariate logistic discriminant analysis. Discriminant analysis allowed statisticians the ability to model binary response data, and the procedure was able to model a categorical response variable with three or more levels. Since the method is still used in some disciplines and since discriminant analysis can, in fact, model data situations that are unavailable to traditional logistic regression, we shall later devote a section to its examination. It should be mentioned, though, that like all other statistical procedures, tests, and models, discriminant analysis carries with it various assumptions. Perhaps the most harsh is the requirement that the data subsumed under each response level have equal variances. A corresponding criterion of multivariate normality for each level is also required. Discriminant model results are fairly robust against criteria violations of variance and normality, but the model lacks certain statistical results that are valuable to researchers. We will elaborate on these relationships when the model is later discussed in full.

1972 was a signature year for statistical modeling. It was in that year that John Nelder and Robert Wedderburn developed the modeling methodology referred to as **generalized linear models** (GLM). Generalized linear models is a covering algorithm allowing for the estimation of a number of otherwise distinct statistical models within a single framework. Any statistical model that is derived from a probability distribution that is a member of the single parameter exponential family of distributions can theoretically be estimated as a generalized linear model. We devote considerable attention to this class of models in Chapter 3 because the binomial distribution is a single parameter exponential family distribution and the logit, or log-odds, function is its canonical or natural link. Given that the binary Bernoulli distribution is a variety of binomial, it is plain to see that logistic regression is a GLM. In fact, both binary and binomial response logistic regression are GLMs.

The value of being able to model logistic regression as a GLM has to do with the simplicity of the **IRLS estimating algorithm.** IRLS estimation is a type of maximum likelihood estimation and shares many of its features. However, its unique characteristics result in few initial value problems and a robust estimation process; that is, GLMs typically suffer few convergence difficulties. When they do it is usually a result of attempting to estimate highly unbalanced data, or models with too few observations compared with predictors, or some other situation where full ML estimation would likewise fail.

In 1974, Nelder, at that time a statistician with Wedderburn at the Rothamsted Experimental Station north of London and a visiting professor at Imperial College in London, coordinated a small cadre of Royal Statistical Society (RSS) statisticians in developing a specifically designed GLM software application. Called GLIM, an acronym for generalized linear interactive modeling, the software was published and maintained by the RSS's Working Party on Statistical Computing. Oxford University Press published the first GLIM manual. With such noted organizations associated with the software, GLIM quickly made it to the marketplace. Because the normal or Gaussian model is also a GLM, the software became a primary statistical package, first throughout the old British Commonwealth nations and then a short time later in the U.S. and the rest of the statistical world. The RSS turned over publication and distribution of GLIM to the Numerical Algorithms Group (NAG), a well-known company with a solid reputation for producing mathematical software, in 1993. After publishing the Version 4 upgrade in 1994, GLIM 4 was discontinued. However, NAG continued to develop an even older statistical application called GenStat, which is described more a bit later.

Wedderburn was not able to continue the work he helped initiate with Nelder, who was some 17 years his senior. Wedderburn, who also developed the concept of quasi-likelihood as an extension to the generalized linear models algorithm, died in 1975 at the early age of 28 while on a canal barge trip with friends in England. He had not known that he was allergic to bee stings, and he succumbed to anaphylactic shock after being bitten. We will discuss quasi-likelihood models later in the text. They play an important role in extended GLMs and form the basis of the important longitudinal models known as **generalized estimating equations** (GEE) (Hardin and Hilbe 2003).

GLIM was easy to use and was programmable. Rather than having to write in FORTRAN, statisticians could expand on the GLIM package by writing their own procedures, called macros, and functions using GLIM's own higher language. Creative programming allowed good programmers to create modeling capabilities that went well beyond basic GLM estimation. For instance, GLIM programs were written for ordered and multinomial logistic regression and even for models as complex as fixed and random effects logistic models. Separate stand-alone FORTRAN programs soon gave way to a general purpose programmable statistical package that was easily expandable, inexpensive to operate, and unified in approach.

When the first IBM personal computer was sold in August 1981, there were several general purpose statistical packages on the market; GLIM, SAS, SPSS, GenStat, Systat, and BMDP were the foremost. Only GLIM and GenStat provided GLM capability. SAS did not develop a GLM procedure until 1994 (GENMOD, in version 6.09). SPSS had no such capability until 2006 (GENLIN,

Version 15), and both Systat and BMDP still do not have it. As previously mentioned, GLIM was discontinued with the release of its fourth version in 1994. NAG put its backing to GenStat instead, which was a general statistical package like SAS, Stata, and SPSS.

Nelder first initiated the idea of GenStat while serving as a visiting professor in Australia from 1965 through 1966. He began to write code with Graham Wilkinson of the University of Adelaide at that time, but actual development of the package did not occur until he returned to Rothamsted Experimental Station (RRES) and took over as head of the Statistics Department in 1968. A GenStat Working Group was formed with RRES statisticians, and in 1970, following Wilkinson's appointment to RRES, the initial GenStat product was published. The earliest GenStat application was primarily devoted to ANOVA and related procedures. A GLM procedure was ported over from GLIM into GenStat in 1976, which signaled the first time logistic regression was made available to its users. The popularity of the application necessitated RRES joining in 1979 with NAG, which marketed the software. Roger Payne took over GenStat development in 1985, following Nelder's retirement. With collaboration from Peter McCullagh, ordered logistic regression was added in 1993. RRES and NAG together created a separate company in 2000 called VSNi, with GenStat as its foremost product. Version 11 was launched in the summer of 2008.

Together, GLIM and Genstat did much to popularize the logistic model and make it available to the statistical community—but as a family within GLM. What is interesting is that Rothamsted statisticians were the first to implement both maximum likelihood and IRLS methods for the estimation of logistic regression models.

Two software applications that were first developed in the mid-1980s ended up having a considerable influence on the future of statistical modeling. S, created at AT&T Laboratories under the direction of Trevor Hastie, was a modular command line statistical programming environment. Hastie authored the glm() command, which was one of S's original models. **glm()** is a traditional GLM algorithm, incorporating many of the options that were in GLIM. No maximum likelihood logistic model was developed for S, even after ownership of the code was passed in 1988 to Statistical Sciences, a commercial software firm inaugurated by R. Douglas Martin, a statistics professor at the University of Washington, for the purpose of developing and selling the software. Commercial S was marketed as S-Plus and quickly became a highly successful enterprise. Academics were particularly fond of the software. S-Plus is currently manufactured by Insightful, has the same **glm()** command as in the original S version, and no separate logistic maximum likelihood implementation.

R is one of the most popular statistical packages used in academia. Being freeware adds to its attraction. R is the descendent of S and its programming code and style are very similar in appearance to its parent language. R was initially developed by Ross Ihaka and Robert Gentleman of the University of Auckland, New Zealand, whose names inspired calling the software R. Their goal was to recapture the freeware concept and open architecture of S, employing programming schema with which many academics were already familiar. Following a period of development and trial/feedback, R was released for general use in 1999 and is available for nearly all operating system formats. The software is now managed by the R Development Core Team and has become a required statistical application at many universities.

Nearly all R statistical procedures, functions, and modules are submitted to the R library by users throughout the world. Although there are a number of extended logistic based models that are estimated using maximum likelihood, R's only basic binary and binomial logistic capability rests with **glm()**.

```
> logitexp<- glm(Y ~ X1*X2, family=binomial(link="logit"))
> summary(logitexp)
```

glm() is a function in the stats library module, which automatically comes with the downloaded software. Its logic is nearly identical with S and S-Plus. We shall use it to show several of the examples later in the text. In addition, R code is provided at the conclusion of each chapter, corresponding to the Stata code used in the text to produce example output. Normally duplicated code is not provided, and quite a few Stata commands have no R equivalent. However, where there are equivalent commands for producing the same or similar output, it is provided. I have also provided R code for creating the majority of graphs found in the text so that they may be compared to Stata graphical output.

The second influential statistical application to be initially developed in the mid-1980s commenced as a commercial package and has remained as such. Following on what was mentioned in the Preface, Stata was founded by Bill Gould, Bill Rogers, and several other econometricians in the proximity of U.C.L.A. With the inauguration of the *Stata Technical Bulletin* in March 1991, users worldwide submitted commands, functions, and data management programs to the bimonthly publication, which, over time, added a host of new capabilities to the commercial package. Those commands that were not made part of the commercial software were still available to users. Today Stata has a prodigious number of statistical commands, including, as previously mentioned, a wide range of logistic regression models.

Since we shall be employing Stata for the majority of examples used in the text, I should mention something of the origin of its basic logistic commands. Stata's binary logistic model, **logit,** was written into the Version 1 executable in 1984. A few years later a binomial, or grouped logistic model, **blogit,** was written with **logit** being embedded in an ADO program that restructured the data as a frequency weighted binary logistic model. **blogit** does not directly estimate a binomial logistic model. In late 1990 I developed an ADO command called **logiodds2,** which was distributed to the Stata community in the January 1991 update, or Support Disk. It was upgraded and published with the first *Stata Technical Bulletin* (STB), which appeared in May of that year. **logiodds2** implemented the newly developed fit statistics designed by Hosmer and Lemeshow. It too utilized **logit,** but produced estimated odds ratios as the default output and provided a number of *m*-asymptotically based fit statistics. **logiodds2** was soon enhanced, renamed **logistic,** and made part of the commercial software with the next release.

The initial **glm** command was added to Stata as a user-authored command in 1993. I initially wrote the command to test other GLM software for an article I was writing. Written entirely as an ADO program and published in the STB, **glm** utilized IRLS as its method of estimation. The command was modified by Royston in 1994 and was later totally rewritten in 2001 by James Hardin and myself as part of our Stata Press book, *Generalized Linear Models and Extensions*. This new implementation of **glm** became part of the commercial Stata package at that time. Until then, only the two user-authored versions of *glm* were available. The Stata Web site, www.stata.com, provides a wealth of information about Stata, Stata's bookstore and publications, tutorials, and links to other statistically related sites.

Finally, I should mention three other software packages: Statistica, SPSS, and LIMDEP. Statistica is a general purpose statistical application with deep roots in the data mining profession. In fact, Statistica, together with its separately sold Data Miner package, is regarded by many as being one of the most comprehensive data mining applications on the market. Statistica's logistic regression command is one of its GLM modeling commands, called GLZ, and except for using the Nonlinear Models module for logistic regression, GLZ is the only route to the model.

SPSS also has extensive data mining capabilities, but is perhaps best known as one of the foremost all-round statistical applications on the market. SPSS is an acronym for Statistical Package for the Social Sciences and is one of the oldest commercial statistical packages. SPSS users can employ logistic models to their data using either of two procedures: Logistic and Genlin. The latter is SPSS's generalized linear models (new with Version 15) procedure, of which both binary and grouped logistic regression are members.

LIMDEP, mentioned earlier, is a popular econometrics package that allows estimation of nearly all of the models we discuss in this text. The term LIMDEP is an acronym for LIMited DEPendent variable models and provides extensive modeling capabilities for discrete response regression models. All of the models we discuss are classified in this manner. The application also allows users to model a variety of continuous response models as well, particularly those of interest to econometricians. LIMDEP and its younger side application, NLOGIT, are authored by William Greene of New York University and are products of Econometrics Software. NLOGIT consists of a suite of commands providing the ability to estimate discrete choice models. Initiated in 1996 as a nested logit extension to LIMDEP's multinomial logistic model, NLOGIT is now considered a superset of LIMDEP.

Logistic regression modeling evolved from the binomial/binary-based maximum likelihood method of estimation. With the implementation of generalized linear models and the iteratively re-weighted least squares (IRLS) algorithm in the early 1970s and the arrival of desktop personal computers a short time thereafter, the road was open to the widespread use of logistic regression models with study data. In 1974 David McFadden expanded the logit model to what he called a discrete choice model (McFadden 1974b). Those in biostatistics and epidemiology prefer the term multinomial over discrete choice, which is still used in the econometric literature, but they are the same model. Other extensions to the logit model were developed in the same article, including the conditional fixed effects logit model, which allowed the logistic modeling of longitudinal and clustered data—also called panel data. In 1975 Zavonia and McElvey developed the initial implementation of ordered response regression models, which was enhanced in 1980 by Peter McCullagh of the University of Chicago. McCullagh named the ordered categorical response logistic model as the proportional odds model (McCullagh 1980). These models were themselves extended to the variety of models we discuss in this text.

All of the models we have alluded to thus far, with the exception of exact logistic regression, are estimated using some variety of maximum likelihood or IRLS, which is, as we shall see, a form of Fisher Scoring. The latter is a simplification of maximum likelihood that can be applied to the estimation of single parameter members of the exponential family of distributions. Maximum likelihood, therefore, is at the base of both these methods of estimation.

Simulation methods have also been used to estimate recalcitrant models that otherwise would be estimated using maximum likelihood, which have been modified into a variety of algorithms. Greene has employed simulated estimation for many of the complex models in LIMDEP. In Chapter 13 we shall discuss the types of models that apparently require this method of estimation.

Finally, when we are faced with small or unbalanced data and are asked to model it using logistic regression, it may be the case that only an exact logistic model will suffice. Although I shall detail its logic a bit more in the next section, it is important to note here that the model is not estimated using maximum likelihood or any of its variations. The algorithms entail multiple permutations and yield exact standard errors and confidence intervals. Current computing power, that is, RAM and storage capacity, prohibit larger models from being calculated; even medium-sized models are questionable. When model specifications exceed user RAM and storage capabilities, Cytel's LogXact shifts to Monte Carlo estimation and Stata to median unbiased estimation (MUE). We shall describe these methods, together with penalized logistic regression, in the final chapter, discovering that their estimation results usually come very close to those of exact methods. SAS and Stata 10 are the only general purpose statistical software applications having exact logistic regression capabilities. LogXact and Stata also provide exact Poisson models, which can be useful for comparison with binomial logistic models.

As the availability of RAM increases as well as storage, many more models will fall within the range required for exact modeling. How this bears on fit analyses will be the subject of many studies during the coming years. It is an extremely powerful method, and takes modeling into a new era. Many of the modeling methods employed by statisticians—including, of course, many of the methods described in this text—far outstrip what those of us who are older thought would be available in our lifetimes.

1.4 Chapter Profiles

In Chapter 2 we discuss the various concepts related to the logistic model. For example, we explicate the concepts of odds, odds ratio, risk, risk ratio, exposure and confounder predictors, and probability, and how to calculate them for each observation in the model. Other concepts and relationships will be developed as well. This chapter is aimed at providing the reader with the basic terminology and concepts that will be used throughout the text. Chapter 2 is an important prerequisite for understanding the discussion of subsequent chapters. This chapter should be carefully read and the concepts understood before continuing.

Chapter 3 is devoted to the examination of methods of model estimation. We discuss in some detail the methods of maximum likelihood (ML) and iteratively re-weighted least squares (IRLS) estimation. The latter is used in the majority of generalized linear models (GLM) procedures, of which both binary and binomial logistic regression are members. Stata's **glm** command, SAS's GENMOD procedure, SPSS's GENLIN, Statistica's GLZ, and S-Plus's and R's **glm()** commands are all based on IRLS methodology. Maximum likelihood estimation is used for Stata's **logit** and **logistic** commands, SAS's PROC LOGISTIC, and SPSS's LOGISTIC procedures (although few SPSS users now employ the command line when modeling). It will be discovered that both ML and IRLS estimation produce identical parameter estimates and standard errors because IRLS is a subset of ML estimation, one in which ML can be simplified as a result of the binomial distribution being a member of the exponential family of distributions.

Those who are not interested in the internal structure of the estimation process may skip this chapter. However, we suggest that everyone read through the discussion, even if it is in a cursory manner. Having a feel for how the estimation process works, although not the same as having a comprehensive understanding of the methodology, will provide an insight as to the limitations of modeling. That is, knowing something of the estimation process helps one know what can go wrong in the modeling process, why difficulties occur, and perhaps how they can be remedied or at least ameliorated.

Chapter 4 is a natural extension of Chapter 3. Once we understand the foremost methods of estimation appropriate for logistic models, we next need to address the actual algorithms themselves. This chapter describes both the Newton–Raphson type of algorithm, used for maximum likelihood-based logistic models, and the structure of the IRLS GLM algorithm, with emphasis on the binomial logit family and link. Schematic algorithms will be provided and discussed. Like Chapter 3, this chapter may be skimmed through for those with little interest in the algorithms used for logistic regression models. Again, though, you should read the chapter, because it will assist in understanding what can go wrong in the modeling process.

In Chapter 5 we discuss how to develop a logistic regression model and how to interpret the results. The chapter focuses on the binary logistic model, but discussion related to standardized coefficients, scaling and other adjustments to the standard errors, handling missing values, constraining coefficients, and stepwise methods all relate to binomial or aggregated response logistic models as well. Also discussed in some detail are the various link specification tests. These inform us about the linear relationship between the logit, which is the linear predictor, and the predictors.

Expanding on the definitions and relationships related to odds and risk ratios discussed in Chapter 2, we more clearly specify how the two terms relate in the context of logistic regression. Many statisticians interpret logistic regression odds ratios as if they were risk ratios. When the outcome prevalence is low (under 10 percent) equating the two may be acceptable. But when prevalence is higher, the relationship is not as clear. In fact, we shall discover examples where the implied use of a risk ratio when interpreting an odds ratio is totally mistaken. We spend considerable time dealing with how logistic regression parameter estimates and odds ratios are most appropriately interpreted.

Interactions are important to modeling. Chapter 6 details the rationale and implementation of interactions in binary logistic models. Interpretation of parameter estimates and odds ratios in the case of interactions is emphasized. The section focuses on binary x binary, binary x categorical, binary x continuous, and categorical x continuous interactions.

A statistical model that has not been evaluated as to its goodness of fit is worthless. This is perhaps strong wording, but it is indeed the case. If the distributional assumptions upon which the model is based are violated, the resulting parameter estimates and standard errors may be mistaken. Unfortunately, model assumptions are many times violated, but some models are fairly robust to certain of the assumptions. In Chapter 7 we discuss assumptions and goodness of fit tests that assess how close the model is to meeting them. Model assessment generally involves two types of analyses: single value summary test statistics and residual analysis. We provide a discussion of each type of these assessment methods.

Chapter 8 is devoted to an examination of binomial logistic regression. Also termed proportional, aggregated, or grouped logistic regression, the binomial logistic model is based on the full binomial probability distribution function (PDF). We discuss how this model differs from the binary response model and how it is at times valuable to convert a model from binary to binomial format. We also examine in some detail the relationship of binomial logistic to rate parameterized Poisson models. This subject is rarely mentioned in texts on logistic regression, but it is nevertheless an important discussion. We shall discover that some types of data situations are better modeled as Poisson than as logistic. This discussion expands on the Chapter 5 distinctions we made on the relationship of odds ratio to risk ratio.

Binomial logistic models, as well as count models such as Poisson and negative binomial, many times have more correlation in the data than allowed by distributional assumptions. When this occurs we call the model overdispersed. Overdispersion results from a variety of causes. Chapter 9 will discuss both apparent and real overdispersion and will detail the methods used to deal with them from the context of logistic regression. We note here that binary response models are not subject to overdispersion unless the model data were collected over groups or clusters. This type of overdispersion is quite different from overdispersion appropriate for binomial models. We distinguish between these different connotations of overdispersion and introduce the notion of implicit overdispersion.

Beginning with Chapter 10 we discuss extensions to the basic binary and binomial logistic regression models. Primarily, extensions are to the Bernoulli or binary response model, estimated using maximum likelihood techniques. In this chapter we discuss the ordered categorical response logistic models. The response term consists of three or more levels, although it may only have two levels. When this occurs, the proportional odds model, which is the standard ordered response parameterization, reverts to the binary logistic model.

Several parameterizations of ordered response are discussed. The proportional odds model assumes that the model coefficients or slopes are proportional, hence the name. Generalized and partial proportional ordered response models have been developed in the case of nonproportionality, which we describe in some detail.

In Chapter 11 we address the unordered response logistic model, also called the multinomial logistic regression, although the term logistic is usually assumed and not specified for the model. As with the proportional odds model, the multinomial model has a response consisting of two or more categorical levels, although a two-leveled response model is nearly always estimated using traditional logistic regression. The results are identical, including the intercept. One may also model a binary response logistic model using an ordered logistic regression procedure, but the intercept term may have the opposite sign from what is estimated when using a standard logistic regression procedure. We discuss this at some length in Chapters 10 and 11.

Multinomial models often succumb to what is commonly referred to as the problem of the independence of irrelevant alternatives (IIA). IIA is examined as well as tests to determine if a model is subject to the problem. We also discuss the multinomial probit, comparing it to the multinomial logit.

Chapter 12 is devoted to an examination of alternative categorical response models. We begin our overview by discussing the continuation ratio model, which assumes that each level of the response, except the first, must have had the immediately previous or lower level selected before the current level. For instance, one must have been a sophomore before being a junior, and a junior prior to being a senior.

We next examine stereotype logistic regression, followed by heterogeneous choice logistic regression and adjacent category logistic regression. Last, we provide an overview of proportional slopes models in general; that is, we discuss ordered probit, ordered complementary loglog, and other proportional sloped models. We emphasize the testing of proportionality, and how we handle a model that fails the proportionality test. Various synthetic models are demonstrated that can be used to develop models with user-selected estimates and cut points.

Many times the data to be evaluated come to us in longitudinal or clustered format. When this occurs there is a distributional violation of maximum likelihood theory, which requires that each observation in the model be independent of the others. Depending on the structure of the violation and the purpose of the modeling, that is, for what we are trying to determine, statisticians have developed a variety of panel, or longitudinal/cluster, models.

This is the subject of Chapter 13. An overview of each major type of logistic panel model is presented with examples and comparisons to other panel models. We provide guidelines for the interpretation of parameter estimates and ancillary model statistics as well as the goodness of fit assessment statistics that are associated with each type of model.

I should mention that conditional fixed effects logistic regression is used as a basis to examine certain other models that are extensions of it. These include matched case-control and rank-ordered logistic regression.

We also examine generalized estimating equations (GEE) at some length, clearly defining and providing examples for the various correlation structures. The Quasilikelihood Information Criterion, or QIC, fit statistic for GEE models is also examined, as well as a new method called quasi-least squares regression (QLS). QLS is used to model certain GEE models that fail to converge due to infeasible correlation values.

Three additional extensions to the fundamental logistic model are examined in Chapter 14. First we discuss survey logistic regression. Second, we describe scobit regression, which is also known as skewed logistic regression. This method allows the probability weight of binomial/Bernoulli models to be centered away from 0.5. Finally, a presentation of three varieties of discriminant analysis is given with comparisons to standard logistic regression. Logistic discriminant analysis, as briefly mentioned in the previous section, was the foremost method used by statisticians during the 1970s and 1980s to predict and classify data in the same manner as we now employ binary and unordered logistic models. To be sure, there are situations when discriminant analysis is even now preferred over logistic regression, for example, when there is perfect prediction in the model. We shall discuss the derivation of the discriminant model and when it should be used, noting that the probabilities calculated by the procedure are based on the logistic distribution.

Finally, we discuss exact logistic regression in Chapter 15. The model is not estimated using maximum likelihood, nor does it employ IRLS. Rather, exact models are estimated using a highly iterative algorithm that calculates repeated nestings of permutations. At this writing only Cytel's LogXact, Stata, and SAS provide its users with the capability of using exact logistic models. This method of estimation is particularly valuable when there are only a few observations in the data to be modeled or when the data are highly unbalanced. In these situations traditional ML and IRLS logistic regression may fail altogether or at best provide incorrect standard errors. Until exact logistic regression was developed by Cytel, such data were simply not able to be properly modeled. Other models have been developed, however, to produce values similar to exact statistics when there is insufficient computer memory to employ exact methods. Monte Carlo, Median Unbiased, and penalized logistic estimates are discussed and compared.

The Preface provided a summary of what is found in the Appendices. Appendix A provides a rather comprehensive overview of Stata commands and functions as well as the logic of how they are used. This appendix is provided to assist those who are not familiar with the software to better understand the commands used for the majority of examples in the book. Information provided in the other appendices may be updated on the book's Web site.

Appendix H presents an overview of marginal effects and discrete change. These statistics are heavily used in econometrics and other disciplines within the social sciences. The methods explained in this Appendix are applicable to nearly all logistic models discussed in the text.

2

Concepts Related to the Logistic Model

The goal of a logistic regression model is to understand a binary or proportional response (dependent variable) on the basis of one or more predictors. For our purposes now, we will exclude discussion of proportional responses or binomial logistic regression models, concentrating instead on binary responses. We shall delay their discussion until Chapter 8, although from time to time we may need to briefly allude to the more general model.

The response variable, which we will refer to many times as simply y, is binary, parameterized in terms of 1/0. One (1) traditionally indicates a success; zero (0) a failure or lack of success. One can also be thought of as having some property, condition, or characteristic and zero as not having that property and so forth. Success is to be thought of in a very wide sense—it can be in terms of yes/no, present/not present, dead/alive and the like. Most software requires the response value to be in terms of 1/0. However, software such as SPSS will accept 0 versus any positive integer (which the algorithm internally reformats to a 1). SAS's PROC LOGISTIC can be confusing because it reverses the relationship. By default, PROC LOGISTIC predicts 0, which can be thought of as a success. Nearly all other applications predict 1 as the success, so keep this in mind when using this procedure. GENMOD, SAS's generalized linear models procedure, of which the binomial-Bernoulli family with logit link is a member, maintains the traditional 1/0 relationship, with a 1 indicating success. We suggest that GENMOD be used for all logistic regression modeling endeavors when using SAS, unless one wishes to apply Hosmer and Lemeshow (2000) fit and classification statistics to the fitted model, which are only available with PROC LOGISTIC. Our caveat is to take care when interpreting your model and when comparing with other software applications.

Stata is somewhat inconsistent with respect to the 0/1 issue. For a binary response logistic model, **logit**, **logistic**, and **glm** all require that one of the levels of the response is 0. For the **glm** command, the second of the binary levels must be a 1. For **logit** and **logistic**, on the other hand, any value greater than or equal to 1 is construed by the algorithm to be 1. For example, if the response has four levels, 0, 1, 2, and 3, levels 1–3 are converted to 1. If the data set on hand as the response is coded as 1, 2, then the statistician must decide on how to recode these values. To convert 1 to 0 and 2 to 1, the following command may be used.

```
. recode response 1=0 2=1
```

One may leave 1 untouched and recode 2 as 0 by:

```
.recode response 2=0
```

or to recode a response having values of 1, 2, and 3 so that level 1 equals 0, and levels 2 and 3 equal 1:

```
. recode response (1=0) (2/3=1)
```

Care must be taken when recoding a response variable, or for that matter any variable; a notation should be made in the study report regarding the operation performed and the reason for so doing.

There are two major uses to which statisticians employ a logistic model. One foremost use is in the interpretation of parameter estimates as odds ratios; the other relates to the calculation of the fitted values, which can be understood as the probability of 1 (success, death, etc.). Both of these uses play important roles in domains such as health and medical research, credit scoring, social science research, and so forth.

Many statisticians tend to interpret odds ratios as approximations of relative risk ratios. We shall define a binary value risk factor, or predictor of interest, as X, with X_1 indicating that $X = 1$ and X_0 indicating $X = 0$. When an odds ratio is used to assert that having X_1 is twice as likely as X_0 to develop some disease, or that patients having X are 40 percent more likely to die within 48 hours of initial symptoms than patients without X, this is risk ratio language, not odds ratio language. Strictly speaking, a logistic regression model that has been parameterized to estimate odds ratios should be interpreted as asserting, for example, that the odds of the outcome are twice as great for X_1 as compared to X_0, not twice as likely. But there are times when a risk interpretation of odds ratio is justified. We shall discuss the relationship of odds to risk and odds ratio to risk ratio at greater length in Chapter 5. For now we shall concentrate on understanding the nature and implications of odds and odds ratios. Keep in mind, however, that there is ongoing debate regarding the appropriate manner on how to interpret odds ratios, which we discuss later.

2.1 2 × 2 Table Logistic Model

We will use the **heart01** data set for an example of how to interpret parameter estimates, using the Stata statistical package. An overview of Stata is presented in Appendix A, and we refer those needing this background to that source. However, since Stata will be used for displaying the majority of the examples in the text, we list here the model commands to be used in the next few chapters. With respect to the basic binary logistic regression model, users have a choice of the following:

logit *y xvars*
logistic *y xvars*
glm *y xvars, fam(bin)*

The binomial or proportional response logistic model can be estimated using:

glm *y xvars, fam(bin denom)*
glogit *y xvars*
binreg *y xvars*

Unlike many texts, we shall provide the commands used to display the various statistical output. This is provided for those readers who do have Stata and who wish to duplicate the examples. Users of other software can interpolate to their own preferred application. However, R code is given at the end of each chapter to duplicate Stata output insofar as possible.

The **heart01** data set includes, among other variables, two that we shall use for our initial example. First is the response variable, *death*; the other is the single explanatory predictor, *anterior*. In epidemiological terminology, a single binary predictor is called the **exposure** or **risk factor** variable.

Response: death   [1 = die; 0 = not die, within 48 hours of admission]
Risk Factor: anterior [1 = anterior MI; 0 = inferior MI]

Note that *anterior* relates to the patient having had a primary heart injury located in the *anterior* or front area of the heart. When this occurs, the patient is coded with a 1 for anterior. A 0 designates an infarct located at the inferior site. Other infarct locations are excluded from the study. This was done so that there would be a clear differentiation of risk. Lumping all non-anterior sites into a single level risks bias. As it is, though, anterior and inferior infarction sites constituted 88 percent of all non-missing observations. The data are from the national Canadian cardiovascular registry. Dates and all identifying variables have been excluded from the data set we use and from the data made available to users on this book's Web site.

First bring the **heart01** data into memory, keeping only death and anterior. Tabulate *death* by *anterior* to observe the relationship between the two variables. Then calculate their odds ratio. Note that 187 out of nearly 4,700 patients (4 percent) died of a primary anterior or inferior site heart attack or myocardial infarction (MI). One hundred and twenty of these deaths resulted from the patient having an anterior MI; 67 were due to an inferior site MI.

```
. use heart01, clear

. keep death anterior

. tab death anterior

           | 0=inferior;1=anterior|
    death |  Inferior    Anterior |      Total
----------+----------------------+----------
        0 |     2,504       2,005 |      4,509
        1 |        67         120 |        187
----------+----------------------+----------
    Total |     2,571       2,125 |      4,696
```

CALCULATION OF ODDS RATIO

When we restructure the above tabulation into a more traditional format, the schematized table appears as:

```
                         Response
                        1        0
                 ---------------
              1 |    a        b |
      Risk      |              |
      Factor  0 |    c        d |
                 ---------------

                    Response (death)
                        1        0
                 ---------------
      Risk      1 |   120    2005 |
(Anterior MI)     |              |
                0 |    67    2504 |
                 ---------------
```

The odds ratio is calculated by A*D/B*C.

```
. di   (120*2504)/(2005*67)
2.2367961                              ← compare with LR output below
```

The odds ratio was calculated by cross-multiplying the diagonals, with a resulting odds ratio of 2.2368. Next, subject the same variables to a logistic regression, with *death* as the response. The regression algorithm calculates the odds ratio of *anterior* by exponentiating its parameter estimate. The output below uses the **logistic** procedure, which is estimated using a full maximum likelihood algorithm.

MAXIMUM LIKELIHOOD LOGISTIC COMMAND

```
. logistic death anterior,

Logistic regression                    Number of obs   =      4696
                                       LR chi2(1)      =     28.14
                                       Prob > chi2     =    0.0000
Log likelihood = -771.92263            Pseudo R2       =    0.0179

-------------------------------------------------------------------
  death | Odds Ratio  Std.Err.    z  P>|z|  [95% Conf. Interval]
--------+----------------------------------------------------------
anterior |  2.236796  .3476527 5.18 0.000  1.649411    3.03336
-------------------------------------------------------------------
```

The estimated odds ratio is identical to that calculated by hand using a cross multiplication method. We shall next use Stata's generalized linear models, or **glm**, command to model the same relationship. Note how the binomial family is supplied to the algorithm. The default link function for the binomial family is the **logit**, or **logistic**; we therefore need not specify it. As mentioned before, probit, complementary loglog and loglog are other possible binomial links. The notion of link will be discussed in more detail later in the text.

The **glm** command uses an iteratively re-weighted least squares algorithm (IRLS) to estimate generalized linear model (GLM) parameters, standard errors, and so forth. IRLS is a simplification of maximum likelihood estimation that can be used when the log-likelihood, which is a reparameterization of the probability function (in this case the binomial PDF), is a member of the exponential family of distributions. The binomial, with Bernoulli as a subset of the binomial, is such an exponential family form member.

GLM COMMAND: BINOMIAL FAMILY; LOGIT LINK DEFAULT

```
<BINOMIAL DENOMINATOR DEFAULT TO 1 -- BERNOULLI DISTRIBUTION>

. glm death anterior, nolog fam(bin) ef /// ef = exponential form

Generalized linear models              No. of obs      =      4696
Optimization      : ML                 Residual df     =      4694
                                       Scale parameter =         1
Deviance        = 1543.845261          (1/df) Deviance = .3288976
Pearson         =        4696          (1/df) Pearson  = 1.000426
```

```
Variance function: V(u) = u*(1-u)       [Bernoulli]
Link function    : g(u) = ln(u/(1-u)) [Logit]

                                        BIC     =         .3296093
Log likelihood   = -771.9226305         BIC     =        -38141.42
------------------------------------------------------------------
           |                  OIM
     death | Odds Ratio  Std.Err.     z  P>|z| [95%Conf.Interval]
---------+--------------------------------------------------------
  anterior |   2.236796  .3476532  5.18 0.000  1.64941   3.033361
------------------------------------------------------------------
```

Note that the odds ratio is the same in all three derivations—calculation from a cross tabulation, maximum likelihood (ML), and IRLS via GLM. The odds ratio can be interpreted, as the exponentiated parameter estimate or coefficient, as "A patient having an anterior site MI has approximately a two and a quarter greater odds of death within 48 hours of admission than does a patient sustaining an inferior site MI." Of course, a true life model would typically include a number of other explanatory predictors. If a predictor affects other predictors as well as the response, epidemiologists call it a **confounder**. In any case, true-life explanatory predictors would very likely alter the odds of *anterior*. However, it is highly doubtful that added predictors would result in the opposite conclusion. It is well known that anterior site infarcts have a higher mortality than do inferior site or other site myocardial infarcts.

We can also determine the probability of death based on having an anterior as well as an inferior MI. We model the data in a normal fashion, but without exponentiating the coefficients. For this we can use Stata's **logit** command, but can equally use the **glm** command as well. Users of other software can make the amendment to the previously given command.

```
. logit death anterior, nolog

Logistic regression                     Number of obs =       4696
                                        LR chi2(1)    =      28.14
                                        Prob > chi2   =     0.0000
Log likelihood = -771.92263             Pseudo R2     =     0.0179
------------------------------------------------------------------
     death|     Coef. Std.Err.      z P>|z| [95% Conf. Interval]
-------+----------------------------------------------------------
anterior| .8050445 .1554244   5.18 0.000   .5004182    1.109671
   _cons|-3.620952 .1237929 -29.25 0.000 -3.863582  -3.378323
------------------------------------------------------------------
```

Stata's **predict** command can be used to obtain, among other statistics, either the model linear predictor, or the fitted probability of success, given the broad sense in which the term can be used. Which statistic is generated depends on the option specified with **predict**. For the **logit, logistic,** and **glm** commands, calculation of the predicted probability of 1 (success) is the default. Adding the option *xb* to the **predict** command calculates the linear predictor. Menu selection applications generally allow the user to specify desired post estimation statistics by clicking on them from a table of choices.

```
. predict fit          /// fit = predicted probability of y=1;
. predict lp,  xb      /// linear predictor, xb. Preferred command:
                           predict xb,  xb
```

The linear predictor is also symbolized as *xb*, η, or g(μ), depending on the context in which the statistics are being used. (Note: *xb* is also written as *x*β, or as β*x*—all meaning the same for our purposes.) The linear predictor is the linear sum of the product of the coefficients and data elements (*x**β). The values may be calculated by hand using the following methods.

ANTERIOR MI

Based on the model output, the linear predictor, *xb*, for an anterior MI is

$$xb = .805*1 + (-3.621)$$

where the value of anterior is 1. The intercept has a value of −3.621. We calculate this by:

```
. di   .805-3.621
-2.816
```

We should interject here that when the logistic model is expressed in this form, *anterior* is the logistic coefficient, which is the natural log of the odds ratio of *anterior*. The constant term, also called the intercept, is the value of *x*β when all predictors have the value of 0. Also to be noted is the absence of an odds ratio for the logistic intercept term when a table of odds ratios has been requested by the model. When logistic parameter estimates are displayed, an intercept or constant term is always displayed together with a standard error and a confidence interval, unless, of course, the user has specifically requested not to have a constant. Odds ratios are obtained by exponentiating the coefficients, yet no odds ratio is displayed for the intercept when using, for example, the **logistic** command. A few software applications have chosen to provide the intercept odds ratio for logistic regression output. The developers at Stata have wisely chosen, however, to exclude it from the table of odds ratios because the exponentiated intercept is neither an odds nor an odds ratio. Nothing is being compared. It is simply an exponentiated value without any meaningful interpretation. In order to avoid confusion, most applications do not display it.

The probability of death on sustaining an anterior MI, which can be expressed as:

$$p \text{ or } \pi,$$
$$mu \text{ or } \mu,$$
$$g^{-1}(\eta) \text{ or } h(\eta)$$

is then calculated as:

```
p  = 1/(1+exp(-xb))   =   1/(1+exp(2.816))   =  .05646567
```

The probability can also be obtained using an alternative transform (inverse link):

```
p = exp(xb)/(1+exp(xb)) = exp(-2.816)/(1+exp(-2.816)) = .05646567
```

INFERIOR MI

The linear predictor for an inferior MI: $xb = .805*0 + (-3.621)$

```
xb = .805*0 + (-3.621)

. di  0 - 3.621
-3.621
```

The probability of death after having an inferior MI is

```
p  = 1/(1+exp(-xb))   =   1/(1+exp(3.621))   =  .02605868
```

Now let's use the **predict** command following GLM.

```
<model using glm>

. predict xb, xb     /* calculates the linear predictor, xb   */

. predict mu, mu     /* calculates the fitted value, p or µ   */

. tab xb

    linear |
 predictor |      Freq.      Percent        Cum.
-----------+-----------------------------------
 -3.620952 |      2,571        54.75       54.75
 -2.815907 |      2,125        45.25      100.00
-----------+-----------------------------------
     Total |      4,696       100.00
```

```
. tab mu

 predicted |
 mean death |       Freq.      Percent         Cum.
------------+-----------------------------------------
  .0260599 |       2,571        54.75          54.75
  .0564706 |       2,125        45.25         100.00
------------+-----------------------------------------
     Total |       4,696       100.00
```

The values calculated by hand and produced by using Stata's built-in post-estimation commands are nearly identical. Any differences are a result of rounding error.

To summarize then, following a logistic modeling estimation, we calculate the linear predictor by summing each parameter estimate times its respective data element across all terms, plus the constant.

$$x_i\beta = \alpha + \beta_1 x_{1i} + \beta_2 x_{2i} + \beta_3 x_{3i} + \ldots + \beta_n x_{ni} \tag{2.1}$$

We then calculate the fitted value, μ, by the formula below:

$$\mu_i = 1/(1 + \exp(-x_i\beta)) \tag{2.2}$$

or

$$\mu_i = (\exp(x_i\beta))/(1 + \exp(x_i\beta)) \tag{2.3}$$

Logit is another concept that is commonly associated with logistic models. It is used with respect to both GLM and full maximum likelihood implementations of the logistic model. The natural log of the odds, or **logit**, is the link function that linearizes the relationship of the response to the predictors. In GLM procedures, it can be determined by calculating $x\beta$ on the basis of μ.

Odds is defined as the relationship of p, the probability of success ($y = 1$), to $1 - p$, or the probability of non-success ($y = 0$). This relationship is shown in Table 2.1. We symbolize this relationship as:

$$\text{odds} = \frac{p}{1-p} = \frac{\mu}{1-\mu} \tag{2.4}$$

Suppose that we have a probability value of 0.5. The odds, given the above formula, is .5/(1 − .5) = .5/.5 = 1. When thought about, this makes perfect sense. If the probability of y is 0.5, it means in colloquial terms, 50:50. Neither binary alternative is favored. The odds are 1. A p of 0.2 gives an odds of 0.2/(1 − 0.2) = 0.2/0.8 = 1/4 = 0.25.

TABLE 2.1

Probability—Odds

Probability	Odds
.1	.1111
.2	.25
.25	.3333
.3	.4286
.4	.6667
.5	1.
.6	1.5
.7	2.3333
.75	3.
.8	4.
.9	9.

The log-odds, or logit, is simply the natural log of the odds.

$$\text{Lnodds} = \text{logit} = \ln(p / (1-p)) = \ln(\mu / (1-\mu)) \tag{2.5}$$

We can then calculate the fit, μ, as:

```
exp[ ln (μ/(1-μ)) ]   = exp(xb)

       (μ/(1-μ))   = exp(xb)

          (1-μ)/μ  = 1/ exp(xb)   = exp(-xb)

        1/μ - 1  = exp(-xb)

          1/μ  = 1 + exp(-xb)

           μ  = 1 / (1 + exp(-xb))
```

which is the value we indicated earlier for μ in terms of *xb*.

Technically, we show the relationship of the log-odds or logit to the fit for a given observation as:

$$x_i\beta = \alpha + \beta_1 x_{1i} + \beta_2 x_{2i} + \beta_3 x_{3i} + \ldots + \beta_n x_{ni} \tag{2.6}$$

$$\ln(\mu_i / (1-\mu_i)) = \alpha + \beta_1 x_{1i} + \beta_2 x_{2i} + \beta_3 x_{3i} + \ldots + \beta_n x_{ni} \tag{2.7}$$

$$\mu_i = 1 / (1 + \exp(-x_i\beta)) \tag{2.8}$$

The variance of the maximum likelihood logistic regression parameter estimate can be obtained for a single binary predictor from the formula:

$$\text{Var}(\beta) = (1/a + 1/b + 1/c + 1/d) \tag{2.9}$$

The letters represent the cells in the 2×2 tabulation of the response on a binary predictor shown earlier in this chapter. The standard error of the same parameter estimate is calculated by taking the square root of the variance. For the death-anterior tabulation we have:

```
. di sqrt(1/120 + 1/2005 + 1/67 + 1/2504)
.15542465
```

Running the logistic model produces:

```
. logit death anterior,  nolog

Logistic regression                      Number of obs   =        4696
                                         LR chi2(1)      =       28.14
                                         Prob > chi2     =      0.0000
Log likelihood = -771.92263              Pseudo R2       =      0.0179
-----------------------------------------------------------------------
    death |     Coef. Std. Err.      z P>|z|  [95% Conf. Interval]
----------+------------------------------------------------------------
  anterior |  .8050445 .1554244   5.18 0.000   .5004182    1.109671
    _cons | -3.620952 .1237929 -29.25 0.000  -3.863582   -3.378323
-----------------------------------------------------------------------
```

We next consider a logistic model based on a $2 \times k$ format. That is, we shall now discuss the modeling of a single categorical predictor.

2.2 $2 \times k$ Table Logistic Model

Describing how a binary response logistic model is used to estimate odds and odds ratios for a categorical predictor is, perhaps, more instructive than modeling a binary predictor. We mean by this that many people find the concepts of odds and odds ratio more understandable with respect to logistic regression when put into the context of a $2 \times k$ table.

We shall again use the **heart01** data, but instead of *anterior* we will use the four-level categorical variable *killip*. Killip levels represent aggregated indicators of cardiovascular and cardiopulmonary severity. Each level designates an ascending degree of damage to the heart with killip level 1 representing patients with normal functioning. Other levels are

Level 2: Moderate congestive heart failure (CHF) and/or evidence of pulmonary edema

Level 3: Severe congestive heart failure and/or pulmonary edema.

Level 4: Myocardial infarction and/or aortal blockage

We expect to see increasing high odds ratios of death within 48 hours of admission for higher killip levels.

First tabulate *death* on killip level to determine that all cells have sufficient counts.

```
. tab death killip

     Death |
  within 48 |                  Killip level
  hrs onset |        1         2         3         4 |     Total
-----------+--------------------------------------------+----------
         0 |    3,671       958       262        37 |     4,928
         1 |       97        71        30        27 |       225
-----------+--------------------------------------------+----------
     Total |    3,768     1,029       292        64 |     5,153
```

A logistic model of *death* on killip level, with level 1 as the base or reference, appears as:

(Note: *kk2* = killip level 2; *kk3* = killip level 3; *kk4* = killip level 4)

```
. tab killip, gen(kk)              /// calculate indicator
                                       variables (1/0) for each
                                       level of killip
. logit death kk2-kk4, nolog or

Logistic regression                    Number of obs   =      5153
                                       LR chi2(3)      =    150.51
                                       Prob > chi2     =    0.0000
Log likelihood = -849.28983            Pseudo R2       =    0.0814
------------------------------------------------------------------
death | Odds Ratio Std. Err.      z   P>|z|  [95% Conf. Interval]
-------+----------------------------------------------------------
   kk2 |   2.804823   .4497351    6.43  0.000   2.048429   3.840521
   kk3 |   4.333438   .9467527    6.71  0.000   2.824005   6.649666
   kk4 |  27.61688   7.545312   12.15  0.000  16.16645  47.17747
------------------------------------------------------------------
```

The parameter estimates have been exponentiated to provide odds ratios. Each level shown is interpreted with respect to the base or reference level; that is, killip level 1.

Building on our explanation of 2 × 2 table logistic models, each level has a calculated odds defined as the ratio of patient deaths to non-deaths for each individual level. The odds ratio is the ratio of the respective odds for each level to the odds of the base or reference level. Therefore, we have:

kk1—reference level

```
. di 97/3671
.02642332                       ODDS KK1
```

kk2

```
. di 71/958
.07411273                       ODDS KK2

. di (71/958)/(97/3671)
2.8048232                       ODDS RATIO OF KK2 RELATIVE TO KK1
```

kk3

```
. di 30/262
.11450382                       ODDS KK3

. di (30/262)/(97/3671)
4.3334383                       ODDS RATIO OF KK3 RELATIVE TO KK1
```

kk4

```
. di 27/37
.72972973                       ODDS KK4

. di (27/37)/(97/3671)
27.616885                       ODDS RATIO OF KK4 RELATIVE TO KK1
```

Notice that each calculated odds ratio is identical to the value displayed in the logistic regression output.

As a side point, we may find it important at times to determine the odds ratio between levels other than one including the reference. What we do is, in effect, specify a new reference. The model may be re-parameterized so that we explicitly define a revised base, or we can simply calculate the odds ratio from the tabulation of *death* on *killip*.

In Stata the reference, or base, can be specified by excluding it from the model. If we wish to determine the odds ratio of level 4 based on level 2, the new referent, we can:

```
. logit death kk1 kk3 kk4, nolog or    [ HEADER STATISTICS NOT
                                         DISPLAYED ]
```

```
-----------------------------------------------------------------
death | Odds Ratio Std. Err.      z  P>|z|  [95% Conf. Interval]
------+----------------------------------------------------------
  kk1 |   .3565287  .0571671   -6.43  0.000   .2603814    .488179
  kk3 |   1.544995  .3532552    1.90  0.057   .9869707   2.418522
  kk4 |   9.846212  2.770837    8.13  0.000   5.671936   17.09256
-----------------------------------------------------------------
```

This informs us that the odds ratio of killip level 4 compared with level 2 is 9.846.

Using the tabulation of *death* on *killip*, we can calculate the odds ratio of the two levels as:

```
. tab death killip

      Death |
  within 48 |                  Killip level
  hrs onset |         1         2         3         4 |     Total
------------+------------------------------------------+----------
          0 |     3,671       958       262        37 |     4,928
          1 |        97        71        30        27 |       225
------------+------------------------------------------+----------
      Total |     3,768     1,029       292        64 |     5,153
```

The odds ratio of *kk4* based on *kk2* is

```
. di (27*958)/(71*37)
9.8462124
```

or, using Stata's **lincom** command, which can be used for complex interrelationships:

```
.
. lincom kk4-kk2, or

 ( 1)  - kk2 + kk4 = 0

-----------------------------------------------------------------
death | Odds Ratio  Std. Err.      z  P>|z|  [95% Conf. Interval]
-----------------------------------------------------------------
  (1) |   9.846212   2.770837    8.13  0.000   5.671936   17.09256
-----------------------------------------------------------------
```

Each of these three methods allows us to easily calculate a desired odds ratio relationship with a single categorical predictor logistic model.

Reverting to the main model, we can view the relationships we developed from a different angle by using the table of logistic coefficients. A logistic regression displaying the respective model coefficients is shown below without the usual ancillary model statistics:

```
. logit death kk2-kk4, nolog        [ HEADER NOT DISPLAYED ]

------------------------------------------------------------------
death |      Coef.  Std. Err.       z  P>|z| [95% Conf. Interval]
------+-----------------------------------------------------------
  kk2 |   1.031341  .1603435      6.43 0.000   .7170731   1.345608
  kk3 |   1.466361  .2184761      6.71 0.000   1.038156   1.894567
  kk4 |   3.318427  .2732137     12.15 0.000   2.782938   3.853916
_cons |  -3.633508  .1028673    -35.32 0.000  -3.835125  -3.431892
------------------------------------------------------------------
```

Using the same logic for the calculation of linear predictors as we showed in the previous section, a formula appropriate for the calculation of a linear predictor for a four-level categorical predictor, with the first level as the base or reference, is given by:

$$x\beta = \beta_0 + \beta_2 x_2 + \beta_3 x_3 + \beta_4 x_4 \tag{2.10}$$

As applied to the four-level killip predictor, we have:

$$x\beta = \beta_0 + kk2(1/0) + kk3(1/0) + kk4(1/0) \tag{2.11}$$

$$x\beta = -3.633508 + 1.031341 * (1/0) + 1.466361 * (1/0) + 3.318427 * (1/0) \tag{2.12}$$

For each level, including killip level 1, we calculate the linear predictor, and, on the basis of the linear predictor, the odds. The odds (not odds ratio) is calculated from the linear predictor as:

$$o = \exp(xb) \tag{2.13}$$

In addition, we calculate the risk, or probability, of the linear predictor by use of the inverse logistic link function:

$$\pi = \frac{1}{1 + \exp(-xb)} \tag{2.14}$$

or

$$\pi = \frac{\exp(xb)}{1 + \exp(xb)} \tag{2.15}$$

In these cases, the odds and risk, or probability, relate to a particular linear predictor, which itself is based on predictor values and coefficients. When there is only a single categorical predictor that can still be regarded as a risk factor or exposure, the terms of the linear predictor consist of the model constant and one coefficient value. All levels but level 1 will have 0 for the level values. When each of these is multiplied by its corresponding coefficient, the value of 0 is retained. Since the lone non-0 level has a value of 1, this means that the coefficient is multiplied by 1 and is not revalued in any manner. The linear predictor is calculated by summing the constant and non-0 valued coefficient.

If *kk2* = 1, the linear predictor, *xb*, will equal:

```
. di 1.031341-3.633508
-2.602167                          log-odds = XB  = linear predictor
```

with an odds of:

```
. di exp(-2.602167)
.0741128                        odds of death when KK2==1
```

and risk value or probability of death within 48 hours of admission.

```
. di 1/(1+exp(2.602167))
.06899909                    ←==    risk of death for a patient
                                    with KK2 with respect to KK1.
```

If *kk3* = 1, the linear predictor, *xb*, will equal:

```
. di 1.466361-3.633508
-2.167147               log-odds = XB = linear predictor
```

with an odds of:

```
. di exp(-2.167147)
.11450383                        odds of death when KK3==1
```

and risk value or probability of death within 48 hours of admission.

```
. di 1/(1+exp(2.167147))
.10273974                    ←==    risk of death for a patient
                                    with KK3 with respect to KK1
```

If *kk4* = 1, the linear predictor, *xb*, will equal:

```
. di 3.318427-3.633508
-.315081                        log-odds = xb  = linear predictor
```

with an odds of:

```
. di exp(-.315081)
.72972976                         odds of death when KK4==1
```

and risk value or probability of death within 48 hours of admission.

```
. di 1/(1+exp(0.315081))
.42187501              ←==    risk of death for a patient with
                             KK4 with respect to KK1
```

Those who are using Stata as their statistical application can produce these same risk values or probabilities of death by using the **predict** command or the same menu option.

The command or menu selection needs to be used directly following model estimation.

```
. predict mu
```

By selectively requesting the corresponding value of mu for each killip level, we can summarize the results as:

kk2 = 1 : .068999
kk3 = 1 : .1027397
kk4 = 1 : .421875

These are the same values as those that we calculated above.

We can also apply the formula:

$$o = \frac{\Pi}{1 - \Pi} \tag{2.16}$$

to convert the above probability or risk values, π, to odds, o.

kk2 – odds

```
. di .068999/(1-.068999)
.0741127
```

kk3 – odds

```
. di .1027397/(1-.1027397)
.11450378
```

kk4 – odds

```
. di .421875/(1-.421875)
.72972973
```

These values are identical to those we produced earlier using cell counts and are identical to the values we later display using Stata's **tabodds** command. Given the odds of *death* for *kk1* and the base or reference level we obtained before as a simple ratio based on the tabulation of *death* and *killip*, we may calculate the odds ratios. We do so here based on the probability values, not from the logistic inverse link function as before.

$$\Pi = \frac{o}{1+o} \tag{2.17}$$

The *kk1* odds = 0.02642, therefore:
probability of death at *kk1*:

```
. di .02642/(1+.02642)
.02573995
```

probability of death at *kk2*:

```
. di .0741128/(1+.0741128)
.06899908
```

probability of death at *kk3*:

```
. di .11450383/(1+.11450383)
.10273974
```

probability of death at *kk4*:

```
. di .72972976/(1+.72972976)
.42187501
```

These all match with the probabilities we calculated based on the logistic inverse link function.

   You should, by now, see how simple it is to calculate the odds ratios based on the odds. Here we have the ratio values that were already calculated. Earlier we used counts from the table of *death* and *killip*. The odds ratio of each non-reference level is the ratio of each respective level odds to the odds of the reference. We symbolize this relationship as:

$$\Psi = \frac{o_k}{o_r} \qquad (2.18)$$

Here the k levels take the values of 2–4 with r as 1.

kk2

```
. di .0741128/.02642
2.8051779
```

kk3

```
. di .11450378/.02642
4.3339811
```

kk4

```
. di .72972973/.02642
27.620353
```

As before, these values are equal to the logistic model table of odds ratios.

Stata has a nice utility called **tabodds** that provides a breakdown of cases, controls, and odds together with their confidence intervals. Accompanying the table are two tests, one for homogeneity of equal odds and the other a score test for the trend of odds. Note that the odds are the same as we calculated using two derivations.

```
. tabodds death killip
```

killip	cases	controls	odds	[95% Conf. Interval]	
1	97	3671	0.02642	0.02160	0.03233
2	71	958	0.07411	0.05824	0.09432
3	30	262	0.11450	0.07848	0.16706
4	27	37	0.72973	0.44434	1.19842

```
Test of homogeneity (equal odds):  chi2(3)   =     288.38
                                    Pr>chi2   =     0.0000

Score test for trend of odds:       chi2(1)   =     198.40
                                    Pr>chi2   =     0.0000
```

Specifying an *or* option yields a table of odds ratios, *chi2* statistics and confidence intervals.

```
. tabodds death killip, or

--------------------------------------------------------------
killip |   Odds Ratio    chi2    P>chi2    [95% Conf. Interval]
--------+-----------------------------------------------------
     1 |    1.000000       .        .            .          .
     2 |    2.804823     44.74   0.0000     2.045340    3.846321
     3 |    4.333438     53.01   0.0000     2.815892    6.668823
     4 |   27.616885    315.30   0.0000    15.789872   48.302629
--------------------------------------------------------------

         [ LOWER TEST STATISTICS NOT DISPLAYED ]
```

The confidence intervals necessitate the use of the variance–covariance matrix. To demonstrate this we first redisplay the table of coefficients, standard errors, and so forth.

```
. logit death kk2-kk4, nolog     [ HEADER STATISTICS NOT
                                   DISPLAYED ]

--------------------------------------------------------------
 death |    Coef.   Std. Err.     z    P>|z| [95% Conf. Interval]
-------+------------------------------------------------------
   kk2 |  1.031341  .1603435    6.43   0.000   .7170731   1.345608
   kk3 |  1.466361  .2184761    6.71   0.000   1.038156   1.894567
   kk4 |  3.318427  .2732137   12.15   0.000   2.782938   3.853916
 _cons | -3.633508  .1028673  -35.32   0.000  -3.835125  -3.431892
--------------------------------------------------------------
```

To display the variance–covariance matrix, we continue as:

```
. corr, _coef co              /// co = variance-covariance matrix

        |      kk2       kk3       kk4      _cons
-------+------------------------------------------
    kk2 |   .02571
    kk3 |  .010582   .047732
    kk4 |  .010582   .010582   .074646
  _cons | -.010582  -.010582  -.010582   .010582
```

As in standard linear regression, we take the square root of the diagonal elements to obtain the respective standard errors.

kk2

```
. di sqrt(.02571)
.16034338
```

kk3

```
. di sqrt(.047732)
.21847654
```

kk4

```
. di sqrt(.074646)
.2732142
```

and even the constant as:

```
. di sqrt(.010582)
.10286885
```

To calculate the 95 percent confidence interval for coefficients, or parameter estimates, we use the same logic as we showed for the binary response model.

$$CI_{95} = \beta +/- 1.96 * SE(\beta) \tag{2.19}$$

We show the calculations for *kk2* only; the other levels, including the constant, can be obtained by substituting the appropriate coefficient and standard error values.

95% CONFIDENCE INTERVAL – *KK2*

```
. di     1.031341 - 1.96 * .1603435
.71706774


. di     1.031341 + 1.96 * .1603435
1.3456143
```

As expected, these values correspond to the values displayed in the results.

Also as shown for the logistic model binary predictor, we can determine the confidence intervals for the odds ratios of each level by exponentiating the formula.

Although we do not use it in our calculations here, the standard error of an odds ratio is determined by using the **delta method**; i.e. $\exp(\beta)*se(\beta)$.

$$\text{Odds Ratio } CI_{95} = \exp\{\beta +/- 1.96 * SE(\beta)\} \tag{2.20}$$

```
. logit, or          [ HEADER NOT DISPLAYED ]

-------------------------------------------------------------------
death | Odds Ratio  Std. Err.     z    P>|z|  [95% Conf. Interval]
------+------------------------------------------------------------
 kk2  |  2.804823   .4497351    6.43   0.000   2.048429   3.840521
 kk3  |  4.333438   .9467527    6.71   0.000   2.824005   6.649666
 kk4  |  27.61688   7.545312   12.15   0.000   16.16645   47.17747
-------------------------------------------------------------------
```

95% CONFIDENCE INTERVAL – ODDS RATIO OF *KK2*

```
. di exp(.71706774)
2.0484179


. di exp( 1.345608)
3.8405209
```

A final note on confidence intervals—many times the traditional 95 percent confidence interval is not sufficiently stringent for a particular study. A 95 percent confidence interval means that we are willing to be mistaken 5 percent of the time; that is, the true parameter or odds ratio can be found within the range of the interval values 95 percent of the time. For some medical studies, this may be overly generous. In nearly all research studies done in the physical sciences, a 99 percent confidence interval is used as the default. A 95 percent confidence interval is considered the default for the vast majority of medical, epidemiological, and social science research. By looking at a standard normal distribution table, or at a critical values chart for the normal or Gaussian distribution, one can determine the appropriate value to substitute for 1.96 in the above formula. For the 99 percent level we find that 2.5758 is the correct value. Substituting the value into the formulae we have used gives us:

99% CONFIDENCE INTERVAL – *KK2*

```
. di    1.031341 - 2.5758 * .1603435
.61832821

. di    1.031341 + 2.5758 * .1603435
1.4443538
```

```
and
```

99% CONFIDENCE INTERVAL – ODDS RATIO OF *KK2*

```
. di    exp(1.031341 - 2.5758 * .1603435)
1.8558229

. di    exp(1.031341 + 2.5758 * .1603435)
4.2391119
```

A look at the same model we have been using, but at the 99 percent confidence interval instead, gives us for the odds ratio output:

```
. logit, level(99) or  [ HEADER STATISTICS NOT DISPLAYED ]
```

```
--------------------------------------------------------------------
death | Odds Ratio  Std. Err.     z   P>|z|  [99% Conf. Interval]
------+-------------------------------------------------------------
  kk2 |   2.804823  .4497351   6.43  0.000   1.855813     4.23913
  kk3 |   4.333438  .9467527   6.71  0.000   2.468484    7.607377
  kk4 |   27.61688  7.545312  12.15  0.000   13.66278    55.82263
--------------------------------------------------------------------
```

Accounting for rounding error, we find that the confidence interval displayed for the *kk2* level is the same as the confidence intervals we calculated by hand.

Care must be taken to remember that when interpreting a logistic model having a categorical predictor, the odds ratio of each level is a relationship of two odds. This section, in particular, attempts to make it clear that the ratios we are talking about are between a numerator—the odds of a given level other than the reference—and a denominator, which is the reference level. We can usually interpret the odds ratio under consideration in light of the context in which the ratio emerged. However, in some similar manner of expression, the interpretation of a categorical predictor level takes a form as in this example:

> Patients entering the hospital with an assigned killip level of 2 have nearly three times the odds of dying in the hospital within 48 hours of admission than do patients assigned a killip level of 1.

When we are considering the parameter estimates, or coefficients, and not the odds ratio, we interpret the relationship as:

> The log-odds of death within 48 hours of admission increases by some 3 percent when a patient enters the hospital with a killip level of 2 compared with a killip level of 1.

As we shall observe when we discuss multivariable logistic models in Chapter 5, we interpret the odds ratio and parameter estimates the same as above, with the exception that all other predictors are held to a constant value.

One final item should be addressed when discussing a categorical predictor. Except when there is a natural division of categories, the number of levels assigned a categorical predictor is arbitrary, or better, essentially arbitrary. If categorization is based on an underlying continuous predictor, we may find that the number of categories is based on inconsistencies in risk across the range of the variable. Ideally, categories represent significant differences in slopes over the variable.

Recall the model we developed above, re-parameterizing our model so that killip level 2 was the reference or base. The table of odds ratios appeared as:

```
--------------------------------------------------------------------
death | Odds Ratio Std. Err.     z   P>|z|  [95% Conf. Interval]
------+-------------------------------------------------------------
  kk1 |  .3565287  .0571671  -6.43  0.000    .2603814    .488179
```

```
kk3 | 1.544995    .3532552  1.90  0.057  .9869707  2.418522
kk4 | 9.846212    2.770837  8.13  0.000  5.671936  17.09256
----------------------------------------------------------------
```

The *p*-value of *kk3* here is not statistically significant at the 0.05 level. Although *kk3* is close to being significant at this level, we may find it valuable to combine levels. If we were concerned about *kk3*'s contribution to the model—that is, to understanding and predicting death based on killip level—we may include *kk3* as part of the reference. We do this by excluding both *kk2* and *kk3*. *kk1* and *kk4* are then interpreted on the basis of this aggregation of levels.

In a case such as this, however, we recommend against combining level *kk3* with *kk2*. It is preferable to keep *kk3* in the model, evaluating it when other confounding predictors are entered into the model. Even if no other predictors are involved, *kk3* still has an effect on the response, albeit not a significant one.

In general, combining levels within a categorical predictor is advisable when:

1. The reference, or base, is either the lowest or highest level of the predictor, and a contiguous level to the base is clearly not significant. The base is then expanded to include the original base and non-significant contiguous level.

2. Contiguous levels have identical or similar parameter estimates, or odds ratios. Even though the two (or more) levels may be significant compared with the base level, having the same slope indicates that they have equal effect on understanding the response. These levels may then be combined.

3. One of two contiguous levels is not significant. The non-significant level may be combined with a contiguous level. Care must be taken, however, when combining with the base if it is not at the extremes, as in our example of killip levels. Be assured that the aggregation results in a new clearly interpretable level and no resulting loss of information.

We have discussed a variety of ways by which the foremost statistics related to a logistic regression, specifically a single categorical predictor model, can be calculated. Understanding how the various statistics all cohere should provide a good sense of the logic of the model.

2.3 Modeling a Quantitative Predictor

Last, we discuss a single predictor model where the predictor of interest, or risk factor, is quantitative, or continuous. Considerable discussion has been given in recent literature to the question of categorizing continuous predictors. Some statisticians believe that continuous predictors may be helpful for graphing purposes and for looking at trends and for interactions. But they argue that for statistical and interpretative purposes, all logistic model predictors should be binary or categorical. Others argue that continuous predictors are fine as long as they are monotonically and smoothly increasing or decreasing. A key consideration involves the loss of individual information when a continuous predictor is categorized.

The problem centers on the fact that a continuous predictor, when modeled, has only one parameter estimate. If there is variability in the underlying odds ratio at different points along the range of the variable, then having a single parameter estimate does not provide sufficient information concerning the predictor to the model. Categorizing the continuous predictor in such a manner as to have separate slopes, or coefficients, with values corresponding to the actual shape of the predictor is preferred to maintaining a single parameter estimate. We describe how transformations of a continuous predictor are made later in this text.

At this point we should discuss interpretation. Consider the model:

```
. logit death age, nolog   [ HEADER STATISTICS NOT DISPLAYED ]


--------------------------------------------------------------------
death |    Coef.   Std. Err.     z   P>|z|  [95% Conf. Interval]
------+-------------------------------------------------------------
  age |  .0565292   .0048193  11.73  0.000   .0470834   .0659749
  ons | -7.076784   .3729006 -18.98  0.000  -7.807656  -6.345912
--------------------------------------------------------------------


. logit, or       [ HEADER STATISTICS NOT DISPLAYED ]


--------------------------------------------------------------------
death | Odds Ratio  Std. Err.    z  P>|z|  [95% Conf. Interval]
------+-------------------------------------------------------------
  age |  1.058157   .0050996  11.73 0.000   1.048209   1.0682
--------------------------------------------------------------------
```

The univariable model informs us that for each 1-year increase in the age of a patient, the odds of dying within 48 hours of admission increases by nearly 6 percent. A 10-year increase in age results in a 76 percent increase in the odds of death. This value can be determined from the parameter estimate of *age* as:

```
. di exp(.0565292 * 10)
1.7599616
```

or from the odds ratio of *age* as:

```
. di 1.058157^10
1.7599531
```

The continuous predictor *age* provides us with the maximal amount of information; that is, we know the exact (within a year) age of each patient in the model. The problem, to reiterate, with having a continuous predictor like this in the model is the fact that we only have a single parameter for all ages, which is a slope. It may be the case that the shape or distribution of age varies so much that several slopes are required in order to best describe the relationship of age to death.

We can obtain a sense of age's distribution by summarizing it and obtaining summary statistics related to its distributional properties.

```
su age, detail
                                  age
-----------------------------------------------------------------
          Percentiles      Smallest
   1%            40             40
   5%            43             40
  10%            46             40        Obs                5388
  25%            54             40        Sum of Wgt.        5388

  50%            66                       Mean           65.80976
                              Largest     Std. Dev.      14.42962
  75%            76            100
  90%            84            100        Variance       208.2138
  95%            93            100        Skewness       .1598778
  99%            99            100        Kurtosis       2.428674
```

Ages range from 40 through 100. The mean and median are both 66. A graphical representation of *age*, may assist us in understanding a bit more how it relates to *death*. (See Figure 2.1.)

```
. histogram age, bin(15) norm
(bin = 15, start = 40, width = 4)
```

<See Figure 2.1 on page 41>

There appears to be a rapid drop of patients after 80 years of age.

```
. sfrancia age

             Shapiro-Francia W' test for normal data
Variable |   Obs        W'           V'           z       Prob>z
---------+-------------------------------------------------------
     age |  5388     0.98438      17.949       3.173    0.00076
```

The percent difference in probabilities and odds between two units or intervals of a continuous predictor is a frequently used measure in research. For example, knowing the percent difference, or effect, in a 1-year

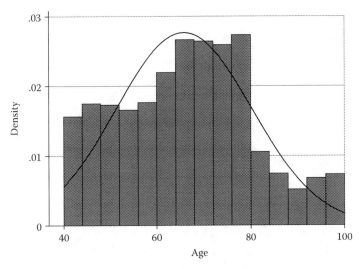

FIGURE 2.1
Histogram of age.

increase in age on the predicted probabilities of a model provides us with an understanding of the effect of age. For an example, we'll use the current model of *death* on *age*.

```
. logit death age, nolog

Logistic regression                          Number of obs  =     5388
                                             LR chi2(1)     =   148.24
                                             Prob > chi2    =   0.0000
Log likelihood = -913.28575                  Pseudo R2      =   0.0751
-------------------------------------------------------------------------
death |    Coef.   Std. Err.      z  P>|z|   [95% Conf. Interval]
------+------------------------------------------------------------------
  age |  .0565292  .0048193   11.73 0.000    .0470834   .0659749
_cons | -7.076784  .3729006  -18.98 0.000   -7.807656  -6.345912
-------------------------------------------------------------------------
```

LINEAR PREDICTOR AT AGE = 50

```
. di -7.075794 + .0565292*50
-4.249334
```

PROBABILITY OF DEATH AT AGE 50

```
. di 1/(1+exp(4.249334))
.01407286
```

LINEAR PREDICTOR AT AGE = 51

```
. di -7.075794 + .0565292*51
-4.1928048
```

PROBABILITY OF DEATH AT AGE 51

```
. di 1/(1+exp(4.1928048))
.01487913
```

PERCENT DIFFERENCE OF PROBABILITY OF DEATH FOR AGES
50 AND 51

```
. di (.01487913 - .01407286)/.01407286
.05729255      5.7% difference.
```

As a check and to give another understanding of the meaning of odds ratios, let us calculate the odds of the probability of death at age 51 and divide it by the odds of the probability of death at age 50. Recalling that the definition of odds is $p/(1-p)$, we calculate and take the ratio of these odds.

ODDS RATIO OF PROBABILITY OF DEATH AT AGE 51 BY
PROBABILITY OF DEATH AT AGE 50

```
. di (.01487913/(1-.01487913) ) / (.01407286/(1-.01407286))
1.0581579
```

Compare the value of 1.058 with the exponentiated value of the coefficient of age in the above model.

```
. di exp(.0565292)
1.0581575
```

Except for a rounding error, the values are identical. If we were to request an odds ratio for the logistic model, it would be the value we calculated by hand.

Categorical and continuous predictors play an important role in assessing model fit. In particular, continuous predictors many times do not have a linear relationship with the logistic link function, the logit. We shall discuss this subject at greater length in Chapter 5. Issues discussed in Chapter 5 will play an important role in determining the fit of a logistic model.

2.4 Logistic Modeling Designs

With a particular study design in mind, researchers employ logistic regression models of nearly every variety to analyze collected data. Mention should be made of the major different types of design and how they relate to logistic regression models.

We may classify study designs into two broad categories: **experimental** and **observational**. Observational may be further categorized into **prospective** and **retrospective** studies. However, as we shall observe in this section, these categories are generalizations that can overlap in certain respects. On the other hand, they are commonly found in statistical and epidemiological literature and a basic familiarity with them is important.

2.4.1 Experimental Studies

Experimental studies are performed by manipulating one factor while controlling others. Repeatedly manipulating one or more factors allows the researcher to establish causal relationships so that specific answers to experimental questions may be given to a set degree of probability. Clinical trials are paradigms of experimental studies, where researchers aim to test the efficacy of various drug treatments. This type of study is prospective, where data are followed over time and evaluated. It is many times a cohort study as well, where, for example, arms of the study are based on randomized subjects who are treated with a drug compared with arms whose randomized patients have been given placebos. Of course, the experimental study design can get quite complicated, but the essential logic of the method is still the same. There are few experimental studies in which logistic regression has been employed as the primary statistical tool.

2.4.2 Observational Studies

2.4.2.1 Prospective or Cohort Studies

In epidemiology, prospective or cohort studies typically begin by collecting data on healthy patients who are exposed, or not exposed, to a specified risk factor and then followed over a long period of time with the aim of determining if, or when, subjects become diseased or contract some specific condition. The data consist of a response variable of interest, a risk factor, and other explanatory predictors, many of which are found to be confounders. Confounders are predictors that are significantly related to the response and to the risk factor but are not themselves of interest to the study.

Also in the cohort data set are observations of subjects who are not exposed to the risk factor. They, too, are followed along with the exposed subjects. The incidence rate of acquiring the disease or condition of interest is compared for both the exposed and non-exposed groups.

The incidence rate is a measure of the proportion of subjects over time acquiring the condition from the total population of the data set. The relative risk, which we discussed earlier, is the ratio of the incidence rate of exposed subjects to non-exposed subjects.

2.4.2.2 Retrospective or Case-Control Studies

Case-control studies are almost always retrospective. That is, the researcher obtains data from a registry, a survey, or from some other data gathering technique. A registry is a collection of data on a specific disease or condition. In the text, three registries are frequently used for examples. The **heart01** data come from the FASTRAK registry, which is the Canadian National Registry of Cardiovascular Disease. It is managed by Hoffman-la Roche Canada and contains data on nearly all cardiovascular events in the nation. Hospitals throughout Canada submit data to the registry. In turn, they receive national and regional reports that can be used to help delivery of services.

The **medpar** data set consists of hospital data on all Medicare patients in the United States. Throughout the 1990s the data consisted of 110 variables. The NRMI-2 registry consists of myocardial infarctions (heart attack) data from hospital reports throughout the United States. It is funded by Genentech Pharmaceuticals and was designed to determine outcomes for patients who were treated with different pharmaceutical agents. Of course, it can also be used to evaluate a host of other outcomes related to having infarctions. I was the lead statistician for each of these registries for a time and found the data to be useful for many of the examples in this text.

Outcomes research is typically done on this type of registry data. It can also be done on cohort data after the data have been collected. When this is done the study is, in fact, a retrospective cohort study. But it is, nevertheless, outcomes analysis.

The idea of case-control relates to a researcher selecting an outcome of interest from the study data and comparing the predictors that are associated with those who have acquired the disease or condition (case) with those who have not (control). Matched case-control and stratified case-control studies as well as cross-sectional studies may also be designed. Some of these will be discussed in detail later. Realize however, that there are a variety of different types of case-control studies, or hybrid case-control studies, but they are generally based on the analysis of data that have already been gathered.

2.4.2.3 Comparisons

All three general types of study design have advantages and disadvantages. Experimental designs are typically used to assess causal relationships. Randomization allows the effects of treatments to be clearly differentiated, which is often not the case for observational studies.

Cohort studies can be used to determine incidence rates or, better, incidence risks and risk ratios. Case-control studies in general cannot. Cohort studies can also be used to assess causal relationships and can study many diseases at a time. It may be that more than one disease can be found to be related to the risk factor, or another risk factor can be assigned retrospectively to evaluate relationships.

Problem areas related to cohort studies involve expense and time. Collecting data for cohort studies may take many years and cost substantial sums of money. The Framingham, Massachusetts, study began in 1948 and has followed patients from all ages to the present, evaluating resulting cardiovascular events for subjects in the study. It is still ongoing.

Case-control studies are much faster to use than longitudinal or prospective studies. The data are already at hand and ready to be analyzed. Registries contain a host of variables that allows researchers to study a wide variety of diseases and relationships. They may be collected prospectively and later treated as if they were case-control data. Studies on time to treatment can also be evaluated using this method. A particular advantage relates to rare events or diseases. In a cohort study, rarely occurring diseases may take many years to obtain sufficient numbers of observations in order to construct a statistically meaningful model. A retrospective data study has no such difficulty.

Problems with case-control studies involve difficulties in being able to claim causal relationships and provide only approximate measures of relative risk. Moreover, the sampling of retrospective case-control data may be biased and not representative of the greater population. In addition, in a retrospective data set, missing values, or mistaken values, are nearly impossible to correct. In a clinical trial or prospective study, researchers can many times go back to the ongoing patient records and correct inaccuracies.

There are a number of associated issues concerning study designs. Many are directly related to using a logistic regression model inappropriately given the data and study design. Excellent additional readings that are recommended on this subject are Mann (2003) and Woodward (2005).

We next turn to the nature of estimation algorithms that are used to determine logistic parameters and associated statistics. Admittedly, most statisticians learning about the logistic model would rather skip this section, but it is necessary to understand if one is truly interested in discerning how estimation works and how and why algorithms must be amended to effect changes necessary to better model a particular data situation.

Exercises

2.1. What is the relationship of the binary logistic link function and the linear predictor of a logistic model?

2.2. In what way does the relationship of the linear predictor and fitted value differ between normal or Gaussian models and logistic models?

2.3. Consider the given tabulation of *died* on HMO membership.

```
          |            HMO
died |         0                    1 |        Total
-----------+-----------------------------+----------
      0 |         825               157 |          982
      1 |         431                82 |          513
-----------+-----------------------------+----------
  Total |       1,256               239 |        1,495
```

Calculate the odds ratio of *died* on HMO. Round to the nearest hundredth.

2.4. Given the following table of binary logistic parameter estimates, calculate the probability of death for a non-white patient who stayed in the hospital for 10 days.

```
-------------------------------------------------------------------
died |    Coef.   Std. Err.      z   P>|z|  [95% Conf. Interval]
-------+-----------------------------------------------------------
white | .2526808   .2065523    1.22  0.221  -.1521542   .6575158
  los |-.0299868   .0077037   -3.89  0.000  -.0450858  -.0148878
_cons |-.5986826   .2132683   -2.81  0.005  -1.016681  -.1806844
-------------------------------------------------------------------
```

2.5. Calculate the standard error of *a3* based on the variance–covariance matrix below.

```
            |        a2         a3       _cons
------------+--------------------------------------
      a2 |    .735256
      a3 |    .533333    1.59199
  _cons |   -.533333   -.533333    .533333
```

2.6. Given the binary logistic regression coefficient of *a3* at –3.344039, calculate its 95 percent confidence intervals using the variance–covariance matrix from Exercise 2.5.

2.7. From the binary logistic regression output below, calculate the odds of surviving given being female.

survived: 1 = survived; 0 = died; sex: 1 = male; 0 = female

```
Logistic regression                    Number of obs    =      2201
                                       LR chi2(1)       =    434.47
                                       Prob > chi2      =    0.0000
Log likelihood = -1167.4939            Pseudo R2        =    0.1569
```

```
--------------------------------------------------------------
survived |      Coef. Std. Err.     z P>|z| [95% Conf. Interval]
---------+----------------------------------------------------
     sex | -2.317175 .1195885 -19.38 0.000 -2.551564  -2.082786
   _cons |  1.00436  .104132   9.65 0.000  .8002648   1.208455
--------------------------------------------------------------
```

2.8. Use the **drugprob** data. Determine if there is a univariable relationship of abusing drugs based solely on being dependent on drugs.

2.9. Research the study design of a cross-sectional study. Is it prospective or retrospective? Can it be both? What is the logic and purpose of this type of study?

R Code

Note: The start of each R command begins directly after >. Lines that do not begin with > are a continuation of code from the previous line. > used for this chapter only.

Section 2.1 2 × 2 Tables

```
> library("foreign") # foreign import package library
> ds <- read.data("heart01.dta") # works for STATA v.5-9
or
> library("Hmisc") # alternative library for STATA import
> ds <- stata.get("heart01.dta") # works for STATA v 10
> #drop all other vars and factor levels except 'death' and
'anterior'
> simple<- data.frame(death=ds$death,
anterior=ds$anterior[,drop=TRUE])
> head(simple)
> #contingency table information
> tsimple<- ftable(simple) #frequency table
> tsimple #show
> rowSums(tsimple)
> colSums(tsimple)
> sum(tsimple) #total number data

> #spiffier format
> library('gmodels')
> CrossTable(simple$death, simple$anterior,
dnn=c('death','anterior'))
> #odds ratio
> tsimple[1,1]*tsimple[2,2]/(tsimple[2,1]*tsimple[1,2])
#compute odds ratio from 2×2 table
> #logistic command (actually IRLS in R):
```

```
 > fit <- glm(death ~ anterior, data=simple,
family=binomial(link='logit'))
 > summary(fit)

 > exp(coef(fit)) #odds ratios corresponding to intercept and
anterior coef

 > confint(fit)
 > exp(confint(fit)) #odds ratios
 > #predictions of probability
 > eta <- coef(fit)[1] + coef(fit)[2]*1 #predictor for
anterior=1 (anterior MI)
 > eta
 > p<- 1/(1+exp(-eta)) #probability
 > p
 > exp(eta)/(1+exp(eta)) #different formula
 > eta<- coef(fit)[1] + coef(fit)[2]*0 #predictor for
anterior=0 (inferior MI)
 > eta
 > p<- 1/(1+exp(-eta)) #probability
 > p
 > #get all fitted values
 > eta<- predict(fit, data.frame(anterior=simple$anterior),
type='link', na.action=na.omit)
 > CrossTable(eta)
 > p<- predict(fit, data.frame(anterior=simple$anterior),
type= 'response', na.action=na.omit)
 > CrossTable(p)
 > sqrt(1/120 + 1/2005 + 1/67 + 1/2504)
[1] 0.1554246
 > fit <- glm(death ~ anterior, data=simple,
family=binomial(link= 'logit'))
 > summary(fit)
```

Section 2.2 2 × k Tables

```
 > #2 x k logistic
 > CrossTable(ds$death, ds$killip, dnn=c('death','killip'))

 > fit2 <- glm(death ~ kk2 + kk3 + kk4, data=ds,
family=binomial(link='logit'))
 > summary(fit2)
 > exp(coef(fit2)) #odds ratios
 > confint(fit2)
 > exp(confint(fit2)) #odds ratios
 > 97/3671
[1] 0.02642332
 > 71/958
[1] 0.07411273
```

```
> (71/958)/(97/3671)
[1] 2.804823
> 30/262
[1] 0.1145038
> (30/262)/(97/3671)
[1] 4.333438
> 27/37
[1] 0.7297297
> (27/37)/(97/3671)
[1] 27.61688
> fit3 <- glm(death ~ kk1 + kk3 + kk4, data=ds,
family=binomial(link='logit'))
> summary(fit3)
> exp(coef(fit3)) #odds ratios
> exp(confint(fit3)) #odds ratios CI
> (27*958)/(71*37)
> vcov(fit2) #covariance matrix
> exp(confint(fit2, level=.99)+ )
```

Section 2.3 Modeling a Quantitative Predictor

```
> #quantitative predictor
> fit4<- glm(death ~ age, data=ds, family=binomial)
> summary(fit4)
> confint(fit4)
> exp(coef(fit4)) #odds ratios
> exp(confint(fit4)) #odds ratios CI
> #histogram and normal tests for age
> layout(1:2)
> hist(ds$age, col='blue', breaks=15, freq=FALSE)
> lines(density(ds$age), col='red')
> qqnorm(ds$age, col='blue')
> qqline(ds$age, col='red')
> shapiro.test(sample(na.omit(ds$age), 1000, replace=TRUE))
#only handles up to 5000 data
```

Section 2.4 Logistic Modeling Designs

No statistical output is given in this section.

.

3

Estimation Methods

As previously mentioned, the logistic regression model may be estimated by using either full maximum likelihood (ML) or a GLM methodology. Maximum likelihood estimation typically uses modified forms of Newton–Raphson estimating equations; GLM uses an iteratively re-weighted least squares (IRLS) algorithm that is a simplification of maximum likelihood estimation but is limited to distributions belonging to the exponential family of distributions. In the case of maximum likelihood, an estimating equation is defined as setting to 0 the derivative of the log-likelihood function of the response distribution with respect to one of the parameters of interest, where there is a single estimating equation for each unknown parameter. The term estimating equation is referred to as such because its solution results in point estimates of its various parameters. In this chapter we discuss the derivation of the maximum likelihood estimating equations upon which logistic regression models are based together with standard errors of the estimates and varieties of ML estimation.

Standard errors (SEs) are based on the inverse negative Hessian matrix. Traditionally, the method used for modeling GLMs employs the Hessian as the **expected information matrix** (EIM). This was generally much simpler than using the **full information matrix,** also called the **observed information matrix** (OIM). The only difference between the two relates to standard errors in models with few observations. OIM is preferable and is used in regular ML estimation. The GLM algorithm can be amended to use OIM for all links, canonical and non-canonical alike. Stata and SAS allow users a choice, but default to the OIM. SAS's first iteration uses EIM and then proceeds to use OIM. It should be noted, though, that for a canonical linked model, the OIM reduces to the EIM. A schematic derivation of the IRLS is given in the following section.

3.1 Derivation of the IRLS Algorithm

Estimation is based on a modification of a two-term Taylor expansion of the log-likelihood function. In its original form, the Taylor expansion appears as:

$$0 = f(X_0) + (X_1 - X_0)f'(X_0) + \frac{(X_1 - X_0)^2}{2!} f''(X_0) + \frac{(X_1 - X_0)^3}{3!} f'''(X_0) + \tag{3.1}$$

The first two terms reduce to:

$$0 = f(X_0) + (X_1 - X_0)f'(X_0) \tag{3.2}$$

which can be recast to:

$$X_1 = X_0 - \frac{f(X_0)}{f'(X_0)} \tag{3.3}$$

The Newton–Raphson method of estimation adopts the above by using the score or gradient of the log-likelihood function as the basis of parameter estimation. The form is

$$\beta_r = \beta_{r-1} - \frac{\partial L(\beta_{r-1})}{\partial^2 L(\beta_{r-1})} \tag{3.4}$$

where

$$\partial L = \frac{\partial L}{\partial \beta} \tag{3.5}$$

and

$$\partial^2 L = \frac{\partial^2 L}{\partial \beta \, \partial \beta'} \tag{3.6}$$

In more traditional form, let:

$$U = \partial L \quad \text{and} \quad H = \partial^2 L \tag{3.7}$$

then,

$$\beta_r = \beta_{r-1} - H^{-1} U \tag{3.8}$$

where

$$H = H_{r-1} \quad \text{and} \quad U = U_{r-1} \tag{3.9}$$

Solving for ∂L or U—The Gradient

In exponential family form, the log-likelihood function is expressed as:

$$L(\theta, \phi; y) = \sum \frac{y\theta - b(\theta)}{\alpha(\phi)} + \sum C(y, \phi) \tag{3.10}$$

Solving for L with respect to β by using the chain rule, we have

$$\frac{\partial L}{\partial \beta} = \sum \frac{\partial L}{\partial \theta} \times \frac{\partial \theta}{\partial \mu} \times \frac{\partial \mu}{\partial \eta} \times \frac{\partial \eta}{\partial \beta} \tag{3.11}$$

Solving for each term:

$$\frac{\partial L}{\partial \theta} = \sum \frac{y - b'(\theta)}{\alpha(\phi)} = \sum \frac{y - \mu}{\alpha(\phi)} \tag{3.12}$$

since:

$$b'(\theta) = \mu \tag{3.13}$$

$$\frac{\partial \mu}{\partial \theta} = \frac{\partial b'(\theta)}{\partial \theta} = b''(\theta) = V(\mu) \tag{3.14}$$

$$\frac{\partial \theta}{\partial \mu} = \frac{1}{V(\mu)} \tag{3.15}$$

$$\frac{\partial \eta}{\partial \beta} = \frac{\partial(x\beta)}{\partial \beta} = x, \quad \text{since } \eta = x\beta \tag{3.16}$$

$$\frac{\partial \mu}{\partial \eta} = \left[g^{-1}(\eta)\right]' = \frac{1}{\partial \eta / \partial \mu} = \frac{1}{g'(\mu)} \tag{3.17}$$

which is the derivative of the link function wrt μ.

Substitution of expressions gives the maximum likelihood estimators of β to be the solution of the vector-based estimating equation:

$$\sum \frac{(y - \mu)x}{\alpha(\phi)V(\mu)g'(\mu)} = \frac{(y - \mu)x}{\alpha(\phi)V(\mu)} \frac{\partial \mu}{\partial \eta} = 0 \tag{3.18}$$

where y and μ are the response and fitted values respectively, x is a $1 \times p$ row vector, and the resulting sum is a $p \times 1$ column vector of 0s.

Solving for $\partial^2 L$—Fisher Scoring

The traditional GLM algorithm substitutes I for H, the Hessian matrix of observed second derivatives. I is the second of two equivalent forms of Fisher information given by:

$$I = -E\left[\frac{\partial^2 L}{\partial\beta\partial\beta'}\right] = E\left[\frac{\partial L}{\partial\beta}\frac{\partial L}{\partial\beta'}\right] \tag{3.19}$$

Solving the above yields:

$$I = \frac{\partial}{\partial\beta_j}\left[\frac{(y-\mu)x_j}{\alpha(\phi)V(\mu)}\frac{\partial\mu}{\partial\eta}\right] * \frac{\partial L}{\partial\beta_k}\left[\frac{(y-\mu)x_k}{\alpha(\phi)V(\mu)}\frac{\partial\mu}{\partial\eta}\right] \tag{3.20}$$

$$I = \frac{(y-\mu)^2\,x_j x_k}{\alpha(\phi)V(\mu)^2}\left[\frac{\partial\mu}{\partial\eta}\right]^2 \tag{3.21}$$

Since:

$$(y-\mu)^2 = \alpha(\phi)V(\mu) \tag{3.22}$$

and letting:

$$V(y) = \alpha(\phi)V(\mu) = (y-\mu)^2 \tag{3.23}$$

I becomes formulated as:

$$I = \frac{x_j x_k}{V(y)}\left[\frac{\partial\mu}{\partial\eta}\right]^2 = \frac{x_j x_k}{V(y)g'^2} \tag{3.24}$$

Putting the various equations together:

$$\beta_r = \beta_{r-1} - \left[\frac{x_j x_k}{V(y)}\left[\frac{\partial\mu}{\partial\eta}\right]^2\right]^{-1}\left[\frac{x_k(y-\mu)}{V(y)}\left[\frac{\partial\mu}{\partial\eta}\right]\right] \tag{3.25}$$

Multiply both sides by I:

$$\left[\frac{x_j x_k}{V(y)}\left[\frac{\partial\mu}{\partial\eta}\right]^2\right]\beta_r = \left[\frac{x_j x_k}{V(y)}\left[\frac{\partial\mu}{\partial\eta}\right]^2\right]\beta_{r-1} + \left[\frac{x_k(y-\mu)}{V(y)}\left[\frac{\partial\mu}{\partial\eta}\right]\right] \tag{3.26}$$

We define W:

$$W(\text{weight}) = \frac{1}{V(y)} \left[\frac{\partial \mu}{\partial \eta} \right]^2 \tag{3.27}$$

and

$$\eta = x_k \beta_{r-1} \tag{3.28}$$

Convert the above algebraic representation to matrix form. This can be done in parts. First, given the definition of W above, the following substitution may be made:

$$\left[\frac{x_j x_k}{V(y)} \left[\frac{\partial \mu}{\partial \eta} \right]^2 \right] \beta_r = \left[X'WX \right] \beta_r \tag{3.29}$$

Second, recalling the definition of V(y) and W:

$$\frac{x_k(y-\mu)}{V(\mu)} \left[\frac{\partial \mu}{\partial \eta} \right] = \frac{x_k(y-\mu)}{\dfrac{1}{W} \left[\dfrac{\partial \mu}{\partial \eta} \right]^2} \left[\frac{\partial \mu}{\partial \eta} \right] \tag{3.30}$$

Third, since:

$$\eta = X_k \, \beta_{r-1} \tag{3.31}$$

$$\left[\frac{x_j x_k}{V(y)} \left[\frac{\partial \mu}{\partial \eta} \right]^2 \right] \beta_{r-1} = X'W\eta \tag{3.32}$$

Combining terms, we have:

$$[X'WX]\beta_r = X'W\eta + \left[\frac{x_k(y-\mu)}{\dfrac{1}{W} \left[\dfrac{\partial \mu}{\partial \eta} \right]^2} \left[\frac{\partial \mu}{\partial \eta} \right] \right] \tag{3.33}$$

$$[X'WX]\beta_r = X'W\eta + \left[X_k W(y-\mu) \left(\frac{\partial \eta}{\partial \mu} \right) \right] \tag{3.34}$$

We let:

$$z = \eta + (y - \mu)\left(\frac{\partial \eta}{\partial \mu}\right) \tag{3.35}$$

$$[X'WX]\beta_r = X'Wz \tag{3.36}$$

so that:

$$\beta_r = [X'WX]^{-1} XWz \tag{3.37}$$

which is a weighted regression matrix used to iteratively update estimates of parameter vector β_r, as well as values for μ and η, and the deviance function, which is defined by Equations 4.20 and 4.21. The iteration threshold for commercial software is typically set at 10^{-6}.

The convergence threshold is based on the minimization of differences between the old and new deviance or log-likelihood statistics, or on the respective values within the vector of parameter estimates. Traditionally, GLM software iterated on the deviance statistic and then calculated the log-likelihood for use in various goodness of fit statistics.

3.2 IRLS Estimation

The IRLS algorithm and related statistical values are based on the formula for the exponential family of distributions.

$$f(y;\theta, \phi) = \prod_{I=1}^{N}\left\{\frac{y_i\theta_i - b(\theta_i)}{\alpha(\phi)} + C(y_i;\phi)\right\} \tag{3.38}$$

where θ is the link function, $b(\theta)$ is the cumulant, $a(\phi)$ is the scale, and $C(y; \phi)$ is the normalization term, which is required to assure that the probabilities sum to 1. For the logistic model, as well as the Poisson and negative binomial count models, the scale is taken as 1 so is not involved in estimation. The first derivative of $b(\theta)$ with respect to θ, or $b'(\theta)$, is the mean; the second derivative is the variance. These are extremely useful relationships and are not found in other distributions. Changing the link gives the user alternate models. For instance, the logit link, $\ln(\mu/(1-\mu))$, is the natural link for the binomial distribution. This type of link is referred to as the canonical link. If the link is changed to the inverse of the cumulative normal distribution, we have a probit model. If the link is changed to $\ln(-\ln(1-\mu))$, based on the Gumbell extreme value distribution, the result is a complementary loglog regression. Links other than the natural link for a given GLM distribution are called non-canonical links.

Families can change as well within the algorithm, including the traditional Gaussian, or linear, regression. GLM algorithms now typically include the Gaussian, gamma, inverse Gaussian, binomial, Poisson, geometric, and negative binomial families, each with the capability of having a range of link functions appropriate to the respective family. Power links may also be employed but are not applicable to binomial models. The usual binomial links include the canonical logit, non-canonical probit, complementary loglog (also referred to as simply cloglog), and loglog links.

GLMs provide a statistician with a wide range of models, including logistic regression. GLM applications typically come with a variety of goodness of fit statistics, residuals, and so forth, to make the modeling process much easier than traditional ML. In fact, this is the advantage of using GLM methods over individual ML implementations. On the other hand, for the logistic model, the ML algorithm can provide easier use of so-called Hosmer–Lemeshow fit statistics, which are based on collapsing observations having the same pattern of covariates. We shall discuss this later, but keep this in mind when selecting a method by which to employ a logistic model.

Table 3.1 below outlines the steps entailed in a typical IRLS fitting algorithm. This will be followed by Table 3.2, a generic IRLS algorithm for binary

TABLE 3.1

IRLS Fitting Algorithm

---

1. Initialize the expected response, μ, and the linear predictor, η, or $g(\mu)$.

2. Compute weights as:

$$W^{-1} = Vg'(\mu)^2$$

where $g'(\mu)$ is the derivative of the link function and V is the variance, defined as the second derivative of the cumulant, or $b''(\theta)$.

3. Compute a working response, a one-term Taylor linearization of the log-likelihood function, with a standard form of (using no subscripts):

$$z = \eta + (y - \mu)g'(\mu)$$

where $g'(\mu)$ is the derivative of the link function.

4. Regress z on predictors $x_1 \ldots x_n$ with weights W to obtain updates on the vector of parameter estimates, β.

5. Compute η, the linear predictor, based on the regression estimates.

6. Compute μ, the model fit, or $E(y)$, as $g^{-1}(\mu)$.

7. Compute the deviance or log-likelihood.

8. Iterate until the change in deviance or log-likelihood between two iterations is below a predicted level of tolerance, or threshold.

TABLE 3.2

Generic Binomial GLM Estimating Algorithm (expected information matrix)

```
Dev   = 0
μ = (y+0.5)/(m+1)                     ///  initialize μ; m=binomial
                                           denom
η = g(μ)                              ///  initialize η; link
                                           function
WHILE (abs(ΔDev)>tolerance)  {  ///  While loop
   w   =  1/(Vg'2)                    ///  weight
   z   =  η + (y-μ)g'                 ///  working response
   β   =  (X'wX)⁻¹ X'wz               ///  estimation of parameters,
                                           β
   η   =  X'β                         ///  calc linear predictor,η
   μ   =  g⁻¹(η)                      ///  calc fit, μ
   Dev0 = Dev
   Dev  = Deviance function           ///  deviance function; or LL
                                           function
   ΔDev = Dev-Dev0                    ///  check for difference in
                                           value
}
Chi2 = Σ (y-μ)²/V(μ)                  ///  calc Pearson χ²
LL   = -2(LL(y;y) – LL(y;μ))
AIC  = (-2LL + 2p)/n                  ///  AIC GOF statistic
BIC  =  -2*LL + p*ln(n)               ///  BIC GOF statistic
```

or Bernoulli models (logit, probit, cloglog, loglog). To fit continuous and count response models, the algorithm need only exchange the initialization of μ with $\mu = (y + \text{mean}(y))/2$: Tables 3.3 and 3.4 show the structure of the Newton–Raphson type maximum likelihood algorithm.

After the tolerance level has been reached (very small difference between old and new deviance statistics), the parameter estimates β, standard errors (SEs), etc. are at their fitted values.

GLM theory typically refers to the link function as $g(\mu)$. The derivative of $g(\mu)$ is used to multiply with the residual term when calculating the working response, z. The derivative is also referred to as simply g'. The inverse link, $g^{-1}(\eta)$ is equal to μ, the fitted value. The inverse link provides μ with its final value.

3.3 Maximum Likelihood Estimation

Maximum likelihood estimation (MLE), or ML, procedures typically use a variety of modified Newton–Raphson algorithm. Continuing the discussion from the last section, full maximum likelihood methodology uses the observed information matrix on which model standard errors are based. We saw how Fisher Scoring allows a simplification of this matrix in the case of distributions belonging to the exponential family. For the sake of completeness, we outline the formulae used for the calculation of the observed information matrix and show how it can be reduced when applied to exponential family models such as GLM.

As mentioned earlier in this chapter, derivation of terms for the estimating algorithm begins as a Taylor linearization and continues through to the calculation of the gradient, or first derivative of the likelihood function. Fisher Scoring, used as the basis of the GLM estimating algorithm, calculates the matrix of second derivatives based on the expected information matrix (EIM), or I, where:

$$I = -E\left[\frac{\partial^2 L}{\partial\beta\,\partial\beta'}\right] = E\left[\frac{\partial L}{\partial\beta}\frac{\partial L}{\partial\beta'}\right] \tag{3.39}$$

Newton–Raphson methodology, or full MLE, calculates the second derivatives of the likelihood on the basis of the observed information matrix (OIM), which allows estimation of likelihood-based algorithms other than the more limited exponential family form. We refer to this matrix as H, signifying Hessian.

$$H = \left[\frac{\partial^2 L}{\partial\beta_j\partial\beta_k}\right] = \sum_{i=1}^{n}\frac{1}{\alpha(\phi)}\left[\frac{\partial}{\partial\beta_k}\right]\left\{\frac{y_i-\mu_i}{V(\mu_i)}\left(\frac{\partial\mu}{\partial\eta}\right)_i x_j x_i\right\} \tag{3.40}$$

Solved, the above becomes:

$$-\sum_{i=1}^{n}\frac{1}{\alpha(\phi)}\left[\underbrace{\frac{1}{V(\mu_i)}\left(\frac{\partial\mu}{\partial\eta}\right)_i^2}_{\text{EIM}} - (\mu_i-y_i)\underbrace{\left\{\frac{1}{V(\mu_i)^2}\left(\frac{\partial\mu}{\partial\eta}\right)_i^2\frac{\partial V(\mu_i)}{\partial\mu} - \frac{1}{V(\mu_i)}\left(\frac{\partial^2\mu}{\partial\eta^2}\right)_i\right\}}_{\text{OIM}}\right]x_{ji}x_{ki}$$

$$\text{(3.41)}$$

Note that in the case of the canonical link, terms from $-(\mu-y)$ above cancel, reducing to the value of the EIM. This happens with the logistic algorithm, as well as with the canonical forms of all GLM models.

Schematizing the above to a generic Newton–Raphson type algorithm gives us the MLE algorithm in Table 3.3.

TABLE 3.3

Generic Maximum Likelihood Algorithm

```
Initialize β
while {||β_new -      β_old|| > tolerance and abs (LLnew - Llold).
                               tolerance)  {
    g = ∂L/∂β                  /// gradient, the derivative of the
                                   LL function
    H = ∂²L/∂β²                 /// Hessian, second derivative of LL
                                   function

    β_old =   β_new
    β_new =   β_old - H⁻¹g      /// update of new vector of
                                   parameter estimates

    L_old =   L_new
    Calc L_new
}
Calc variance matrix           /// for use in calculating standard
                                   errors
```

TABLE 3.4

A Newton–Raphson Type Estimating Algorithm

```
g   = g(μ)    = link
g'  = g'(μ)   = 1ˢᵗ derivative of link, wrt μ
g'' = g''(μ)  = 2ⁿᵈ derivative of link, wrt μ
V   = V(μ)    = variance
V'  = V(μ)    = derivative of variance
m   = binomial denominator
φ   = phi, a constant scale parameter
off = offset
μ   = (y + mean(y))/2   : binomial = (y+0.5)/(m+1)
η   = g
βn  = 0
while MAX(ABS(βn-βc))> tolerance  {
    βc  = βn
    z   = p(y-μ)/(Vg'Φ)                        <a column vector>
    s   = X'z                                  <gradient>
    We  = p/(ΦVg'2)                            <weight: expected IM>
    Wo  = We+p(y-μ){(Vg''+V'g')/(V²g'³Φ)}      <observed IM>
    Wo  = diag(Wo)                             <diagonalize Wo>
    H   = -X'WoX                               <Hessian>
    βn  = βc - H⁻¹s   :==: βc+(X'WoX)⁻¹ X'(p(y-μ))
                      :==: (X'WoX)⁻¹ X'W[η+(y-μ)g']
                      :==: (X'WoX)⁻¹ X'Wz <if z=η+(y-μ)g'>
    η       = X'βn + off                       <linear predictor>
    μ       = g⁻¹(η)                           <inverse link>
}
```

In more detail, the Newton–Raphson algorithm takes the form as above. In the algorithm, the line beginning the while loop needs explanation.

MAX(ABS($\beta n - \beta c$)) is the maximum value of the element-wise absolute differences between βn (the new β) and βc (the current β). The intercept is considered as just another parameter and is included in the comparison. The threshold set specifies that the elements of the parameter vector β must not change (much) from one iteration to another in order for the algorithm to converge. As with IRLS algorithms, the usual numeric threshold value is 10^{-6}.

We should also point out the generic score and Hessian terms used in the MLE approach, such that they are applicable to MLE logistic regression:

$$\text{Score } s = X'(y - \mu)/(Vg'\phi) \quad <\text{a column vector}> \quad (3.42)$$

$$\text{Hessian } H = -X'[diag[1/(\phi Vg'^2) + (y-\mu)\{(Vg'' + V'g')/(V^2 g'^3)\}]]X \quad (3.43)$$

MLE parameter estimates are calculated by iteratively updating the following relationship until convergence is achieved:

$$\beta n = \beta c - H^{-1}s \quad (3.44)$$

The parameter estimates, standard errors, and so forth of the logistic regression model, being the canonical form of the binomial family, are identical regardless of the method of estimation, that is, full maximum likelihood (ML) or iterative re-weighted least squares (IRLS). Again, it should be kept in mind that IRLS estimation is also referred to as Fisher Scoring.

To summarize, since the logistic model is the canonical form of the general binomial family of models, the more complicated observed information matrix used to calculate standard errors reduces to the expected information matrix. The expected information matrix is used only in generalized linear models (GLM). It therefore follows that the binary logistic regression model can be estimated using either GLM or MLE methodology. Both methods produce identical results. Many statisticians prefer using the GLM approach, though, due to the host of fit statistics that are typically associated with the procedure as implemented in commercial software. We shall discuss these considerations later in the text.

Exercises

3.1. Why is there a difference in standard errors when a non-canonical linked generalized linear model is estimated using IRLS instead of using a full maximum likelihood method of estimation?

3.2. What is a score function?

3.3. What is the working response in a generalized linear model's algorithm? How does it differ from the model response?

3.4. In exponential family form, the cumulant of a probability function is defined as b(θ). What exactly is θ, and what statistic is produced by taking the derivative of the cumulant with respect to θ? What statistic is produced taking the second derivative?

3.5. Maximum likelihood and iterative re-weighted least squares (IRLS), or Fisher Scoring, are the two foremost methods of estimation discussed in this chapter. In what sense is IRLS a variety, or subset, of maximum likelihood estimation?

3.6. List several other methods of estimation aside from maximum likelihood and IRLS. When are these alternatives primarily used?

3.7. Complete research on and explain the differences between frequentist and Bayesian estimation. Which of the two are we emphasizing in this text? Can logistic regression modeling be performed using the method we do not employ in this text?

R Code

No statistical output is given in this chapter.

4

Derivation of the Binary Logistic Algorithm

4.1 Terms of the Algorithm

It is important to have an overview of how the logistic model is derived from the Bernoulli probability distribution function, which we have already seen is a subset of the general binomial PDF. In this chapter we shall derive the statistical functions required for the logistic regression algorithm from the Bernoulli PDF. In the process we address functions that are important to model convergence and the estimation of appropriate parameter estimates and standard errors. Moreover, various statistics will emerge that are important when assessing the fit or worth of the model.

p, or π, are standard terms that are used to indicate the probability of success, ranging from near 0 to near 1. We shall prefer p rather than π, but know that both are common in the literature dealing with logistic regression. The variable y usually represents the actual model response, with a value of 1 or 0. A value of $y = 1$ indicates a success, or satisfaction of a given criterion; $y = 0$ represents a lack of success, which may be thought of as a failure to satisfy the defined criterion. Given these terms, the Bernoulli PDF can be expressed as:

Given these terms, the Bernoulli PDF can be expressed as:

$$f(y_i; p_i) = \Pi_{i=1}^{n} \; p_i^{y_i}(1 - p_i)^{1-y_i} \tag{4.1}$$

expressed in terms of the exponential family form, without subscripts:

$$f(y; \theta, \phi) = \Pi\exp\{(y\theta - b(\theta)) / \alpha(\phi) + c(y; \phi)\} \tag{4.2}$$

we have:

$$f(y; p) = \Pi\exp\{y\ln(p / (1 - p)) + \ln(1 - p)\} \tag{4.3}$$

Equation 4.3 may be re-parameterized to estimate p given y, rather than the reverse as is the case for probability functions. The resultant function is called the likelihood function, which appears as:

$$L(p;y) = \Pi\exp\{(y\ln(p/(1-p)) + \ln(1-p)\} \qquad (4.4)$$

The single parameter exponential family of distributions has a unique and extremely helpful feature. If a distribution is parameterized to reflect the above structure, then the link, mean, and variance of the distribution can be calculated based on the following relationships.

LINK

$$\theta = \ln(p/(1-p)) = \eta = \ln(\mu/(1-\mu)) \qquad (4.5)$$

MEAN

The mean is calculated as the first derivative of the cumulant, $b(\theta)$, wrt θ:

$$b(\theta) = -\ln(1-p) \qquad (4.6)$$

$$b'(\theta) = p = \mu \qquad (4.7)$$

where $b'(\theta)$ is the derivative of $b(\theta)$ with respect to θ. Notice that we can now define μ as identical to p. μ is the traditional term used in generalized linear models for the mean. In the actual GLM algorithm, μ, the fitted value, is calculated from the linear predictor as the inverse link.

INVERSE LINK

$$\mu = g^{-1}(\eta) \qquad (4.8)$$

$$= 1/(1+e^{-\eta}) \qquad (4.9)$$

or

$$= e^{\eta}/(1+e^{\eta}) \qquad (4.10)$$

The link, defined in this manner, is usually signified as $g(\mu)$. We therefore have:

$$g(\mu) = \ln(\mu / (1 - \mu)) \tag{4.11}$$

and the derivative of $g(\mu)$ as:

$$g'(\mu) = 1 / (\mu / (1 - \mu)) \tag{4.12}$$

VARIANCE

The variance is calculated as the second derivative of $b(\theta)$ wrt θ:

$$b''(\theta) = p(1 - p) = \mu(1 - \mu) \tag{4.13}$$

which is simply the second derivative of the cumulant.

LOG-LIKELIHOOD

The log-likelihood is determined by taking the natural log of the likelihood function so that it appears in exponential family form as:

$$L(\mu; y) = \Sigma\{y\ln(\mu / (1 - \mu)) + \ln(1 - \mu)\} \tag{4.14}$$

Note the summation of observations across the model rather than the product employed. Logging the function allows summation, which is much easier to use in calculations than multiplication. For this reason, the log-likelihood is always used in preference to the likelihood for maximization purposes.

Many formulations of the log-likelihood function, when used as the measure of convergence in a binary logistic regression, are partitioned into two functions—one when $y = 1$, and the other when $y = 0$. This dual representation of the log-likelihood appears as:

$$L(y = 1) = \Sigma\{\ln(\mu / (1 - \mu)) + \ln(1 - \mu)\} = \Sigma \ln(\mu) \tag{4.15}$$

and

$$L(y = 0) = \Sigma \ln(1 - \mu) \tag{4.16}$$

Given that the logistic linear predictor, xb, is η, we can parameterize the log-likelihood in terms of xb rather than μ. In fact, maximum likelihood algorithms require estimation in terms of xb. We have seen above that the relationship of xb to μ is

$$\mu = \exp(xb) / (1 + \exp(xb)) \tag{4.17}$$

As a consequence, algebraic conversion results in the binary logistic log-likelihood function being expressed as:

$$L(xb;y) = \Sigma\{y(xb) - \ln(1+\exp(xb))\} \tag{4.18}$$

The above is also observed in the following form, using β rather than b, as:

$$L(x\beta;y) = \Sigma\{y(x\beta) - \ln(1+\exp(x\beta))\} \tag{4.19}$$

DEVIANCE

The deviance function, which is also used as a goodness of fit statistic for logistic models, is defined as:

$$D = 2\Sigma\{LL(y;y) - LL(\mu;y)\} \tag{4.20}$$

which calculates to:

$$D = 2\Sigma\{y\ln(y/\mu) + (1-y)\ln(1-y)/(1-\mu))\} \tag{4.21}$$

Notice that the full log-likelihood function replaces a y for each instance of μ in the standard log-likelihood.

As with the log-likelihood, the deviance function can also be partitioned into two formulations. These appear as:

$$D(y = 1) = 2\Sigma \ln(1/\mu) \tag{4.22}$$

and

$$D(y = 0) = 2\Sigma \ln(1/(1-\mu)) \tag{4.23}$$

The first and second derivatives of the log-likelihood are also important when estimating logistic regression by means of maximum likelihood. We give them as:

$$\frac{\partial(L)}{\partial\mu} = \frac{y}{\mu} - (1-y)(1-\mu)^{-1} = \frac{y-\mu}{\mu(1-\mu)} \tag{4.24}$$

$$\frac{\partial^2(L)}{\partial\mu^2} = \frac{y}{\mu^2} - \frac{1-y}{(1-\mu)^2} \tag{4.25}$$

4.2 Logistic GLM and ML Algorithms

We are now at the point that the basics have been covered. We have discussed the nature of odds, odds ratio, logit, the maximum likelihood and GLM IRLS estimating algorithms, the estimation algorithms for the Bernoulli logistic model, and the derivation of the main terms in the logistic estimation model. We now turn to a fully worked-out example.

The binary response logistic regression GLM algorithm may be expressed as illustrated in Table 4.1.

TABLE 4.1

GLM Binary Logistic Regression Algorithm

```
Dev=0
μ = (y+.5)/2
η = ln(μ/(1-μ))
WHILE (abs(ΔDev) > tolerance level) {  ///  tolerance
                                            usually .000001

        w = μ(1-μ)
        z = η + (y-μ)/w
        β = (X'WX)⁻¹ X'Wz
        η = Xβ
        μ = 1/(1+exp(-η))
        OldDev = Dev
        Dev = 2Σ ln(1/μ) if y==1
        Dev = 2Σ ln(1/(1-μ)) if y==0
        ΔDev = Dev - OldDev
}
X² = Σ (y-μ)² / (μ/(1-μ))
LL = Σ{ y*ln(μ/(1-μ)) + ln(1-μ) }
```

The Pearson *Chi2* statistic is shown in the above algorithm. It is displayed in nearly all statistical output, no matter the software package. It is the sum of squared raw residuals divided by the variance. The *Chi2* statistic is used as a goodness of fit statistic but plays a more important role as the basis for defining the model *Chi2* dispersion statistic. We shall discuss this statistic in more detail when addressing the issue of overdispersion.

Recall, for the logistic model, the less complex GLM approach, as displayed in Table 4.1, produces identical results to the more computer intensive full maximum likelihood estimation. Using the **heart01** data of Chapter 2, we show both approaches, modeling *death* on *anterior*. Note that rounding errors show minor differences at the one-millionth place (0.000001).

GLM ESTIMATION

```
. glm death anterior, fam(bin)

Generalized linear models              No. of obs       =       4696
Optimization  : ML                     Residual df      =       4694
                                       Scale parameter  =          1
Deviance       =   1543.845261         (1/df) Deviance  =   .3288976
Pearson        =          4696         (1/df) Pearson   =   1.000426

Variance function   : V(u) = u*(1-u)
Link function: g(u) = ln(u/(1-u))

                                       AIC              =   .3296093
Log likelihood = -771.9226305          BIC              =  -38141.42
-------------------------------------------------------------------
            |              OIM
    death|   Coef.   Std. Err.     z   P>|z|  [95% Conf. Interval]
--------+----------------------------------------------------------
anterior| .805044    .15542    5.18  0.00   .500417      1.10967
   _cons|-3.62095    .12379  -29.25  0.00  -3.86358     -3.37832
-------------------------------------------------------------------
```

MLE ESTIMATION

```
. logit death anterior,  nolog

Logistic regression                    Number of obs    =       4696
                                       LR chi2(1)       =      28.14
                                       Prob > chi2      =     0.0000
Log likelihood = -771.92263            Pseudo R2        =     0.0179
-------------------------------------------------------------------
    death|    Coef. Std.Err.     z    P>|z|  [95%Conf. Interval]
--------+----------------------------------------------------------
anterior| .8050445 .1554244   5.18  0.000   .5004182     1.109671
   _cons|-3.620952 .1237929 -29.25  0.000  -3.863582    -3.378323
-------------------------------------------------------------------
```

4.3 Other Bernoulli Models

Although this is a text on logistic based regression models and at this point in our discussion an examination of binary response or Bernoulli models, we should perhaps give a brief mention of the other commonly used Bernoulli models. An overview of these models, in addition to other major models based on the binomial distribution, is given in Table 4.2.

TABLE 4.2

Bernoulli Functions

	Link	I-link
Logit	$\ln(\mu/(1-\mu))$	$1/(1+\exp(-\eta))$
Probit	$\Phi^{-1}(\mu)$	$\Phi(\eta)$
Cloglog	$\ln(-\ln(1-\mu))$	$1-\exp(-\exp(\eta))$
Loglog	$-\ln(-\ln(\mu))$	$\exp(-\exp(-\eta))$

The foremost Bernoulli models can be listed as follows:

> Logit
> Probit
> Complementary loglog
> Loglog

In terms of generalized linear models (GLM), each of the above represents a separate link in the binomial/Bernoulli family. Each is parameterized in both individual and grouped format. The logit link, the link of the logistic model, is the canonical link, meaning that it stems directly from the binomial/Bernoulli PDF.

The complementary loglog model was the first of the above models to be developed, in 1922 by R.A. Fisher, perhaps the leading figure in twentieth century statistics. The probit model was designed by Bliss in 1934, 10 years before Berkson's logit model. If these models had been estimated using a generalized linear models approach, it is quite likely that the logit model would have preceded the others. But GLMs were not derived until 1972. In any case, probit and complementary loglog regression models have a solid history in the modeling of binary response data and are still widely used in research. This is particularly the case with the probit model.

Conceived of as a GLM, a binary logit model may be amended to become a probit, complementary loglog, or loglog model by simply exchanging link and inverse link functions.

Φ is based on the cumulative normal or Gaussian distribution defined as:

$$\mu = \int_{-\infty}^{\eta} \phi(u)du \tag{4.26}$$

where $\Phi(u)$ is the standard normal density function:

$$\exp(-.5u^2)/\sqrt{2\pi} \tag{4.27}$$

or

$$\exp(-.5u^2) / 2.5066283 \qquad (4.28)$$

Non-logit binomial/Bernoulli links are all termed non-canonical links. When estimating non-canonical linked Bernoulli models, it must be understood which information matrix is being used to calculate the parameter estimates and standard errors. In the case of logit, it makes no difference if the observed or expected information matrix is used, since the former reduces to the latter. However, this is not necessarily the case for non-canonical linked Bernoulli models. If the probit, for instance, is being modeled using a traditional GLM algorithm, it is likely that the expected information matrix is used. This is called Fisher Scoring. If a full maximum likelihood algorithm is used to estimate a probit model, most likely the observed information matrix has been used.

The problem is that when there are relatively few observations in the data, standard errors based on the observed information are preferred to those based on the simpler expected information. It is possible to convert an algorithm from using the expected to the observed, and this has been done for the default GLM commands in SAS and Stata software. The reader should be cautioned, though, that other software may not implement this conversion for their GLM procedures.

The results of a probit and logit model are generally not too different from one another. If one prefers having a model estimate odds ratios, then the proper selection is a logit model. If the model response is based on a latent normally distributed variable, then a probit may be more appropriate. Complementary loglog and loglog links are asymmetrically sigmoidal and are typically used to model extreme values. Both logit and probit are symmetric about 0.5.

Those interested in learning more about these models, the link functions, and their relationships should read the relevant chapters in Hardin and Hilbe (2007).

Exercises

4.1. Given the binary logistic link function, $\ln(\mu/(1 - \mu))$, derive the inverse link function, $1/(1 + \exp(-\eta))$. Show each step in the derivation.

4.2. For the binary logistic model, what is the difference in meaning between the terms θ, and $x\beta$? Does the same relationship in meanings exist for the binary probit model? Why?

4.3. What is the primary difference in probabilities produced by a logit model compared to the probabilities produced by a complementary loglog model?

4.4. List reasons for using the deviance statistic as a criterion of convergence rather than the log-likelihood. Likewise, list when it is preferable to use the log-likelihood statistic as the criteria of convergence rather than the deviance.

4.5. What is the relationship of a probability function and likelihood function?

4.6. Design an IRLS fitting algorithm using the programming language of your favorite statistical software.

R Code

Note: R's logistic regression command is based on a generalized linear models approach, and is part of the **glm()** function. There is no difference between GLM and maximum likelihood logistic estimation. Stata has three different logistic regression commands: **logit, logistic**, and **glm** (with binomial family and default logit link). Again, logistic regression is performed in R using the **glm()** function with the binomial family and default logit link.

No example output is given in Sections 1 and 3.

Section 4.2 Logistic GLM and Logistic Algorithms

```
library('foreign') #foreign import package library
heart01<- read.dta('heart01.dta') #works for STATA v.5-9
head(heart01) #show a few records
nrow(heart01) #show number of rows
fit1<- glm(death ~ anterior, data=heart01, family=binomial
(link='logit'))
summary(fit1)
```

5

Model Development

5.1 Building a Logistic Model

We shall include more variables from the **heart01** data to use as an example of a multivariate, or multivariable, **logistic** regression in Chapter 2. Recall that the response and predictor from the earlier univariable model were:

Response : *death* [1 = die; 0 = not die, within 48 hours of initial symptoms].

Predictor : *anterior* [1 = anterior MI; 0 = inferior MI]

The following predictors will be added to the model:

hcabg	History of 1 = CABG; 0 = PTCA ///	
	CABG = Coronary Artery Bypass Graft	
	PTCA = Percutaneous Transluminal Coronary Angioplasty	
killip	killip level 1–4	/// increasing order of severity
agegrp	agegrp	/// 4 age groups
center	Provider or Center	/// id number

We will also want to model *age* as a continuous predictor. However, the data are currently stored as a four-level categorical predictor. We shall create discrete ages from 40 to 100, with random numbers being created for each age group. The following code can be used. Again, we do this for pedagogical purposes; the actual ages of the patients are unknown.

```
. tab hcabg

   Hist of |
1=CABG;0=PT |
        CA |      Freq.     Percent        Cum.
-----------+-----------------------------------
         0 |      5,207       96.64       96.64
         1 |        181        3.36      100.00
-----------+-----------------------------------
     Total |      5,388      100.00
```

73

```
. tab killip

    Killip |
     level |        Freq.        Percent         Cum.
-----------+-------------------------------------------
         1 |        3,768         73.12         73.12
         2 |        1,029         19.97         93.09
         3 |          292          5.67         98.76
         4 |           64          1.24        100.00
-----------+-------------------------------------------
     Total |        5,153        100.00


. tab agegrp

    agegrp |        Freq.        Percent         Cum.
-----------+-------------------------------------------
      =<60 |        1,907         35.39         35.39
     61-70 |        1,390         25.80         61.19
     71-80 |        1,425         26.45         87.64
      > 80 |          666         12.36        100.00
-----------+-------------------------------------------
     Total |        5,388        100.00
```

. gen byte age = 40+trunc(21 * uniform()) if age ==1 /// 40-60

. replace age = 61+trunc(10 * uniform()) if age ==2 /// 61-70

. replace age = 71+trunc(10 * uniform()) if age ==3 /// 71-80

. replace age = 81+trunc(20 * uniform()) if age ==4 /// 81-100

Every software package handles categorical predictors in different ways. Some allow an easy creation of separate dummy or indicator variables from the levels of a categorical variable, while others prefer that users not specifically create indicator variables. Rather, the software prefers to automatically factor designated categorical variables for modeling purposes, but not maintain the indicator variables as separate predictors. Stata uses the first of these alternatives, allowing the user to easily create indicator variables. This way, the user has more direct control over how indicators are used in the model. We shall find that having this type of control can make the modeling process more efficient in the long run. Care must be taken, however, when creating indicator predictors from a categorical variable with a missing level. See Appendix A for a full discussion.

In any case, indicator predictors for the levels of *killip* and *agegrp* can be created by:

. tab killip, gen(kk)   /// Note: *kk* indicators were created earlier in
                                         text as well

. tab agegrp, gen(age)

Levels of indicators will take the form of *kk1*, *kk2*, ... and *age1*, *age2*, and so forth. Ascending numbers are attached to the ending of user-defined names, as specified within the parentheses of the **generate** function.

The variables now in the **heart01** data are

```
. d                /// abbreviation for "describe"

Contains data from heart01.dta
  obs:       5,388
  vars:         15                               18 Oct 2007 16:30
  size:    107,760      (99.0% of memory free)
-------------------------------------------------------------------
variable    storage   display   value
name        type      format    label     variable label
-------------------------------------------------------------------
death       byte      %4.0g               Death within 48 hrs onset
anterior    byte      %9.0g     anterior  1=anterior; 0=inferior
hcabg       byte      %4.0g               Hist of 1=CABG;0=PTCA
kk1         byte      %8.0g               killip==       1.0000
kk2         byte      %8.0g               killip==       2.0000
kk3         byte      %8.0g               killip==       3.0000
kk4         byte      %8.0g               killip==       4.0000
age1        byte      %9.0g               =<60
age2        byte      %9.0g               61-70
age3        byte      %9.0g               71-80
age4        byte      %9.0g               >80
center      int       %12.0g              Provider or Center
killip      byte      %8.0g               Killip level
agegrp      byte      %8.0g     age
age         byte      %9.0g
-------------------------------------------------------------------
Sorted by:   center
```

When one is deciding which explanatory predictors best explain the death of a patient within 48 hours of admission, a useful tactic, suggested by Hosmer and Lemeshow (2000), is to model each potential predictor on the response to determine which predictors clearly fail to contribute to the model. Probability value (*p*-value) criteria for keeping predictors for additional analysis generally range from 0.10 to 0.5, based in large part on the traditions of the discipline of the statistician or researcher and in part on what is reasonable for the study data being modeled. Unless there is clinical information or data from previous studies indicating otherwise, we shall retain predictors having parameter *p*-values of 0.25 or less.

The **ulogit** command can be used to perform univariable **logistic** regression models on a list of binary, continuous, and categorical predictors. However, it is preferable to separate the categorical predictors from the other predictors. If this is not done, the correct *p*-values will not be calculated for the categorical predictors due to a reference level not being identified. Note that the **ulogit** command is not part of Stata's commercial package. It was written to accompany short courses I was teaching on logistic regression in 1991. I have since updated the code to have it work satisfactorily with Stata version 8 and higher. It is posted at http://www.stata.com/users/jhilbe/.

```
. ulogit death anterior hcabg age

                Univariable Logistic Regression Models

Intercept LL = -785.9942                       1 degree of freedom

Variable      OR         LL         Chi2      Prob    95 percent CI

anterior    2.2368    -771.923     28.143    0.0000   1.6494   3.0334
hcabg       1.6821    -986.051      2.710    0.0997   0.9422   3.0031
age         1.0582    -913.286    148.241    0.0000   1.0482   1.0682
```

The two binary predictors, *anterior and hcabg,* have positive effects with respect to the response, although *hcabg* is not significant at the $p < 0.05$ level. The continuous predictor, *age,* has a slightly positive relationship and is significant at $p < 0.05$.

We may interpret the results of each predictor separately as illustrated in Section 5.1.1 (Boxes 5.1 through 5.3.)

5.1.1 Interpretations

Box 5.1

Patients with a history of having had an anterior infarct have some 2.25 times greater odds of dying within 48 hours of admission than those patients having had an inferior infarct.

Patients with a history of having had a CABG have a 68 percent greater odds of dying within 48 hours of admission than those patients having had a PTCA.

For continuous variables, such as age, the odds ratio represents the effect of a one unit change in the variable. For every year increase in age, the odds of death within 48 hours of admission increase by approximately .06, or 6 percent. Odds ratios less than 1 represent a net decrease in odds; odds ratios greater than 1 represent a net increase in odds. For a 10-year increase in age, we multiply the age increase by the coefficient and exponentiate. The parameter estimate of age is 0.0565292. Therefore, the odds ratio of a 10-year increase in age is exp(10*.0565292), or 1.7599616. There is a 76 percent increase in odds for a 10-year increase in age.

The three categorical predictors are now modeled individually.

```
. logistic death kk2-kk4

Logistic regression                    Number of obs   =        5153
                                       LR chi2(3)      =      150.51
                                       Prob > chi2     =      0.0000
Log likelihood = -849.28983            Pseudo R2       =      0.0814

------------------------------------------------------------------
death | Odds Ratio  Std. Err.    z    P>|z|  [95% Conf. Interval]
------+-----------------------------------------------------------
  kk2 |   2.804823   .4497351   6.43  0.000   2.048429    3.840521
  kk3 |   4.333438   .9467527   6.71  0.000   2.824005    6.649666
  kk4 |  27.61688    7.545312  12.15  0.000  16.16645    47.17747
------------------------------------------------------------------
```

The more severe the *killip* level, and the older the age group, the greater the odds of death. These are all unadjusted values, and all make sense. All predictors are to be kept in the model at this stage in development.

Note that the above **ulogit** table is found in many research journal articles where the goal is to develop a multivariable logistic model. It is the first stage in determining if a variable has any impact on the value of the dependent or response variable. In this case, the response is *death*: 1 = patient died within 48 hours; 0 = patient alive after 48 hours of admission. Note also that any patient who dies after 50 hours of admission will be coded as 0.

An additional point needs to be made at this stage. Recall that when the **heart01** data were listed above, the number of observations was presented as 5388. Yet the univariable model of *death* on *killip* levels displays *Number of obs = 5153*. This is 235 observations less than the expected 5388. However, such differences are common in real data and are a result of missing values. A tabulation of *killip*, including missing values, is given as:

```
. tab killip, miss

  Killip |
   level |      Freq.     Percent        Cum.
---------+---------------------------------------
       1 |      3,768       69.93       69.93
       2 |      1,029       19.10       89.03
       3 |        292        5.42       94.45
       4 |         64        1.19       95.64
       . |        235        4.36      100.00
---------+---------------------------------------
   Total |      5,388      100.00
```

The matrix inversion algorithm that is an essential part of the maximum likelihood estimation process requires that there are no missing values in the data. Statistical software is written to drop observations in which any variable in the model—a response variable as well as predictors—has a missing value. This is done prior to matrix inversion. Therefore, it is not uncommon to find differing numbers of model observations when adding or dropping predictors from a model. Unfortunately, this has a bearing on comparative model fit analysis, a point we discuss later in the text.

We return now to the interpretation of the univariable categorical model of *death* on *killip* level.

Box 5.2

The odds of dying within 48 hours of admission increase by a factor of 2.8 for killip level 2 patients over that of killip level 1 patients. Patients with a killip level 3 have an increase in the odds of death within 48 hours of hospital admission of some 4.33 times that of killip level 1 patients. Patients with killip level 4 have an increase in odds of death by over 27 times that of patients with killip level 1.

If the underlying incidence of outcome, or probability, value is low (for example, less than 10 percent), the odds ratios in the above table approximate risk ratios, and may be interpreted in terms of the following:

Patients with a killip level 2 are 2.8 times more likely to die within 48 hours of admission than patients with killip level 1. Patients with killip level 3 are 4.33 times more likely to die than killip level 1 patents. Patients with killip level 4 are over 27 times more likely to die than patients with killip level 1.

Most texts use risk language to express all odds ratio relationships. The concept of likeliness is more understandable to some than is odds ratio, and for many models is justified on the basis of the underlying probabilities. We shall generally not employ risk language, though, since it is not technically appropriate. However, the reader should understand that many statisticians tend to prefer the use of probabilities than odds when interpreting logistic model odds ratios. This subject is discussed at length later in the text.

```
. logistic death age2-age4

Logistic regression                      Number of obs   =        5388
                                         LR chi2(3)      =      168.68
                                         Prob > chi2     =      0.0000
Log likelihood = -903.06717             Pseudo R2       =      0.0854

------------------------------------------------------------------
death | Odds Ratio Std. Err.    z    P>|z|  [95% Conf. Interval]
------+-----------------------------------------------------------
 age2 |  2.086041   .5547185   2.77  0.006   1.238717  3.512965
```

```
age3 |   5.604167   1.295943    7.45   0.000    3.561835   8.817558
age4 |  11.78908    2.775262   10.48   0.000    7.43188   18.70083
-----------------------------------------------------------------
```

Box 5.3

Patients 61–70 years of age have 2 times greater odds of dying within 48 hours of admission than patients 60 and younger. Patients 71–80 have 5.5 times greater odds of dying than patients 60 and younger. Patients 80 and older have over 11.5 times greater odds of dying than patients 60 and younger.

Categorical predictors must always be evaluated in terms of a reference. You may change the reference if you wish, but in all cases, the reference level, which has been made into a separate indicator variable, is left out of the model. For instance, we can change the reference for *age**, which was derived from *agegrp*, by leaving out the level we select. If *age4* is selected as the reference, the model appears as:

```
. logistic death age1-age3

Logistic regression                    Number of obs   =       5388
                                       LR chi2(3)      =     168.68
                                       Prob > chi2     =     0.0000
Log likelihood = -903.06717            Pseudo R2       =     0.0854
-----------------------------------------------------------------
death | Odds Ratio  Std. Err.     z    P>|z|  [95% Conf. Interval]
------+----------------------------------------------------------
age1 |   .0848243   .0199685  -10.48   0.000   .0534736   .1345554
age2 |   .176947    .0361498   -8.48   0.000   .1185619   .2640835
age3 |   .4753695   .0744059   -4.75   0.000   .3497835   .6460456
-----------------------------------------------------------------
```

Notice the inverse relationship of the odds ratios. The *age1* level, patients 60 and younger, have a 92 percent reduction in the odds of dying [1 − 0.08] within 48 hours of admission than patients over the age of 80 (*age4*). Patients age 71–80 (*age3*) have half the odds of dying as those over 80.

Because no univariable model has a *p*-value higher than 0.25, we may enter all of the predictors into a single multivariable logistic model.

5.1.2 Full Model

We model all predictors—one model with *age* as a continuous predictor, one as a categorical predictor.

AGE CONTINUOUS

```
. logistic death anterior hcabg age kk2-kk4
```

Logistic regression   Number of obs   =   4503
                                                 LR chi2(6)   =   198.58
                                                 Prob > chi2   =   0.0000
Log likelihood = -643.81958   Pseudo R2   =   0.1336

death	Odds Ratio	Std. Err.	z	P>|z|	[95% Conf. Interval]	
anterior	1.91492	.3201291	3.89	0.000	1.379908	2.657366
hcabg	2.005966	.7017128	1.99	0.047	1.01056	3.98185
age	1.053013	.0060901	8.93	0.000	1.041144	1.065017
kk2	2.281317	.4105659	4.58	0.000	1.60323	3.2462
kk3	2.297846	.6170106	3.10	0.002	1.35756	3.889403
kk4	14.90713	5.258332	7.66	0.000	7.466923	29.76092

AGE CATEGORICAL

```
logistic death anterior hcabg kk2-kk4 age2-age4
```

Logistic regression   Number of obs   =   4503
                                                 LR chi2(8)   =   212.57
                                                 Prob > chi2   =   0.0000
Log likelihood = -636.82619   Pseudo R2   =   0.1430

death	Odds Ratio	Std. Err.	z	P>|z|	[95% Conf. Interval]	
anterior	1.901714	.3186377	3.84	0.000	1.369379	2.640988
hcabg	2.103391	.7423324	2.11	0.035	1.053204	4.200757
kk2	2.254419	.4069454	4.50	0.000	1.582647	3.211331
kk3	2.174442	.5851167	2.89	0.004	1.283215	3.684649
kk4	14.31629	5.09811	7.47	0.000	7.123816	28.77054
age2	1.6548	.5132771	1.62	0.104	.9009995	3.039252
age3	4.580415	1.219386	5.72	0.000	2.718318	7.718083
age4	8.9873	2.443214	8.08	0.000	5.275074	15.31193

A continuous predictor, such as *age*, has a parameter estimate that gives us only one value across the entire range of values. It may well be—and usually is—that the odds change as one gets older. We see that this is exactly what happens in this case. The odds of death for each of the four age divisions are substantially different from one another. We would not be able to know this if we only had a single parameter estimate. Of course, we may graph *age* against *death* and observe changes in risk, or rather odds, but we do not have a quantitative differentiation. For this reason, many researchers—and referees—prefer to categorize or factor continuous predictors, particularly for medical research. They may, however, use the continuous predictor for graphing purposes. There may be times, though, when a continuous predictor should be retained in the model. I advise doing so when there is a uniformity of odds across levels. Continuous predictors should also remain as such if the researcher is duplicating or testing previous research in which a continuous predictor(s) was entered in the model. There is no single criterion for making the decision.

The full model, where each predictor is adjusted by the others, tells us the same basic story as the univariable models. We should combine age categories 1 and 2; that is, we need to make our reference age larger—all ages less than 71. We do this because the *p*-value of *age2* is 0.104, giving evidence that it does not contribute to the model when adjusted by the other predictors.

5.1.3 Reduced Model

```
. logistic death anterior hcabg kk2-kk4 age3-age4

Logistic regression                 Number of obs   =        4503
                                    LR chi2(7)      =      209.90
                                    Prob > chi2     =      0.0000
Log likelihood = -638.15957         Pseudo R2       =      0.1412
---------------------------------------------------------------------
   death | Odds Ratio  Std. Err.   z    P>|z|  [95% Conf. Interval]
---------+-----------------------------------------------------------
anterior |   1.894096   .3173418  3.81  0.000   1.363922    2.630356
   hcabg |   2.195519   .7744623  2.23  0.026   1.099711    4.383246
     kk2 |   2.281692   .4117012  4.57  0.000   1.602024    3.249714
     kk3 |   2.218199   .5971764  2.96  0.003   1.308708    3.759743
     kk4 |  14.63984   5.218374  7.53  0.000   7.279897   29.44064
    age3 |   3.549577   .7119235  6.32  0.000   2.395823    5.258942
    age4 |   6.964847   1.44901   9.33  0.000   4.632573   10.47131
---------------------------------------------------------------------
```

The likelihood ratio test is a much-preferred method of assessing whether a predictor or groups of predictors significantly contribute to a model. Most statisticians simply use the *p*-value based on the *z*-statistic, or alternatively the Wald statistic with statistical packages such as SAS to determine whether a predictor should be retained in a model. However, it has been demonstrated by simulation tests that the likelihood ratio test is preferred. When the *p*-value of a predictor is close to the criterion selected for retention, it is preferred to use the likelihood ratio test rather than rely on the table of parameter estimates.

The likelihood ratio statistic is defined as:

$$G = -2(\text{LLr} - \text{LLf}) \tag{5.1}$$

where LLr represents the log-likelihood of the reduced model and LLf is the log-likelihood of the full or original model. Using Stata, we calculate the likelihood ratio test by:

```
. qui logistic death anterior hcabg kk2-kk4 age2-age4
                              /// Full model

. estimates store A
                              /// store model log-likelihood

. qui logistic death anterior hcabg kk2-kk4 age3-age4
                              /// reduced model

. estimates store B

. lrtest A B

Likelihood-ratio test                LR chi2(1)   =      2.67
(Assumption: B nested in A)          Prob > chi2 =    0.1025
```

The likelihood ratio test confirms what we learned by the full model *p*-value for *age2* ($p = 0.104$). *age2*, when adjusted by the other predictors, does not significantly contribute to the model, that is, to an understanding of the response. Note that the predictor *p*-value and *Chi2* value of the likelihood ratio test do not always cohere as well as in this case.

The model appears to be well fitted; in other words, all *p*-values are significant. However, the values may only appear to be significant. It may be the case that the model is not appropriate for the data, and that the *p*-values only coincidentally appear to contribute to the model. Or, it may be that the predictors contribute to the model, but that the model is not suitable for the data. We can assess the fit of the model by using a number of techniques developed by statisticians over the past 20 years. In some cases, however, the fit tests we use are fairly new.

5.2 Assessing Model Fit: Link Specification

Logistic models are essentially nonlinear, but by virtue of a link function, become linear. The logistic link function is called the *logit*, defined as the natural log of the odds of the response variable having the value of 1. We test the assumption of the linearity of the logit with respect to the model predictors using various link specification tests.

We shall describe five types of link specification tests appropriate for logistic regression: the Box–Tidwell test, the Tukey–Pregibon link test, a test using partial residuals, a linear slopes test, and a test using generalized additive models. These are not tests to normalize the continuous predictor but rather tests of the relationship of linearity between the logit and predictors.

5.2.1 Box–Tidwell Test

The Box–Tidwell test is a simple method that can be used to assess the linear relationship between the logit, defined as $\ln(\mu/(1-\mu))$ for binary logistic models, or as $\ln(\mu/(1-\mu/m))$ for binomial or aggregated logistic models, and the model covariates on the right-hand side of the regression equation:

$$\ln(\mu/(1-\mu)) = \beta_0 + \beta_1 X_1 + \beta_2 X_2 + \ldots + \beta_n X_n \tag{5.2}$$

As we recall from our earlier discussions, the logit function is the logistic linear predictor, which in generalized linear models terminology is also referred to as η. A more generic symbol for the linear predictor is xb. Both of these terms are commonly found in related statistical literature.

If the relationship between the logit and right-hand terms is not linear—that is, that the model is not "linear in the logit"—then the estimates and standard errors are biased, and therefore, cannot be trusted. The linearity of the relationship is an essential assumption of the logistic model. It is, therefore, important to test the assumption during the model building process.

The Box–Tidwell test is performed by adding an interaction of each continuous predictor with its log transform and modeling the predictors and interactions using logistic regression. If an interaction term or terms are statistically significant, then the assumption of linearity has been violated. In fact, the violation affects the ability of the model to avoid a Type II error.

Let us use the **heart01** data to construct a single predictor logistic model, assessing its linearity using the Box–Tidwell test. We shall use *age*, the only continuous variable in the **heart01** data, as our predictor, with *death* as the response.

First, examine the range of *age*, assuring ourselves that it is indeed continuous. We can, however, recall our earlier discussion of *age* in Section 2.3 and our attempts to normalize it for the purposes of categorization.

```
. su age

Variable |     Obs        Mean    Std. Dev.       Min        Max
---------+------------------------------------------------------
     age |    5388    65.80976    14.42962         40        100

. logit death age, nolog
```

```
Logistic regression                        Number of obs   =        5388
                                           LR chi2(1)      =      148.24
                                           Prob > chi2     =      0.0000
Log likelihood = -913.28575               Pseudo R2       =      0.0751
------------------------------------------------------------------------
death |      Coef.    Std. Err.     z    P>|z|  [95% Conf. Interval]
------+-----------------------------------------------------------------
  age |   .0565292    .0048193   11.73  0.000   .0470834    .0659749
_cons | -7.076784    .3729006  -18.98  0.000  -7.807656   -6.345912
------------------------------------------------------------------------
```

We now create an interaction of *age* with the natural log of *age*, and model. As with any regression model interaction, *age* must also be entered in the model, but we are not interested in its significance.

```
. gen lnageage = age * ln(age)

. logit death age lnageage, nolog

Logistic regression                        Number of obs   =        5388
                                           LR chi2(2)      =      148.99
                                           Prob > chi2     =      0.0000
Log likelihood = -912.91306               Pseudo R2       =      0.0754
------------------------------------------------------------------------
   death |    Coef.    Std. Err.     z   P>|z|  [95% Conf. Interval]
---------+--------------------------------------------------------------
     age |  .2433231   .2199722    1.11  0.269  -.1878145  .6744608
lnageage | -.0352509   .0414788   -0.85  0.395  -.1165478   .046046
   _cons | -9.626208   3.035231   -3.17  0.002 -15.57515  -3.677264
------------------------------------------------------------------------
```

Lnageage is not significant, indicating that the relationship of logit to the predictor, *age*, is linear. If it were significant, many statisticians attempt a transformation of *age*, for example, creating the log of *age*, or the square or square root of *age*, and so forth. If the transformed variable is entered into the model as the predictor, and is found not to be significant, they conclude that the model is now linear. However, this is an incorrect application of the method. The correct response is to model *age* using another binomial family (or Bernoulli) link function, such as probit, complementary loglog, or loglog model. Two caveats should be given regarding the test: it fails to inform the statistician of the shape of the nonlinearity, and it is rather insensitive to small departures from linearity.

5.2.2 Tukey–Pregibon Link Test

The Tukey–Pregibon link test is so named as a result of the contributions of John Tukey and Daryll Pregibon (some 30 years later) in designing methods of fit analysis for logistic models. Pregibon (1980) actually devised the method, basing it on his previous work that had been heavily influenced by Tukey (1949).

I created the name for this test since it did not have an established name. Stata employs the test in its software, calling it **linktest**. The logic of the test is the same as for the Box–Tidwell; that is, it is to be used to assess the appropriateness of the link, not as a method for suggesting variable transforms.

The test is used as a post-estimation command; in other words, one first estimates a model and then uses the test. Since the test is based on a continuous predictor, we shall use *age*, as we did for the Box–Tidwell test. The conclusion should be the same—that there is a linear relationship between the logit and predictor(s).

We first quietly estimate the logistic model, then use Stata's linktest to determine if the assumption of linearity has been violated.

```
. quietly logit death age

. linktest, nolog

Logistic regression                    Number of obs   =       5388
                                       LR chi2(2)      =     149.37
                                       Prob > chi2     =     0.0000
Log likelihood = -912.72187            Pseudo R2       =     0.0756
------------------------------------------------------------------
   death |    Coef.   Std. Err.    z     P>|z|  [95% Conf. Interval]
---------+--------------------------------------------------------
    _hat |  .4512473  .5308716   0.85   0.395  -.5892419   1.491737
  _hatsq | -.0963071  .0922522  -1.04   0.297  -.2771181   .0845039
   _cons | -.7241011  .7349082  -0.99   0.324  -2.164495   .7162924
------------------------------------------------------------------
```

The predictor of interest is _hatsq, which is the square of the *hat* matrix diagonal. If it is significant, the assumption has been violated. The test shows us that the above model is well specified, in other words, that there is a linear relationship between the logit and *age*.

We shall discuss the *hat* matrix diagonal in Section 7.4.1, where the formula and explanation are provided. We also describe how the statistic can be calculated without having to resort to the matrix programming capability of one's statistical application.

The next tests we discuss involve more calculations on the part of the statistician and are based on interpretation rather than on a *p*-value.

5.2.3 Test by Partial Residuals

Partial residuals are the basis for the well-known partial residual plot, as well as a more sophisticated family of models called generalized additive models. The value of a partial residual graph, or plot, is that a continuous predictor can be evaluated as to its linearity in the logit on the basis of other adjusters. The methods of assessing linearity that we have thus far examined are essentially univariable; they are unaffected by other predictors in the model. The partial residual, on the other hand, allows a continuous predictor to be examined in its role as a copredictor in the model, that is, as it is adjusted by other predictors.

Partial residuals are post-estimation calculations, followed by their associated graphical representation. The equation for partial residuals appears as:

$$PR_i = (y_i - \mu_i)/(\mu_i(1 - \mu_i)) + \beta_j x_{ij} \tag{5.3}$$

You may recognize the first term here as a raw residual divided by the logistic variance function. The product of the coefficient of the continuous predictor and its value is added to the variance-adjusted residual to create the partial residual. Partial residuals are used in more complex analyses as well, for example, with generalized additive models, which we shall discuss later in this chapter.

The key to partial residuals is the manner in which they are graphed. The partial residual graph or plot is a two-way graph of the response on the residual, smoothed by a cubic spline. Sometimes a *lowess* smoother is employed instead, but the traditional method of smoothing is by means of a cubic spline.

Using the **heart01** data, we model *death* on *anterior*, *hcabg*, *killip* levels, and the continuous predictor *age*. A default cubic spline is used as the smoother, with a default bandwidth of 4. The author-created program, called **lpartr** (Hilbe, 1992b), is used to estimate, graph, and smooth the data. Note the fairly linear smooth of *age*, even with the default setting.

Continuous predictors to be estimated and smoothed are placed to the right of the **lpartr** command. Binary and categorical predictors that are used in the logistic regression model are all placed within the *with()* option. *Smooth()* directs the algorithm to use a cubic spline, with the default. If another bandwidth is desired, the alternate value is placed within the *smooth()* parentheses. Figure 5.1 illustrates the command below.

```
. lpartr death age, with(ant]erior hcabg kk2-kk4)   smooth
```

A *lowess* smooth with the default of .8 is provided as another option. The smooth appears quite similar to the cubic spline smooth. Together, the relatively flat line indicates that the smooths provide evidence of linearity. On the other hand, a more careful look at the smooth gives the impression that there may be three different slopes: the first running from age 40 to about 60, another from 60 to 80, and a final slope from 80 to 100. The middle slope appears steeper than the other contiguous slopes. The manner by which we can determine if this appearance is real is to categorize *age*, run the model

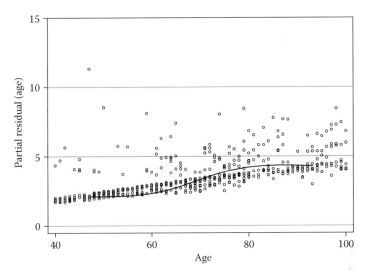

FIGURE 5.1
Cubic spline, bandwidth = 4.

with a factored *age* at these cut points, and see if the slope differences actually exist in the data. This takes us to a test of linearity that is associated with the partial residuals approach.

5.2.4 Linearity of Slopes Test

This test can be regarded as either a follow-up on the previous test using partial residuals, or as an independent test of the linearity of a continuous predictor. When incorporated in a multivariate or multivariable logistic model, a continuous predictor is not independent but is adjusted by the other predictors in the model.

Continuing the above examination of partial residual graphs, we categorize *age*. Due to findings based on analysis we perform later in the text, four levels of *age* will first be generated based on the middle 10-year age divisions. The Stata code below, generating four indicator variables, is clear. Recall that an indicator variable has values of 1 or 0 only. In addition, one indicator is used as a reference and is not directly estimated by the model. The same logic may be used to create indicators with other command-line applications.

```
. gen byte  agegrp=1  if age<=60
. replace   agegrp=2  if age>60 & age <=70
. replace   agegrp=3  if age>70 & age <=80
. replace   agegrp=4  if age>80
```

The newly created indicator variables can be labeled in Stata using the following code:

```
. label define agegrp 1 "=<60" 2 "61-70" 3 "71-80" 4 ">80"
```

The label name *agegrp*, can be attached or associated with the predictor *agegrp*, using the code:

```
. label values agegrp agegrp
```

A label name does not have to be the same as its associated predictor.

A tabulation of the newly created *agegrp* displays the groups, their labels, frequency, and percentages.

```
tab agegrp

    agegrp |      Freq.     Percent        Cum.
-----------+-----------------------------------
     =<60 |      1,907       35.39       35.39
    61-70 |      1,390       25.80       61.19
    71-80 |      1,425       26.45       87.64
      >80 |        666       12.36      100.00
-----------+-----------------------------------
     Total |      5,388      100.00
```

The middle two levels can be combined as a separate indicator or dummy variable called *agegrp23*. *agegrp23* takes the value of 1 if either *agegrp2* or *agegrp3* has values of 1.

This is done so that each age group represents 20 years.

```
. gen age23 = age2==1 | age3==1

. tab age23

    age23 |      Freq.     Percent        Cum.
-----------+-----------------------------------
        0 |      2,573       47.75       47.75
        1 |      2,815       52.25      100.00
-----------+-----------------------------------
    Total |      5,388      100.00
```

The observations are consistent. Now we model the same data as used in the partial residuals plot application.

```
. logit death anterior hcabg kk2-kk4 age23 age4, nolog
                              /// header not displayed
```

```
------------------------------------------------------------------
   death |     Coef. Std. Err.      z  P>|z|  [95% Conf. Interval]
---------+--------------------------------------------------------
anterior |  .6634689 .1669831    3.97  0.000  .3361881   .9907497
   hcabg |  .6705911 .3507393    1.91  0.056 -.0168452  1.358027
     kk2 |  .8208859 .1797935    4.57  0.000   .468497  1.173275
     kk3 |  .8243462 .2690031    3.06  0.002  .2971098  1.351583
     kk4 |  2.644704 .3498393    7.56  0.000  1.959031  3.330376
   age23 |  1.129668 .2569174    4.40  0.000  .6261191  1.633217
    age4 |  2.189783 .2718955    8.05  0.000  1.656877  2.722688
   _cons | -5.077218 .2588188  -19.62  0.000 -5.584494 -4.569943
------------------------------------------------------------------
```

We see a clear linear trend in the age groups, with slopes in parentheses: *age1* (0), *age23* (1.13), and *age4* (2.19). The results indicate that the variances in the smooth are largely artifacts of the data, and may not represent factual differences in the underlying or latent slope. Again, the age divisions modeled represent the 20-year age groups 40–60, 60–80, and 80–100.

Next, we model the data, breaking the middle years of *age* into 10-year age divisions: under 60, 60–70, 70–80, and 80+. The reference level has some 1,900 observations, and levels two and three are approximately 1,400 observations each. The highest level, *age4*, has only 666 observations, or 12 percent of the data. When we combined the middle two levels, the resulting level had 52 percent of the observations. Separating them into 10-year divisions may help determine if the coefficients were unduly biased due to the large differences in group size. Modeling in this manner produces the partial table of coefficients below. Note that the *s indicate estimated predictors that we do not display.

```
. logit death anterior hcabg kk2-kk4 age2 age3 age4, nolog

death |  Coef.      Std. Err.     z   P>|z|    [95% Conf. Interval]
------+------------------------------------------------------------
         ***          ***        ***    ***         ***         ***
 age2 | .5036804   .3101746    1.62   0.104   -.1042506    1.111611
 age3 | 1.52179    .2662173    5.72   0.000    1.000013    2.043566
 age4 | 2.195812   .2718518    8.08   0.000    1.662993    2.728632
```

The trend is clear, but the *age2* group, representing ages 61–70, is not significant at even the .1 level. In effect, this means that the reference is in fact a combined *age1–age2*, since non-significance in this situation indicates that *age2* is not statistically different from *age1*. Re-modeling the data excluding *age2* should produce coefficients and standard errors that are not too different from those displayed in the table above. If the *p*-value were close to 1.0, we would expect very little difference in values.

```
. logit death anterior hcabg kk2-kk4 age3 age4, nolog

death |    Coef.   Std. Err.    z   P>|z|   [95% Conf. Interval]
-------------------------------------------------------------------
            ***        ***     ***    ***        ***        ***
 age3 | 1.266828   .2005658   6.32  0.000    .8737267   1.65993
 age4 | 1.940876   .2080462   9.33  0.000    1.533112   2.348639
```

The differences in values between *age4* and *age3* in both models are near identical.

```
. di 2.195812    - 1.52179
.674022


. di 1.940876 - 1.266828
.674048
```

The conclusion we draw from the above discussion is that we find no appreciable violation in the assumption of linearity between the sole continuous predictor, *age*, and the logit function. This is consistent with what we found using the Box–Tidwell and Tukey–Pregibon tests.

5.2.5 Generalized Additive Models

The generalized additive model, or GAM, is a method originally devised by Stone (1985), but which was substantially enhanced to its standard formulation by Hastie and Tibshirani (1990), both of whom are currently at Stanford University. At the time of their seminal text on the subject, Trevor Hastie was at AT&T Laboratories in New Jersey and Robert Tibshirani was at the University of Toronto. In association with William Cleveland and J.M. Chambers, Hastie was one of the developers of the S statistical language, which was later licensed to a commercial firm, Statistical Sciences in Seattle, and sold as S-Plus software. Hastie was responsible for authoring both **glm()** and **gam()** in S, which became primary functions in S-Plus. After publication of their text on the subject, Tibshirani and Hastie developed a separate software GAM modeling application, marketed as G.A.I.M.—Generalized Additive Interactive Modeling. For a comprehensive review of this software, as well as other GAM implementations, see Hilbe (1993b).

Recall that the structure of a linear model appears as:

$$y_i = \sum_{i=1}^{n} \{\beta_0 + \beta_{1i}X_{1i} + \ldots + \beta_{ji}X_{ji} + \varepsilon_i\} \tag{5.4}$$

where β_0 is the linear constant, $\beta_1 X_1$ through $\beta_j X_j$ are the j products of the respective parameter estimates and predictor value, and ε is the error.

The *n* observations in the model are indicated for each *i* observation in the data.

The above formula can be parameterized to appear as, without reference to individual observations,

$$y = \beta_0 + \sum_{j=1}^{J} \beta_j X_j + \varepsilon \qquad (5.5)$$

where *j* is a subscript indicating the *J* predictors in the model.

The generalized additive model, however, amends the linear structure of the model by smoothing the values of specified continuous predictors in the model. The parameter estimate is therefore a function of the smooth. This relationship is commonly given as:

$$y = \beta_0 + \sum_{j=1}^{J} f_j \left(X_j \right) + \varepsilon \qquad (5.6)$$

with *f* as a smooth function. In the standard formulation of generalized additive models, the partial residuals of a continuous predictor are smoothed by a cubic spline, or alternatively by a *lowess* or alternative smoother. This is similar to the partial residual graphs discussed in Section 5.2.3. The parameters of the smooths are related to the bandwidth used for the smooth.

It should be mentioned that although a nonparametric smooth is nearly always used in GAM modeling, the statistician may alternatively employ a parametric function, such as a polynomial function.

One should not interpret a generalized additive model in the same manner as a generalized linear model. The reason this type of model is employed with data is to better determine the underlying shape of a continuous predictor, as adjusted by the remainder of the model. GAM software produces an adjusted partial residual graph of the smooth, which can then be used by the statistician to determine the best transformation required to effect linearity in the associated generalized linear model. In fact, the GAM algorithm uses a generalized linear model's algorithm for estimation of the model. However, at each iteration, the partial residuals of identified continuous predictors are calculated and smoothed as defined by the user. Partial residuals are used because, for each predictor, they remove the effect of the other predictors from the response. They can therefore be implemented to model the effects for predictors, X_j.

R will be used for modeling a generalized additive logistic model of the **heart01** data. The goal will be to evaluate the structure of *age* as adjusted by the other predictors in the binary logistic model. We should expect to observe a graph showing properties similar to the partial residual graph displayed earlier in this chapter.

We shall use the R GAM function that is in the **mgcv()** library, which comes with the default download of the software. **mgcv()** was authored by S. Wood in 2000. He subsequently wrote an excellent text on GAM titled *Generalized Additive Models: An Introduction with R* (2006).

R can read data files created by other leading software packages, including Stata. However, one must use the **foreign()** library prior to loading such a file into memory. With respect to Stata files, at the time of this writing, only Stata versions 5 through 9 can be read into R. The current version of Stata is 10. Stata users must load the file into Stata memory and save it into version 8/9 format using the *saveold()* command:

```
. use heart01
. saveold "c:\heartr"
```

The older version file was saved by another name to keep it from overwriting the higher version file. The file was saved directly into the C: directory.

The following R commands load the Stata **heartr** file into memory, load the **mgcv()** package or library of files, including **gam()**, and model the data using the **gam()** function. The variable *age* is smoothed by a cubic spline and is entered as a term in the GAM logistic regression command. We only include the factored *killip* levels as adjusters into the model.

```
> library(foreign)
> data("c:/heartr.dta")   /// OR USE heartr<-read.dta("c:/heartr.dta")
> library(mgcv)
> heartgam <- gam(death ~ s(age) + kk2 + kk3 + kk4,
family=binomial,data=heartr)
```

To obtain a summary of the parameter estimates and other related GAM statistics, type:

```
> summary(heartgam)

Family: binomial
Link function: logit

Formula:
death ~ s(age) + kk2 + kk3 + kk4

Parametric coefficients:
            Estimate Std. Error z value Pr(>|z|)
(Intercept)  -3.8072     0.1121 -33.952  < 2e-16 ***
kk2           0.8744     0.1630   5.364 8.14e-08 ***
kk3           1.0988     0.2244   4.897 9.73e-07 ***
kk4           3.0820     0.2873  10.726  < 2e-16 ***
---
Signif. codes:  0 '***' 0.001 '**' 0.01 '*' 0.05 '.' 0.1 ' ' 1

Approximate significance of smooth terms:
```

```
          edf Est.rank Chi.sq p-value
s(age) 1.003     3.000   95.39  <2e-16 ***
---
Signif. codes:   0 '***' 0.001 '**' 0.01 '*' 0.05 '.' 0.1 ' ' 1

R-sq.(adj) =   0.0687   Deviance explained = 13.4%
UBRE score = -0.68742   Scale est.  = 1        n = 5153
```

A plot of the smoothed predictor *age* can be displayed using the *plot()* function. Residuals are indicated by periods, and the standard error of the smooth is appropriately shown, with wider ranges at the extremes. The smooth appears to linearize *age*. Figure 5.2 illustrates the command below.

```
> plot(heartgam, residuals=TRUE, se=TRUE, pch=".")
```

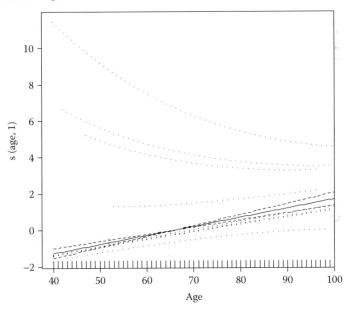

FIGURE 5.2
GAM smooth of age.

We should see if adding other predictors to the model aside from *killip* level changes the apparent relationship of linearity. Adding *anterior* and *hcabg* to the model gives us the following output and graph.

```
> heartgam <- gam(death~s(age)+anterior+hcabg+kk2+kk3+kk4,
family=binomial,data=heartr)

> summary(heartgam)

Family: binomial
Link function: logit
```

```
Formula:
death ~ s(age) + anterior + hcabg + kk2 + kk3 + kk4

Parametric coefficients:
                   Estimate Std. Error z value Pr(>|z|)
(Intercept)         -4.2431     0.1592 -26.652  < 2e-16 ***
anteriorAnterior     0.6503     0.1673   3.887 0.000102 ***
hcabg                0.7203     0.3512   2.051 0.040232 *
kk2                  0.8195     0.1802   4.548 5.42e-06 ***
kk3                  0.8228     0.2686   3.063 0.002188 **
kk4                  2.7089     0.3538   7.656 1.92e-14 ***
---
Signif. codes:  0 '***' 0.001 '**' 0.01 '*' 0.05 '.' 0.1 ' ' 1

Approximate significance of smooth terms:
        edf Est.rank Chi.sq  p-value
s(age) 3.553        8  90.73 3.31e-16 ***
---
Signif. codes:  0 '***' 0.001 '**' 0.01 '*' 0.05 '.' 0.1 ' ' 1

R-sq.(adj) =  0.0563   Deviance explained = 13.8%
UBRE score = -0.71114  Scale est. = 1          n = 4503

> plot(heartgam)
```

Figure 5.3 illustrates the above output. The graph—or plot—is much more clear, and indicates that linearity is compromised at the extremes. This is

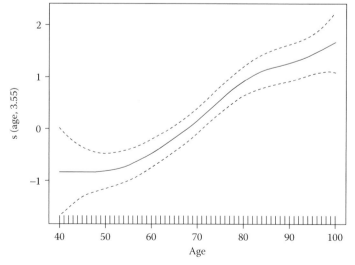

FIGURE 5.3
GAM smooth of adjusted age.

not at all uncommon since the fewer values at the extremes result in greater variability. Note that the confidence intervals are considerably wider at the extremes as well. The graph appears somewhat different from the earlier one, but nevertheless does not lead us away from asserting the linear relationship between the model logit and explanatory predictors.

It must be remembered that although the GAM plot function uses partial residuals, the plot will be different from a simple partial residual plot. The GAM residuals are based on GAM-based fit and variance functions, whereas the partial residual graph we developed earlier is based on a logistic regression model.

5.2.6 Fractional Polynomials

Fractional polynomials, initially developed by Royston and Altman (1994), is a combined regression and graphical method designed to inform the statistician of the best power to use for transforming a continuous predictor so that it is linear to the link function used in the model of which it is a part. As in the case with additive models, continuous predictors in fractional polynomial procedures may be adjusted by other predictors in the model. The method can become quite complex, and for our purposes, we only present a brief summary of how fractional polynomials can be employed to assist in assessing linearity in the logit.

The basic structure of a fractional polynomial can be expressed by a formula that appears similar to that used in the previous section for generalized additive models. Parameterized for a single continuous predictor and its adjusters, the model can be given as:

$$y = \beta_0 + \sum_{j=1}^{J} f_j(X_j)\beta_j \qquad (5.7)$$

Unlike GAMs, here $f_j(X_j)$ designates a power function, generally ranging from -2 to $+2$. For those readers familiar with generalized linear model power links, the same meaning is attributed to the values assigned to the powers, x^p.

Expanding the above formula to the form actually modeled, we have:

$$y = \beta_0 + \beta_1 X^{P1} + \beta_2 X^{P2} + \ldots + \beta_m X^{Pm} \qquad (5.8)$$

with degree m, and powers, p, equal to $(p_1, \ldots, p_m)$.

Fractional polynomial modeling is similar to traditional polynomial regression except that the former allows non-integer as well as negative powers.

Additionally, fractional polynomials can be used by a variety of models other than the normal or Gaussian. They were designed to be used with members of the family of generalized linear models (including Gaussian, binomial/ Bernoulli, Poisson and negative binomial, gamma, and inverse Gaussian), but can also be applied to Cox proportional hazards models.

A listing of the primary powers, p, used in fractional polynomials is displayed in Table 5.1. A power of 1/3, not listed in the table, could be used for modeling volumes, but statisticians rarely engage in such modeling tasks. It should also be pointed out that powers can be specified for values intermittent to those displayed in the table. However, in actual practice, fine-tuning a model for a power such as 1.75 adds little to the viability of the model. Software implementations, now only in Stata, specify the range of powers to be tested for a continuous predictor as: $(-2, -1, -0.5, 0, 0.5, 1, 2, 3)$

TABLE 5.1

Primary Powers Used in Fractional Polynomials

$p = -2$	inverse quadratic	2	square	3	cubic
-1	inverse	1	linear		
$-.5$	inverse square root	.5	square root	0	log

The degree, m, of a fractional polynomial model refers to the number of terms in powers of X that occurs in the model. Values of m at 1 or 2 generally account for most, if not all, of the transformation needed to appropriately adjust a continuous predictor. For example, adding the square of X to a model where X takes a parabolic form usually linearizes X with respect to the model.

The power or powers, p, of a fractional polynomial regression equation are listed within parentheses, or alternatively, braces. For example:

$$y = \beta_0 + \beta_1 X^2 \qquad\qquad p = \{2\}$$
$$y = \beta_0 + \beta_1 X^2 + \beta_2 X^{-1} \qquad p = \{2, -1\}$$

The fractional polynomial algorithm searches the powers designated by the user, combining and comparing them with the aim of providing the values that can be used to provide the predictor with an optimal transformation.

Stata has two basic commands for modeling fractional polynomials: **fracpoly** and **mfp**. The latter command searches for the best, or, the most parsimonious statistically significant, model from alternative parameterizations. **fracpoly** provides the user with the information required to make an appropriate decision on the preferred model. A choice is provided to the user to account for circumstances that may affect the type of model desired.

We model the **heart01** data of *death* on the continuous predictor *age*, adjusted by *killip* level. *Age* is partitioned into four groups, and various powers are attempted for given configurations. Slopes for each mean-centered age division are displayed, together with the parameter estimates of the three nonreference *killip* levels, and a constant. Below the initial table of parameter estimates is a line giving the value of the deviance for the preferred model, and a listing of its associated power values. Note that *degree(2)* is the default; we used *degree(4)* to capture more of the variability.

```
. fracpoly logit death age kk2-kk4, degree(4) compare

-> gen double Iage__1 = X^3-285.2272038 if e(sample)
-> gen double Iage__2 = X^3*ln(X)-537.490326 if e(sample)
-> gen double Iage__3 = X^3*ln(X)^2-1012.862191 if e(sample)
-> gen double Iage__4 = X^3*ln(X)^3-1908.666571 if e(sample)
   (where: X = age/10)

Iteration 0:   log likelihood = -924.54235
               -    -    -
Iteration 5:   log likelihood =  -797.7554

Logistic regression                   Number of obs   =        5153
                                      LR chi2(7)      =      253.57
                                      Prob > chi2     =      0.0000
Log likelihood =  -797.7554           Pseudo R2       =      0.1371
------------------------------------------------------------------
   death |     Coef. Std. Err.     z   P>|z|  [95% Conf. Interval]
---------+--------------------------------------------------------
Iage__1 | -2.501229 1.605217  -1.56 0.119 -5.647397    .6449384
Iage__2 |  3.113422 2.023847   1.54 0.124 -.8532445    7.080089
Iage__3 | -1.313447 .8692188  -1.51 0.131 -3.017085    .3901905
Iage__4 |  .1871408 .1265591   1.48 0.139 -.0609104    .4351921
    kk2 |  .8655287  .163073    5.31 0.000  .5459115    1.185146
    kk3 | 1.088649 .2242206    4.86 0.000  .6491849    1.528114
    kk4 | 3.082218  .289309   10.65 0.000  2.515183    3.649253
   _cons | -3.910871 .1407253 -27.79 0.000 -4.186687   -3.635054
------------------------------------------------------------------
Deviance: 1595.51. Best powers of age among 494 models fit: 3 3 3 3.

Fractional polynomial model comparisons:
------------------------------------------------------------------
age             df     Deviance   Dev. dif.  P (*)   Powers
------------------------------------------------------------------
Not in model     0     1698.580   103.069    0.000
Linear           1     1600.745     5.234    0.631   1
m = 1            2     1600.745     5.234    0.514   1
m = 2            4     1597.240     1.729    0.785   -2 -2
m = 3            6     1595.786     0.276    0.871   -1 -1 -.5
m = 4            8     1595.511      --       --     3 3 3 3
------------------------------------------------------------------
(*) P-value from deviance difference comparing reported model
with m = 4 model
```

The above table of comparative deviances shows that there is no significant difference between any two levels—except between a constant only and linear model. Figure 5.4 illustrates the command below.

```
. fracplot age, msize(small)
```

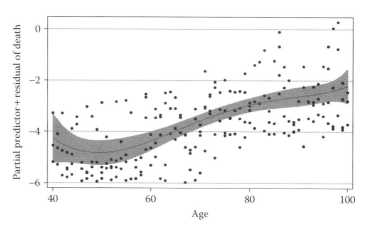

FIGURE 5.4
Fractional polynomial (3 3 3 3), adjusted for covariates.

Modeling the same data using the automatic multivariable fractional polynomial (MFP) gives the same results.

```
. mfp logit death age kk2-kk4
Deviance for model with all terms untransformed =  1600.745,
5153 observations

[kk4 included with 1 df in model]

age          lin.    FP2    1600.745    3.505  0.320    1      -2 -2
             Final          1600.745                    1

[kk2 included with 1 df in model]
[kk3 included with 1 df in model]

Fractional polynomial fitting algorithm converged after 1 cycle.
Transformations of covariates:

-> gen double Iage__1 = age-65.82592664 if e(sample)

Final multivariable fractional polynomial model for death
```

```
--------------------------------------------------------------------
Variable |     ----Initial-----           -----Final-----
         |  df    Select   Alpha    Status    df    Powers
---------+----------------------------------------------------------
     age |   4    1.0000   0.0500      in      1      1
     kk2 |   1    1.0000   0.0500      in      1      1
     kk3 |   1    1.0000   0.0500      in      1      1
     kk4 |   1    1.0000   0.0500      in      1      1
--------------------------------------------------------------------


Logistic regression                  Number of obs   =       5153
                                      LR chi2(4)      =     248.34
                                      Prob > chi2     =     0.0000
Log likelihood = -800.37241          Pseudo R2       =     0.1343
--------------------------------------------------------------------
   death |    Coef. Std. Err.     z   P>|z|  [95% Conf. Interval]
---------+----------------------------------------------------------
Iage__1 |  .0497254 .0051881   9.58  0.000   .0395569      .059894
     kk2 |   .874362 .1630111   5.36  0.000   .554866     1.193858
     kk3 |  1.098789 .2243805   4.90  0.000   .6590115    1.538567
     kk4 |  3.081978  .287323  10.73  0.000   2.518835    3.645121
   _cons | -3.807229 .1121425 -33.95  0.000  -4.027024   -3.587434
--------------------------------------------------------------------
Deviance: 1600.745.
```

The only difference between the final model selected by the fractional polynomial algorithm and a standard logistic regression is with the constant. It has a different value due to the centering of *age*. The conclusion is to leave *age* untransformed. See Section 6.4.1 for a discussion on centering.

```
. logit death age kk2-kk4, nolog    [HEADER INFO DELETED]

--------------------------------------------------------------------
  death |    Coef. Std. Err.     z   P>|z|  [95% Conf. Interval]
--------+-----------------------------------------------------------
    age |  .0497254 .0051881   9.58  0.000   .0395569      .059894
    kk2 |   .874362 .1630111   5.36  0.000   .554866     1.193858
    kk3 |  1.098789 .2243805   4.90  0.000   .6590115    1.538567
    kk4 |  3.081978  .287323  10.73  0.000   2.518835    3.645121
  _cons | -7.080451 .3980213 -17.79  0.000  -7.860558   -6.300343
--------------------------------------------------------------------
```

5.3 Standardized Coefficients

Standardizing coefficients is a common method used in ordinary least squares regression to put all explanatory predictors on the same metric so that they may be compared. The goal is to see which predictors, of all significant predictors, have the most influence on the response. The unstandardized coefficient specifies that a one-unit change in the value of the predictor results in the coefficient value change in the response. However, the metrics employed by predictors may be vastly different. For example, a predictor may be measured in a natural metric, such as centimeters, inches, kilograms, and so forth, or measured using categorical levels, including poor, average, good, excellent. The metric might also be binary, 1/0, or yes/no. Instead of having predictors based on different metrics, standardizing converts all predictors into standardized units. Each predictor is measured in standard deviation units and may therefore be compared equally. A standardized coefficient of 2 is one standard deviation higher than a standardized coefficient of 1.

The same logic prevails for logistic regression coefficients. However, the fact that the response is binary instead of continuous poses various problems. In a similar manner as to how statisticians have dealt with the R^2 statistic, several different versions of standardized logistic coefficients can be found in the statistical literature.

The version of standardized logistic regression coefficient we shall use here is defined as:

$$\text{Std}(\beta_i) = \beta_i \frac{\text{SD}(X_i)}{\sqrt{\frac{\pi^2}{3}}} \tag{5.9}$$

or, given

```
. di sqrt((_pi*_pi)/3)
1.8137994
```

$$\text{Std}(\beta_i) = \beta_i * \text{SD}(X_i) / 1.8137994 \tag{5.10}$$

where β is the parameter estimate of a predictor, X, and SD(X) is the standard deviation of X. $\pi^2/3$, or 3.29, is the fixed variance of the logistic distribution. Taking the square root of this variance results in the standard deviation of the logistic distribution.

The unstandardized coefficient for *anterior* is 0.6387 (refer to output below). The standard deviation of *anterior* may be calculated by:

```
. su anterior

    Variable |        Obs        Mean    Std. Dev.       Min        Max
-------------+--------------------------------------------------------
    anterior |       4696   .4525128    .4977929          0          1
```

The value is .4977929. Calculating the values in the formula above, we have:

```
. di .6387*(.4977929)/1.8137994
.17528968
```

which is identical to the value displayed in the output below.

```
. lstand              /// lstand must be performed after logistic

Table of Predictor Estimates:
Standardized Coefficients and Partial Correlations

No.  Var          Coef        OR    St.Coef    PartCorr    Prob(z)
==================================================================
0    Constant   -4.8159
1    anterior    0.6387    1.8941   0.1753     0.0918      0.000
2    hcabg       0.7864    2.1955   0.0781     0.0447      0.026
3    kk2         0.8249    2.2817   0.1818     0.1128      0.000
4    kk3         0.7967    2.2182   0.1016     0.0674      0.003
5    kk4         2.6837   14.6398   0.1639     0.1918      0.000
6    age3        1.2668    3.5496   0.3081     0.1597      0.000
7    age4        1.9409    6.9648   0.3522     0.2392      0.000
==================================================================
```

Based on standardized coefficients, the *age* predictors have the most influence on the response.

Statisticians standardized model predictors to assess their comparative worth to the model when they are expressed in differing metrics. Interpretation of standardized binary and categorical predictors may not always be clear, but most statisticians who employ standardized logistic coefficients in their research do so to assess predictor influence in the model. Standardization is much more common when working with normal or ordinary least squares models.

We may also mention partial correlation. SPSS is the only commercial package that displays the partial correlation of logistic predictors, and it is rarely used now in research, but we provide the formula for completeness.

The Wald statistic is the square of *t*, which is used in many software applications, such as S-Plus. Wald is a popular statistic upon which the determination of the predictor *p*-value is based. Stata uses the *z*-statistic to calculate generalized linear models (GLM) as well as all maximum likelihood standard errors other than OLS. *z* has a normal distribution, Wald has a Chi2 (χ^2) distribution. Both yield similar results, especially as the number of observations grows beyond 30.

In any case, the Wald statistic for a given predictor is defined as the square of the predictor coefficient divided by its standard error. The partial correlation is defined as:

$$R_{partial} = \text{sqrt}\left\{(W-2)/(2*\text{abs}(\text{LLo}))\right\}$$ (5.11)

where W is a predictor's Wald statistic and *abs(LLo)* is the absolute value of the constant-only log-likelihood. The partial correlation may be interpreted as the correlation of the predictor with the response, holding other predictor values constant. Note that the partial correlations for the two *age* predictors are among the highest, as they were for their standardized coefficients.

5.4 Standard Errors

5.4.1 Calculating Standard Errors

Thus far, we have simply assumed we know what are meant by standard errors. We know that standard errors tell us about the variability of the model parameter estimates and are the basis for calculating parameter estimate confidence intervals. It can be said with little reservation that parameter confidence intervals, based on standard errors, are substantially more important to the understanding of the response than are the estimated parameters themselves. The estimation of a true parameter tells us that the latter can be found anywhere in the range of the confidence interval, which, for logistic models, is defined in terms of the normal distribution.

Standard errors are calculated as the square root of the model diagonal of the variance-covariance matrix, or inverse Hessian. The latter can be obtained after one has modeled the data.

We use the **heart01** data, modeling *death* on *anterior, hcabg,* and the levels of the categorical predictor, *killip*.

```
. logit death anterior hcabg kk2-kk4, nolog

Logistic regression                      Number of obs   =       4503
                                         LR chi2(5)      =     113.64
                                         Prob > chi2     =     0.0000
Log likelihood = -686.28751              Pseudo R2       =     0.0765
------------------------------------------------------------------------
    death |      Coef. Std. Err.      z   P>|z|  [95% Conf. Interval]
----------+-------------------------------------------------------------
 anterior |    .7226527 .1646645    4.39 0.000 .3999162   1.045389
    hcabg |    .7185074 .3460598    2.08 0.038 .0402428   1.396772
      kk2 |    .9453414 .1771164    5.34 0.000 .5981997   1.292483
      kk3 |   1.176885 .2621423    4.49 0.000 .6630952   1.690674
      kk4 |   2.920631 .3300369    8.85 0.000  2.27377   3.567491
    _cons |  -4.072202  .149476  -27.24 0.000 -4.36517  -3.779235
------------------------------------------------------------------------
```

To request the inverse Hessian, or model variance-covariance matrix, the following command is used in Stata. Most software applications have a similar means to display the matrix.

```
. corr, _coef covariance

            | anterior    hcabg      kk2       kk3       kk4     _cons
------------+--------------------------------------------------------
  anterior  |  .027114
     hcabg  |  .003167   .119757
       kk2  | -.002213  -.002022   .03137
       kk3  | -.004683  -.002488  .012959   .068719
       kk4  | -.001471   .003044   .01262    .01275   .108924
     _cons  | -.016048  -.007831  -.01115  -.009679 -.011838  .022343
```

Again, standard errors are calculated from the square root of each element of the primary diagonal of the matrix. Since only the lower half of the matrix is displayed above, standard errors are obtained by taking the square root of the top terms of the matrix. The matrix is symmetric, meaning that the upper half is identical in reverse to the lower half.

SE—ANTERIOR

```
. di sqrt(.027114)
.16466329
```

SE—HCABG

```
. di sqrt(.119757)
.34605924
```

SE—KK2

```
. di sqrt( .03137)
.17711578     /// rounding error may give different results at
                  5th or 6th place
```

We see that the calculated standard errors are nearly identical to those displayed in the table. Except for rounding error, they are the same.

5.4.2 The *z*-Statistic

The *z*-statistic is simply the ratio of the parameter estimate and standard error. For *anterior* this becomes:

Z—ANTERIOR

```
. di .7226527/.1646645
4.3886369
```

Z—HCABG

```
. di .7185074/.3460598
2.0762521
```

and so forth.

A z is used here to indicate that the normal distribution is used to determine the predictor p-values. Some software applications use t rather than z, which indicates the use of the t-distribution. t is used for traditional ordinary linear regression models and was carried over to other members of the generalized linear models family by packages such as S-Plus and others. However, the statistic is normally distributed and the probability or p-value is properly indicated as z.

5.4.3 *p*-Values

The p-value is calculated as:

$$[1 - [\text{normal distribution}]] * 2 \qquad (5.12)$$

for a two-sided value. We use the *normprob()* function in Stata to calculate the p-value of *anterior* and *hcabg*. I'll show how to calculate the value directly from the stored values of the parameter estimate and standard error. These values have many more significant values than do the values displayed in the table of estimates.

ANTERIOR   VALUE DISPLAYED IN TABLE

```
. di _b[anterior]/_se[anterior]
4.3886372                                             4.39

. di (1-normprob(4.3886372))*2
.00001141                                             0.000
```

HCABG

```
. di _b[hcabg]/_se[hcabg]
2.0762524                                             2.08

. di (1-normprob(2.0762524))*2
.03787061                                             0.038
```

5.4.4 Confidence Intervals

The confidence intervals are calculated using the parameter estimate and standard error using the formula:

$$\text{coef} +/- \alpha/2 * \text{SE} \qquad (5.13)$$

For the 95 percent confidence level, which is $\alpha = 0.05$, we have:

$$\text{coef} +/- 1.96 * SE \qquad (5.14)$$

The upper and lower values of the significance level may be calculated using either of the following two methods:

```
. di invnorm(.05/2)     /// which is .025 with reference to 0.00
-1.959964
```

```
. di invnorm(.975)      /// which is half of .95 with reference to 1.00
1.959964
```

We are now prepared to calculate the lower and upper confidence levels at the 95 percent level. The table of parameter estimates is again displayed for ease of comparison:

```
------------------------------------------------------------------
   death |    Coef. Std. Err.     z P>|z| [95% Conf. Interval]
---------+--------------------------------------------------------
anterior | .7226527 .1646645   4.39 0.000 .3999162   1.045389
   hcabg | .7185074 .3460598   2.08 0.038 .0402428   1.396772
     kk2 | .9453414 .1771164   5.34 0.000 .5981997   1.292483
     kk3 | 1.176885 .2621423   4.49 0.000 .6630952   1.690674
     kk4 | 2.920631 .3300369   8.85 0.000  2.27377   3.567491
   _cons | -4.072202 .149476 -27.24 0.000 -4.36517  -3.779235
------------------------------------------------------------------
```

ANTERIOR—LOWER

```
. di .7226527 - 1.96*.1646645
.39991028
```

ANTERIOR—UPPER

```
. di .7226527 + 1.96*.1646645
1.0453951
```

HCABG—LOWER

```
.   di  .7185074 - 1.96*  .3460598
.04023019
```

HCABG—UPPER

```
. di  .7185074 + 1.96*  .3460598
1.3967846
```

5.4.5 Confidence Intervals of Odds Ratios

Recall that the odds ratio of a logistic regression parameter estimate is obtained by exponentiating the coefficient.

We do the same thing to obtain the odds ratio confidence intervals. The table of odds ratios is displayed as:

```
------------------------------------------------------------------
   death | Odds Ratio  Std. Err.    z  P>|z|  [95% Conf. Interval]
---------+--------------------------------------------------------
anterior |   2.05989   .3391907  4.39  0.000    1.4917    2.844505
   hcabg |  2.051369   .7098963  2.08  0.038  1.041063    4.042131
     kk2 |  2.573692   .4558429  5.34  0.000  1.818841    3.641818
     kk3 |  3.244251   .8504555  4.49  0.000   1.94079    5.423135
     kk4 |  18.55298    6.12317  8.85  0.000  9.715963     35.4276
------------------------------------------------------------------
```

We simply exponentiate the entire formula for calculating confidence intervals:

ANTERIOR—LOWER ODDS RATIO

```
. di exp(.7226527 - 1.96*.1646645)
1.4916909
```

ANTERIOR—UPPER ODDS RATIO

```
. di exp(.7226527 + 1.96*.1646645)
2.8445222
```

The same procedure holds for the other predictors. Note that if the odds ratio confidence interval contains the value of 1.0, the predictor is not significant. Moreover, a value of zero in the coefficient confidence interval indicates a non-significant predictor. Refer to Section 6.4.5 for additional discussion.

5.5 Odds Ratios as Approximations of Risk Ratios

5.5.1 Epidemiological Terms and Studies

The majority of statisticians use the odds ratios produced from a logistic regression model as an estimate of risk ratio, or of relative risk. In fact, John Cornfield (1951) originally proposed the notion of odds ratio as a good approximation to relative risk when the latter could not be determined.

This is not always justified. Because of this common interpretation of odds ratio, we shall discuss the relationship of odds ratio and risk ratio in more detail. The relationship is based on a paradigm format taken from epidemiology, to which we now turn.

The notion of risk relates to what in epidemiology is called a *risk factor,* or *exposure.* A risk factor is the predictor of primary interest in a logistic (or rate parameterized Poisson) model for which the data have been collected prospectively. For instance, consider a response e.g. *death,* that we are attempting to understand, and suppose we have a single risk factor, *smoking.* The relationship of *death* to *smoking* is a relationship of response to risk or exposure. In a sense, smoking creates a risk of death. How much of a risk is quantified as the coefficient, or rather the exponentiated coefficient, of the risk factor. The exponentiated coefficient of a logistic model, as we already know, is an odds ratio. However, for cohort studies where the incidence of the response is small, it is justified to call the exponentiated coefficient of the exposure a *risk ratio.*

It should be mentioned that some authors prefer to call the primary predictor of interest the exposure, considering it is the primary risk factor among others. In this text I shall maintain the identity of the terms exposure and risk factor when applied in this context. Bear in mind, however, that the term exposure can be used differently in other contexts, and be aware of which meaning is used in a particular study.

The exposure, or risk factor, is associated with an outcome, or response variable. When additional predictors are entered into the model as adjusters, many, perhaps most, will also be confounders. *Confounders* are binary, categorical, or continuous predictors that are statistically associated with both the outcome and risk factor, without being a consequence of either. Adding or dropping a confounder affects the coefficient value of the exposure. If a confounder significantly interacts with the risk factor/expousre, it is said to be an *effect modifier.*

Typically, but not always, the exposure term is binary, with a value of 1 indicating that the subject is exposed to some risk, and 0 indicating that the subject is not exposed. Epidemiologists generally symbolize the above relationship as $\Pr(D\,|\,E)$ and $\Pr(D\,|\,{\sim}E)$, respectively, indicating the probability of disease given the subject is exposed to some risk, and the probability of disease given that the subject is not exposed to the risk. The ratio of these two risks, or probabilities, is called the *risk ratio,* or *relative risk.* This relationship is commonly expressed as:

$$RR = \frac{\Pr(D\,|\,E)}{\Pr(D\,|\,{\sim}E)} \tag{5.15}$$

Prior to examining the relationship of odds ratios and risk ratios, it may be wise to reiterate our discussion from Section 2.4 on the types of study design appropriate to the use of odds and risk ratios. In particular, we are interested in the type of study using a logistic model for which resulting odds ratios can be interpreted as if they were risk ratios. Recall from earlier discussion that risk language allows us to assert, for example, that males are twice as likely to die of an anterior-site heart attack than females, holding the values of other model predictors as constant. In general, only prospective study designs permit us to use this type of language, and then only if the outcome is rare. In other cases, we must generally interpret odds ratio as the term implies—as the ratio of odds. We would then say that the odds of males dying of an anterior-site heart attack are twice that of females, holding constant the values of other predictors.

Recall from our earlier discussion (Section 2.4) that, in a paradigm case, *prospective cohort studies* begin by defining a risk factor for a given disease, then follow a defined group of subjects over time to determine the incidence rate of acquiring the disease. The incidence rate can clearly be obtained. We simply count the number of subjects acquiring the disease of interest and compare the rates occurring among subjects exposed to a specified risk factor from those not exposed. If the study is large and the outcome rare, true risk ratios may be obtained.

Retrospective case-control studies, on the other hand, begin with a fixed number of subjects from which the binary response is pre-established by stratification. The cases represent a sample from a greater population of subjects who have the condition of interest, for example, a disease. The controls are also sampled from the population, but from the group not having the condition of interest. One may not normally use logistic models based on case-control studies to estimate risk ratios based on estimated odds ratios. The foremost reason is that the proportion of cases having the condition of interest in the study may be considerably different from the proportion of cases having the condition of interest in the greater population. Case-control data are fixed from the start; we discover the proportion of cases and controls based upon it. Prospective studies, on the other hand, if representative of the greater population, can provide true estimates of the probability of acquiring the condition.

It is important to mention that many well-respected authors have argued that, under certain conditions, retrospectively collected data may be treated as if it were prospective. The logic of the argument is beyond this text, but it is based on data collected in the retrospective study as reflecting the true sampling fraction of cases and controls in the population. If a case-control study reflects the underlying population, or the appropriate sampling fractions are added to the study, then case-control models can be used to estimate relative risk, given that the incidence of the outcome is rare.

It should be noted that authors tend to differ in their views as to when a case-control study may be used as if it were a prospective cohort study. Hosmer and Lemeshow (2000), for example, speak of pretending that the case-control data were collected in a cohort study with the implication being that "analysis of data from case-control studies via logistic regression may proceed in the same way and using the same computer programs as cohort studies." They still recommend that the outcome must be rare and that data are a fair sample from the population.

The approach taken in this text, and the position we take with respect to this discussion, is that any interpretation of odds ratios as estimates of risk ratio needs to be specifically noted in a study report. That is, if the data from a case-control study are used as if they were prospectively collected for the purposes of estimating risk and risk ratios, a notation needs to be given in the study report that this approach has been followed. We shall interpret exponentiated logistic regression coefficients in this text as if they were odds ratios only, and not as if they are approximations of risk ratios. Relative risk models are discussed later in this section, together with recommendations on which type of model best estimates true risk ratios inherent in the population data. Although these models are not themselves logistic regression models, they do play an important role in the modeling of data that are many times improperly estimated using logistic models.

5.5.2 Odds Ratios, Risk Ratios, and Risk Models

It will perhaps be helpful to explain the relationship of odds ratio and risk ratio, and describe when an odds ratio can be used to approximate a risk ratio, by schematizing the relations. We shall use epidemiological terminology in describing the relationships, but realize that these terms are applicable for any prospective study. Those in other disciplines may prefer to use Y in place of D and X instead of E. The relationships obtain regardless of which symbols are used.

Given a univariable relationship between the outcome, D (disease) and E (risk factor/exposure) from a prospective cohort study, we have the table:

		E		
		1	0	
D	1	A	B	$A+B$
	0	C	D	$C+D$
		$A+C$	$B+D$	$A+B+C+D$

A is a count of the number of subjects, or patients, who were exposed to the risk and then later acquired the disease. B is a count of patients who were not exposed to the risk, but who nevertheless became diseased. C symbolizes a count of exposed subjects who did not acquire the disease, and D a count of subjects who neither were exposed nor became diseased. The *reference* level for E, although not usually identified as such when dealing with 2×2 relationships, is $E = 0$. This simply means that the level of E we are primarily interested in is exposed $E = 1$, compared to nonexposed $E = 0$. The relationship or reference level is essential when discussing ratios. However, a reference level is perhaps more apparent when dealing with multilevels, for example, a 2×4 table. We discuss multileveled relationships in considerable detail later in the text. We turn now to defining the odds and risk ratios on the basis of the above table, and to assessing when the odds ratio approximates the risk ratio.

The odds ratio and risk ratio may be calculated, respectively, as:

$$\text{OR} = \frac{A/C}{B/D} = \frac{AD}{CB} \tag{5.16}$$

$$\text{RR} = \frac{A/(A+C)}{B/(B+D)} = \frac{A(B+D)}{B(A+C)} \tag{5.17}$$

Given that $\simeq$ symbolizes "approximately equal to," we have:

$$\text{OR} \simeq \text{RR}$$

$$\frac{AD}{BC} \simeq \frac{A(B+D)}{B(A+C)} \tag{5.18}$$

Divide A/B from both sides

$$\frac{D}{C} \simeq \frac{(B+D}{(A+C} \tag{5.19}$$

A and B are the numerators of both the OR and RR formulae. The smaller their values, the closer are the values of the numerator and denominator for each side of the approximation, in other words, the closer $B + D$ is to D and the closer $A + C$ is to C. When the terms are close in value to one another, we may conclude that the odds ratio approximates the risk ratio.

Epidemiologists and statisticians traditionally agree that the odds ratio is an approximation to the risk ratio when the incidence of the outcome, or response, is less than 10 percent. The outcome comprises effects from both the exposed and unexposed groups. Therefore, the incidence of outcome is expressed as:

INCIDENCE OF OUTCOME

$$\frac{A+B}{A+B+C+D} \tag{5.20}$$

We can give the main 2×2 table values for each cell, which should help make the above schematics more meaningful. Suppose we substitute the following values into the main table:

		E		
		1	0	
D	1	9	14	23
	0	100	150	250
		109	164	273

Here $(A + B) = 23$ and $(A + B + C + D) = 273$. Dividing, we have the incidence of outcome rate of 0.084, and are therefore justified in using risk language when expressing the relationship of the outcome to the risk factor or exposure. If the value of A were 19 and the value of B were 24 instead, the incidence of outcome rate would be $43/293 = 0.1468$. It is evident that the number of subjects with the outcome must be small to use risk language. Risk language is expressed in terms of probabilities, not odds, and appears in phrases such as, "E1 is twice as likely to sustain D as E0." This language may be used when the incidence of outcome rate is less than 10 percent, or 0.1; otherwise, odds ratio language must be used.

The odds and risk ratios of the above table may be calculated as follows:

ODDS RATIO

```
. di (9/100)/(14/150)    or   . di (150*9)/(100*14)
.96428571                     .96428571
```

RISK RATIO

```
. di (9/109)/(14/164)
.9672346
```

In this example, we may assert that $E = 1$ is .964 times more likely to have D than is $E = 0$. This relationship may be more clearly expressed by saying that subjects exposed to a specified risk are 3.5 percent more likely not to have D than are unexposed subjects, or likewise, subjects exposed to a specified risk are 3.5 percent less likely to have D than are unexposed subjects.

Even though .964 is an odds ratio, we may use risk language to express the relationship. Of course, the true risk ratio is .967, some three one-thousandths greater than the odds ratio. We verify the calculations using Stata output. The **glm** command is used since it has an option to exclude the header statistics. Note that the data used to generate the output below use y and x in place of the D and E.

```
. glm y x [fw=cnt], fam(bin) eform nolog nohead
```

```
-----------------------------------------------------------------
       |              OIM
y | Odds Ratio   Std. Err.    z    P>|z|   [95% Conf. Interval]
--+--------------------------------------------------------------
x |   .9642857   .4303852   -0.08   0.935   .4020619    2.312696
-----------------------------------------------------------------
```

As shall be discussed later in this section, the risk ratio for a univariable relationship can be obtained using the Poisson command, with robust standard errors. We can use the same command as above, substituting *poi* in place of *bin* in the *fam()* option.

```
. glm y x [fw=cnt], fam(poi) eform nolog nohead robust

------------------------------------------------------------------------
        |                Robust
    y |        IRR    Std. Err.      z    P>|z|   [95% Conf. Interval]
----+-------------------------------------------------------------------
    x |   .9672346    .3963099   -0.08   0.935   .4332761       2.15923
------------------------------------------------------------------------
```

The values are identical to those we calculated by hand. Note also the fairly close approximation of the odds ratio confidence intervals compared to the confidence intervals for the risk ratio.

How is the relationship affected when 10 is added to the values of *A* and *B* in the table? We determined earlier that the incidence of outcome is 0.1468. The table we displayed earlier now appears as:

		E		
		1	0	
D	1	19	24	43
	0	100	150	250
		119	174	293

The odds and risk ratios may be calculated as:

ODDS RATIO

```
. di (19/100)/(24/150)
1.1875
```

RISK RATIO

```
. di (19/119)/(24/174)
1.157563
```

When *A* and *B* had values of 9 and 14, respectively, subjects exposed to the risk were more likely not to have *D* than were unexposed patients. Here the effect is reversed. But, we cannot use risk language; we cannot use the term "likely." The incidence of outcome is 0.147, greater than the criterion of 0.1. [Note: We will discuss the viability of this criterion later in this section.] Rather, we most appropriately express the relationship by asserting that subjects exposed to the risk have a near 19 percent increase in the odds of having *D* over subjects not so exposed. Note also the magnitude of the difference between the odds and risk ratios—a 3 percent difference. Raising *A* and *B* by another 10 count results in an incidence of outcome of 0.20—20 percent, or double the established criterion.

```
. di 63/313
.20127796
```

The odds ratio and risk ratios are now 1.279 and 1.217, respectively. The difference in effect is greater, and the magnitude of the difference in the two statistics is greater. As a general principle, in cohort prospective studies, the more common the outcome of interest, the greater the calculated odds ratio overestimates the relative risk, or risk ratio.

We may demonstrate the above discussion with an example taken from the **heart01** data set, which is a prospective design. For an example, we model *death* within 48 hours of admission to the hospital on having an *anterior*-site infarct.

```
. logit death anterior, nolog or

Logistic regression                    Number of obs   =        4696
                                       LR chi2(1)      =       28.14
                                       Prob > chi2     =      0.0000
Log likelihood = -771.92263            Pseudo R2       =      0.0179
-----------------------------------------------------------------
    death | Odds Ratio Std. Err.    z  P>|z| [95% Conf. Interval]
----------+------------------------------------------------------
 anterior |   2.236796 .3476527  5.18 0.000 1.649411      3.03336
-----------------------------------------------------------------
```

We know from having previously described these data that *anterior* == 0 refers to patients having had an inferior site myocardial infarction. Let us change this definition so that *anterior* == 1 indicates having had an anterior infarct, but *anterior* == 0 specifies having had any other type of infarct. The risk factor is having an anterior-site heart attack.

Let us denote the risk of death occurring for those sustaining an anterior infarct as p_e and the risk of death among those having an other-site infarct as p_u. The subscripts e and u signify *exposed* (1) and *unexposed* (0). The risk ratio is defined as:

$$\rho = \frac{p_e}{p_u} \tag{5.21}$$

and represents how much more or less likely it is for an individual sustaining an anterior infarct to die within 48 hours of admission compared to those who died having had an infarct at another site. If ρ has a value of 1.5, we can say that a patient having had an anterior infarct is 50 percent more likely to die within 48 hours of admission than is a patient having an infarct at another cardiac site. Values of ρ greater than 1.0 specify that exposed patients—that is, patients with an anterior infarct site—are more at risk than patients with an other-site infarct. Values of ρ less than 1.0 specify that a patient sustaining an other-site infarct is more at risk than the anterior-site patient. Or, we can say that values of ρ less than 1.0 indicate that an anterior-site patient is less at risk than a patient not having had an anterior-site infarct.

You may recall that the odds ratio is defined, for a binary exposure term, as:

$$OR = \frac{p_1/(1-p_1)}{p_0/(1-p_0)} \tag{5.22}$$

For the example model we have:

$$OR = \frac{\Pr(\text{Anterior}=1)/\Pr(1-\text{Anterior}=1)}{\Pr(\text{Anterior}=0)/\Pr(1-\text{Anterior}=0)} \tag{5.23}$$

In terms of the risk factor as we have defined it, the odds ratio is:

$$OR = \frac{p_e/(1-p_e)}{p_u/(1-p_u)} \tag{5.24}$$

When p_e and p_u are small compared to $(1-p_e)$ and $(1-p_u)$, respectively, then the values of ρ, the risk ratio, and OR, the odds ratio, are nearly the same. As the incidence rate grows, the odds ratio will overestimate the risk ratio when the latter is greater than 1.0 and underestimate the risk ratio when less than 1.0.

Most authors assert that the majority of epidemiological studies entail incidence of outcome rates small enough to allow for the interpretation of an odds ratio as an estimate of risk ratio. This is certainly the case for many cohort studies. Moreover, some epidemiologists have argued that certain types of case-control studies, which are retrospective rather than prospective, can be viewed as if they were prospective, and as such, the odds ratios from these studies can be interpreted as if they were risk ratios—but only if the incidence of outcome rate is less than 10 percent. Although this argument is commonly employed in some disciplines, we take the more conservative position here that it is likely best to interpret odds ratios as risk ratios for only prospective studies with rare outcomes.

Let us look at the difference between the odds and risk ratios on the **heart01** data. For simplicity, we'll model the response, *death*, on the risk factor, *anterior*. Recall, though, that *anterior* has been redefined for this analysis to mean 1 = anterior-site infarct; 0 = infarct at another site. We tabulate the two variables to set up the calculations.

```
. tab death anterior

           |         0          1
    death | Inferior   Anterior |      Total
-----------+----------------------+----------
        0 |     2,504      2,005 |      4,509
        1 |        67        120 |        187
-----------+----------------------+----------
    Total |     2,571      2,125 |      4,696
```

The odds ratio can be obtained by cross-multiplying 0,0 * 1,1 and 1,0 * 0,1, dividing the results as shown below:

ODDS RATIO

```
. di (2504*120)/(2005*67)
2.2367961
```

The resulting value is identical to the odds ratio displayed in the logistic model given earlier in this section.

The risk ratio is calculated as the ratio of two incidence rates; that is, the incidence of p_e divided by the incidence rate of p_u. We calculate this as the ratio of *death* == 1 to the total for *anterior* to the ratio of *death* == 1 to the total of other site. The following calculation produces the risk ratio.

RISK RATIO

```
. di (120/2125)/ (67/2571)
2.1669535
```

A Poisson regression can produce incidence rate ratios, which we interpret for this type of study as a relative risk ratio. The latter is calculated as the exponentiation of the Poisson parameter estimate. See Hilbe (2007a) and Zou (2004).

```
. poisson death anterior, irr robust nolog

Poisson regression                        Number of obs    =       4696
                                          Wald chi2(1)     =      26.69
                                          Prob > chi2      =     0.0000
Log pseudolikelihood =  -776.2572         Pseudo R2        =     0.0171
------------------------------------------------------------------------
             |              Robust
    death |       IRR   Std. Err.     z   P>|z|  [95% Conf. Interval]
-----------+------------------------------------------------------------
anterior | 2.166953   .3243484  5.17   0.000  1.616003    2.905741
------------------------------------------------------------------------
```

We observe that the two values are identical.

Likewise, a log-binomial model produces risk ratios as well. The model is a generalized linear model where the binomial family is used with a log link. The canonical form is the logit link, which results in a logistic regression. The problem with using a log-binomial model is that the predicted values may fall outside the probability range of 0 to 1. For the **binreg** program, fitted values exceeding 0/1 are truncated at each iteration. This method allows the model to converge and produce risk ratios. Many data situations do not allow themselves to be modeled as a log-binomial. No such problem exists with the current data. The log-binomial model produces the same value of risk ratio, as did the robust Poisson model, and as did our hand calculations. Note that both of the exponentiated coefficients of the two models are identical, and the standard errors are the same to the one-thousandth.

```
. binreg death anterior, rr n(1) nolog

Generalized linear models                No. of obs     =       4696
Optimization       : MQL Fisher scoring  Residual df    =       4694
                     (IRLS EIM)          Scale parameter =          1
Deviance           =   1543.845261       (1/df) Deviance =   .3288976
Pearson            =   4695.998849       (1/df) Pearson  =   1.000426

Variance function: V(u) = u*(1-u)        [Bernoulli]
Link function     : g(u) = ln(u)         [Log]

                                         BIC            = -38141.42
-------------------------------------------------------------------
             |                EIM
    death |  Risk Ratio Std. Err.   z  P>|z| [95% Conf. Interval]
---------+---------------------------------------------------------
anterior |   2.166953 .3243138 5.17 0.000 1.616054      2.90565
-------------------------------------------------------------------
```

Peterson and Deddens (2008) have demonstrated that the robust Poisson method of calculating prevalence ratios, a term they use instead of risk ratios, produce less-biased estimates than the log-binomial for moderate sample sized models when the prevalence or incidence rates are very high. On the other hand, for both moderate-valued prevalences and moderate-sized models, the log-binomial is less biased than the robust Poisson. In all of the simulations, the log-binomial method was slightly more powerful in determining the correct risk or prevalence ratio, as well as producing smaller standard errors than the robust Poisson.

We defer the discussion of robust standard errors to later in this chapter (see Section 5.7). We can say at this point that robust standard errors are not based on an underlying probability distribution, but rather on the data. Nearly all major software applications offer robust standard errors as an option when estimating Poisson models—as well as other models based on generalized linear models. The problem we are addressing relates to employing odds ratios as an approximation of risk ratio. So doing allows the statistician to use a probabilistic interpretation of the relationship of the response to the risk factor, as in our example of *death* to *anterior*. The incidence of outcome for the *death-anterior* relationship is 0.04, well below the criterion for allowing us to interpret the logistic odds ratios as if they were risk ratios.

```
. di 187/4696
.03982112
```

One may conclude, for this example, that a patient is 2.2 times more likely to die within 48 hours of admission if he has an anterior-site infarct than if he has primary damage to his heart at another site. Again, when the incidence rate of death is low in the model (not population), then we may be able to make such a conclusion.

The following table from Gould (2000) presents actual risk ratio values for given probability levels and odds ratios. Probability values ($Pr(y|x)$, are on the vertical axis; odds ratios are listed on the horizontal axis.

Table 5.2 shows us that *p*-values, or incidence rates, of .01 result in fairly equal odds and risk ratios. The traditional criterion of .1, or 10 percent, incidence provides good support for using risk language with odds ratios. The greater the odds ratio, however, the less odds ratios and risk ratios are the same. For probability values of .2, only odds ratios of 1 or less approximate risk ratios. Note that when the odds ratio is equal to 1, we can use risk ratio language no matter the probability.

Zhang and Yu (1998) proposed a formula to convert binary response logistic regression odds ratios in prospective cohort studies to estimated risk ratios. Their method has been used in a host of research studies in the medical and public health domains. The authors argued that the formula is most useful when the odds ratio is greater than 2.5 or less than 0.5. Odds ratios exceeding these extremes tend to increasingly overestimate or underestimate, respectively, as the values become more extreme. Note in advance, however, that both the risk ratios and the associated confidence intervals

TABLE 5.2

Table of Risk Ratios Given Probability of Event and Odds Ratios

$Pr(y|x)$	.25	.50	.75	1.00	1.50	2.00	4.00
.2	.2941	.5556	.7895	1.000	1.364	1.667	2.500
.1	.2702	.5263	.7692	1.000	1.429	1.818	3.077
.01	.2519	.5025	.7519	1.000	1.493	1.980	3.883
.001	.2502	.5003	.7502	1.000	1.499	1.998	3.988
.0001	.2500	.5000	.7500	1.000	1.500	2.000	4.000

produced by this method are biased, and tend to become more so as the number of model confounders increase (McNutt et al., 2003; Wolkewitz et al., 2007). Because the formula has been used to such an extent, it is important to discuss its logic, as well as why it does not provide the approximation it intends.

Derived from the formula for odds ratio:

$$OR = \frac{p_e/(1-p_e)}{p_u/(1-p_u)} \tag{5.25}$$

where p_e is the incidence rate for exposed patients (*anterior* $== 1$), and p_u is the incidence rate for unexposed patients (*anterior* $== 0$), the formula for risk ratio in terms of odds ratio is

$$RR > \frac{OR}{(1-p_u)+(p_u*OR)} \tag{5.26}$$

To use the formula, one must first calculate the incidence rate of the unexposed group, that is, when the risk factor equals 0.

CALCULATE INCIDENCE RATE OF DEATH FOR UNEXPOSED RISK FACTOR

```
. su death if anterior==0

Variable |        Obs        Mean    Std. Dev.        Min        Max
---------+-----------------------------------------------------------
   death |       2571    .0260599    .1593445          0          1
              =========
```

Using the odds ratio of *anterior* at 2.2368, the estimated risk ratio is as follows.

CALCULATE ESTIMATED RISK RATIO

```
. di 2.2368/((1-.02606)+(.02606*2.2368))
2.1669568
```

Note that this value is identical to the risk ratio produced by both the Poisson model and hand calculations from the tabulation of *death* and *anterior*.

The **oddsrisk** command, a user-authored Stata command, was developed to calculate the estimated risk ratios from odds ratios produced in a logistic regression model. We use it to calculate the same value as above.

```
. oddsrisk death anterior
```

```
------------------------------------------------------------------------
Incidence for unexposed risk group =        0.0261
------------------------------------------------------------------------
Predictor      Odds Ratio    Risk Ratio      [95% Conf. Interval]
------------------------------------------------------------------------
anterior           2.2368        2.1670        1.6220        2.8807
------------------------------------------------------------------------
```

The rounded risk ratio value is identical to the value we calculated using Stata's calculator. The confidence intervals, however, differ from the robust Poisson and log-binomial models by some two one-hundredths. We consider this difference negligible. However, when more confounders are entered into the algorithm, the values of the predictor risk ratios and their associated confidence intervals vary considerably more from the true risk ratios; in other words, they are biased.

Modeling *death* on *anterior* and associated confounders using the **oddsrisk** algorithm provides the following output:

```
. oddsrisk death anterior hcabg kk2-kk4 age3 age4
```

```
------------------------------------------------------------------------
Incidence for unexposed risk group =        0.0251
------------------------------------------------------------------------
Predictor      Odds Ratio    Risk Ratio      [95% Conf. Interval]
------------------------------------------------------------------------
anterior           1.8941        1.8525        1.3516        2.5269
hcabg              2.1955        2.1316        1.0970        4.0401
kk2                2.2817        2.2106        1.5782        3.0760
kk3                2.2182        2.1524        1.2986        3.5162
kk4               14.6398       10.9059        6.2886       17.1776
age3               3.5496        3.3361        2.3147        4.7510
age4               6.9648        6.0578        4.2455        8.4600
------------------------------------------------------------------------
```

Modeling the same data using a robust Poisson command gives us:

```
. glm death anterior hcabg kk2-kk4 age3 age4, nolog fam(poi)
eform robust nohead
```

```
------------------------------------------------------------------------
           |             Robust
    death  |      IRR  Std. Err.     z    P>|z|  [95% Conf. Interval]
-----------+------------------------------------------------------------
  anterior | 1.775475  .2841127   3.59   0.000   1.297492   2.429542
     hcabg | 2.025651   .623352   2.29   0.022   1.108218   3.702578
       kk2 | 2.155711  .3575015   4.63   0.000     1.5575   2.983687
```

```
      kk3 | 2.103697   .5493533   2.85   0.004   1.260964    3.509647
      kk4 | 8.178385   2.253432   7.63   0.000   4.765755   14.03471
     age3 | 3.324076   .6332468   6.31   0.000   2.288308    4.82867
     age4 | 5.906523   1.194716   8.78   0.000   3.973368   8.780211
--------------------------------------------------------------------
```

Log-binomial results are

```
. binreg death anterior hcabg kk2-kk4 age3 age4, rr
                                       /// header statistics
                                       /// deleted
--------------------------------------------------------------------
            |                 EIM
      death | Risk Ratio  Std. Err.   z   P>|z|  [95% Conf. Interval]
---------+----------------------------------------------------------
   anterior |  1.765091   .2735107  3.67  0.000  1.302772   2.391473
      hcabg |  1.986219   .6013924  2.27  0.023  1.097231   3.595473
        kk2 |  2.163118   .3571723  4.67  0.000  1.565058   2.989716
        kk3 |  2.007106   .4902613  2.85  0.004  1.243522   3.239569
        kk4 |  6.655951  1.597871   7.90  0.000  4.157815   10.65504
       age3 |  3.336484   .6318699  6.36  0.000  2.301902   4.836055
       age4 |  5.640176  1.089631   8.95  0.000   3.86232   8.236392
--------------------------------------------------------------------
```

We can observe from the above output that the Zhang–Yu algorithm and robust-Poisson and log-binomial model values differ. In fact, the true risk ratios for these predictors differ from all three outputs. Research has shown that Zhang–Yu algorithm risk ratio values and confidence intervals become increasingly biased with added predictors/confounders (McNutt et al., 2003). This is due to the formula used in the algorithm being incorrect. McNutt et al. argue that the algorithm fails because "the formula, used as one summary value, fails to take into consideration the more complex relation in the incidence of disease related to exposure for each covariate pattern." In other words, for the purposes of converting an odds ratio to a risk ratio, calculations must be done at the covariate pattern level rather than at a global one. In fact, Yu and Wang (2008) attempt such a method, but the calculations are tedious and difficult to employ. [Note: Currently SAS's Proc NLP is required for using the Yu-Wang algorithm. As of this writing, the method has not undergone evaluation by other statisticians.] However, research is still ongoing as to whether robust Poisson or log-binomial estimates give more proximate values to the true risk ratios for given data. Wolkewitz et al. (2007) prefer using robust Poisson; McNutt et al. (2003) prefer the log-binomial. We prefer the log-binomial since estimation is based on the binomial distribution, and we believe that simulation studies have demonstrated it to yield close approximations to the true risk ratios. On the other hand, the log-binomial algorithm frequently fails to converge (Hardin and Hilbe, 2007), in which case use of the robust Poisson method is reasonable.

Next, we turn to an examination of univariable models comparing logit, Poisson, and log-binomial with the aim of identifying appropriate estimates of risk. We shall also discuss the calculation of standard errors and confidence intervals as well as the development of full models for comparative purposes.

5.5.3 Calculating Standard Errors and Confidence Intervals

In the previous section we developed initial concepts of odds ratio and risk ratio based on 2×2 cell relationships and subsequently on the tabulation of *death* and *anterior*. The data from which the tabulation was calculated consist of 4,696 separate observations. Specifically, we shall now use an aggregated form of the logistic model of *death* on *anterior* (**heart01**) to demonstrate how standard errors, confidence intervals, and several other statistical relationships can be calculated. We commence by displaying the observation-based tabulation of *death* on *anterior*, which was used in the previous subsection.

```
. tab death anterior

          |         0          1
  death   |  Inferior   Anterior  |     Total
----------+------------------------+----------
     0 |     2,504      2,005  |     4,509
     1 |        67        120  |       187
----------+------------------------+----------
  Total  |     2,571      2,125  |     4,696
```

The tabulated data from the above table were stored separately in a file called **anterior**, consisting of three variables: *anterior, death,* and *count*.

```
. use anterior

. list

     +----------------------------+
     | count    death    anterior |
     |----------------------------|
  1. |   120        1           1 |
  2. |    67        1           0 |
  3. |  2005        0           1 |
  4. |  2504        0           0 |
     +----------------------------+
```

We now repeat the schema we used for the 2×2 table of response and risk factor displayed in Chapter 2, but now with marginals, that is, summation values for the respective columns and rows. Note that we reversed the terms in the table from our earlier discussion in this section.

```
                        Response
                        1        0
                  ---------------
    Risk      1 |    a        b  | a+b
    or          |                |
    Exposure  0 |    c        d  | c+d
                  ---------------
                      a+c      b+d    N
```

Inserting the appropriate values into the schema, we have:

```
                    Response (death)
                      1        0
                  ---------------
    Risk or       1 |  120     2005 | 2125
    Exposure        |               |
    (Anterior MI) 0 |   67     2504 | 2571
                  ---------------
                      187      4509    4696
```

Modeling *death* on *anterior* using the **logit** command gives us:

LOGIT MODEL PARAMETER (COEFFICIENT) ESTIMATES

```
. logit death anterior [fw=count], nolog

Logistic regression                    Number of obs   =        4696
                                       LR chi2(1)      =       28.14
                                       Prob > chi2     =      0.0000
Log likelihood = -771.92263            Pseudo R2       =      0.0179
------------------------------------------------------------------
   death |    Coef.   Std. Err.     z    P>|z|  [95% Conf. Interval]
---------+--------------------------------------------------------
anterior |  .8050445 .1554244    5.18  0.000   .5004182   1.109671
   _cons | -3.620952 .1237929  -29.25  0.000  -3.863582  -3.378323
------------------------------------------------------------------
```

Note that, when aggregated data are being modeled, one must enter the counts as a frequency weight. The results are identical with the observation-based model.

As we have previously discussed, the parameter estimate odds ratio may be calculated directly from the 2 × 2 table. Keeping in mind that the table has been reversed, the odds ratio is defined on the basis of the above table as:

CALCULATE ODDS RATIO

Odds of Response given Exposure = 1 : a/b
Odds of Response given Exposure = 0 : c/d

$$OR = \psi = (a/b)/(c/d) \tag{5.27}$$

Where a/b determines the odds of *death* for *anterior* infarct (risk factor = 1), c/d determines the odds of *death* for *other-site* infarct (risk factor = 0). $(a/b)/(c/d)$ is the ratio of the two odds—or odds ratio.
    As shown earlier in Section 5.5.1, the calculation may be simplified as:

$$OR = \psi = (a*d)/(b*c) \tag{5.28}$$

```
. di (120*2504) / (2005*67)
2.2367961
```

CALCULATE PARAMETER ESTIMATE

The logistic parameter estimate may be calculated from the odds ratio by taking its natural log. Generally, logistic regression algorithms determine the model odds ratios by first modeling the data, producing a table of parameter estimates, and then exponentiating the coefficients. Here we do the reverse.

$$\text{Parameter estimate} = \ln(\psi)$$

```
. di ln(2.2368)
.80504627
```

which we observe is identical, taking rounding error into consideration, to the coefficient displayed in the model output.
    Standard errors may be calculated using methods developed by Woolf (1955).

CALCULATE STANDARD ERROR OF PARAMETER ESTIMATE

$$SE(\ln(\psi)) = \text{sqrt}\left(1/a + 1/b + 1/c + 1/d\right) \tag{5.29}$$

```
. di sqrt(1/120 + 1/67 + 1/2005 + 1/2504)
.15542465
```

This calculated value matches the value displayed in the model output: .1554244
    Confidence intervals are calculated by standard methods, as we next demonstrate.

CALCULATE 95 PERCENT CONFIDENCE INTERVALS

$$\ln(\psi) +/- 1.96 * SE(\ln(\psi)) \tag{5.30}$$

LOWER CI

```
. di .80504627 - 1.96*.15542465
.50041396
```

UPPER CI

```
. di .80504627 + 1.96*.15542465
1.1096786
```

which correspond very well to the model output of (.5004182, 1.109671). Remember, the parameter estimate, standard error, and confidence interval are being estimated by maximum likelihood. The hand calculations prove to be near identical.

Exponentiating each of the confidence interval limits results in the confidence intervals for the odds ratio.

ODDS RATIO CONFIDENCE INTERVAL

```
. di exp(.50041396)
1.6494039

. di exp(1.1096786)
3.0333833

. logit death anterior [fw=count], nolog or  [ NO HEADER
DISPLAYED]
```

```
------------------------------------------------------------
    death | Odds Ratio Std. Err.   z   P>|z| [95% Conf. Interval]
---------+--------------------------------------------------
anterior |  2.236796 .3476527 5.18  0.000 1.649411    3.03336
------------------------------------------------------------
```

Again, the coherence is maintained.

We now turn to similar calculations for the risk ratio. This will allow us to better relate the two statistics and to perhaps better understand when risk ratios may be estimated using odds ratios.

RISK RATIO

When calculated by hand, the risk ratio is based on the same table as was the calculation of the odds ratio and related model statistics.

```
                 Response (death)
                   1      0
                 ---------------
Risk   or    1 |  120    2005 | 2125
Exposure       |              |
             0 |   67    2504 | 2571
                 ---------------
                  187    4509   4696
```

RELATIVE RISK/RISK RATIO

The risk of dying if one has an anterior MI is calculated as the count of those patients who do die out of all patients in the study who have had an anterior MI. The risk of dying if one has an MI at another site on the heart is calculated as the count of those patients who do die out of all patients in the study who have had an MI at a nonanterior site. The ratio of these two risks is the risk ratio. The calculations from the table appear below.

$$RR = \frac{a/(a+b)}{c/(c+d)} = \frac{a(c+d)}{c(a+b)} \tag{5.31}$$

RISK: ANTERIOR = 12/2125

```
. di 120/2125
.05647059
```

RISK: OTHER = 67/2571

```
. di 67/2571
.0260599
```

RISK RATIO

```
. di .05647059 / .0260599
2.1669534
```

The risk ratio is 2.16695. Therefore, the coefficient or parameter estimate is determined as:

Parameter estimate = ln(RR)

```
. di ln(2.1669534)
.77332222
```

Now we calculate the 95 percent confidence intervals of each risk, in other words, the numerator and denominator of the risk ratio.

95 PERCENT CONFIDENCE INTERVAL FOR THE RISK OF DEATH FOR ANTERIOR MI PATIENTS

```
. di .05647059 - 1.96*sqrt(.05647059 * (1-.05647059)/4696)
.04986851
```

```
. di .05647059 + 1.96*sqrt(.05647059 * (1-.05647059)/4696)
.06307267
```

95 PERCENT CONFIDENCE INTERVAL FOR THE RISK OF DEATH FOR OTHER SITE MI PATIENTS

```
. di  .0260599 - 1.96*sqrt(.0260599 * (1- .0260599)/4696)
.02150326
. di  .0260599 + 1.96*sqrt(.0260599 * (1- .0260599)/4696)
.03061654
```

We can determine the standard error, though, using the formula:

STANDARD ERROR OF COEFFICIENT ESTIMATE OF ln(RISK RATIO)

$$SE(\ln(RR)) = sqrt(1/a - 1/(a+b) + 1/c - 1/(c+d)) \tag{5.32}$$

```
. di sqrt( 1/120- 1/2125 + 1/67 - 1/2571)
.1496635
```

COEFFICIENT STANDARD ERRORS

The coefficient confidence intervals are calculated from the parameter estimate (.7733222) and coefficient standard error (.1496635).

$$\ln(RR) +/- 1.96 * SE(\ln(RR)) \tag{5.33}$$

```
. di  .7733222 - 1.96 * .1496635
.47998174
```

```
. di  .7733222 + 1.96 * .1496635
1.0666627
```

RISK RATIO 95 PERCENT CONFIDENCE INTERVAL

As we did with the odds ratio, exponentiating the respective values provides us with the 95 percent confidence interval of the risk ratio.

```
. di exp(.47998174)
1.6160449

. di exp(1.0666627)
2.9056662
```

To summarize, we have:

OR = 2.2367961	RR = 2.1669534
CI = (1.6494039, 3.0333833)	CI = (1.6160449, 2.9056662)

5.5.4 Risk Difference and Attributable Risk

Two other statistics are commonly found in the literature dealing with logistic models aimed at describing epidemiological relationships: risk difference and attributable risk. We briefly describe each.

Risk difference, also called absolute risk difference, is the difference between the two levels in a binary risk factor, or exposure. Prospective follow-up study results should serve to help make this concept more clear. Suppose a study of the effects of smoking on lung cancer. Two groups of subjects were involved in a longitudinal study. Smokers and nonsmokers were followed over a given time period to observe the incidence rate of contracting lung cancer. A total of 112 out of the 1,612 smokers in the study developed lung cancer, while only 98 of 3,218 nonsmokers were so diagnosed. The incidence of outcome is 210/4,830, or 0.0435—4.3 percent.

```
              Contract Lung Cancer
                   1       0
              ---------------
  Smoker    1 |   112    1500 | 1612
             |               |
Nonsmoker   0 |    98    3120 | 3218
              ---------------
                  210    4620   4830
```

The list of data as depicted in the editor is displayed as:

```
. l

       +--------------+
       | cnt    x    y |
       |--------------|
   1.  |  112    1    1 |
   2.  | 1500    1    0 |
   3.  |   98    0    1 |
   4.  | 3120    0    0 |
       +--------------+
```

```
RISK X=1                    /// PREVALENCE RISK
. di 112/1612
.06947891
```

```
RISK X=0                    /// PREVALANCE RISK
. di 98/3218
.0304537
```

```
RISK RATIO                  /// PREVALENCE RELATIVE RISK
. di .06947891/.0304537
2.2814604
```

```
RISK DIFFERENCE
. di .06947891 -  .0304537
.03902521
```

The risk of contracting lung cancer is .039, or 39 chances in 1000, greater for smokers compared to nonsmokers.

A 95 percent confidence interval may be calculated for the risk difference (Woodward, 2005, p. 55):

95 PERCENT RISK DIFFERENCE CONFIDENCE INTERVAL

$$p_1 - p_2 \pm 1.96 \sqrt{(p_1(1-p_1)/n_1) + (p_2(1-p_2)/n_2)} \qquad (5.34)$$

```
. di .06947891 -  .0304537 - 1.96*sqrt(  ((.06947891*
(1-.06947891))/1612) + ((.0304537*(1-.0304537))/3118))
.0252248
```

```
. di .06947891 -  .0304537 + 1.96*sqrt(  ((.06947891*
(1-.06947891))/1612) + ((.0304537*(1-.0304537))/3118))
.05282562
```

The risk difference is, therefore, .039 with 95 percent confidence intervals of .025 and .053.

```
. glm y x [fw=cnt], nolog nohead fam(poi) eform robust

----------------------------------------------------------------
         |              Robust
   y |       IRR    Std. Err.     z    P>|z|   [95% Conf. Interval]
---+------------------------------------------------------------
   x |   2.28146   .3078309    6.11   0.000    1.751309    2.972098
----------------------------------------------------------------


. binreg y x [fw=cnt], rr nolog   /// header statistics deleted

----------------------------------------------------------------
         |                EIM
   y | Risk Ratio   Std. Err.     z    P>|z|   [95% Conf. Interval]
---+------------------------------------------------------------
   x |    2.28146    .307799    6.11   0.000   1.751357    2.972016
----------------------------------------------------------------
```

Attributable risk is a mixture of relative risk and prevalence of the risk. By itself, a relative risk does not tell us of the impact of the risk in the population. If few people in a study, or population, are affected by a risk factor compared with another risk factor, the second risk factor will have much more impact on, for example, contracting lung cancer, than the first risk factor. The attributable risk, in effect, weights the relative risk by prevalence.

The formula for the attributable risk is

$$\theta = \frac{r - r_{Eu}}{r} \tag{5.35}$$

where r is the overall risk of the outcome and r_{Eu} is the risk in the unexposed group. Given the table given above,

$$r = (a+c)/n = (a+c)/(a+b+c+d) \tag{5.36}$$

and

$$r_{Eu} = c/(c+d) \tag{5.37}$$

For the lung cancer example we have:

```
             Contract Lung Cancer
                 1        0
             ---------------
  Smoker   1 |   112     1500 |  1612
            |                 |
Nonsmoker   0 |    98     3120 |  3218
             ---------------
                 210     4620    4830
```

OVERALL RISK OF LUNG CANCER

```
. di (112+98)/4830
.04347826
```

RISK OF LUNG CANCER FOR UNEXPOSED GROUP

```
. di 98/3218
.0304537
```

ATTRIBUTABLE RISK

```
. di (.04347826 - .0304537)/ .04347826
.29956489
```

The attributable risk for lung cancer given smoking is 0.3, or 30 percent.

The 95 percemt confidence interval of the attributable risk is given as follows (Woodward, 2005, p. 148).

95 PERCENT CONFIDENCE INTERVAL OF ATTRIBUTABLE RISK

$$\frac{(ad - bc)\exp(\pm u)}{nc + (ad - bc)\exp(\pm u)} \tag{5.38}$$

with u defined as:

$$\frac{1.96(a + c)(c + d)}{ad - bc}\sqrt{\frac{ad(n - c) + c^2 b}{nc(a + c)(c + d)}} \tag{5.39}$$

The + or − in front of u in the top expression indicates the lower and upper bounds, respectively.

With $a = 112$, $b = 1500$, $c = 98$, $d = 3120$, and $n = 4830$:

```
.
. local a=112
. local b=1500
. local c=98
. local d=3120
. local n=4830

. local u = (1.96*('a'+'c')*('c'+'d'))/('a'*'d' - 'b'*'c') *
sqrt( ('a'*'d'*('n'-'c') + 'c'*'c' *'b') / ('n'*'c'*('a'+'c')*
('c'+'d')))

. local arl = (('a'*'d' - 'b'*'c') * exp(-'u')) / ('n'*'c' +
('a'*'d' - 'b'*'c')*exp(-'u'))
```

```
. di 'arl'
.21051337          ← LOWER BOUND

. local aru = (('a'*'d' - 'b'*'c') * exp('u')) / ('n'*'c' +
('a'*'d' - 'b'*'c')*exp('u'))

. di 'aru'
.40687309          ← UPPER BOUND
```

The attributable risk is therefore .30 with 95 percent confidence intervals of .21 and .41. Note that attributable risk confidence intervals are only meaningful for the study population. Projection to the population as a whole may not be warranted.

5.5.5 Other Resources on Odds Ratios and Risk Ratios

Understanding the relationship of odds ratio and risk ratio is not as clearcut as many would like to have it. Below are references to seven journal articles that deal with this area in different manners and present a clear analysis of the subject.

Deddens, J.A. and M.R. Petersen (2004). Estimating the relative risk in cohort studies and clinical trials of common outcomes, *American Journal of Epidemiology*, 159(2): 213–214.

Deeks, J. (1998). When can odds ratios mislead? Odds ratios should be used only in case-control studies and logistic regression analysis [letter], *British Medical Journal*, 317(716); 1155–1156; discussion 1156–1157.

Greenland, S. (1987). Interpretation and choice of effect measures in epidemiological analysis, *American Journal of Epidemiology* 125(5): 761–768.

Lee, J. (1994). Odds ratio or relative risk for cross sectional data? *International Journal of Epidemiology*, 23(1): 201–203.

McNutt, L.A., C. Wu, X. Xue, and J.P Hafner (2003). Estimating the relative risk in cohort studies and clinical trials of common outcomes, *American Journal of Epidemiology*, 157(10): 940–943.

Pearce, N. (1993). What does the odds ratio estimate in a case-control study? *International Journal of Epidemiology*, 22(6): 1189–1192.

Zocchetti, C., D. Consonni, and P.A. Bertazzi (1997). Relationship between prevalence rate ratios and odds ratios in cross-sectional studies, *International Journal of Epidemiology*, 26(1): 220–223.

Sources that provide a readable support for the position that certain types of case-control data may be treated as if they were prospective include:

Collett (1991), Weinberg and Wacholder (1993), Hosmer and Lemeshow (2000), Woodward (2005), Lumley, Kronmal, and Ma (2006), and van der Laan (2008).

5.6 Scaling of Standard Errors

Scaling of standard errors is perhaps the oldest method used to adjust for extra correlation in the data. In the case of binary response data, we might want to adjust standard errors to account for the clustering effect of how the data have been gathered. For instance, in the case of our **heart01** data, it seems reasonable to believe that patients are treated more similarly within a provider or hospital than between hospitals. Recall that likelihood theory, upon which all maximum likelihood estimation is based, requires that each observation, or covariate pattern, is independent of the others. The GLM-based IRLS method of estimation, called Fisher Scoring, is a subset of maximum likelihood estimation, and therefore also requires independence of observations. When clustering occurs, this important assumption upon which estimation of these models is founded is violated. In the simplest case, only standard errors might be affected. More substantial violations may affect parameter estimates and related model statistics as well. The problem is that a model may appear to be well specified and fitted, but is not. Scaling attempts to adjust standard errors due to clustering effects, as well as for any other unaccounted-for source of extra-model correlation.

Scaling adjusts the weights, *W*, in:

$$\beta = (X'WX)^{-1}X'Wz \tag{5.40}$$

which is the GLM estimating equation that results in the production of parameter estimates, standard errors, and the remaining regression statistics. Refer to Chapter 3 for details. Scaling replaces *W* by the inverse square root of the Pearson or deviance dispersion statistic. So doing in effect adjusts the model standard errors to the value that would have been calculated if the dispersion statistic had originally been 1.0. Statisticians have used the deviance-based dispersion statistic for the purposes of scaling, but simulation studies have demonstrated that the Pearson dispersion is preferable.

We shall discuss simulation methods in more detail in Chapter 6. However, it may be instructive here to provide a simulation to provide the rationale for using the Pearson dispersion for the scaling of binomial or grouped logistic models.

First, we create a simulated binomial logistic model with a constant denominator value of 100. We set the number of observations in the data as 50,000, and create two random variates, assigning them specific values of $x1 = 0.25$ and $x2 = -1.5$. The model intercept is assigned the value of 1.0. The code producing the simulated data just described is given as:

CREATE 50000 OBSERVATIONS

```
. set obs 50000
```

CREATE TWO RANDOM VARIATES

```
. gen x1 = abs(rnormal())
. gen x2 = abs(rnormal())
```

CREATE A LINEAR PREDICTOR WITH CONSTANT = 1, x1 = 0.25, x2 = –1.5

```
. gen xb = 1 + .25*x1 - 1.5*x2
```

CREATE BINOMIAL LOGISTIC RESPONSE WITH DEFINED DATA

```
. gen d = 100
. gen exb = 1/(1+exp(-xb))
. gen by = rbinomial (d, exb)

. glm y x1 x2, fam(bin 100) nolog
```

```
Generalized linear models              No. of obs        =         50000
Optimization        : ML               Residual df       =         49997
                                       Scale parameter   =             1
Deviance        =   50707.00439        (1/df) Deviance   =      1.014201
Pearson         =   50318.12465        (1/df) Pearson    =      1.006423

Variance function: V(u) = u*(1-u/100)
Link function     : g(u) = ln(u/(100-u))

                                       AIC               =       5.87019
Log likelihood   = -146751.7419        BIC               =     -490249.5
-------------------------------------------------------------------------
            |                OIM
        y |      Coef.   Std. Err.      z    P>|z|   [95% Conf. Interval]
------------+------------------------------------------------------------
       x1 |   .2496956   .0016227   153.88  0.000    .2465152   .2528761
       x2 |  -1.499944    .001913  -784.07  0.000   -1.503694  -1.496195
     _cons |   1.001436   .0021072   475.25  0.000    .9973058   1.005566
-------------------------------------------------------------------------
```

Note that the parameter estimates are nearly identical to the values we defined. Rounding error will give slightly different results each time this method is followed. Setting a random seed number, however, will guarantee that the same values will always be obtained.

Observe the values of the deviance and Pearson dispersion statistics. The Pearson deviance is equal to 1.0 to six one-thousandths. The deviance dispersion is some 1.5 percent higher than the theoretically ideal value of 1.0. Any value greater than 1 indicates binomial overdispersion. Overdispersed models, as we shall discuss, yield biased standard errors, and therefore may appear to be significant when in fact they are not.

The Pearson dispersion statistic maintains this same function for count response models as well as for binomial. When a model displays a Pearson dispersion statistic (or what we shall generally refer to as simply the dispersion) greater than 1.0, then the model requires adjusting. Scaling by the Pearson dispersion amends the standard errors so that they approximate the values that would have been estimated were the model not overdispersed.

Note that there is no esoteric rule to help us decide how much greater than 1.0 the dispersion must be for us to assert it as an indicator of overdispersion. As a general rule, the larger the data set, the smaller the difference needs to be between the dispersion statistic and 1.0 for us to claim likely overdispersion in the data. Data sets with smaller numbers of observations usually require values with a substantially greater difference between the dispersion and 1.0 to be classifiable as likely overdispersed. Chapter 9 is devoted to the problem of overdispersion, which will provide additional insight about the criteria used to determine whether a model is truly overdispersed, or whether the dispersion statistic reflects random variation in otherwise independent data.

We use the **heart01** data to provide an example of scaling by the Pearson dispersion:

```
. glm death anterior hcabg kk2-kk4 age3 age4, fam(bin) eform
nolog

Generalized linear models            No. of obs       =       4503
Optimization        : ML             Residual df      =       4495
                                     Scale parameter  =          1
Deviance         =  1276.319134      (1/df) Deviance  =   .283942
Pearson          =  4212.631591      (1/df) Pearson   =  .9371817

Variance function: V(u) = u*(1-u)    [Bernoulli]
Link function     : g(u) = ln(u/(1-u))  [Logit]

                                     AIC              =   .2869907
Log likelihood   = -638.1595669      BIC              =  -36537.86
```

```
--------------------------------------------------------------------
            |                    OIM
    death   | Odds Ratio  Std. Err.    z    P>|z|  [95% Conf. Interval]
--------+-----------------------------------------------------------
anterior  |   1.894096   .3173418   3.81  0.000  1.363922    2.630356
   hcabg  |   2.195519   .7744623   2.23  0.026  1.099711    4.383246
     kk2  |   2.281692   .4117012   4.57  0.000  1.602024    3.249714
     kk3  |   2.218199   .5971764   2.96  0.003  1.308708    3.759743
     kk4  |  14.63984   5.218374   7.53  0.000  7.279897    29.44064
    age3  |   3.549577   .7119235   6.32  0.000  2.395823    5.258942
    age4  |   6.964847   1.44901    9.33  0.000  4.632573    10.47131
--------------------------------------------------------------------
```

Note that the model is underdispersed, with a Pearson dispersion of 0.94. The scaled model appears as:

```
. glm death anterior hcabg kk2-kk4 age3 age4, fam(bin) irls
scale(x2) eform
```

[HEADER STATISICS THE SAME]

```
--------------------------------------------------------------------
            |                    EIM
    death   | Odds Ratio  Std. Err.    z    P>|z|  [95% Conf. Interval]
--------+-----------------------------------------------------------
anterior  |   1.894096   .3072126   3.94  0.000  1.378293    2.60293
   hcabg  |   2.195519   .7497424   2.30  0.021  1.124249    4.287577
     kk2  |   2.281692   .3985602   4.72  0.000  1.62021     3.213237
     kk3  |   2.218199   .5781152   3.06  0.002  1.330936    3.696952
     kk4  |  14.63984   5.05181    7.78  0.000  7.444058    28.7914
    age3  |   3.549577   .6891997   6.52  0.000  2.426073    5.193369
    age4  |   6.964847   1.402759   9.64  0.000  4.693262    10.3359
--------------------------------------------------------------------
```

(Standard errors scaled using square root of Pearson X2-based dispersion)

Note that when using Stata for the estimation, the *irls* option must accompany the *scale()* option. This is because the *irls* option specifies the use of the expected information matrix for production of standard errors. Scaling is traditionally based on using this matrix rather than the observed information matrix. But, since the logit link is canonical, the two matrices are identical, and theoretically, one should not have to use the *irls* option. It is important and makes a difference in standard errors, when noncanonical links are used, such as probit and complementary loglog.

Notice also that the predictors all have similar *p*-values to the unadjusted model. This tells us that there is little unexplained correlation in the data. However, the data have been collected from a number of different facilities. There may be extra correlation in the data that is not revealed by scaling. Another commonly used method of adjustment, especially when clustering is evident and can be identified, is the use of a robust variance estimator. We turn to that discussion next.

5.7 Robust Variance Estimators

heart01 data were collected from various hospitals, or providers, throughout Canada. We know that, in many cases of medical treatment, the way treatment is given, and to whom it is given, can vary far more between providers than from within providers, that is, it is likely that patients are treated in a more similar manner within a facility than between facilities. Within a facility, patients tend to see the same group of physicians, the hospital environment is the same, the nursing care is similar, and so forth. Therefore, it is reasonable to consider that there is a higher correlation of treatment within a provider than between providers. This situation violates the independence of observations criterion underlying maximum likelihood theory. Statisticians have developed a number of methods to deal with—or adjust for—the extra correlation found in this type of data. We refer to this variety of correlation as overdispersion. The scaling of standard errors that we discussed in the previous section is one means of handling overdispersion. In fact, we shall observe in Chapter 9 that scaling is perhaps the most elementary way of handling real overdispersion. But, although it is a relatively simple method to implement, it can also be effective.

We have the same problem with longitudinal data as with simple clustering. Patient treatment over time, for example, is correlated, which violates the assumption that each patient's treatment is independent of another. Levels of correlation can exist in the data as well. Individual patients within a hospital are themselves clustered on the basis of their identity, and treatment of patients in general within a facility is clustered by provider. Cluster units can be nested within each other. Statistical models have been developed to control for various levels of clustering, and hence, levels of correlation. Termed mixed models, they take us beyond the scope of this chapter, but we will address them later in the text. What we can do here, though, is to subsume various levels of clustering, if they exist, under one cluster unit. Doing so still accommodates the overdispersion; it just does not break it out into components.

The simplest manner of dealing with clustering, whether it be of groups of subjects or items or of the same subjects over time, is to ignore it. The data are pooled together without consideration of any underlying clusters or groups to which the subjects may belong—hence the name *pooled estimator*. When we model in this manner, the results are called consistent, but not efficient. The standard errors are biased.

Statisticians have developed various "robust" variance estimators, which can be used to adjust for the clustering effect in the longitudinal and clustered data. The most common type of adjustments are called sandwich estimators, because the term used to adjust the overall variance is sandwiched between two model variance matrices. The value of the robust estimator is that it provides a consistent estimate of the variance-covariance matrix of parameter estimates, even when the model variance function has been misspecified.

Unlike the standard variance estimator, $-H(\beta)^{-1}$, the robust estimator does not need the log-likelihood, $LL(\beta;x)$, to be the distribution function for x. For this, we redefine the standard errors as a measurement of the standard errors of the calculated β if we indefinitely resample and estimate the model. This gives rise to the notion of robust standard errors as being empirical standard errors. We can formally interpret robust standard errors as:

> If the sampling of data were continually repeated and estimated, we would expect the estimated coefficient to be in a specific range 95 percent of the time, using a 95 percent CI. (Hardin and Hilbe, 2001, 2007)

Robust standard errors are implemented as follows:

1. Estimate the model.
2. Calculate the linear predictor, $x\beta$.
3. Calculate the score vector: $g' = \partial g(\beta;x) = x\partial LL(x\beta)/\partial(\beta) = ux$.
4. Abstract the variance-covariance matrix.
5. Calculate the degree of freedom for the adjustment: $n/(n-1)$.
6. Combine terms: $V(b) = H^{-1}(n/(n-1))\Sigma(gg')\,H^{-1}$.
7. Replace the model variance-covariance (V – C) matrix with the robust estimator.
8. Proceed with an additional iteration; a new robust variance results.

Adjustment for clusters is performed by calculating and summing the scores within each respective cluster. Then the data set is collapsed so that there is only one observation per cluster. The robust variance estimator is determined in the same manner as in the noncluster case, except n is now the number of clusters and u consists of cluster sums.

The robust score equations for binary logistic regression are given as:

$$y = 1 : \exp(-x\beta) / (1 + \exp(-x\beta)) \tag{5.41}$$

$$y = 0 : -\exp(x\beta) / (1 + \exp(x\beta)) \tag{5.42}$$

If the scores are negatively correlated, their sums will be correspondingly small, and the robust variance estimator will produce standard errors that are smaller than their associated naïve estimated standard errors. The reverse is the case with positively correlated scores, producing standard errors that are greater than their associated model standard errors.

The pooled estimator for the logistic model is

$$L(x_{it}\beta; y_{it}) = \sum_{i=1}^{n} \sum_{t=1}^{ni} \{y_{it}(x_{it}\beta) - \ln(1 + \exp(x_{it}\beta))\} \tag{5.43}$$

which is the same as the standard logistic regression model, except there are loops summing within and across clusters. The within-cluster variance is adjusted by the robust variance estimator, while individual groups or clusters are considered to be independent of one another.

We are now prepared to model the **heart01** data, employing a robust variance estimator clustered on *center*, which is the hospital where treatment occurs. Note that there are now only 47 "observations" in the data—since there are 47 centers. Model GOF statistics are now based on $n = 47$ rather than $n = 4483$. However, the number of observations, independent of clustering, is still displayed at the top of the output.

```
. glm death anterior hcabg kk2-kk4 age3 age4, fam(bin)
cluster(center) eform

Generalized linear models              No. of obs     =       4503
Optimization      : ML                 Residual df    =       4495
                                       Scale parameter =         1
Deviance       =   1276.319134         (1/df) Deviance =   .283942
Pearson        =   4212.631591         (1/df) Pearson  =  .9371817

Variance function: V(u)  = u*(1-u)      [Bernoulli]
Link function    : g(u)  = ln(u/(1-u))  [Logit]

                                       AIC            =  .2869907
Log pseudolikelihood = -638.1595669    BIC            = -36537.86
                   (Std. Err. adjusted for 47 clusters in center)
-----------------------------------------------------------------
         |               Robust
   death | Odds Ratio  Std. Err.    z   P>|z|  [95% Conf. Interval]
---------+-------------------------------------------------------
anterior |   1.894096  .3866808  3.13  0.002  1.269488    2.82602
   hcabg |   2.195519  .6870412  2.51  0.012  1.188972   4.054177
     kk2 |   2.281692  .4066383  4.63  0.000  1.609006   3.235611
     kk3 |   2.218199  .7249049  2.44  0.015  1.169038   4.208934
     kk4 |   14.63984  4.396843  8.94  0.000  8.126271   26.37431
    age3 |   3.549577  .5975619  7.53  0.000   2.55199   4.937125
    age4 |   6.964847  1.668158  8.10  0.000  4.355513    11.1374
-----------------------------------------------------------------
```

Again, model predictors have *p*-values similar to those of the original unadjusted model. We therefore have good evidence that there is little to no center or hospital effect, that is, the probability of *death* within 48 hours of admission, explained by site of infarct, history of a CABG versus PTCA, *killip* level, and *age*, is not affected by the hospital at which a patient is treated. This is, in fact, valuable information.

There are two other important methods by which standard errors are adjusted. These methods are valuable when the model is misspecified, such as when the underlying PDF and likelihood upon which the standard errors are based is not appropriate for the data being modeled. These two methods calculate bootstrapped and jackknifed standard errors.

5.8 Bootstrapped and Jackknifed Standard Errors

Many logistic regression applications allow the user to apply bootstrapping to standard errors. Essentially, bootstrapping is data resampling; in other words, subsamples are repeatedly drawn from the data with replacement and compared with the original full sample estimate. The default bootstrap method used by Stata is called nonparametric variance estimation. Parametric bootstrapping resamples residuals, not cases as the nonparametric method does. A more detailed description of the varieties of bootstrapping can be found in Hardin and Hilbe (2007). We will use the default bootstrap settings to produce bootstrapped standard errors. A total of fifty replications are performed on the entire dataset.

BOOTSTRAPPED STANDARD ERRORS

```
glm death anterior hcabg kk2-kk4 age3 age4, vce(boot) fam(bin)
eform
(running glm on estimation sample)

Bootstrap replications (50)
----+--- 1 ---+--- 2 ---+--- 3 ---+--- 4 ---+--- 5
..................................................    50

Generalized linear models          No. of obs      =        4503
Optimization      : ML             Residual df     =        4495
                                   Scale parameter =           1
Deviance          = 1276.319134    (1/df) Deviance =     .283942
Pearson           = 4212.631591    (1/df) Pearson  =    .9371817

Variance function: V(u) = u*(1-u)      [Bernoulli]
Link function    : g(u) = ln(u/(1-u)) [Logit]

                                   AIC             =    .2869907
Log likelihood    = -638.1595669   BIC             =   -36537.86
```

```
-----------------------------------------------------------------
         |  Observed  Bootstrap                       Normal-based
   death | Odds Ratio Std. Err.   z   P>|z|  [95% Conf. Interval]
---------+-------------------------------------------------------
anterior |  1.894096  .2685635  4.50  0.000  1.434532    2.500885
   hcabg |  2.195519  .9653016  1.79  0.074  .9274496    5.197376
     kk2 |  2.281692  .3916942  4.81  0.000  1.629794    3.194342
     kk3 |  2.218199  .7176781  2.46  0.014  1.176527    4.182144
     kk4 |  14.63984  5.396134  7.28  0.000  7.108693    30.14968
    age3 |  3.549577   .783638  5.74  0.000  2.302806    5.471367
    age4 |  6.964847  1.380083  9.79  0.000  4.723307    10.27015
-----------------------------------------------------------------
```

Based on bootstrapped standard errors, all predictors still significantly contribute to the model except for *hcabg*, the history of a patient having had a CABG rather than a PTCA. *hcabg* now has a *p*-value of 0.074. With so many observations in the data, and the fact that *hcabg* showed itself to be significant at the $\alpha = 0.05$ level for each of the previous methods we have used thus far, we can feel confident that it likely contributes to an understanding of death.

We shall next try jackknifed standard errors. Stata allows one to calculate a jackknifed estimate of the variance using different methods. One may form it by estimating the vector of coefficients for $n - 1$ observations n separate times. Each of the n estimates is calculated for a different subsample of the original dataset. Stata also allows the user to calculate a one-step jackknife estimate, a weighted jackknife, and a variable subset-size jackknife estimate of the variance. We will use the standard jackknife to see how *anterior* fares. Some 4,503 replications of the data are performed, with one observation excluded each time. If you wish to replicate the example, be warned that it takes quite a while to estimate 4,503 separate regressions, making comparisons at each replication. Expect about 100 replications per minute. Of course, faster machines and greater RAM increases calculation speed.

JACKKNIFED STANDARD ERRORS

```
. glm death anterior hcabg kk2-kk4 age3 age4, fam(bin)
vce(jack) eform
(running glm on estimation sample)

Jackknife replications (4503)
----+--- 1 ---+--- 2 ---+--- 3 ---+--- 4 ---+--- 5
..................................................     50
..................................................     100

..................................................     4450
..................................................     4500
...
```

```
Generalized linear models              No. of obs        =       4503
Optimization      : ML                 Residual df       =       4495
                                       Scale parameter   =          1
Deviance          =   1276.319134      (1/df) Deviance   =    .283942
Pearson           =   4212.631591      (1/df) Pearson    =   .9371817

Variance function: V(u) = u*(1-u)      [Bernoulli]
Link function    : g(u) = ln(u/(1-u))  [Logit]

                                       AIC               =   .2869907
Log likelihood    =  -638.1595669      BIC               =  -36537.86
-------------------------------------------------------------------------
            |                 Jackknife
   death    | Odds Ratio  Std. Err.    t   P>|t|  [95% Conf. Interval]
------------+------------------------------------------------------------
 anterior   |  1.894096    .329821   3.67  0.000  1.346299    2.664787
    hcabg   |  2.195519   .8166423   2.11  0.035  1.058864    4.552334
      kk2   |  2.281692   .4133476   4.55  0.000  1.599607    3.254624
      kk3   |  2.218199    .659657   2.68  0.007  1.238222    3.973766
      kk4   |  14.63984   6.264284   6.27  0.000  6.327251    33.87329
     age3   |  3.549577   .7169922   6.27  0.000  2.388872    5.274243
     age4   |  6.964847   1.513508   8.93  0.000  4.548728    10.66432
-------------------------------------------------------------------------
```

Employing jackknifed standard errors to the model shows *hcabg* to contribute significantly to the model, but little else has little bearing on the outcome. The standard errors have been consistent for each method we have used.

It should also be reiterated that of the various adjustments given to the model standard error, no predictor exceeded the given criterion of $p < 0.05$. When this happens, we can feel statistically confident that there are no outside clustering effects.

We have purposefully overlooked a point of considerable importance. We can observe that the coefficients, or equally, the odds ratios, on killip levels 2 and 3 are very close in value. In fact, there is an odds difference of only .06 between the two levels. When this happens, regardless if both levels significantly contribute to the model, merging them to one level is likely a good idea. In model building, we want to effect the most parsimonious group of predictors that optimally fit the data. We expect fuzzy edges, but not so fuzzy that overlapping can occur. If we do merge *kk2* and *kk3*, we keep the same reference, but now have three levels.

We will model the data with merged *kk2 – kk3* levels. But first, we must create a new predictor that takes the value of 1 if either *kk2* = 1 or *kk3* = 1. We also need to make certain that if the *killip* predictor is missing for a particular observation, entailing that both *kk2* and *kk3* are missing for that observation, that *kk23* is also missing. All of these conditions can be satisfied by two commands.

```
. gen byte kk23=kk2==1 | kk3==1
                        /// kk23=1/0; missing made 0 as well

. replace kk23=. if kk2==. | kk3==.
                        /// kk23 missing if either is missing

. glm death anterior hcabg kk23 kk4 age3 age4, fam(bin) nolog
eform

Generalized linear models                  No. of obs       =        4503
Optimization      : ML                     Residual df      =        4496
                                           Scale parameter  =           1
Deviance          =  1276.329389           (1/df) Deviance  =   .2838811
Pearson           =  4208.029511           (1/df) Pearson   =   .9359496

Variance function: V(u) = u*(1-u)          [Bernoulli]
Link function    : g(u) = ln(u/(1-u))      [Logit]

                                           AIC              =   .2865488
Log likelihood   = -638.1646945            BIC              = -36546.27

-------------------------------------------------------------------------
             |                OIM
    death | Odds Ratio  Std. Err.    z    P>|z|  [95% Conf. Interval]
---------+---------------------------------------------------------------
anterior |   1.892058  .3163837   3.81  0.000   1.363325   2.625848
   hcabg |   2.196883  .7747551   2.23  0.026   1.100579   4.385232
   kk23 |   2.265916  .3786814   4.89  0.000   1.633016   3.144106
    kk4 |   14.64305  5.218625   7.53  0.000   7.282363   29.44357
   age3 |   3.546141  .7104483   6.32  0.000   2.394544   5.251571
   age4 |   6.952852  1.441791   9.35  0.000   4.630758   10.43936
-------------------------------------------------------------------------

. abic
AIC Statistic =      .2865488                AIC*n       = 1290.3293
BIC Statistic =      .2894897                BIC(Stata)  = 1335.2169
```

To anticipate discussion in the following chapter, the AIC and BIC goodness-of-fit tests have been calculated for the model. Both are comparative tests, with the lower value associated with the preferred model. We quietly run the model having both *kk2* and *kk3*, then run the author-created **abic** command, which calculates two of the more popular versions of both commands.

```
. qui glm death anterior hcabg kk2 kk3 kk4 age3 age4, fam(bin)
nolog eform

. abic
AIC Statistic =      .2869907                AIC*n       = 1292.3191
BIC Statistic =      .2908262                BIC(Stata)  = 1343.6191
```

Both the AIC and BIC statistics of the model having $kk23$ are lower than the model with both $kk2$ and $kk3$. This gives us evidence that the merged-kk model is preferred over the nonmerged.

5.9 Stepwise Methods

Stepwise methods have seemed to play a greater role in the social sciences than in epidemiology and medical research. Primarily, this may be because those in these latter areas of research many times know exactly what they are evaluating and usually have few predictors in their model. Predictors that have already been proven by past research to contribute to an understanding of the response are left in the model regardless of their significance value. Some medical researchers, of course, may decide that they would rather drop predictors that fail tests of significance, but usually this is after some reflection. They rarely make such decisions on the basis of an algorithm.

Those in social science research, on the other hand, many times have scores of potential predictors, which they wish to pare down to use in a parsimonious model. Of course, there are always examples where these generalizations do not hold, but this observation does seem to be the case for many studies.

Health outcomes research typically bases its research on large registries of observational data sets. This includes state and national patient file data, health provider financial information, and so forth. Health outcomes researchers usually handle large data sets as do social theorists.

One of the reasons why there is a difference in the type of data analyzed deals with the goal of the study. The primary type of model we have thus far discussed has been the smaller more epidemiological type. With this model paradigm, we typically have a single exposure or risk factor, together with one or more confounders, which as such must influence both the response and each other. We have also used examples where there are three or four predictors, any of which can be regarded as a primary predictor of interest. Actually, many researchers put three to seven or eight predictors that they think might have a bearing on the outcome in a model and run it to determine what comes out as significant. This is a crass manner of putting it, but this method is in fact used more often than some would like to think. As far as the logistic model is concerned, either tactic is fine.

There are also researchers—more in the social and health outcomes research domains—who put 50 or more predictors in a logistic model and use a stepwise algorithm to drop variables that do not contribute to the understanding of the response. This happens when the response, or subject related to the response, is somewhat new and there have not been many prior studies determining the relationships between selected variables. Stepwise methods can provide a quick way to drop variables from consideration when they show no evidence of being related to the response or other predictors. This can save a great deal of research time and money.

There are several types of stepwise methods. The foremost are

- Forward selection
- Backward selection
- Forward and backward selection

There are also mechanisms to force specified predictors into the model. This may happen when you have a variable or group of variables that you want to stay in the model—at least until the final stage. One may wish to keep predictors one is interested in, or that one believes on the basis of past studies to have a bearing on the response, and only after stepwise methods have eliminated other variables will the maintained predictors be further evaluated for inclusion in the final model. Moreover, for particular groups of predictors, you may want to have them entered or deleted from the model as a group. Past research may have indicated that certain variables cannot be meaningfully separated, at least for the purposes of the study under consideration. Categorical variables that have been made into indicator predictors are almost always entered into the model as a group. This is a bit dangerous, however, because it may turn out that categorical indicators may be deleted as a group due to the p-values of certain levels, when all that was needed was to combine levels.

The key to using stepwise methods is the criterion used for inclusion, or exclusion. Linear regression methods typically use either the R^2 or F statistics as selection criteria. Probably the most popular criterion for logistic regression models is the p-value. At any stage in model development, predictors having a p-value exceeding some user specified level are excluded from the model. Likewise, predictors having a p-value less than some level may be retained in the model. The results are the same, but the algorithm may stress a particular retention/exclusion method.

Forward selection methods begin with a constant-only model, then a predictor is added and the resultant model re-evaluated with an aim of deciding whether to keep or exclude the predictor. Then another predictor is added, evaluated together with a previously accepted predictor, and either retained or excluded. This procedure continues until all variables, or potential predictors, have been evaluated.

A backward selection method starts from a model with predictors having $p < 0.25$. Those predictors having the highest p-value are excluded from the model one at a time. At each iteration, the algorithm checks the p-values of all other predictors as well, tossing the one that is highest until a model is arrived at in which all predictors significantly contribute to the model—at the specified p-value criterion.

p-values can be determined by the z, t, or Wald statistic on the one hand, or by calculating a likelihood ratio test on the other. p-values based on these statistics or by these methods are usually similar, but, as mentioned earlier, the likelihood ratio test statistic is preferred. Although the calculations involved require more computer time, the results are worth it. However, if one has 100 variables to search, using the likelihood ratio test may take longer than acceptable. If the goal is to use the selection method as a simple variable reduction technique, but not to determine a final model, then the use of z (Stata, Genstat), t (S-plus), or Wald (SAS, SPSS) statistics as the basis of the p-value may be a wiser tactic.

Most statisticians prefer a forward and backward selection algorithm. The problem with the forward and backward selection methods is that a variable, once dropped, cannot be brought back into the model. Predictors are typically related, and we don't know in advance how the exclusion or retention of one may affect the p-value of another. An excluded variable may well significantly contribute to the model after another variable later has been excluded. A forward and backward selection method seeks to eliminate this situation.

We shall use the **lr_model** data to demonstrate how the forward selection method works. The data consist of a number of patient characteristics and physiological test results. We need not concern ourselves with the specifics of the particular tests involved; rather, we are interested in observing how the algorithm selects model predictors.

```
. use lr_model

. stepwise, pe(.1): logistic heartatk (reg2-reg4) (race2
race3) gender height weight bpsystol tcresult tgresult
hdresult hgb diabetes albumin (hsi2-hsi5)
                      begin with empty model
p = 0.0000 <   0.1000   adding   hsi2 hsi3 hsi4 hsi5
p = 0.0000 <   0.1000   adding   gender
p = 0.0000 <   0.1000   adding   albumin
p = 0.0000 <   0.1000   adding   tgresult
p = 0.0001 <   0.1000   adding   height
p = 0.0095 <   0.1000   adding   diabetes
p = 0.0476 <   0.1000   adding   hdresult
p = 0.0264 <   0.1000   adding   tcresult
p = 0.0718 <   0.1000   adding   reg2 reg3 reg4
p = 0.0811 <   0.1000   adding   bpsystol
```

```
Logistic regression                   Number of obs   =      4119
                                      LR chi2(15)     =    175.68
                                      Prob > chi2     =    0.0000
Log likelihood = -617.34036           Pseudo R2       =    0.1246
---------------------------------------------------------------------
heartatk | Odds Ratio  Std. Err.    z    P>|z|  [95% Conf. Interval]
---------+-----------------------------------------------------------
   hsi2 |   1.371909   .3251834    1.33  0.182  .8621147    2.18316
   hsi3 |   .5750433   .1858951   -1.71  0.087  .3051627   1.083602
   hsi4 |   .5576161   .1950999   -1.67  0.095  .2808763    1.10702
```

```
     hsi5 |   .3950171   .1481593   -2.48   0.013   .1893892    .8239041
   gender |   5.322067   1.308023    6.80   0.000   3.287589    8.615554
  albumin |   .2800112   .0738735   -4.82   0.000   .1669585    .4696153
 tgresult |   1.002251   .0009789    2.30   0.021   1.000334    1.004171
   height |   .9555865   .0114476   -3.79   0.000   .9334109    .9782889
 diabetes |   1.982298   .5769437    2.35   0.019   1.120543    3.506787
 hdresult |   .9824571   .0072327   -2.40   0.016    .968383    .9967358
 tcresult |   1.003877   .0018959    2.05   0.040   1.000168      1.0076
     reg2 |   1.344029   .3509895    1.13   0.258   .8056004     2.24232
     reg3 |   1.385065   .3669276    1.23   0.219   .8240857     2.32792
     reg4 |   1.898037   .4779219    2.54   0.011   1.158708    3.109106
 bpsystol |   1.006073   .0034925    1.74   0.081   .9992516    1.012942
-----------------------------------------------------------------------
```

It is clear that when all levels of a categorical predictor are entered into a stepwise model as a group that further analysis may be necessary. All levels are entered if one level meets the defined criterion. In the case of region (*reg1* as the reference), only the highest level is significant. We would likely want to dichotomize the predictor such that level 4 is assigned the value of 1, and all others are 0. Likewise, for *hsi**, we can combine levels 1 and 2 as the reference, with other levels being evaluated in terms of their combined effect. The resultant model appears as:

```
. logistic heartatk reg4 gender height tcresult tgresult
hdresult diabetes albumin hsi3 hsi4 hsi5

Logistic regression                    Number of obs   =        4119
                                       LR chi2(11)     =      169.06
                                       Prob > chi2     =      0.0000
Log likelihood = -620.64962            Pseudo R2       =      0.1199
-----------------------------------------------------------------------
heartatk | Odds Ratio  Std. Err.    z   P>|z| [95% Conf. Interval]
---------+-------------------------------------------------------------
     reg4 |   1.510363   .2609739    2.39   0.017   1.076473    2.119139
   gender |   5.602316   1.364958    7.07   0.000   3.475196    9.031416
   height |    .955444   .0113642   -3.83   0.000   .9334282    .9779791
 tcresult |   1.004242   .0018669    2.28   0.023    1.00059    1.007908
 tgresult |   1.002454   .0009449    2.60   0.009   1.000604    1.004308
 hdresult |   .9819033   .0071793   -2.50   0.012   .9679325    .9960757
 diabetes |   2.067756     .59605    2.52   0.012   1.175256    3.638029
  albumin |   .2604338   .0677299   -5.17   0.000   .1564332    .4335765
     hsi3 |   .4375065    .114448   -3.16   0.002   .2620097    .7305527
     hsi4 |    .416755   .1218746   -2.99   0.003   .2349405     .739271
     hsi5 |    .293678   .0948859   -3.79   0.000   .1559022    .5532107
-----------------------------------------------------------------------
```

Using a stepwise method for the array of potential predictors with which we began certainly resulted in a faster resolution of the model than if we tested each variable individually. The downside is that one of the variables excluded by the algorithm may now be contributory. If there is a variable we think might significantly contribute to the model, it is always possible to evaluate it following stepwise results. In the case of these data, however, the model selected by the algorithm, with our follow-up amendment, appears to be the optimal model. The algorithm worked well, but this is not always the case. Care must be taken when using an automatic selection process like this, and it may be necessary to fine-tune the model before declaring it to be final. Evaluating the model using the methods outlined in Chapter 7 is an absolute prerequisite to accepting a model—be it a small, two-predictor model or one entailing many predictors. A listing of caveats on using stepwise methods can be found at: http://www.stata.com/support/faqs/stat/stepwise.html.

A caveat is in order regarding fitting a stepwise model with missing observations. Recall that a maximum likelihood procedure drops all observations with missing values in the model. That is, if one term has a missing value for predictor $x2$ at observation number 30, then observation number 30 is deleted from the model. When employing stepwise methods, if a predictor with a missing value is dropped as part of the stepwise procedure, the remaining predictors are left without the observation that was missing in the excluded predictor. If the final model is remodeled without using stepwise methods, the parameter estimates will differ in the two models.

Consider the low birth weight data, **lbw**, used in Hosmer and Lemeshow (2000). We model *low* on *age, smoke* and two levels of *race* defined as:

low   : low birth weight baby; < 2500 gr

age   : continuous predictor

smoke : 1 = smoked during pregnancy; 0 = not a smoker during pregnancy

race2   : black /// race1 (white, is reference level)

race3   : other

Five missing values are created for *age*, at *age* > 34.

```
. replace age=. if age>34
(5 real changes made, 5 to missing)
```

Now model the data using stepwise methods.

```
. sw logit low age smoke race2 race3, pr(.2)
                  begin with full model
p = 0.5733 >= 0.2000   removing age

Logistic regression                    Number of obs   =        184
                                       LR chi2(3)      =      13.93
```

```
                                        Prob > chi2      =      0.0030
Log likelihood = -108.46928             Pseudo R2        =      0.0603
    -------------------------------------------------------------------
    low |       Coef.   Std. Err.     z    P>|z|   [95% Conf. Interval]
  ------+------------------------------------------------------------
  race3 |     1.07341   .4056996    2.65   0.008    .2782536    1.868567
  smoke |    1.113056   .3755059    2.96   0.003    .3770781    1.849034
  race2 |    1.152815    .500782    2.30   0.021    .1713003     2.13433
  _cons |   -1.804816   .3617762   -4.99   0.000   -2.513884   -1.095748
    -------------------------------------------------------------------
```

Next, model the same predictors as the final model, but not using stepwise methods.

```
. logit low smoke race2 race3, nolog

Logistic regression                     Number of obs    =          189
                                        LR chi2(3)       =        14.70
                                        Prob > chi2      =       0.0021
Log likelihood = -109.98736             Pseudo R2        =       0.0626
    -------------------------------------------------------------------
    low |       Coef.   Std. Err.     z    P>|z|   [95% Conf. Interval]
  ------+------------------------------------------------------------
  smoke |    1.116004   .3692258    3.02   0.003    .3923346    1.839673
  race2 |    1.084088   .4899845    2.21   0.027    .1237362     2.04444
  race3 |    1.108563   .4003054    2.77   0.006    .3239787    1.893147
  _cons |   -1.840539   .3528633   -5.22   0.000   -2.532138   -1.148939
    -------------------------------------------------------------------
```

The reason for the difference is that the stepwise method excludes the five missing values of *age*. Note that the displayed number of observations for the stepwise method is 184 compared with 189 of the standard method.

5.10 Handling Missing Values

The problem with missing values is always a concern to someone building a well-fitted statistical model. Some types of data frequently have missing values, such as survey data. But, they find their way into data sets of nearly every type.

The first time many researchers become aware of missing values and how they can affect models is when they compare models after adding or excluding a predictor from the model. One normally expects that adding a predictor to one's model has nothing to do with the number of observations in the model. The expectation is that the observations stay the same. However, this is not always the case, and for many large observational data sets, is a rarity. The reason is that no missing elements are allowed in an inverted matrix, which the estimating algorithm calculates to produce parameter estimates and standard errors. Therefore, nearly all commercial estimating algorithms drop an observation if there is a missing value for any of its predictors, or for the response. No message is given; the only way the statistician knows that observations have been dropped is by noting the change of values in the model header display: *No. of Obs.* This method of dropping observations is termed *listwise deletion*.

One of the primary problems with having two models with different numbers of observations is the fact that they cannot be evaluated using a likelihood ratio test or a deviance test. In fact, the number of observations must be identical when comparing the coefficients of two nested or otherwise different models. The solution is to make certain that no missing values exist in the models being compared, or if there are, that deleting a predictor or predictors from the model does not affect the number of observations. We shall briefly discuss methods that can be used to assure the latter situation.

If we are not comparing models, but are only concerned with the model itself, we need to evaluate the manner in which missing values are distributed in the data. Are missing values created by the software storing the original data, such as when an entered value exceeds the criterion for acceptance? If the variable in question is gender, and database instructions require that 1 be assigned for *males* and 0 for *females*, then if the software finds a 2 entered into the cell, it may assign it a missing value status. A few software packages recode the value to some identifier, such as −1. Other software simply prohibit entry of any value, leaving it blank. We don't know if this was an instance of miscoding, or if it was simply unknown. There are a variety of ways in which missingness is understood.

How missing values are distributed can play an important role in how we deal with them. If missing values in a data set are **missing at random,** commonly referred to by the acronym MAR, they are not randomly distributed across observations in the data, but rather are distributed within one or more of the subsamples within the data. The structure of the observations may allow values to be predicted. If missing values are MCAR, or **missing completely at random,** they are randomly distributed across all observations in the data. Information or trends in the data have no bearing on the missingness of the observations. When observations are MCAR, we may be able to employ missing value software that impute values to the missing cells. Finally, NIM, or **non-ignorable missingness,** exists in the data when missing values are not distributed randomly across observations, yet their missingness cannot be predicted on the basis of observations or structures in the data. NIM elements are the most difficult type of missingness to deal with. See Hardin and Hilbe (2003) for an elaboration of these types of missingness.

If there is a pattern to the missingness of observations, we must address the cause underlying their generation. Using imputation software is usually not available for this type of missingness. It is wise, therefore, to scan the data to see which variables being considered for inclusion in the model have missing values. For those that do, we need to know how many terms are involved and their percentage, and if they are random. Many statistical software applications have procedures that give users this information. We shall use Stata's missing value commands for examples on how to evaluate and deal with missing values. For additional information and guidance on handling missingness, see the classic text by Little and Rubin (1987), or more recent texts such as Allison (2001).

It should be noted from the outset that, depending on the data and number of missing values, many statisticians do not attempt imputation of missing values and simply allow the algorithm to delete observations with identified missing values. The important caveat, however, is that comparisons between nested models based on the likelihood ratio or deviance tests still require identical numbers of model observations. Goodness-of-fit comparisons based on the AIC or BIC tests do not mandate equal numbers.

We shall use the **felon** data for an example with missing values. The data are from a hiring profile legal case for a mid-sized U.S. city over the course of a year. We want to determine if being a felon has a bearing on being hired for a city job. Other predictors of interest include *gender*, if the applicant is Native American (the city has many Native American residents), if they were employed within 6 months of applying for this job, and their level of education. The type of jobs associated with this study were low level, not requiring a college degree.

```
Contains data from felon.dta
  obs:           733
  vars:           10                                17 Feb 2008 12:04
  size:       10,262 (99.9% of memory free)
-------------------------------------------------------------------
              storage  display  value
variable name   type   format   label        variable label
-------------------------------------------------------------------
hired          byte    %8.0g                 1=hired; 0=not hired
felony         byte    %9.0g    felony       1=felon; 0=not felon
gender         byte    %9.0g                 1=male; 0=female
natamer        byte    %12.0g   natamer      1=Native Amer;
                                             0= Not N. Amer
mo6            byte    %22.0g   mo6          1=Worked >6 mo prev job
educ           byte    %15.0g   educ         education level
educ1          byte    %8.0g                 educ==Not HS Grad
educ2          byte    %8.0g                 educ==HS Grad
educ3          byte    %8.0g                 educ==JC
educ4          byte    %8.0g                 educ==2+ yrs College
-------------------------------------------------------------------
```

We know from past experience that city data such as this study generally have a considerable number of missing values. An overview can be obtained by summarizing the variables, observing the number of valid observations.

```
. count
  733
```

There are 733 observations in the data, as can be observed from both the data set header above and from a count of observations just issued. Per-variable observations are listed as:

```
. su

  Variable |       Obs        Mean    Std. Dev.       Min        Max
-----------+-----------------------------------------------------------
     hired |       733    .3124147    .4637944         0          1
    felony |       682    .1143695    .3184931         0          1
    gender |       732    .6584699    .4745471         0          1
   natamer |       721     .112344    .3160084         0          1
       mo6 |       616    .6396104    .4805036         0          1
-----------+-----------------------------------------------------------
      educ |       639    1.967136    .8855421         1          4
     educ1 |       639    .3442879    .4755077         0          1
     educ2 |       639    .4084507    .4919324         0          1
     educ3 |       639    .1830986    .3870503         0          1
     educ4 |       639    .0641628    .2452346         0          1
```

Variables *mo6* and *education levels* have some 100 missing values.

Again, a major problem when modeling with predictors having missing values rests in the inability to compare nested models using likelihood ratio and deviance tests, unless we make adjustments to the model data.

```
. logistic hired felony gender natamer mo6 educ2-educ4

Logistic regression                             Number of obs   =        564
                                                LR chi2(7)      =      27.05
                                                Prob > chi2     =     0.0003
Log likelihood = -317.12951                     Pseudo R2       =     0.0409
------------------------------------------------------------------------------
     hired | Odds Ratio  Std. Err.      z    P>|z|    [95% Conf. Interval]
-----------+------------------------------------------------------------------
    felony |  .4846043   .1770132   -1.98   0.047    .2368458    .9915367
    gender |  2.056871   .4394179    3.38   0.001    1.353198    3.126459
   natamer |  .4449538   .1614659   -2.23   0.026    .2184886     .906152
       mo6 |  1.137472   .2300916    0.64   0.524    .7651682    1.690926
     educ2 |  .7121205   .1601832   -1.51   0.131    .4582322    1.106678
     educ3 |  .6532553   .1861892   -1.49   0.135    .3736591    1.142064
     educ4 |  .8873214   .3377706   -0.31   0.753    .4207846    1.871122
------------------------------------------------------------------------------
```

There are only 564 observations in the model. The reason for this is due to missing values being spread throughout the data. It is fairly clear that time from previous employment and educational level have no bearing on being hired for these jobs in this city. We can test for being a Native American though, but when we model without the predictor, we find that the number of model observations has changed. We cannot therefore directly perform a likelihood ratio test.

```
. logistic hired felony gender mo6 educ2-educ4

Logistic regression                     Number of obs   =        572
                                        LR chi2(6)      =      21.16
                                        Prob > chi2     =     0.0017
Log likelihood = -322.60337             Pseudo R2       =     0.0318
-------------------------------------------------------------------
  hired | Odds Ratio  Std. Err.     z    P>|z|   [95% Conf. Interval]
--------+----------------------------------------------------------
 felony |   .4700817   .1707425  -2.08   0.038    .2306752   .9579565
 gender |   2.013741   .4278946   3.29   0.001    1.327804   3.054029
    mo6 |   1.166732   .2338304   0.77   0.442    .7877307   1.728083
  educ2 |   .6991336   .1558118  -1.61   0.108    .4517081   1.082088
  educ3 |   .6728011   .1902775  -1.40   0.161     .386505   1.171165
  educ4 |   .9483138   .3590638  -0.14   0.889    .4515005   1.991801
-------------------------------------------------------------------
```

This problem may be solved using Stata by employing its **mark** and **markout** commands. First, we define a binary variable that will identify missing values. It has become nearly a standard among Stata users to call this constant *nomiss*. Second, we designate the variables in the data that we wish to mark as missing. The **markout** command is used for this purpose. Tabulating *nomiss* allows us to see how many missing values exist in the specified variables.

```
. mark nomiss

. markout nomiss hired felony gender natamer educ educ2-educ4

. tab nomiss

   nomiss |      Freq.     Percent        Cum.
----------+-----------------------------------
        0 |        111       15.14       15.14
        1 |        622       84.86      100.00
----------+-----------------------------------
    Total |        733      100.00
```

There are 111 missing values—622 are *not missing*. Using *nomiss* to exclude missing observations for models based on the variables specified in **markout**, we can perform a likelihood ratio test.

```
. logistic hired felony gender natamer mo6 educ2-educ4 if
nomiss==1, nolog

Logistic regression                    Number of obs    =        564
                                        LR chi2(7)       =      27.05
                                        Prob > chi2      =     0.0003
Log likelihood = -317.12951            Pseudo R2        =     0.0409
-----------------------------------------------------------------------
  hired | Odds Ratio  Std. Err.    z   P>|z|  [95% Conf. Interval]
--------+--------------------------------------------------------------
 felony |  .4846043   .1770132  -1.98  0.047   .2368458   .9915367
 gender |  2.056871   .4394179   3.38  0.001   1.353198   3.126459
natamer |  .4449538   .1614659  -2.23  0.026   .2184886   .906152
    mo6 |  1.137472   .2300916   0.64  0.524   .7651682   1.690926
  educ2 |  .7121205   .1601832  -1.51  0.131   .4582322   1.106678
  educ3 |  .6532553   .1861892  -1.49  0.135   .3736591   1.142064
  educ4 |  .8873214   .3377706  -0.31  0.753   .4207846   1.871122
-----------------------------------------------------------------------

. estimates store A1

. logistic hired felony gender mo6 educ2-educ4 if nomiss==1,
nolog

Logistic regression                    Number of obs    =        564
                                        LR chi2(6)       =      21.33
                                        Prob > chi2      =     0.0016
Log likelihood = -319.99137            Pseudo R2        =     0.0323
-----------------------------------------------------------------------
  hired | Odds Ratio  Std. Err.    z   P>|z|  [95% Conf. Interval]
--------+--------------------------------------------------------------
 felony |  .4714159    .171547  -2.07  0.039   .2310225   .9619538
 gender |     2.042   .4346923   3.35  0.001   1.345413   3.099245
    mo6 |  1.174303    .236055   0.80  0.424   .7919067   1.741351
  educ2 |  .7116346   .1591963  -1.52  0.128   .4590277   1.103253
  educ3 |  .6693647   .1897315  -1.42  0.157   .3840517   1.166637
  educ4 |  .9329941    .353542  -0.18  0.855   .4439466   1.960772
-----------------------------------------------------------------------

. lrtest A1

Likelihood-ratio test                  LR chi2(1)    =       5.72
(Assumption: . nested in A1)           Prob > chi2 =      0.0167
```

The predictor, *natamer*, is indeed a significant predictor of *hired*, adjusted by the other predictors. We may also test for the significance of *felony*, our predictor of interest.

```
. qui logistic hired felony gender natamer mo6 educ2-educ4 if
nomiss==1, nolog
```

```
. estimates store B1

. qui logistic hired felony gender natamer mo6 educ2-educ4 if
nomiss==1, nolog

. lrtest B1

lrtest B1

Likelihood-ratio test                    LR chi2(1)   =       4.44
(Assumption: . nested in B1)             Prob > chi2 =     0.0352
```

Felony is also a significant contributor to the model, with a likelihood ratio test value of $p = 0.0352$. As previously mentioned, this test is preferable to the z or Wald tests we use when checking the table of model parameter estimates, etc., that are displayed when modeling.

Long and Freese (2006) developed another means for identifying missing values and for observing if any patterns exist in how missingness is distributed in the data. The **misschk** command calculates the number and percentage of missing values for each variable specified by the user. *educ* was left as a categorical variable and not factored into separate indicator variables. Therefore, six variables are evaluated by the utility.

```
. misschk hired felony gender natamer mo6 educ


Variables examined for missing values


     #  Variable          # Missing    % Missing
    ------------------------------------------------
     1  hired                    0          0.0
     2  felony                  51          7.0
     3  gender                   1          0.1
     4  natamer                 12          1.6
     5  mo6                    117         16.0
     6  educ                    94         12.8


Missing for |
      which |
  variables? |      Freq.       Percent         Cum.
 ------------+------------------------------------------
    _2_4__ _ |         1          0.14         0.14
    _2__5 6  |        44          6.00         6.14
    _2___ 6  |         1          0.14         6.28
    _2___ _  |         5          0.68         6.96
    __34__ _ |         1          0.14         7.09
    __45_ _  |         2          0.27         7.37
    ___4_ _  |         8          1.09         8.46
    ____5 6  |        13          1.77        10.23
```

```
      5   |          58           7.91          18.14
      6   |          36           4.91          23.06
          |         564          76.94         100.00
----------+-----------------------------------------
   Total  |         733         100.00


Missing for |
  how many  |
 variables? |     Freq.        Percent           Cum.
------------+---------------------------------------------
       0  |          564          76.94          76.94
       1  |          107          14.60          91.54
       2  |           18           2.46          94.00
       3  |           44           6.00         100.00
----------+----------------------------------------------
   Total  |          733         100.00
```

The middle table needs an explanation. The top line indicates that one observation has missing values for the second and fourth variables specified in the list of variables presented after the **misschk** command:

1: hired 2: felony 3: gender 4: natamer 5: mo6 6: educ

Therefore, one observation has missing values for *felony* and *natamer*, forty-four observations have missing values for *felony* (2), *mo6* (5), and *educ* (6), one observation for *felony* (2) and *educ* (6), and so forth. The bottom line indicates that 564 observations had no missing values.

The final table given by **misschk** simply indicates the number of observations having 0, 1, 2, or 3 missing values.

The utility also has several other features that assist in working with missing values, including one that sets up the necessary information to create a graphic of missing variables. The interested reader should access the software and accompanying documentation.

Methods exist to impute values where there were once missing values. Multiple imputation is perhaps the most sophisticated, cycling through repeated regressions aimed at providing the appropriate values where missing ones exist. For a binary predictor such as *felony*, the algorithm uses a logit model for predicting values.

It is clear that *mo6* and *educ* are not significant predictors. Imputing values for these variables will likely not increase their worth to the model. Moreover, both variables have a sizeable number of missing values, 16 percent and 12.8 percent, respectively, so imputation is questionable. However, for the three predictors that are significant, their percentage of missing values is relatively small, and we can use multiple imputation techniques to convert them to binary values.

First, for comparison purposes, we model *hired* on *felony, gender,* and *natamer*.

```
. logistic hired felony gender natamer

Logistic regression                          Number of obs   =        671
                                             LR chi2(3)      =      23.30
                                             Prob > chi2     =     0.0000
Log likelihood = -383.46986                  Pseudo R2       =     0.0295
-------------------------------------------------------------------------
   hired | Odds Ratio  Std. Err.     z    P>|z|  [95% Conf. Interval]
---------+---------------------------------------------------------------
  felony |   .5557088   .1740779  -1.88  0.061   .3007497    1.026808
  gender |   1.968143   .3846429   3.46  0.001   1.341846     2.88676
 natamer |   .4416102   .1454782  -2.48  0.013   .2315424    .8422628
-------------------------------------------------------------------------
```

Stata's user-created multiple imputation command is called **ice**. The command used to generate imputed values for the three predictors is

```
. ice felony gender natamer using imputed, m(5) cycles(20)
cmd(felony:logistic) replace

   #missing |
     values |       Freq.      Percent          Cum.
------------+-----------------------------------------
          0 |         671        91.54         91.54
          1 |          60         8.19         99.73
          2 |           2         0.27        100.00
------------+-----------------------------------------
      Total |         733       100.00

   Variable | Command | Prediction equation
------------+---------+--------------------------------
     felony | logistic| gender natamer
     gender | logit   | felony natamer
    natamer | logit   | felony gender

Imputing 1..2..3..4..5..file imputed.dta saved
```

The new imputed data set is stored in **imputed.** Bringing it into memory, we remodel the data using the first 733 observations:

```
. use imputed,clear

. logistic hired felony gender natamer in 1/733

Logistic regression                          Number of obs   =        733
                                             LR chi2(3)      =      24.66
                                             Prob > chi2     =     0.0000
Log likelihood = -442.87604                  Pseudo R2       =     0.0271
```

```
-------------------------------------------------------------------
 hired | Odds Ratio  Std. Err.     z   P>|z|  [95% Conf. Interval]
-------+-----------------------------------------------------------
felony |   .5463534   .1578574  -2.09  0.036   .3101249   .9625219
gender |   1.867168   .3341425   3.49  0.000   1.314786   2.651623
natamer |  .4657953   .1398872  -2.54  0.011    .258561   .8391261
-------------------------------------------------------------------
```

The model is based on a full 733 observations and shows improved *p*-values. We need to employ the various fit statistics detailed elsewhere in this text to assess the worth of the model, either considered by itself or in comparison to the original model.

Comparisons of the tabulations of original and imputed data for each predictor are

```
. tab felon, miss

     1=felon; |
0=not felon |      Freq.       Percent        Cum.
------------+-----------------------------------------
  No Felony |        604         82.40        82.40
     Felony |         78         10.64        93.04
          . |         51          6.96       100.00
------------+-----------------------------------------
      Total |        733        100.00
```

```
. tab felony in 1/733, miss

     felony |
imput. (51 |
    values) |      Freq.       Percent        Cum.
------------+-----------------------------------------
  No Felony |        652         88.95        88.95
     Felony |         81         11.05       100.00
------------+-----------------------------------------
      Total |        733        100.00
```

```
. tab gender, miss

     1=male; |
  0=female |      Freq.       Percent        Cum.
------------+-----------------------------------------
         0 |        250         34.11        34.11
         1 |        482         65.76        99.86
         . |          1          0.14       100.00
------------+-----------------------------------------
      Total |        733        100.00
```

```
. tab gender in 1/733, miss

    gender |
  imput. (1 |
    values) |      Freq.       Percent        Cum.
------------+-----------------------------------
         0 |        250         34.11       34.11
         1 |        483         65.89      100.00
------------+-----------------------------------
     Total |        733        100.00

. tab natamer, miss

   1=Native |
Amer; 0= Not |
    N. Amer |      Freq.       Percent        Cum.
------------+-----------------------------------
non Nat Amer |       640         87.31       87.31
Native Amer |        81         11.05       98.36
          . |         12          1.64      100.00
------------+-----------------------------------
     Total |        733        100.00

. tab natamer in 1/733, miss

    natamer |
  imput. (12 |
    values) |      Freq.       Percent        Cum.
------------+-----------------------------------
non Nat Amer |       651         88.81       88.81
Native Amer |        82         11.19      100.00
------------+-----------------------------------
     Total |        733        100.00
```

5.11 Modeling an Uncertain Response

Many times the data we have gathered for a study have been measured with uncertainty. It may be that one or more predictors, or the response, have an associated uncertainty about them. We consider the latter case of the response having a known uncertainty associated with it. The uncertainty is expressed in two ways—degrees of specificity and of sensitivity. At times, we know these values due to previous studies. At other times, we may have to base the values on our understanding and experience with the data.

For a binary response logistic model, *sensitivity* is defined as the percent or fraction of correctly classified positive responses. *Specificity* is the percent of correctly classified negative responses. Both statistics are calculated in classification tables, which can be generated following a logistic regression model. For an example, we shall use the **medpar** data, which consists of Medicare inpatient hospital records as maintained by the Health Care Financing Administration (HCFA) in Baltimore, Maryland. Medicare is a national health care system for U.S. citizens who are age 65 and older or who have been classified as disabled. These data are from the 1991 Arizona Medpar files. We have kept only a few variables from the then 110 variables, which were associated with each patient hospitalization.

For our example, we model *died* on *los* and *type2* and *type3*. *died* indicates that the patient died while in the hospital. *los* records the total length of stay in the hospital for that patient. *type* refers to the type of admission: 1 = elective, 2 = urgent, 3 = emergency. Type has been made into binary indicator variables. We leave *type1* out as the reference.

```
. logit died los type2 type3, nolog or

Logistic regression                    Number of obs   =       1495
                                       LR chi2(3)      =      39.48
                                       Prob > chi2     =     0.0000
Log likelihood = -941.69422            Pseudo R2       =     0.0205
------------------------------------------------------------------------
   died | Odds Ratio  Std. Err.      z    P>|z|   [95% Conf. Interval]
--------+---------------------------------------------------------------
    los |  .9629564   .0074959   -4.85   0.000    .9483762   .9777607
  type2 |  1.488538   .2133153    2.78   0.006    1.124031   1.971249
  type3 |  2.521626   .5756899    4.05   0.000    1.611946   3.944672
------------------------------------------------------------------------
```

We now issue the post estimation **estat** *class* command to calculate a classification of predicted values by observed or given values (0 and 1).

```
. estat class

Logistic model for died

              -------- True --------
Classified |         D            ~D  |      Total
-----------+----------------------------+-----------
     +     |        28            24  |         52
     -     |       485           958  |       1443
-----------+----------------------------+-----------
   Total   |       513           982  |       1495

Classified + if predicted Pr(D) >= .5
True D defined as died != 0
----------------------------------------------------------
Sensitivity                     Pr( +| D)      5.46%
Specificity                     Pr( -|~D)     97.56%
```

SENSITIVITY

```
. di 28/513
.0545809
```

Given being in True level = Yes (D), the number of positive cases is 28. The percent of 28 of the total number of True = Yes (513) is 5.46 percent.

SPECIFICITY

```
. di 958/982
.97556008
```

Given being in True level = No (~D), the number of negative cases is 958. The percent of 958 of the total number of True = No (982) is 97.56 percent.

Statisticians usually compare the sensitivity to one minus the specificity. The graph of this relationship is referred to as a **receiver operating characteristic** (ROC) curve, which we discuss in Section 7.2.2.

These values inform us about the sensitivity and specificity statistics based on comparing the observed versus predicted values. They are rarely the values we use for adjusting the model for observed specificity and sensitivity in the collection or gathering of data.

Magder and Hughes (1997) suggested a method to adjust the response of a binary logistic model given user-supplied values of the sensitivity and specificity. They used a method of estimation called **expectation-maximization** (EM), which is a technique commonly used when traditional maximum likelihood methods are not suitable. We use the values obtained in the previous classification table with a Stata command authored by Mario Cleves (Stata Corp.) and Alberto Tosetto (St. Bortolo Hospital, Vicenza, Italy) (1997).

Suppose that the data were collected with uncertainty, having an estimated specificity of 97.5 percent and sensitivity of 95 percent. In other words, we assume that the data are mostly valid, but there is still some residual error in the collection and recording of data. These values are inserted into the model.

```
. logitem died los type2 type3, nolog spec(.975) sens(.95)

logistic regression when outcome is uncertain
                                      Number of obs   =         1495
                                      LR chi2(3)      =        -1.04
Log likelihood =  -941.1761           Prob > chi2     =       1.0000
-------------------------------------------------------------------
       | Odds Ratio   Std. Err.     z    P>|z|   [95% Conf. Interval]
-------+-----------------------------------------------------------
   los |    .956208    .0094359  -4.54   0.000    .9378916    .974882
 type2 |   1.547785    .2412247   2.80   0.005    1.14038    2.100737
 type3 |    2.71013    .6798863   3.97   0.000    1.657489    4.431286
-------------------------------------------------------------------
```

The odds ratios have been adjusted as indicated below:

los	.963	to	.956
type2	1.49	to	1.55
type3	2.52	to	2.71

The predictor *p*-values remain significant in both models, but the value of z is on average slightly inflated in the adjusted model. The log-likelihood functions differ only negligibly (0.5).

The response values are adjusted using the following logic. For cases where the response, *y*, is equal to 1, and has been correctly classified,

$$y_{adj1} = \frac{Pr(y = 1 \mid x\beta) * \text{sens}}{Pr(y = 1 \mid x\beta) * \text{sens} + Pr(y = 0 \mid x\beta) * (1 - \text{spec})} \qquad (5.44)$$

Where *y* has been incorrectly classified,

$$y_{adj0} = \frac{Pr(y = 1 \mid x\beta) * (1 - \text{spec})}{Pr(y = 1 \mid x\beta) * (1 - \text{spec}) + Pr(y = 0 \mid x\beta) * \text{spec}} \qquad (5.45)$$

Additional discussion on this type of model and how the EM algorithm is implemented in this case can be found in Cleves and Tosetto (2000).

5.12 Constraining Coefficients

There are times that a statistician will want to constrain the logistic coefficients of a model to a given value. This value may be one that they wish to use because there are known coefficients that have been calculated based on the entire population, or because they simply wish to use coefficients developed for one model with another. Whatever the reason, this section will demonstrate how offsets can be used to constrain predictor coefficients to a given constant value.

To demonstrate how coefficients may be constrained for use with the same data, we shall use the **medpar** data using a logistic model with *died* as the response.

```
. logit died white hmo los age80, nolog
```

```
Logistic regression                    Number of obs    =        1495
                                        LR chi2(4)       =       43.44
                                        Prob > chi2      =      0.0000
Log likelihood = -939.71071            Pseudo R2        =      0.0226
```

```
------------------------------------------------------------------
  died |      Coef.  Std. Err.     z   P>|z|   [95% Conf. Interval]
-------+----------------------------------------------------------
 white |   .2064595  .2080442   0.99  0.321  -.2012997    .6142186
   hmo |  -.0139475  .1510154  -0.09  0.926  -.3099322    .2820372
   los |  -.0295671  .0078148  -3.78  0.000  -.0448838   -.0142505
 age80 |   .6255816  .1282944   4.88  0.000   .3741293    .877034
 _cons |  -.7054139  .2169202  -3.25  0.001  -1.13057    -.280258
------------------------------------------------------------------
```

The coefficient of *hmo*, for example, may be constrained to the value estimated by the above model by creating a new covariate that is the product of *hmo* and its associated coefficient.

```
. gen coefhmo= -.0139475*hmo
```

The model may be estimated as before, but leaving out *hmo* from the list of predictors, and placing it into the model as an offset.

```
. logit died white los age80, nolog offset(coefhmo)
```

```
Logistic regression                    Number of obs   =        1495
                                        Wald chi2(3)    =       40.23
Log likelihood = -939.71071            Prob > chi2     =      0.0000
------------------------------------------------------------------
  died |      Coef.  Std. Err.     z   P>|z|   [95% Conf. Interval]
-------+----------------------------------------------------------
 white |   .2064595   .207772   0.99  0.320  -.2007662    .6136852
   los |  -.0295671  .0078065  -3.79  0.000  -.0448676   -.0142666
 age80 |   .6255816  .1281731   4.88  0.000   .3743671    .8767962
 _cons |  -.7054139   .216081  -3.26  0.001  -1.128925    -.281903
coefhmo |   (offset)
------------------------------------------------------------------
```

Notice that all of the other predictor values are the same. We entered *hmo*, but constrained it to have a specified coefficient. The same may be done for any of the predictors. All predictors but *white* may be constrained to the values from the above model by generating new covariates, as defined above, for each predictor.

```
. gen coefage80=.6255816*age80
```

```
. gen coeflos=-.0295671*los
```

The sum of all predictors for which you wish to have coefficients constrained is calculated and saved as a new covariate. In effect, a partial linear predictor is being created.

```
. gen constr = coefhmo + coefage80 + coeflos
```

Enter the summed value as an offset as before. Notice that the coefficient of *white* and the intercept are the same as before.

```
. logit died white, nolog offset(constr)
```

Logistic regression				Number of obs	=	1495
				Wald chi2(1)	=	0.99
Log likelihood = -939.71071				Prob > chi2	=	0.3192

died	Coef.	Std. Err.	z	P>|z|	[95% Conf. Interval]	
white	.2064595	.2072682	1.00	0.319	-.1997786	.6126977
_cons	-.7054142	.1991244	-3.54	0.000	-1.095691	-.3151375
constr	(offset)					

Again, notice that the coefficient of *white* and the value of the intercept are identical to their values in the original model.

If you wish to use these constrained coefficients in another model with additional predictors, simply add the other predictors and include the constrained predictors as an offset.

```
. logit died white type2 type3, nolog offset(constr)
```

Logistic regression				Number of obs	=	1495
				Wald chi2(3)	=	23.39
Log likelihood = -928.65304				Prob > chi2	=	0.0000

died	Coef.	Std. Err.	z	P>|z|	[95% Conf. Interval]	
white	.2710414	.2091607	1.30	0.195	-.1389061	.6809889
type2	.4176459	.1438319	2.90	0.004	.1357406	.6995512
type3	.9136077	.2227881	4.10	0.000	.476951	1.350264
_cons	-.8992056	.2064119	-4.36	0.000	-1.303765	-.4946458
constr	(offset)					

The coefficient for *hmo, los,* and *age80* cannot be directly observed, but they are present in the model nonetheless. We know that they are since we specifically defined them based on the initial model.

The above method can be used for any statistical package. Stata has a constraint command allowing the user to define a coefficient for a given predictor. For example, given the first model we used in this section, the value of *hmo* is assigned a constraint at that value. The constraint is called 1 and can be used for many Stata commands. It cannot, however, be used with **logit**. We therefore use the **glm** command to model the same logistic model as above, setting *hmo*'s coefficient at −0.0139475.

```
. constrain 1 hmo=-.0139475
```

The constraint can be used in the following manner:

```
. glm died white hmo los age80, nolog fam(bin) constraint(1)
```

```
Generalized linear models            No. of obs      =       1495
Optimization      : ML               Residual df     =       1490
                                     Scale parameter =          1
Deviance          =   1879.421413    (1/df) Deviance =   1.261357
Pearson           =   1514.66544     (1/df) Pearson  =   1.016554

Variance function: V(u) = u*(1-u)
Link function     : g(u) = ln(u/(1-u))

                                     AIC             =   1.263827
Log likelihood    = -939.7107063     BIC             =  -9012.302

 ( 1)   [died]hmo = -.0139475
-----------------------------------------------------------------
        |               OIM
  died  |      Coef.  Std. Err.      z   P>|z|  [95% Conf. Interval]
--------+--------------------------------------------------------
 white  |   .2064595  .2077722    0.99  0.320  -.2007666    .6136856
   hmo  |  -.0139475          .       .      .          .           .
   los  |  -.0295671  .0078066   -3.79  0.000  -.0448677   -.0142666
 age80  |   .6255816  .1281731    4.88  0.000    .374367    .8767963
 _cons  |  -.7054139  .2160812   -3.26  0.001  -1.128925   -.2819026
-----------------------------------------------------------------
```

Notice that the other coefficient values are identical to the original unconstrained model.

Suppose that we have a known coefficient on *white* coming from a national study that we wish to incorporate into our own study model. As a consequence, we constrain the coefficient of *white* to .25. We first drop the old constraint and redefine it. Of course, it was possible to simply assign it to constraint 2.

```
. constraint drop 1
```

```
. constraint 1 white= .25
```

```
. glm died white age80 hmo los, nolog fam(bin) constraint(1)
```

```
Generalized linear models            No. of obs      =       1495
Optimization      : ML               Residual df     =       1490
```

```
                                       Scale parameter =            1
Deviance          =    1879.464996     (1/df) Deviance  =    1.261386
Pearson           =    1515.821841     (1/df) Pearson   =    1.01733

Variance function: V(u)  = u*(1-u)
Link function    : g(u)  = ln(u/(1-u))

                                       AIC              =    1.263856
Log likelihood    = -939.7324981       BIC              =   -9012.258

 ( 1)   [died]white = .25
---------------------------------------------------------------------
            |               OIM
   died |     Coef.   Std. Err.      z    P>|z|    [95% Conf. Interval]
--------+------------------------------------------------------------
  white |      .25         .        .        .         .           .
  age80 |  .6243386   .1281721     4.87   0.000    .3731258    .8755513
    hmo | -.0155519   .1508309    -0.10   0.918   -.3111749    .2800712
    los | -.0294859   .0078065    -3.78   0.000   -.0447864   -.0141853
  _cons | -.7457952    .099352    -7.51   0.000   -.9405216   -.5510689
---------------------------------------------------------------------
```

Other more complex constraints may be applied. Stata users may find additional information in the Reference manuals. Users of other software should refer to the appropriate manuals to determine if similar capabilities exist.

Exercises

5.1. Under what conditions should odds ratios be interpreted as if they were risk ratios?

5.2. The following binary logistic regression output uses the **affairs** data set, with *affair* (1=had affair; 0=never had affair) as the response and predictors:

male: 1 = male; 0 = female

kids: 1 = children in family; 0 = no children in family

and a five-level categorical predictor, *religious*, which has been factored into indicator variables. Level 1, *anti-religious*, is the reference.

notrel: 2—not religious

slghtrel: 3—slightly religious

smerel : 4—somewhat religious

vryrel: 5—very religious

```
Logistic regression                    Number of obs   =        601
                                       LR chi2(6)      =      34.36
                                       Prob > chi2     =     0.0000
Log likelihood = -320.50848            Pseudo R2       =     0.0509
---------------------------------------------------------------------
   affair | Odds Ratio Std. Err.    z   P>|z|  [95% Conf. Interval]
---------+-----------------------------------------------------------
     male |    1.2409 .2424715    1.10 0.269  .8460812   1.819958
     kids |  2.346751 .5681122    3.52 0.000  1.460172   3.771636
   notrel |   .470281 .1651811   -2.15 0.032  .2362555   .9361227
 slghtrel |  .6688562 .2381866   -1.13 0.259  .3328214   1.344171
   smerel |  .2633311 .0942904   -3.73 0.000  .1305312   .5312389
   vryrel |  .2792907 .1209288   -2.95 0.003   .119536   .6525508
---------------------------------------------------------------------
```

a. Which predictors are statistically significant at the .05 level?

b. How can the significance of a predictor be determined from the confidence intervals?

c. Interpret the odds ratio of *kids*.

d. Interpret the odds ratio of *vryrel*.

e. Are the levels *smerel* and *vryrel* statistically different? The levels *slghtrel* and *smerel*? The levels *notrel* and *slghtrel*? In view of what is determined from evaluating these relationships, what, if anything, should be recommended about amending the model?

f. What is the value of the parameter estimate of *male*?

g. To the nearest thousandth, calculate the 95 percent confidence intervals for the coefficient of *kids*.

h. Calculate the deviance statistic of the above model.

i. The likelihood ratio test of the model, displayed in the output above, is 34.36. Given the *p*-value of the statistic, what does the statistic indicate about the model?

j. If the model were re-estimated excluding the predictor *male*, what would you expect to happen to the odds ratio of *kids*? Why?

5.3. Below is the **kyp** data set, which is from a study of kyphosis, a disease of the spine. The data come from Bell et al. (1989) as found in Hastie and Tibshirani (1990). *kyp* is the binary response with 1 indicating that the patient has kyphosis, 0 if the patient does not. Patients in the study underwent corrective spine surgery.

Predictors include:

start : continuous—indicating the first disk or vertebrae level of the surgery (1–18) numb: continuous—indicating the number of disks involved in the surgery (2–14)

age: continuous—age in months

```
Logistic regression                 Number of obs    =           83
                                    LR chi2(3)       =        21.79
                                    Prob > chi2      =       0.0001
Log likelihood = -32.508134         Pseudo R2        =       0.2510
```

kyp	Coef.	Std. Err.	z	P>|z|	[95% Conf.	Interval]
start	-.1981754	.0657199	-3.02	0.003	-.3269841	-.0693667
numb	.2981585	.1778791	1.68	0.094	-.0504781	.6467951
age	.0059798	.0055195	1.08	0.279	-.0048383	.0167979
_cons	-1.213598	1.234168	-0.98	0.325	-3.632523	1.205327

a. Interpret the coefficient of *start*.

b. Should *numb* be excluded from a final model of the data? Why?

c. What does the intercept_*cons* indicate?

5.4. The models below use the **absent** data set, consisting of data from a study on slow learners in the Australian school system. The response variable is students who have many absences (overmonth: 1 = >30 days; 0 = <30 days) are modeled on:

slow: 1 = slow learner; 0 = standard learner

aborig: 1 = aboriginal ethnic background; 0 = other ethnic background

girl: 1 = girl; 0 = boy

grade2: 9th grade <grade1 (reference; 8th grade)>

grade3: 10/11th grade

grade4: 12th grade

```
Logistic regression                 Number of obs    =          140
                                    LR chi2(6)       =        22.25
                                    Prob > chi2      =       0.0011
Log likelihood = -54.567247         Pseudo R2        =       0.1693
```

ovrmonth	Coef.	Std. Err.	z	P>|z|	[95% Conf.	Interval]
slow	.9606643	.6538048	1.47	0.142	-.3207696	2.242098
aborig	1.591901	.5246611	3.03	0.002	.563584	2.620218
girl	-.4236748	.5076044	-0.83	0.404	-1.418561	.5712115
grade2	-.6426341	.9233743	-0.70	0.486	-2.452415	1.167146
grade3	1.042728	.7633262	1.37	0.172	-.4533638	2.53882
grade4	1.558996	.8516379	1.83	0.067	-.1101832	3.228176
_cons	-3.309335	.8176474	-4.05	0.000	-4.911894	-1.706775

The predictor *girl* is excluded from the model below. Note, however, that there are more observations in the new model than in the one above. Why would that be the case?

```
Logistic regression                    Number of obs    =          146
                                       LR chi2(5)       =        21.38
                                       Prob > chi2      =       0.0007
Log likelihood = -56.155735            Pseudo R2        =       0.1599
-------------------------------------------------------------------
ovrmonth |     Coef. Std. Err.    z   P>|z|  [95% Conf. Interval]
---------+---------------------------------------------------------
   slow  |  .9200935 .6300386  1.46 0.144 -.3147594    2.154946
  aborig |  1.556833 .5221025  2.98 0.003  .5335306    2.580135
  grade2 | -.8742673 .9102697 -0.96 0.337 -2.658363    .9098286
  grade3 |  .9245047 .7564289  1.22 0.222 -.5580688    2.407078
  grade4 |  1.348837  .817174  1.65 0.099 -.2527951    2.950468
   _cons | -3.399795 .8087578 -4.20 0.000 -4.984931   -1.814659
-------------------------------------------------------------------
```

5.5. Given the following tabulation, perform the calculations as instructed. y is assumed as the response variable, x is the risk factor, or predictor.

```
                        y
                   1        0
               ---------------------
           1 |   8        48    |
        x      |                 |
           0 |   4        32    |
               ---------------------
```

   a. Calculate the risk ratio of y on x, in other words, the relative risk of y given x.

   b. Calculate the odds ratio of y on x.

   c. If modeled using a logistic regression, can the odds ratios be interpreted as if they are risk ratios? Why?

   d. Calculate the 95 percent confidence interval of the risk ratio.

   e. Calculate the 95 percent confidence interval of the logistic coefficient.

5.6. After scaling the standard error of a logistic coefficient by the Pearson dispersion, how is the standard error to be interpreted?

5.7. What is the problem with using the likelihood ratio test to evaluate the statistical significance of a predictor in a model if the predictor has missing values and the remaining predictors do not? Can it be remedied? If so, how? If not, why?

5.8. The following binary logistic model was developed to determine risk factors for coronary heart disease (*chd*). It is from the **chd** data. Included as predictors are

cat :	Serum catecholamine level	1 = high; 0 = low
exer :	Regular exercise	1 = yes; 0 = no
smk :	Smoker	1 = yes; 0 = no
age :	Age	continuous : 40–76
chl :	Serum cholesterol	continuous : 94–357
sbp :	Systolic blood pressure	continuous : 92–300

```
Logistic regression                 Number of obs   =        609
                                     LR chi2(6)      =      35.89
                                     Prob > chi2     =     0.0000
Log likelihood = -201.33493          Pseudo R2       =     0.0818
```

chd	Odds Ratio	Std. Err.	z	P>|z|	[95% Conf. Interval]	
cat	2.146958	.8111049	2.02	0.043	1.023872	4.501956
exer	.6618895	.1973543	-1.38	0.166	.3689643	1.187372
smk	2.287023	.6962681	2.72	0.007	1.259297	4.153488
age	1.032983	.0157079	2.13	0.033	1.00265	1.064233
chl	1.009384	.0032874	2.87	0.004	1.002962	1.015848
sbp	1.000367	.0054368	0.07	0.946	.9897676	1.01108

a. What are the odds of having CHD for a 1-year increase in *age*? A 10-year increase?

b. What are the odds of having CHD for someone who does not exercise compared to someone who does?

c. What is the probability of having CHD for a 63-year-old with a systolic blood pressure of 120, serum cholesterol of 234, a high serum catecholamine level, who exercises very little, but does not smoke?

d. What is the percent difference in the probability of having CHD for the same person at age 70 compared with age 63, with the other characteristics the same as in item c above?

e. Both *cat* and *smk* are significant predictors of *chd*. Are they also statistically different from one another?

f. Estimate a relative risk ratio model of *chd* on the same predictors. Which predictors have near identical odds and risk ratios?

Which predictor(s) no longer significantly contribute to the model?

g. Remodel the data constraining *cat* to have an odds ratio of 2.00. How are other predictor odds ratios affected?

5.9. The following is a constructed 2 × 4 table. Using A as the reference, calculate a binary logistic regression of *happy* on *grade*. Calculate the probability of a student being *happy* if earning a *grade* of A, then of D.

```
                   Grade
Happy        A      B      C      D
      +--------------------------+
  No  |    158    198     89     28 |
 Yes  |     12     30     18      5 |
      +--------------------------+
```

5.10. The following tables were developed following a study on the importance of consultants in a jury trial. The tables are given as:

VIOLENT CRIMES

Outcome	Consultant	No Consultant	Total
Guilty	21	10	31
Not Guilty	12	3	15
Total	23	13	46

NON-VIOLENT CRIMES

Outcome	Consultant	No Consultant	Total
Guilty	18	11	29
Not Guilty	1	2	3
Total	19	13	32

a. Develop a logistic regression model of *verdict* on *type* of crime and *consultant*.

b. Does the *type* of crime significantly contribute to the *verdict*?

c. Does having a *consultant* at the trial make a difference in the *verdict*?

d. Does scaling the standard errors by *Chi2* make a difference?

R Code

Section 5.1 Building a Logistic Model

```
#preliminaries to get dataset
#DATA: 'heart01.dta'
#death: 1=yes, 0=no
#anterior: Anterior, Inferior
#age: in years
#hcabg: History of 1=CABG, 0=PTCA
#killip: Level 1-4 (4 more severe)
#agegrp: 4 age groups
#center: Provider or Center ID
library('foreign')
heart<- read.dta('heart01.dta') #read Stata format file
heart$anterior <- heart$anterior[,drop=TRUE] #drop empty
  levels that complicate tabulations
heart$center<- factor(heart$center) #convert to factor from
  numeric levels
heart$killip<- factor(heart$killip) #convert to factor from
  numeric levels
head(heart) #show some records
attach(heart) #make fields available directly as names

#Section 5.1:
t1<-table(hcabg) #frequency table
t1 #show
cat(format(t1/sum(t1), digits=3),'\n') #show fractions in each
  level
sum(t1) #total number

t2<-table(killip)
t2 #show
cat(format(t2/sum(t2), digits=3),'\n') #show fractions in each
  level
sum(t2)

t3<-table(agegrp)
t3 #show
cat(format(t3/sum(t3), digits=3),'\n') #show fractions in each
  level
sum(t3)

#construct age variable from categorized age
#(demo only, to keep 'age' from data intact for comparisons)
ageNew<- ifelse(age1==1, 40+trunc(21*runif(1)),
  ifelse(age2==1, 61+trunc(10*runif(1)),
    ifelse(age3==1, 71+trunc(10*runif(1)),
      ifelse(age4==1, 81+trunc(20*runif(1)),NA))))
```

```
#construct dummy variables for killip and age as 4 categories
#(demo only, to keep kk1-kk4 and age1-age4 already in dataset)
#NOTE: Dummy variables are generated automatically by glm().
kk1New<- ifelse(killip=='1', 1, 0)
kk2New<- ifelse(killip=='2', 1, 0)
kk3New<- ifelse(killip=='3', 1, 0)
kk4New<- ifelse(killip=='4', 1, 0)

age1<- ifelse(agegrp=='<-60', 1, 0)
age2<- ifelse(agegrp=='60-69', 1, 0)
age3<- ifelse(agegrp=='70-79', 1, 0)
age4<- ifelse(agegrp=='>=80', 1, 0)

#describe dataset heart01:
summary(heart)
str(heart) #show fields and encoding

#ulogit function to show univariate fit
ulogit <- function (y,x) {
  f<- glm(y ~ x, family=binomial())
  if (is.factor(x) && length(levels(x))>2) {
    print(anova(f, test='Chisq'))
  } else {
    b<- coef(f)[2]
    ci<- confint(f)[2,]
    cat('NullDeviance Deviance OR P-value LCL UCL\n')
    cat(format(f$null.deviance,nsmall=2,width=10),format
      (f$deviance,nsmall=2,width=10),

format(exp(b),nsmall=4,width=10),format(pchisq(b*b/vcov(f)
  [2,2],1,lower.tail=FALSE),
      nsmall=4,width=10),format(exp(ci[1]),nsmall=4,width=10),
         format(exp(ci[2]),
      nsmall=4,width=10),'\n')
  }
}

ulogit(death,anterior)
ulogit(death,hcabg)
ulogit(death,age)

#helper function to summarize logistic fit
showLogistic<- function (fit) {
  s<-summary(fit)
  print(s)
  b<- coef(fit) #coefficients
  o<- exp(b) #odds ratios
  CI<- confint(fit) #confidence intervals for coeff
  oCI<- exp(CI) #confidence intervals for odds ratios
```

```
  cat('\n')
  cat('Variable Coefficient LCL UCL\n')
  for (i in 1:length(b)) {
    cat(format(names(o)[i],width=15),formatC(b[i],digits=5,
        width=10),
        format(CI[i,1],digits=5,width=10),formatC(CI[i,2],
          digits=5,width=10),'\n')
  }
  cat('\n')
  cat('Variable Odds Ratio LCL UCL\n')
  for (i in 1:length(o)) {
    cat(format(names(o)[i],width=15),formatC(o[i],digits=5,
        width=10),
        formatC(oCI[i,1],digits=5,width=10),formatC(oCI[i,2],
          digits=5, width=10),'\n')
  }
  cat('\n')
}


#killip factor fits:
fitk<- glm(death ~ kk2+kk3+kk4, data=heart, family=binomial())
showLogistic(fitk)
fitk2<- glm(death ~ killip, data=heart, family=binomial())
  #note how dummy vars are created
summary(fitk2) #same fit as above

tk<-table(factor(killip, exclude=NULL)) #include NA's
tk #show
cat(format(tk/sum(tk), digits=3),'\n') #show fractions in each
  level
sum(tk)


#agegrp factor fits:
fitage<- glm(death ~ age2+age3+age4, data=heart,
  family=binomial())
showLogistic(fitage)
fitage2<- glm(death ~ agegrp, data=heart, family=binomial())
  #dummy vars auto-created
summary(fitage2) #slight differences

fitage3<- glm(death ~ age1+age2+age3, data=heart,
  family=binomial())
showLogistic(fitage3)

#full model
fitFull<- glm(death ~ anterior+hcabg+age+kk2+kk3+kk4,
  data=heart, family=binomial())
showLogistic(fitFull)
```

```
#Full model with categorical age
fitFull2<- glm(death ~ anterior+hcabg+kk2+kk3+kk4+age2+age3+
  age4, data=heart, family=binomial())
showLogistic(fitFull2)

#drop age2 (combine with age1)
fitRed<- glm(death ~ anterior+hcabg+kk2+kk3+kk4+age3+age4,
  data=heart, family=binomial())
showLogistic(fitRed)

#likelihood ratio test (deviance difference)
anova(fitRed,fitFull2,test='Chisq')
```

Section 5.2 Assessing Fit

```
#Box-Tidwell
summary(age) #show some descriptive stats for age
sd(age)
length(age)

fitage<- glm(death ~ age, family=binomial(link=logit))
summary(fitage)

lnageage<- age*log(age) #Box-Tidwell interaction variable
fitlnageage<- glm(death ~ age + lnageage, family=binomial)
summary(fitlnageage)

#Tukey-Pregnibon
hatage<- hatvalues(fitage) #get hat values for death ~ age fit
summary(glm(death ~ age + hatage + I(hatage^2), family=binomial))

#Partial Residuals
lpartr<- function (glmModel, iVar)
{
  xlab <- attributes(glmModel$terms)$term.labels[iVar]
    #construct x-axis label for var
  x<- model.matrix(glmModel)[, iVar + 1] #get values for var
  y<- coef(glmModel)[iVar + 1]*x + residuals(glmModel,
    type='pearson') #partial residual
  plot(x, y, col='blue', xlab = xlab, ylab = 'Partial Residual')
  lines(lowess(x,y), col='red')
  invisible()
}
fitPartial<- glm(death ~ age + anterior + hcabg + kk2 + kk3 +
  kk4, family=binomial)
lpartr(fitPartial, 1)

#Linearity of slopes test
```

```
#construct dummy 'agegrp' factor as 'agegrpNew' to avoid
  changing 'agegrp' in heart01
agegrpNew<- rep(1,length(age)) #start with all level 1 (age <= 60)
agegrpNew<- ifelse(age>60 & age<=70, 2, agegrpNew)
agegrpNew<- ifelse(age>70 & age<=80, 3, agegrpNew)
agegrpNew<- ifelse(age>80, 4, agegrpNew)
agegrpNew<- factor(agegrpNew, labels=c('<=60', '61-70',
  '71-80', '>80'))

tagegrp<-table(agegrp) #frequency table for agegrp
tagegrp #show
cat(format(tagegrp/sum(tagegrp), digits=3),'\n') #show
  fractions in each level
sum(tagegrp) #total number

age23<- as.numeric(age2==1)age3==1) #middle ages (numeric, not
  factor)
tage23<-table(age23) #frequency table for agegrp
tage23 #show
cat(format(tage23/sum(tage23), digits=3),'\n') #show fractions
  in each level sum(tage23) #total number

fit1<- glm(death ~ anterior + hcabg + kk2 + kk3 + kk4 +
  age23 + age4, family=binomial)
summary(fit1)

fit2<- glm(death ~ anterior + hcabg + kk2 + kk3 + kk4 + age2 +
  age3 + age4, family=binomial)
summary(fit2)

fit3<- glm(death ~ anterior + hcabg + kk2 + kk3 + kk4 + age3 +
  age4, family=binomial)
summary(fit3)
2.1958 - 1.5218
1.9409 - 1.2668

#GAM
#NOTE: 'heart01' dataset already loaded as 'heart'.
library('mgcv') #get package for GAM functions

heartgam<- gam(death ~ s(age) + kk2 + kk3 + kk4,
  family=binomial)
summary(heartgam)
plot(heartgam, residuals=TRUE, se=TRUE, col='blue', pch='.')

heartgam2<- gam(death ~ s(age) + anterior + hcabg + kk2 +
  kk3 + kk4, family=binomial)
summary(heartgam2)
plot(heartgam2, col='blue')
```

```
#Fractional polynomials for GLM not available yet in a R
  package!
#But see: Amber G. (1999) Fractional polynomials in S-plus,
# http://lib.stat.cmu.edu/S/fracpoly.
```

Section 5.3 Standardized Coefficients

```
sqrt((pi*pi)/3)
sd(as.numeric(anterior), na.rm=TRUE) #std. dev. dropping
  missing data
.6387*.4977929/1.8137994

lstand<- function (lfit) {
#compute standardized coefficients per Hilber
  f<- sqrt(pi*pi/3) #factor for logistic
  b<- coef(lfit)
  o<- exp(b)
  v<- as.character(attributes(lfit$terms)$variables) #get
    names of vars
  cat('Seq. Variable Coef OR Std. Coef.','\n')
  cat(0, format(v[2], width=10), format(b[1], width=8,
    digits=4), format(o[1], width=8, digits=4), '\n')
  for (i in 1:(length(lfit$model)-1)) {
    x<- lfit$model[i+1][,1] #get values of i-th variable
    if (is.numeric(x)) {
      s<- sd(x, na.rm=TRUE) #get std. dev.
    } else { #factor
      s<- sd(as.integer(x), na.rm=TRUE) #std. dev. for factor
    }
    bs<- b[i+1]*s/f #std. coef.
    cat(i, format(v[i+2], width=10), format(b[i+1], width=8,
      digits=4), format(o[i+1], width=8, digits=4),
      format(bs, width=8, digits=4), '\n')
  }
  invisible() #suppress return print
}
lstand(fit3)
```

Section 5.4 Standard Errors

```
fit4<- glm(death ~ anterior + hcabg + kk2 + kk3 + kk4,
  family=binomial)
summary(fit4)
v<- vcov(fit4) #variance-covariance matrix
v
sqrt(v[2,2]) #anterior
sqrt(0.027113125) #and again
sqrt(v[3,3]) #hcabg
sqrt(v[4,4]) #kk2
```

```
0.7227 / 0.1646606 #z-value for anterior
0.7185 / 0.3460558 #z-value for hcabg

se<- sqrt(diag(v)) #S.E.s of coefficients in fit4 model
se
coef(fit4)[2]/se[2] #could also use
coef(fit4)['anteriorAnterior']/se['anteriorAnterior']
coef(fit4)[3]/se[3] #also coef(fit4)['hcabg']/se['hcabg']

coef(fit4)[2]/se[2] #anterior value in table is 4.389
2*pnorm(4.38874, lower.tail=FALSE) #complementary normal
   probability x 2
coef(fit4)[3]/se[3] #hcabg z value in table is 2.076
2*pnorm(2.076255, lower.tail=FALSE) #complementary normal
   probability x 2

qnorm(0.05/2) #normal quantile for p = 0.025
qnorm(1-0.05/2) #normal quantile for p= 0.975

matrix(c(coef(fit4)-1.96*se, coef(fit4)+ 1.96*se),ncol=2)
   #Wald 95% C.I.s for coefficients
confint(fit4) #more accurate 95% C.I.s
0.7227 - 1.96*0.1646606 #95% Wald LCL for anterior
0.7227 + 1.96*0.1646606 #95% Wald UCL for anterior
0.7185 - 1.96*0.3460558 #95% Wald LCL for hcabg
0.7185 + 1.96*0.3460558 #95% Wald LCL for hcabg

exp(coef(fit4)) #odds ratios for coefficients
exp(matrix(c(coef(fit4)-1.96*se, coef(fit4)+ 1.96*se),ncol=2))
   #Wald 95% C.I.s for odds ratios
exp(confint(fit4)) #more accurate 95% C.I.s for odds ratios
exp(0.7227 - 1.96*0.1646606) #95% Wald LCL for anterior odds ratio
exp(0.7227 + 1.96*0.1646606) #95% Wald UCL for anterior odds ratio
```

Section 5.5 Odds Ratio ~ Risk Ratios

```
#SECTION 5.5: ODDS RATIOS
(9/100)/(14/150) #OR
(150*9)/(100*14) #OR
(9/109)/(14/164) #RR
y<- c(1,0, 1,0) #Exposure variable
x<- c(1,1, 0,0) #Outcome variable
cnt<- c(9,100, 14,150) #counts
fit5a<- glm(y ~ x, weights=cnt, family=binomial(link=logit))
summary(fit5a)
exp(coef(fit5a)) #ORs
exp(confint(fit5a)) #C.I. for ORs
```

```
fit5b<- glm(y ~ x, weights=cnt, family=poisson)
fit5b
exp(coef(fit5b)) #RRs
exp(confint(fit5b)) #C.I. for RRs

(19/100)/(24/150) #OR
(19/119)/(24/174) #RR
63/313 #incidence

fit5<- glm(death ~ anterior, data=heart,
  family=binomial(link=logit))
fit5 #show fit
exp(coef(fit5)) #show odds ratios for coefficients (unit
  change in covariates)

library('gmodels') #package for CrossTable()
CrossTable(death, anterior, prop.r=FALSE, prop.c=FALSE, prop.
  t=FALSE, prop.chisq=FALSE)
(2504*120)/(2005*67) #calc. odds ratio
(120/2125)/(67/2571) #calc. risk ratio

fit6<- glm(death ~ anterior, data=heart, family=poisson)
fit6
exp(coef(fit6)) #show the IRR

fit7<- glm(death ~ anterior, data=heart,
  family=binomial(link=log))
fit7
exp(coef(fit7)) #show RR

187/4696

mean(death[anterior=='Inferior'], na.rm=TRUE) #get risk for
  unexposed
2.2368/(1-0.02606 + 0.02606*2.2368)

library('epitools') #package for oddsratio() and riskratio()
oddsratio(anterior, death, method='wald')
riskratio(anterior, death, method='wald')

oddsratio(anterior,hcabg, kk2, kk3, kk4, age3, age4,death,
  method='wald')
riskratio(anterior,hcabg, kk2, kk3, kk4, age3, age4, death,
  method='wald')

fit9<- glm(death ~ anterior + hcabg +kk2 + kk3 + kk4 + age3 +
  age4, data=heart, family=poisson) #poisson loglinear fit
summary(fit9)
exp(coef(fit9)) #show the IRRs
exp(confint(fit9)) #95% C.I.s for IRRs
```

```
fit9_5<- glm(death ~ anterior + hcabg +kk2 + kk3 + kk4 + age3
  + age4, data=heart, family=binomial(link=log)) #log-
  binomial fit
summary(fit9_5)
exp(coef(fit9_5)) #show the RRs
exp(confint(fit9_5)) #95% C.I.s for RRs

CrossTable(death, anterior, prop.r=FALSE, prop.c=FALSE,
  prop. t=FALSE, prop.chisq=FALSE) #show 2x2 table again

library('reshape')
dframe<- data.frame(death=death, anterior=anterior) #new,
  simple data frame
dfcounts<- data.frame(cast(melt(dframe, measure="death"),
  anterior~ ., function(x) { c(Lived=sum(x==0),
  Died=sum(x==1)) })) #create data frame of counts
dfcounts #show

fit8<- glm(cbind(Died,Lived)~ anterior, data=dfcounts,
  family=binomial)
summary(fit8) #show fit based on counts instead of binary data
confint(fit8) #Confidence intervals on coefs
exp(confint(fit8)) #C.I. for odds ratios
(120*2504)/(2005*67) #compute OR
log(2.2368) #coefficient
sqrt(1/120 + 1/67 + 1/2005 + 1/2504) #S.E. of
  log(OR)=coefficient
0.8050463 - 1.96*0.1554246 #Wald 95% LCL
0.8050463 + 1.96*0.1554246 #Wald 95% UCL
exp(0.5004141) #OR LCL
exp(1.109679) #OR UCL
#NOTE: confint() returns more accurate C.I.'s than the simple
  Wald C.I.s exp(coef(fit8)) #get ORs

120/2125 #risk for anterior
67/2571 #risk for inferior
0.05647059/0.0260599 #risk ratio
log(2.166953) #ln(RR)

0.05647059 - 1.96*sqrt(0.05647059*(1-0.05647059)/4696) #Wald
  95% LCL for proportion
0.05647059 + 1.96*sqrt(0.05647059*(1-0.05647059)/4696) #Wald
  95% UCL for proportion
0.0260599 - 1.96*sqrt(0.0260599*(1-0.0260599)/4696) #Wald 95%
  LCL for proportion
0.0260599 + 1.96*sqrt(0.0260599*(1-0.0260599)/4696) #Wald 95%
  UCL for proportion

sqrt(1/120 - 1/2125 + 1/67 -1/2571) #S.E. for ln(RR)
0.773322 - 1.96*0.1496635 #Wald 95% LCL for ln(RR)
```

```
0.773322 + 1.96*0.1496635 #Wald 95% UCL for ln(RR)
exp(0.4799815) #Wald 95% LCL for RR
exp(1.066662) #Wald 95% UCL for RR

112/1612 # pervalence risk x=1
98/3218 # prevalence risk x=0
.06947891/.0304537 #prevalence relative risk
.06947891 - .0304537 # risk diff

# risk difference
.06947891 - .0304537 - 1.96*sqrt( ((.06947891*(1-
  .06947891))/1612) + ((.0304537*(1-.0304537))/3118))
.06947891 - .0304537 + 1.96*sqrt( ((.06947891*(1-
  .06947891))/1612) + ((.0304537*(1-.0304537))/3118))

fit10_5<- glm(y ~ x, weights=cnt, data=heart,
  family=poisson(link=log)) #poisson
summary(fit10_5)
exp(coef(fit10_5)) #show the RRs
exp(confint(fit10_5)) #95% C.I.s for RRs

fit10_6<- glm(y ~ x, weights=cnt, data=heart,
  family=binomial(link=log)) #log-binomial fit
summary(fit10_6)
exp(coef(fit10_6)) #show the ORs
exp(confint(fit10_6)) #95% C.I.s for ORs

(112+98)/4830 #Risk Lung Cancer
98/3218 # Risk unexposed grp
(.04347826 - .0304537)/ .04347826 # attrib risk

aru <- (1.96*(112+98)*(98+3120))/(112*3120 - 1500*98) *
  sqrt((112*3120*(4830-98) + 98*98 *1500) /
  (4830*98*(112+98)*(98+3120)))
((112*3120 - 1500*98) * exp(-aru)) / (4830*98 + (112*3120
  - 1500*98)*exp(-aru))
((112*3120 - 1500*98) * exp(aru)) / (4830*98 + (112*3120
  - 1500*98)*exp(aru))
```

Section 5.6 Scaling Standard Errors

```
#SECTION 5.6: SCALING OF STANDARD ERRORS
x1<- rnorm(50000, mean=0, sd=1) #unit normal covariate
x2<- rnorm(50000, 0, 1) #unit normal covariate
xb<- 1 + 0.25*x1 - 1.5*x2 #model logit in x1 and x2
p<- 1/(1+exp(-xb)) #response probabilities
y<- rbinom(50000, size=100, prob=p) #random binomial counts
fit11<- glm(cbind(y, 100-y) ~ x1 + x2, family=binomial)
  #logistic fit on counts
summary(fit11)
```

```
fit11$deviance/fit11$df.residual #residual deviance per d.f.
sum(residuals(fit11, type='pearson')^2)/fit11$df.residual
  #Pearson X2 per d.f.
sum(residuals(fit11, type='deviance')^2)/fit11$df.residual
  #residual deviance per d.f.
rm(x1, x2, xb, p, y) #clean up large arrays

fit12<- glm(death ~ anterior + hcabg +kk2 + kk3 + kk4 + age3 +
  age4, data=heart, family=binomial(link=logit)) #ordinary
  logistic fit
showLogistic(fit12)
sum(residuals(fit12, type='pearson')^2)/fit12$df.residual
  #Pearson X2 per d.f.
sum(residuals(fit12, type='deviance')^2)/fit12$df.residual
  #residual deviance per d.f.

fit13<- glm(death ~ anterior + hcabg +kk2 + kk3 + kk4 + age3 +
  age4, data=heart, family=quasibinomial) #quasi-binomial
  logistic fit
showLogistic(fit13)
```

Section 5.7 Robust Variance

```
h<- na.omit(heart) #drop incomplete rows
nrow(h)
fit16<- glm(death ~ anterior + hcabg +kk2 + kk3 + kk4 + age3 +
  age4, data=h, family=binomial(link=logit)) #ordinary
  logistic fit
summary(fit16) #show ordinary logistic fit
exp(coef(fit16)) #odds ratios

library('sandwich') #package for variance sandwich estimation
vcovHC(fit16)
sqrt(diag(vcovHC(fit16)))
sqrt(diag(vcovHC(fit16, type="HC0")))

library('haplo.ccs') #package for sandcov()
sandcov(fit16, h$center)
sqrt(diag(sandcov(fit16, h$center)))
```

Section 5.8 Bootstrap and Jackknife STD ERR

```
summary(fit16) #show ordinary logistic fit again
exp(coef(fit16)) #odds ratios
sqrt(diag(vcov(fit16))) #show SEs of coefs

library('boot')
ffit<- function (x, i) {
  xx<- x[i,]
```

```
  fglm<- glm(death ~ anterior + hcabg +kk2 + kk3 + kk4 + age3
    + age4, data=xx, family=binomial(link=logit)) #ordinary
    logistic fit on resample
  return(sqrt(diag(vcov(fglm)))) #return SE of coeffs
}
bse<- boot(h, ffit, R=50) #do 50 resamples by cases (rows)
sqrt(diag(vcov(fit16))) #show SEs of coefs
apply(bse$t, 2, mean) #mean estimate of SEs

#jackknife is preferred for small numbers of data, bootstrap
  for large numbers
#Here, jackknife will require 5,403 model fits, but the
  bootstrap above
#only required 50 (although 100 might be surer).
#The following code will generate jackknife estimates of the
  SEs, but is not
# recommended due to the length of time required to complete:
library('bootstrap')
theta<- function (i) {
  xx<- h[i,] #resample
  fglm<- glm(death ~ anterior + hcabg +kk2 + kk3 + kk4 + age3
    + age4, data=xx, family=binomial(link=logit)) #ordinary
    logistic fit on resample
  return(sqrt(diag(vcov(fglm))[iCoef])) #SE of coef[iCoef]
}
sqrt(diag(vcov(fit16))) #show SEs of coefs again
for (iCoef in 1:8) { #repeat these time-consuming calculations
  for each coef jSE<- jackknife(1:nrow(h), theta) #perform
  jackknife for coef[iCoef] cat(iCoef, jSE$jack.se, '\n')
}

kk23New<- as.numeric(kk2==1 ~ kk3==1) #new variable equiv to
  merging kk2 and kk3
sum(kk23New!=kk23, na.rm=TRUE) #Equal to kk23 in dataset?

fit17<- glm(death ~ anterior + hcabg + kk23 + kk4 + age3 +
  age4, data=heart, family=binomial(link=logit))
summary(fit17) #show fit
exp(coef(fit17)) #odds ratios

abic<- function (glmfit) {
  d <- glmfit$deviance
  nobs <- length(glmfit$y)
  k <- glmfit$rank #number predictors, incl. intercept
  aic <- glmfit$aic
  bic <- d + 2*k*log(k)
  cat('AIC per datum: ',format(aic/nobs, digits=8),' AIC total:
',format(aic,digits=8),'\n')
  cat('BIC per datum: ',format(bic/nobs, digits=8),' BIC total:
```

```
',format(bic,digits=8),'\n')
  return(c(aic,bic))
}

abic(fit17) #show AIC and BIC for kk23 fit
abic(fit16) #show AIC and BIC for kk2 and kk3 fit
```

Section 5.9 Stepwise

```
detach(heart) #remove this data frame from easy access

library('foreign')
lrdata<- read.dta('lr_model.dta') #read Stata format file
head(lrdata) #show some rows

fit18<- glm(heartatk ~ reg2 + reg3 + reg4 + race2 + race3 +
  gender + height + weight + bpsystol + tcresult + tgresult +
  hdresult + hgb + diabetes + albumin + hsi2 + hsi3 + hsi4 +
  hsi5, data=lrdata, family=binomial) #set up overall model
showLogistic(fit18)

fit19<- step(fit18, direction='both') #stepwise fits
showLogistic(fit19) #final fit

fit20<- glm(heartatk ~ reg4 + gender + height + tcresult +
  tgresult + hdresult + diabetes + albumin + hsi3 + hsi4 +
  hsi5, data=lrdata, family=binomial) #set up overall model
showLogistic(fit20)

lbw<- read.dta('lbw.dta') #read Stata format file
head(lbw) #show some rows
lbw$age<- ifelse(lbwdata$age>34, NA, lbwdata$age) #mark age>34
  values as missing

fit21<- glm(low ~ age + smoke + race2 + race3, data=lbw,
  family=binomial)
fit22<- step(fit21, direction='both') #stepwise fits have
  problem due to missing data

lbw2<- na.omit(lbw) #remove missing data rows
fit23<- glm(low ~ age + smoke + race2 + race3, data=lbw2,
  family=binomial)
fit24<- step(fit23, direction='both') #stepwise fits
summary(fit24) #show final fit

fit25<- glm(low ~ smoke + race2 + race3, data=lbw,
  family=binomial)
summary(fit25) #show fit on original dataset lbw
```

Section 5.10 Missing Values

```
library('foreign')
felon<- read.dta('felonx.dta') #read Stata format file
head(felon) #show some rows
str(felon) #show structure
nrow(felon) #show number of records
summary(felon) #descriptive stats on variables

fit26<- glm(hired ~ felony + gender + natamer + mo6 + educ2 +
  educ3 + educ4, data=felon, family=binomial)
summary(fit26)

fit27<- glm(hired ~ felony + gender + mo6 + educ2 + educ3 +
  educ4, data=felon, family=binomial)
summary(fit27)

nomiss<- !with(felon, is.na(hired) | is.na(felony) |
  is.na(gender) is.na(natamer) | is.na(educ) | is.na(educ2) |
  is.na(educ3) | is.na(educ4))
table(nomiss)

fit28<- glm(hired ~ felony + gender + natamer + mo6 + educ2 +
  educ3 + educ4, data=felon, subset=nomiss, family=binomial)
  #note use of subset = summary(fit28)
fit29<- glm(hired ~ felony + gender + mo6 + educ2 + educ3 +
  educ4, data=felon, subset=nomiss, family=binomial)
summary(fit29)
anova(fit29, fit28, test='Chisq') #LRT for natamer

fit30<- glm(hired ~ gender + natamer + mo6 + educ2 + educ3 +
  educ4, data=felon, subset=nomiss, family=binomial)
anova(fit30, fit28, test='Chisq') #LRT for felony

library('mice') #missing data imputation library for
  md.pattern(), mice(), complete()
names(felon) #show variable names
md.pattern(felon[,1:4]) #show patterns for missing data in 1st
  4 vars

library('Hmisc') #package for na.pattern() and impute()
na.pattern(felon[,1:4]) #show patterns for missing data in 1st
  4 vars

fit31<- glm(hired ~ felony + gender + natamer, data=felon,
  family=binomial)
summary(fit31)

#no '.ice' equivalent in R that will give exactly the same
  imputed values as Stata.
```

```
#simple imputation can be done by
felon2<- felon #make copy
felon2$gender<- factor(felon2$gender) #convert to factor
felon2$felony<- impute(felon2$felony) #impute NAs (most
  frequent)
felon2$gender<- impute(felon2$gender) #impute NAs
felon2$natamer<- impute(felon2$natamer) #impute NAs
na.pattern(felon2[,1:4]) #show no NAs left in these vars
fit32<- glm(hired ~ felony + gender + natamer, data=felon2,
  family=binomial)
summary(fit32)

#better, multiple imputation can be done via mice():
imp<- mice(felon[,1:4]) #do multiple imputation (default is 5
  realizations)
for (iSet in 1:2) { #show results from 1st two imputation
  datasets
  fit<- glm(hired ~ felony + gender + natamer,
    data=complete(imp, iSet), family=binomial) #fit to
    iSet-th realization
  print(summary(fit))
}

felon3<- read.dta('imputedx.dta') #read Stata format file
head(felon3) #show some rows
str(felon3)  #show structure
nrow(felon3) #show number of records
summary(felon3) #descriptive stats on variables
fit33<- glm(hired ~ felony + gender + natamer,
  data=felon3[1:733,], family=binomial)
summary(fit33)

c(table(felon$felony), sum(is.na(felon$felony))) #frequency in
  original
c(table(felon3$felony[1:733]), sum(is.
  na(felon3$felony[1:733]))) #frequency in imputed
c(table(felon$gender), sum(is.na(felon$gender))) #frequency in
  original
c(table(felon3$gender[1:733]), sum(is.
  na(felon3$gender[1:733]))) #frequency in imputed
c(table(felon$natamer), sum(is.na(felon$natamer))) #frequency
  in original
c(table(felon3$natamer[1:733]), sum(is.
  na(felon3$natamer[1:733]))) #frequency in imputed
```

Section 5.11 Uncertain Response

```
library('foreign')
medpar<- read.dta('medparx.dta')
```

```
head(medpar) #show some rows
str(medpar)  #show structure
nrow(medpar) #show number of records
summary(medpar) #descriptive stats on variables

fit34<- glm(died ~ los + type2 + type3, data=medpar,
  family=binomial)
summary(fit34)
exp(coef(fit34)) #show odds ratios
table(medpar$died, fitted(fit34, type='response')>=0.5) #show
  confusion matrix
(28/513) #sensitivity
(958/982) #specificity

library('PresenceAbsence') #package for sensitivity/
  specificity
d<- cbind(1:nrow(medpar), medpar$died, fitted(fit34,
  type='response')) #construct data
confmat<- cmx(d, threshold = 0.5, which.model = 1) #get
  confusion matrix
confmat #show confusion matrix
sensitivity(confmat, st.dev = TRUE) #calc. sensitivity
specificity(confmat, st.dev = TRUE) #calc. specificity

#Magder-Hughes methodology not yet implemented in R
```

Section 5.12 Constrained Coefficients

```
fit35<- glm(died ~ white + hmo + los + age80, data=medpar,
  family=binomial)
summary(fit35)

coefhmo<- coef(fit35)['hmo']*medpar$hmo #effect of hmo in fit
fit36<- glm(died ~ white + los + age80 + offset(coefhmo),
  data=medpar, family=binomial) #fix effect of hmo via
  offset()
summary(fit36)

coefage80<- coef(fit36)['age80']*medpar$age80 #effect of age80
  in fit
coeflos<- coef(fit36)['los']*medpar$los #effect of age80 in fit
constr<- coefhmo + coefage80 + coeflos #combined effects
fit37<- glm(died ~ white + offset(constr), data=medpar,
  family=binomial) #fix effects of hmo, age80, los via
  offset()
summary(fit37)
```

```
fit38<- glm(died ~ white + type2 + type3 + offset(constr),
   data=medpar, family=binomial) #fix effects of hmo, age80,
   los via offset()
summary(fit38)

#There is no equivalent to constraint() in R other than
   offset().
```

6

Interactions

6.1 Introduction

Interactions play an important role in modeling. They are used when the levels of one predictor influence the response differently according to the levels of another predictor. For example, consider the continuous predictor *age* and a binary risk factor, *sex*. The response is *death*. If we model *death* on *age* and *sex*, we assume that the coefficients for both *age* and *sex* (male/female) are independent of one another. Moreover, the assumption of such a model, also called a main effects model, is that the coefficient, or slope, of *age* is constant across all values of *age* in the data, that is, the difference in odds or risk between males and females does not change with age. We know, though, that women tend to live longer than men do, so the odds of death at a certain age should be different between the two levels of *sex*. Creating an interaction of *age* and *sex* aims to reflect this relationship.

We shall discuss four types of interactions in this chapter, each of which is commonly used in research. Graphical representations of each interaction type are also developed. We shall also discuss how interactions are to be constructed and interpreted, and develop a means to calculate appropriate standard errors and confidence intervals. The four types of interactions we discuss are

1. Binary × Binary
2. Binary × Categorical
3. Binary × Continuous
4. Categorical × Continuous

Other types not discussed in detail are

- Continuous × Continuous
- Categorical × Categorical
- 3-Level interactions

Continuous × Continuous interactions are seldom used in modeling. Inter-action effects measure the differential effect of different levels of the covariates involved in the interaction. For a continuous predictor, an interaction occurs when the slope of the predictor changes at different points along the range of its values, based on the value or level of another predictor. Different slopes are developed for each of these levels. In the case of a Continuous × Continuous interaction, the simple multiplication of the two predictors results in a single continuous predictor, having a single coefficient. However, this tells us little about any interactive effect between the two predictors. With both terms in an interaction being continuous, it is difficult to access the specific contribu-tion of each predictor to an understanding of the response. To meaningfully assess an interactive effect of a continuous predictor on another predictor, it is preferable that the other predictor is either binary or categorical.

Categorical × Categorical interactions are occasionally found in research, but can be difficult to interpret when there are more than three levels for each predictor. The number of interactions increases as M × N grows. Problems develop for smaller data sets due to the possibility of very small observation numbers in certain cells. Consider the **medpar** data. Two of the predictors are *age* and *type*, where *age* is a 9-level categorical variable and *type* is a 3-level predictor. Tabulating *age* on *type* gives us:

```
. tab age type                  /// numeric age levels added

           |         Admission type
Age Group  | Elective    Urgent  Emergency |     Total
-----------+--------------------------------+----------
1     <55  |        4         2          0  |         6
2   55-59  |       46         9          5  |        60
3   60-64  |      133        22          8  |       163
4   65-69  |      206        63         22  |       291
5   70-74  |      240        55         22  |       317
6   75-79  |      247        60         21  |       328
7   80-84  |      148        31         12  |       191
8   85-89  |       73        17          3  |        93
9     90+  |       37         6          3  |        46
-----------+--------------------------------+----------
    Total  |    1,134       265         96  |     1,495
```

Note how few observations are in age level group 1 (< 55), and for higher type levels of age, in groups 2, 3, and 7 to 9. The majority of Medicare patients are 65 and over, with a percentage choosing to be part of Medicare at age 63, even though they earn substantially reduced benefits. Level 1 patients are typically disabled, as are an undetermined number of level 2 patients.

It is typical that the cells of most discrete predictors like *age* will have substantial differences in values, making meaningful interactions problem-atic. My suggestion is that when considering a Categorical × Categorical interaction, limit levels to three, checking to make certain that sufficient observations exist in all cells. Keep in mind, however, that a Categorical × Categorical interaction is simply not obtained by multiplying the two predictors. Rather, interactions must be created for every possible combination of the nonreference levels of the two categorical predictors. Again, meaningful interpretation can be problematic.

On the other hand, we often find Binary × Binary, Binary × Continuous, and Categorical × Continuous interactions in research literature. We shall, therefore, emphasize these types of relationships. Also discussed are Binary × Categorical interactions.

6.2 Binary × Binary Interactions

We use the **Titanic** data set to demonstrate Binary × Binary interactions. The data were gathered on the survival status of the 1316 passengers and 885 crew members on the ill-fated *Titanic*, which sunk in the North Atlantic in April 1912, due to an impact with an iceberg. The relevant variables are

Response: survived	1 = yes, 0 = no
Predictor: sex	1 = male, 0 = female
age	1 = adult, 0 = child
class	1 = 1st class; 2 = 2nd class; 3 = 3rd class;
	4 = crew (to be used later)

Model *survived* on *sex* and *age*. For the purpose of examining the nature of Binary × Binary interactions, I do not include the other categorical variable, *class*. If included though, it would make no difference to our results. However, I drop the *Titanic* crew from the data since we are only interested in passengers. All crew were adults. If retained, the corresponding values for *sex* and *age* would bias the relationships we are constructing.

```
. use titanic
. drop of class==4
. keep survived sex age
```

Estimate a main effects model to begin:

```
. logit survived sex age, nolog

Logistic regression                    Number of obs   =        1316
                                       LR chi2(2)      =      355.13
                                       Prob > chi2     =      0.0000
Log likelihood = -695.81171            Pseudo R2       =      0.2033
```

```
------------------------------------------------------------------
survived |    Coef. Std. Err.    z  P>|z| [95% Conf. Interval]
---------+--------------------------------------------------------
     sex | -2.345788 .1360596 -17.24 0.000  -2.61246 -2.079116
     age | -.6358321 .2332456  -2.73 0.006 -1.092985 -.1786791
   _cons |  1.548988 .2408192   6.43 0.000  1.076991  2.020985
------------------------------------------------------------------
```

The above model may be parameterized as odds ratios by adding an *or* to the command line, or we may use the **glm** command with *nohead* as an option to display only the table of odds ratios:

. glm survived age sex, nolog nohead fam(bin) eform

```
------------------------------------------------------------------
         |              OIM
survived | Odds Ratio Std. Err.    z  P>|z| [95% Conf. Interval]
---------+--------------------------------------------------------
     age |  .5294947 .1235025  -2.73 0.006 .3352142  .8363748
     sex |  .0957717 .0130307 -17.24 0.000 .0733539  .1250408
------------------------------------------------------------------
```

A possible interaction can often be identified if we exclude a predictor from a model and note that the odds ratio or the coefficient of another predictor is greatly affected. However, if the two predictors each have somewhat opposite effects, excluding one from the model will give no indication that an interaction may be required. Such is the case with this example:

. logit survived age, nolog

```
Logistic regression                   Number of obs   =        1316
                                      LR chi2(1)      =       10.12
                                      Prob > chi2     =      0.0015
Log likelihood = -868.31746           Pseudo R2       =      0.0058


------------------------------------------------------------------
survived |    Coef. Std. Err.    z  P>|z| [95% Conf. Interval]
---------+------- -------------------------------------------------
     age | -.6403735 .2008588  -3.19 0.001  -1.03405 -.2466975
   _cons |  .0918075 .1917671   0.48 0.632 -.2840491  .4676642
------------------------------------------------------------------
```

Excluding *sex* gives us—on the surface—no reason to believe that there may be a significant interactive effect between *sex* and *age* with respect to *survived*. Regardless, we shall create such an interaction, if only to show how Binary × Binary interactions are constructed and interpreted.

. gen byte sexage = sex * age

Remember, as we include lower-level terms when constructing ANOVA interactions, the main effects terms should be included in the regression model.

```
. logit survived sex age sexage, nolog

Logistic regression                     Number of obs   =        1316
                                        LR chi2(3)      =      372.85
                                        Prob > chi2     =      0.0000
Log likelihood = -686.95067             Pseudo R2       =      0.2135

------------------------------------------------------------------------
survived |      Coef. Std. Err.     z   P>|z|  [95% Conf. Interval]
---------+--------------------------------------------------------------
     sex | -.6870434 .3969786  -1.73 0.084 -1.465107   .0910203
     age |  .5279292 .3276433   1.61 0.107 -.1142399  1.170098
  sexage | -1.846994 .4228132  -4.37 0.000 -2.675692 -1.018295
   _cons |  .4989912   .30747   1.62 0.105  -.103639  1.101621
------------------------------------------------------------------------
```

If the above model is parameterized in terms of odds ratios, we have:

```
. glm survived sex age sexage, nolog fam(bin) eform nohead

------------------------------------------------------------------------
         |              OIM
survived | Odds Ratio Std. Err.     z   P>|z|  [95% Conf. Interval]
---------+--------------------------------------------------------------
     sex |  .5030612 .1997045  -1.73 0.084 .2310532   1.095291
     age | 1.695418  .5554923   1.61 0.107 .8920439   3.222309
  sexage |  .1577106 .0666821  -4.37 0.000 .0688591    .3612103
------------------------------------------------------------------------
```

We have used the generalized linear models command, **glm**, to estimate the odds ratios. We could have done the same using the **logit** command. However, **glm** allows the user to exclude associated statistics that are typically displayed together with the table of parameter estimates or odds ratios, as well as their corresponding standard errors, z-statistic and p-values, and confidence intervals.

The terms that are relevant for interpreting the *sexage* interaction are as follows:

β_1 = coefficient on *sex*   −0.6870434

β_2 = coefficient on *age*   0.5279292

β_3 = coefficient on *sexage*   −1.846994

sex

age

We will not need the actual interaction term, only its coefficient.

To determine the meanings of the four interactions entailed in the above model, use the formulae:

$$\beta_{B \times B} = \beta_1 + \beta_3 x \tag{6.1}$$

and

$$\beta_{B \times B} = \beta_2 + \beta_3 x \tag{6.2}$$

The formulae for parameterizing as odds ratios are

$$OR_{B \times B} = \exp[\beta_1 + \beta_3 x] \tag{6.3}$$

and

$$OR_{B \times B} = \exp[\beta_2 + \beta_3 x] \tag{6.4}$$

$\exp[\beta_1 + \beta_3 x]$ calculates the two interactions based on *age* (1 = adult; 0 = children).

$\exp[\beta_2 + \beta_3 x]$ calculates the two interactions based on *sex* (1 = male; 0 = female).

6.2.1 Interpretation—as Odds Ratio

1     $OR_{B \times B} = \exp[-0.6870434 - 1.846994 \text{*age} = 1]$

```
. di exp(-.6870434  - 1.846994*1)
.07933805
```

Among adults, the odds of surviving are 0.08 times as great for males as they are for females.

Or, inverting,

```
. di 1/.07933805
12.604293
```

Among adults, the odds of surviving are 12 times less for males than they are for females.

2     $OR_{B \times B} = \exp[-0.6870434 - 1.846994 \text{*age} = 0]$

```
. di exp(-.6870434  - 1.846994*0)
.50306122
```

Among children, the odds of surviving for males are half the odds for females, or
Among children, the odds of surviving for females are twice the odds for males.

$$3 \qquad OR_{B \times B} = \exp[0.5279282 - 1.846994*\text{sex} = 1]$$

```
. di exp(.5279292  - 1.846994*1)
.26738524
```

Among males, adults have .26 times the odds of surviving as children do. Among males, adults are three times less likely to survive than children are. Among males, children have some three times greater odds of surviving than adults do.

$$4 \qquad OR_{B \times B} = \exp[0.5279282 - 1.846994*\text{sex} = 0]$$

```
. di exp(.5279292  - 1.846994*0)
1.6954178
```

Among females, adults have a 70 percent greater odds of surviving than do children.

The above relationships may also be understood in terms of tabulations. This is accomplished by tabulating the response on each main effect separately by the other main effect term. Note that when we interpreted the results based on the interaction formula, the first two interactions began with "among adults" and "among children," respectively. This clearly corresponds to the manner in which we set up the tabulations below. The reason that the formula-based interpretations and the tabulations below are in reverse order is because we began with age = 1 (adults) above, whereas sorted tabulations begin with the level 0.

Begin by tabulating on *age*.

```
. sort age
. by age: tab survived sex

----------------------------------------------------

-> age = child

           |        Gender
  Survived |     women       man |      Total
-----------+--------------------+-----------
        no |        17        35 |         52
       yes |        28        29 |         57
-----------+--------------------+-----------
     Total |        45        64 |        109
----------------------------------------------------
```

```
-> age = adults

                |          Gender
   Survived |       women          man  |       Total
-----------+------------------------------+----------
        no  |          106         659  |        765
       yes  |          296         146  |        442
-----------+------------------------------+----------
     Total  |          402         805  |      1,207
```

The ratio of odds for each of the two levels in each of the above tabulations can be calculated by using the following schemata:

$$\frac{X_1 = 1 / X_1 = 0}{X_0 = 1 / X_0 = 0} \qquad (6.5)$$

Or by cross-multiplying:

$$\frac{X_1 = 1 * X_0 = 0}{X_1 = 0 * X_0 = 1} \qquad (6.6)$$

For the first tabulation, the odds ratio may be calculated as:

```
. di (29/35)/(28/17)
.50306122
```

or, cross-multiplying:

```
. di (29*17)/(28*35)
.50306122
```

The second tabulation entails the odds ratio, as calculated below:

```
. di (146/659)/(296/106)
.07933806
```

These odds ratios may be recognized as being identical to calculations #2 and #1, respectively, done using the formula for interactions.

Sorting on *sex* and engaging the same type of tabulations give us the odds ratios we earlier calculated for #4 and #3.

```
. sort sex
. by sex: tab survived age

----------------------------------------------------
-> sex = women
```

```
          | Age (Child vs Adult)
Survived  |    child      adults  |      Total
----------+------------------------+----------
     no   |      17         106  |        123
    yes   |      28         296  |        324
----------+------------------------+----------
   Total  |      45         402  |        447
```

```
-> sex = man
```

```
          | Age (Child vs Adult)
Survived  |    child      adults  |      Total
----------+------------------------+----------
     no   |      35         659  |        694
    yes   |      29         146  |        175
----------+------------------------+----------
   Total  |      64         805  |        869
```

The odds ratio of the first of the above two tabulations is

```
. di (296/106)/(28/17)
1.6954178
```

The final odds ratio results from the last tabulation.

```
. di (146/659)/(29/35)
.26738528
```

It is clear now that there are four interpretations for a Binary × Binary interaction. Moreover, the coefficient of an interaction term is only one aspect of the interaction and does not by itself specify the interactive relationship.

It should be noted, though, that the main effects terms used in an interaction have a meaning independent of their use in the interaction. Specifically, given main effects *sex* and *age,* and their interaction *sexage,* the odds ratio or coefficient of *sex* reflects the effect of *sex* on *survived* when *age* == 0. Likewise, the odds ratio or coefficient of *age* reflects the effect of *age* on *survived* when *sex* == 0. Moreover, the z-value (or *t;* Wald) for *sex* tests whether the effect of *sex* significantly differs from zero when *age* == 0. Likewise, the z-value (or *t;* Wald) for *age* tests if the effect of *age* significantly differs from zero when *sex* == 0. However, it is understood that regardless of *sex* having an insignificant *p*-value if *age* == 0, this has no bearing on the *p*-value of *sex* for other values of *age,* that is, *sex* may be insignificant if *age* = 0, but significant if *age*! = 0.

6.2.2 Standard Errors and Confidence Intervals

Since the values of interaction coefficients and odds ratios are not directly displayed in the model output, but must be calculated separately from a combination of the main effects and interaction coefficients, the standard errors and confidence intervals for the various interactions must also be calculated. I delay a discussion of how to construct the appropriate standard errors and confidence intervals for the various interactions until Section 6.3, Binary × Continuous Interactions. The same method is used for all types of interactions we consider, but it may be preferred to discuss this when dealing with more complicated interactions.

6.2.3 Graphical Analysis

Graphical representation of interactions can play an important role in better understanding both the interaction and model as a whole. We can use graphs to determine if an interaction even exists between two predictors and the response, or we can use graphs to give a visual import to the interactive relationship. Graphs of interaction effects are frequently used with Binary and Categorical × Continuous interactions. In fact, in Section 7.4.3, we use what is frequently called a conditional effects plot to view the relationship of levels of a binary predictor throughout the range of a continuous predictor. Those engaged in statistical consulting will find that graphs like the conditional effects plot can add greatly to the presentation of a statistical model. Nonstatistical clients tend to appreciate these types of graphs more than they do numbers.

Using Stata, a graph of the interaction of *age* and *sex* can be developed using the following commands (see Figure 6.1):

```
. qui xi3: logit survived i.sex*i.age
. postgr3 sex, by(age) title("Interaction of Age and Gender")
```

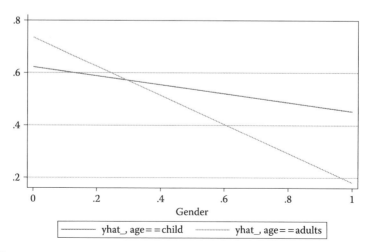

FIGURE 6.1
Interaction of age and gender.

We can clearly see the interaction of *age* and *sex* (gender) in the above graph. We can also identify the probability levels at which the interaction takes place. **xi3** is a user-authored command that allows graphing of both two-way and three-way interactions. **xi3** is used with the **postgr3** command to graph the interactions (Long and Freese, 2006a). A more complete explanation follows shortly.

It should be mentioned that using *i.sex*i.age* in a command automatically includes the main effects of *sex* and *age* when either or both *sex* and *age* are categorical variables. This is not the case when both variables involved in the interaction are continuous. A variable needs to be generated that is the product of the two continuous variables, with all three variables included in the **logit** or estimation command.

We can also determine if adding passenger *class*, a categorical variable, to the model changes the interactive effect, and if so to what degree. We first create dummy or indicator variables for each of the three levels of *class*. We then enter it into the model, leaving out first-class passengers as the reference. We are using levels of class as predictors, which possibly contribute to the model. The term *class* is not being used in an interaction.

```
. tab class, gen(class)
. logit survived sex age sexage class2 class3, nolog
```

```
Logistic regression                    Number of obs    =        1316
                                       LR chi2(5)       =      487.73
                                       Prob > chi2      =      0.0000
Log likelihood = -629.51262            Pseudo R2        =      0.2792
```

survived	Coef.	Std. Err.	z	P>|z|	[95% Conf. Interval]
sex	-.7148625	.4061458	-1.76	0.078	-1.510894 .0811685
age	.0885981	.3359649	0.26	0.792	-.569881 .7470771
sexage	-1.862079	.4350085	-4.28	0.000	-2.71468 -1.009478
class2	-1.02726	.1989846	-5.16	0.000	-1.417262 -.6372569
class3	-1.800574	.1753357	-10.27	0.000	-2.144225 -1.456922
_cons	2.062627	.3525665	5.85	0.000	1.371609 2.753644

We find that the levels of class appear to contribute significantly to the model. We now see if the effect has changed with the addition of the new predictor.

We can again let Stata automatically assign interaction terms and main effect levels for categorical predictors by using a prefatory **xi:** and a leading *i.* for each predictor that needs to be factored. This method is used to produce the output below.

```
. xi: logit survived i.sex*i.age i.class
```

```
i.sex          _Isex_0-1      (naturally coded; _Isex_0 omitted)
i.age          _Iage_0-1      (naturally coded; _Iage_0 omitted)
```

```
i.sex*i.age  _IsexXage_#_#  (coded as above)
i.class       _Iclass_1-3   (naturally coded; _Iclass_1 omitted)

Logistic regression                     Number of obs   =       1316
                                        LR chi2(5)      =     487.73
                                        Prob > chi2     =     0.0000
Log likelihood = -629.51262             Pseudo R2       =     0.2792
-------------------------------------------------------------------
   survived |    Coef. Std. Err.    z  P>|z| [95% Conf. Interval]
------------+------------------------------------------------------
    _Isex_1 | -.7148625 .4061458  -1.76 0.078 -1.510894   .0811685
    _Iage_1 |  .0885981 .3359649   0.26 0.792  -.569881   .7470771
_IsexXage_~1 | -1.862079 .4350085  -4.28 0.000  -2.71468 -1.009478
  _Iclass_2 |  -1.02726 .1989846  -5.16 0.000 -1.417262 -.6372569
  _Iclass_3 | -1.800574 .1753357 -10.27 0.000 -2.144225 -1.456922
       _cons |  2.062627 .3525665   5.85 0.000  1.371609  2.753644
-------------------------------------------------------------------
```

I should note that the **postgr3** command requires the saved results associated with **xi3** in order to graph the interaction. We use the Stata command **quietly** (or **qui**) to prepare use of **postgr3** without displaying the logit model results.

```
. qui xi3: logit survived i.sex*i.age i.class
. postgr3 sex, by(age) title("Interaction of Age and Gender;
  add Class") lpattern(longdash_dot shortdash_dot)
```

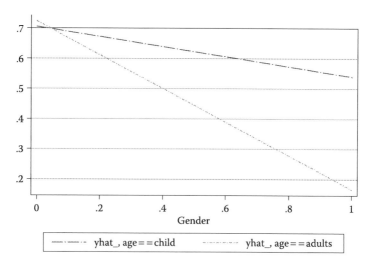

FIGURE 6.2
Interaction of age and gender; add class.

Figure 6.2 shows that the effect has changed, with the interaction occurring higher to the upper left. Recall that the probabilities are probabilities of *age* being adult and *sex* being male.

Respective values over 0.5 indicate a higher probability of being adult rather than a youth and a higher probability of being male than female.

For information and download of **postgr3** and **xi3**, see:

http://www.ats.ucla.edu/stat/stata/ado/analysis

http://www.ats.ucla.edu/stat/stata/faq/xi3.htm

6.3 Binary × Categorical Interactions

In the main effects model of *survived* on *age* and *sex*, we found that both *age* and *sex* were significant predictors. Their respective odds ratios are .5295 and .0958. However, when *class* is added to the model, we find the following:

```
. logit survived age sex class2 class3, or nolog   [header
info not displayed]
```

```
-------------------------------------------------------------------
survived | Odds Ratio Std. Err.     z  P>|z| [95% Conf. Interval]
---------+---------------------------------------------------------
    age  |    .3479809 .0844397   -4.35 0.000 .2162749   .5598924
    sex  |    .0935308 .0135855  -16.31 0.000 .0703585   .1243347
  class2 |    .3640159 .0709594   -5.18 0.000 .2484228   .5333952
  class3 |    .1709522 .0291845  -10.35 0.000 .1223375   .2388853
-------------------------------------------------------------------
```

The odds ratio of *sex* is nearly unchanged, but the value of *age* has appreciably changed—from .529 to .348. We may reasonably suspect that there may be an interactive effect between *age* and *class*.

Simply for pedagogical purposes, let us take a closer look at the sex-class relationship.

```
. sort survived
. by survived: tab sex class
```

```
---------------------------------------------------------------
-> survived = no
```

		class (ticket)		
Gender	1st class	2nd class	3rd class	Total
women	4	13	106	123
man	118	154	422	694
Total	122	167	528	817

```
------------------------------------------------------------
-> survived = yes

             |           class (ticket)
     Gender  |  1st class  2nd class  3rd class  |     Total
-----------+------------------------------------+----------
     women  |        141         93         90  |       324
       man  |         62         25         88  |       175
-----------+------------------------------------+----------
     Total  |        203        118        178  |       499
```

The two interaction terms are created using the code:

```
. gen byte sc2 = sex * class2
. gen byte sc3 = sex * class3
```

Looking at the women's *class1* and *class2* levels for those who died (*survived* = 0) we find very few observations compared to other levels. It is likely that when an interaction is constructed involving these levels, a significant relationship will not be possible.

```
. logit survived sex class2 class3 sc2 sc3, nolog

Logistic regression                        Number of obs   =        1316
                                           LR chi2(5)      =      515.16
                                           Prob > chi2     =      0.0000
Log likelihood = -615.79775                Pseudo R2       =      0.2949
------------------------------------------------------------------------
survived |     Coef. Std. Err.     z   P>|z|  [95% Conf. Interval]
---------+--------------------------------------------------------------
    sex  | -4.206016 .5307502  -7.92 0.000 -5.246267   -3.165764
  class2 | -1.594815 .5871694  -2.72 0.007 -2.745646    -.4439844
  class3 | -3.726095  .526913  -7.07 0.000 -4.758825   -2.693365
    sc2  |  .4202889  .644876   0.65 0.515 -.8436449    1.684223
    sc3  |  2.801977 .5621158   4.98 0.000  1.70025     3.903703
   _cons |  3.562466 .5070426   7.03 0.000  2.56868     4.556251
------------------------------------------------------------------------
```

The terms that are relevant for interpreting the *sc2* and *sc3* interactions are as follows:

β_2 = coefficient on *class2*   −1.594815

β_3 = coefficient on *class3*   −3.726095

β_4 = coefficient on *sc2*   0.420289

β_5 = coefficient on *sc3*   2.801977

Sex   1/0

The appropriate formulae for calculating the odds ratios for the interactions entailed in creating *sc2* and *sc3* are

$$OR_{B \times C2} = \exp[\beta_2 + \beta_4 x] \qquad (6.7)$$

and

$$OR_{B \times C3} = \exp[\beta_3 + \beta_5 x] \qquad (6.8)$$

where x is the value of *sex*.
We have, therefore:

MEN × CLASS 2

$$\exp(-1.594815 + .4203889^* sex = 1)$$

$$\exp(-1.594815 + .4203889) = 0.30899626$$

Among males, the odds of surviving are .3 times as great for second-class passengers as they are for first-class passengers, or
Among males, second-class passengers have 1/3 the odds of surviving as do first-class passengers.

WOMEN × CLASS 2

$$\exp(-1.594815 + .4203889^* sex = 0)$$

$$\exp(-1.594815) = 0.20294607$$

Among females, the odds of surviving are .2 as great for second-class passengers as they are for first-class passengers, or
Among females, second-class passengers have 1/5 the odds of surviving as do first-class passengers.

MEN × CLASS 3

$$\exp(-3.726095 + 2.801977^* sex = 1)$$

$$\exp(-3.726095 + 2.801977) = 0.39688131$$

Among males, the odds of surviving are .4 as great for third-class passengers as they are for first-class passengers, or,
Among males, third-class passengers have .4 the odds of survival as do first-class passengers.

WOMEN × CLASS 2

$$\exp(-3.726095 + 2.801977^*\text{sex} = 0)$$

$$\exp(-3.726095) = 0.02408671$$

Among females, the odds of surviving are .02 as great for third-class passengers as they are for first-class passengers, or,

Among females, third-class passengers have 2 percent the odds of survival as do first-class passengers.

The interaction of *sex* and *class* may be graphed using the same methods employed in the previous section. See Figure 6.3.

```
. qui xi3: logit survived age i.sex*i.class
. postgr3 sex, by(class) title("Interaction of Class Level
and Gender") lpattern(longdash_dot shortdash_dot longdash_
star)
```

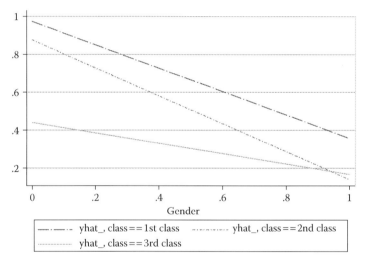

FIGURE 6.3
Interaction of class level and gender.

We find that second- and third-class passengers interact with *sex* at the high probability male level. A likelihood ratio test (not shown) of the main effects and interaction model result in a *Chi2* of 377.23 and a *p*-value of $p > chi2 = 0.0000$. Note that we did not statistically look at the relationship of levels 2 and 3 in this section, but rather of levels 2 and 1 and 3 and 1. Changing level 2 to be the reference level will allow us to quantify the interaction.

Near the beginning of this section, we found that an interaction between *age* and *class* is most likely significant. When making such a statement, however, it means that both levels of *class* significantly interact with the binary predictor, *age*. We cannot tell if a significant interaction exists between a binary and categorical predictor simply by looking at the *p*-value significance of the interaction term. Each level needs to be examined individually and assessed for its significance. We shall discuss this aspect of handling interactions in the following section. In any case, it should be apparent that we use the same methods for the *age* × *class* interaction as used for the *sex* × *class* interaction.

Recall that we modeled *survived* on *sex class2 class3* and the two *sex-class* interactions. For constructing an *age-class* interaction, we can instead model *survived* on *age class2 class3* and two *age-class* interactions. Moreover, whereas we did not include *age* in the previous model, it makes no difference to the interactive statistics if we had. We first create the *age-class* interactions.

```
. gen byte acl2 = age*class2
. gen byte acl3 = age*class3
```

Now we model *survived* on the main effects terms and interactions.

```
. logit survived age class2 class3 acl2 acl3, nolog
```

```
Logistic regression                    Number of obs   =         1316
                                        LR chi2(5)      =       187.48
                                        Prob > chi2     =       0.0000
Log likelihood = -779.63885            Pseudo R2       =       0.1073
```

survived	Coef.	Std. Err.	z	P>|z|	[95% Conf. Interval]
age	-17.8419	.	.	.	. .
class2	2.71e-06	.1729138	0.00	1.000	-.3389021 .3389075
class3	-18.97649	.2637056	-71.96	0.000	-19.49334 -18.45963
acl2	-1.053884	.	.	.	. .
acl3	17.34916	.2549336	68.05	0.000	16.8495 17.84883
_cons	18.32108	.1152079	159.03	0.000	18.09528 18.54688

```
Note: 0 failures and 30 successes completely determined.
```

A problem! From the note, it appears that the model perfectly predicts a response level. To determine the source of the problem, we sort by the response variable, *survived*, and tabulate the two main effect terms, *age* and *class*.

```
. sort survived
. by survived: tab age class
```

```
----------------------------------------------------------------
-> survived = no

Age (Child |                class (ticket)
vs Adult)  | 1st class   2nd class   3rd class |     Total
-----------+-----------------------------------+----------
    child  |         0           0          52 |        52
   adults  |       122         167         476 |       765
-----------+-----------------------------------+----------
    Total  |       122         167         528 |       817


----------------------------------------------------------------
-> survived = yes

Age (Child |                class (ticket)
vs Adult)  | 1st class   2nd class   3rd class |     Total
-----------+-----------------------------------+----------
    child  |         6          24          27 |        57
   adults  |       197          94         151 |       442
-----------+-----------------------------------+----------
    Total  |       203         118         178 |       499
```

We can immediately observe the source of the problem. There were no children in first or second class who died. The only children who died were in third class—and 52 of them died. A total of 57 children from all classes survived. Our initial suspicion that *age-class* would be a significant interaction proves to be mistaken. However, if there were sufficient values in the problem cells, we would construct and interpret interactions as we did for *sex-class*.

We now turn to Binary × Continuous interactions, which perhaps appear more often in research reports and journal articles than any other interaction.

6.4 Binary × Continuous Interactions

6.4.1 Notes on Centering

Before discussing the construction and interpretation of Binary × Continuous interactions, it is important to elaborate a bit on centering. Some authors argue that a continuous predictor should be centered if it makes no sense to have a 0 value as its minimum value. For instance, if a study is being performed on secondary school teachers and one of the predictors is "years of education," we do not expect to have 0 as a realistic value. In addition, a patient's systolic blood pressure will not have a 0 value—unless the patient is dead, of course. On the other hand, temperature values in terms of the Kelvin scale begin with 0 (absolute zero). Years of education among the population as a whole can also begin with 0, indicating those who have undergone no formal education. Height and weight are common variables, which statisticians center before constructing an interaction with these terms included. Both are also commonly centered, regardless of being made part of an interaction.

Aside from assisting in the interpretation of a continuous predictor, some statisticians have argued that centering reduces multicollinearity in a model, particularly when the continuous predictor does not take values of 0 (Aiken and West, 1991). Collinearity exists in a model when one or more predictors are highly correlated; interaction terms are indeed correlated with their main effects values. On the other hand, Echambadi and Hess (2007) have demonstrated that mean-centering fails to ameliorate the multicollinearity problem and that any collinearity resulting from interactions remains, regardless of any centering adjustment made to the predictors.

The benefit, therefore, for centering rests in the interpretation of interactions. Not all statisticians employ centering when constructing interactions, but some believe it is important. Before addressing the viability of centering, though, we first need to define what is meant by the term.

There are various types of centering. The most common is mean-centering. Median-centering is also used in research, as is centering on some predefined value. Mean-centering is emphasized in this section.

Creating a mean-centered continuous variable is simple. First, determine the mean value of the variable and subtract it from every value of the variable. Using *weight* as an example, we would center it using the following logic:

$$center\_weight = weight - mean(weight) \tag{6.9}$$

In Stata, using the **auto** data set that comes with the software, we can center and model the resultant mean-adjusted *weight*, which we name *c_weight*, using the following code.

```
. use auto
. sum weight

Variable |     Obs        Mean    Std. Dev.      Min        Max
---------+--------------------------------------------------------
  weight |      74    3019.459    777.1936       1760       4840

. di r(mean) /// this command is not needed for centering, but is
3019.4595    /// displayed showing r(mean) = mean of weight
. gen c_weight = weight - r(mean)

. l weight c_weight in 1/5
```

```
        +--------------------+
        | weight   c_weight  |
        |--------------------|
   1.   |  2,930   -89.45946 |
   2.   |  3,350    330.5405 |
   3.   |  2,640   -379.4595 |
   4.   |  3,250    230.5405 |
   5.   |  4,080    1060.541 |
        +--------------------+
```

Modeling *foreign* by *weight* and by *c_weight* produce the same coefficient and standard errors, but differing intercepts. *foreign* is a binary variable with 1 = foreign-made car, 0 = domestic-made car. We shall also calculate the linear predictor for each model, comparing them afterward.

. glm foreign weight, nolog nohead fam(bin)

```
-----------------------------------------------------------------------
foreign |      Coef.   Std. Err.      z   P>|z|   [95% Conf. Interval]
--------+--------------------------------------------------------------
 weight | -.0025874   .0006094   -4.25  0.000   -.0037817  -.001393
  _cons |  6.282599   1.603967    3.92  0.000    3.138882  9.426316
-----------------------------------------------------------------------
```

. predict xb_w, xb

. glm foreign c_weight, nolog nohead fam(bin)

```
-----------------------------------------------------------------------
 foreign  |      Coef.   Std. Err.      z   P>|z|   [95% Conf. Interval]
----------+------------------------------------------------------------
c_weight  | -.0025874   .0006094   -4.25  0.000  -.0037817   -.001393
   _cons  | -1.529913   .4248365   -3.60  0.000  -2.362577  -.6972483
-----------------------------------------------------------------------
```

. predict xb_cw, xb
. l xb* in 1/5

```
        +----------------------+
        |     xb_w      xb_cw  |
        |----------------------|
   1.   | -1.298446   -1.298446 |
   2.   | -2.385149   -2.385149 |
   3.   | -.5481037   -.5481037 |
   4.   |  -2.12641    -2.12641 |
   5.   | -4.273942   -4.273942 |
        +----------------------+
```

The linear predictors of the two models are identical, even with different intercepts. The mean-adjustment to *weight* balances the difference in intercept values.

Recall that the meaning of an intercept relates to the estimated linear predictor if the model predictor(s) are equal to 0. For such a model, the intercept is the only remaining value contributing to the value of the linear predictor. For the centered predictor, though, the intercept is the value of the linear predictor if the predictor (*weight*) is at its mean. Interpretation of the coefficient and odds ratio is also in terms of its relationship to the mean. To demonstrate that this is the case, we calculate the mean value of the linear predictors from the mean-centered model.

```
. su xb_cw
```

```
Variable |     Obs       Mean   Std. Dev.       Min        Max
---------+-------------------------------------------------------
   xb_cw |      74  -1.529912   2.010901  -6.240356   1.728797
```

Note that the value –1.5299 is the intercept from the mean-centered model. The value of the mean-centered intercept is, therefore, the linear predictor for *weight* at its mean value. When serving as a predictor in a logistic model, *c_weight* is a measure of how far a subject's weight is from the mean weight of subjects in the model. *weight*, on the other hand, is simply a measure of a subject's weight based on a theoretical weight of 0. When these terms are in interaction, the same meaning attaches to the terms. If basing the interactive relationship makes more sense when values are related to the mean of the continuous predictor rather than to one that assumes a 0 value, then centering may be preferred. If no such relationship is obtained, then it may be preferred to keep the predictor unadjusted. Otherwise, there is apparently no statistical reason to prefer centering to non-centering, only interpretative ones.

For the examples we use in the text, I refrain from mean-adjusting continuous predictors. If we were actually modeling data for the purpose of publication or research, I would seriously consider doing so for several of the example models used throughout the text. For the sake of focusing on the issues under discussion, however, I decided against their inclusion.

We now return to considering the Binary × Continuous interaction, which, as I previously indicated, is perhaps the most used type of interaction. It is also generally used as the paradigm interaction for displaying examples of graphing interactions.

6.4.2 Constructing and Interpreting the Interaction

We shall use the **medpar** data to discuss the constructing and interpretation of Binary × Continuous interaction. First, we model the main effects terms, with *died* as the response and *white* and *los* as the binary risk factor and continuous predictor, respectively.

```
. logit died white los

Logistic regression                           Number of obs   =       1495
                                              LR chi2(2)      =      19.85
                                              Prob > chi2     =     0.0000
Log likelihood = -951.50591                   Pseudo R2       =     0.0103
------------------------------------------------------------------------------
    died |       Coef.   Std. Err.     z    P>|z|    [95% Conf. Interval]
---------+--------------------------------------------------------------------
   white |   .2526808   .2065523   1.22  0.221  -.1521542    .6575158
     los |  -.0299868   .0077037  -3.89  0.000  -.0450858   -.0148878
   _cons |  -.5986826   .2132683  -2.81  0.005  -1.016681   -.1806844
------------------------------------------------------------------------------
```

Parameterized as odds ratios we have:

```
------------------------------------------------------------------------------
    died | Odds Ratio Std. Err.     z    P>|z|    [95% Conf. Interval]
---------+--------------------------------------------------------------------
   white |   1.287472 .2659303   1.22  0.221  .8588558    1.929992
     los |   .9704583 .0074761  -3.89  0.000  .9559155    .9852224
------------------------------------------------------------------------------
```

Regardless of the nonsignificance of *white*, it still may be the case that it plays a significant role in the model when in interaction with *los*.

Having a continuous predictor in the model assumes that the slope or log-odds of the predictor is the same across the entire range of its values. In the case of *los*, the range is from 1 through 116. If there is a difference in how long patients stay in the hospital based on whether they are white or non-white, the slope of the continuous predictor is in fact not constant. We can check for this by categorizing *los* into four levels and then observing whether there is a constant odds ratio across all levels. If there is not, then an interaction term should be included in the model.

```
. gen byte losgrp = 1 if los<5
. replace losgrp = 2 if los>=5 & los<10
. replace losgrp = 3 if los>=10 & los<20
. replace losgrp = 4 if los>=20 & los!=.

. tab losgrp, gen(lgp)


     losgrp |       Freq.      Percent        Cum.
------------+-----------------------------------------
          1 |         376       25.15       25.15
          2 |         502       33.58       58.73
```

```
          3 |          481         32.17              90.90
          4 |          136          9.10             100.00
------------+------------------------------------------------
      Total |        1,495        100.00
```

```
. logit died lgp2-lgp4, nolog

Logistic regression                    Number of obs    =         1495
                                       LR chi2(3)       =        86.58
                                       Prob > chi2      =       0.0000
Log likelihood = -918.14114            Pseudo R2        =       0.0450

----------------------------------------------------------------------
    died |      Coef.  Std. Err.     z  P>|z|  [95% Conf. Interval]
---------+------------------------------------------------------------
    lgp2 |  -1.211274  .1452134  -8.34  0.000  -1.495887   -.9266607
    lgp3 |  -1.111399  .1450896  -7.66  0.000   -1.39577    -.827029
    lgp4 |  -.8310827  .2087024  -3.98  0.000  -1.240132   -.4220334
   _cons |   .1599144   .103472   1.55  0.122   -.042887    .3627158
----------------------------------------------------------------------
```

```
. glm died lgp2-lgp4, nolog fam(bin) eform nohead

----------------------------------------------------------------------
    died | Odds Ratio  Std. Err.     z  P>|z|  [95% Conf. Interval]
---------+------------------------------------------------------------
    lgp2 |  .2978177  .0432471  -8.34  0.000  .2240498     .3958734
    lgp3 |  .3290981  .0477487  -7.66  0.000  .2476423     .4373467
    lgp4 |  .4355774  .0909061  -3.98  0.000  .2893461     .6557121
----------------------------------------------------------------------
```

The slopes are clearly not the same across levels. Therefore, we should include an interaction of *white* and *los*.

As an aside, perhaps a more preferable way to check for an interactive effect of *white* and *los* is to stratify the above table of slopes by *white*.

WHITE = 1

```
. glm died lgp2-lgp4 if white==1, nolog fam(bin) nohead

----------------------------------------------------------------------
         |              OIM
    died |      Coef.  Std. Err.     z  P>|z|  [95% Conf. Interval]
---------+------------------------------------------------------------
    lgp2 |  -1.266584  .1512953  -8.37  0.000  -1.563117   -.9700507
    lgp3 |  -1.165982  .1513857  -7.70  0.000  -1.462693   -.8692721
```

```
  lgp4 |   -.8494984    .2213825   -3.84 0.000    -1.2834  -.4155967
  _cons |    .2192651    .1078562    2.03 0.042    .0078708   .4306593
-----------------------------------------------------------------
```

WHITE = 0

```
. glm died lgp2-lgp4 if white==0, nolog fam(bin) nohead

-----------------------------------------------------------------
          |                OIM
    died  |     Coef.    Std. Err.    z   P>|z|  [95% Conf. Interval]
----------+------------------------------------------------------
   lgp2  |  -.5108256    .5374838  -0.95 0.342  -1.564275   .5426234
   lgp3  |  -.4155154    .5289593  -0.79 0.432  -1.452257   .6212258
   lgp4  |  -.3677248    .6576311  -0.56 0.576  -1.656658   .9212084
  _cons  |  -.5877867    .3944053  -1.49 0.136  -1.360807   .1852336
-----------------------------------------------------------------
```

It is even more evident that an interaction term is in order.

We create an interaction term named *whlo*, including it in the model with *white* and *los*.

```
. gen byte whlo = white*los
. logit died white los whlo

Logistic regression                      Number of obs   =       1494
                                         LR chi2(3)      =      24.95
                                         Prob > chi2     =     0.0000
Log likelihood = -948.53533              Pseudo R2       =     0.0130
-----------------------------------------------------------------
    died  |     Coef.    Std. Err.    z   P>|z|  [95% Conf. Interval]
----------+------------------------------------------------------
  white  |   .7703072    .2956605    2.61 0.009   .1908233   1.349791
    los  |   .0100189    .0161884    0.62 0.536  -.0217098   .0417475
   whlo  |  -.0476866    .018308    -2.60 0.009  -.0835696  -.0118036
  _cons  |  -1.048343    .280243    -3.74 0.000  -1.597609  -.4990767
-----------------------------------------------------------------
```

The odds ratios are

```
-----------------------------------------------------------------
    died  | Odds Ratio  Std. Err.    z   P>|z|  [95% Conf. Interval]
----------+------------------------------------------------------
  white  |    2.16043   .6387538    2.61 0.009  1.210246    3.85662
    los  |   1.010069   .0163514    0.62 0.536  .9785242   1.042631
   whlo  |    .9534326   .0174554   -2.60 0.009  .9198271    .9882658
-----------------------------------------------------------------
```

We see that there is an apparently significant interactive effect between *white* and *los*. As observed, however, when handling Binary × Binary interactions, the coefficient of the interaction is only part of the meaning of the interaction, and the interaction term itself plays no role in the interpretation itself.

6.4.3 Interpretation

Again, the value listed in the table of odds ratios above is **not** the odds ratio of the interaction term, nor is the value given as the (log-odds) coefficient of *whlo* in fact **the** coefficient.

The model, with the *whlo* interaction coefficient included, appears as:

```
----------------------------------------------------------------------
    died |     Coef.   Std. Err.     z   P>|z|    [95% Conf. Interval]
---------+------------------------------------------------------------
   white |   .7703072   .2956605   2.61   0.009    .1908233    1.349791
     los |   .0100189   .0161884   0.62   0.536   -.0217098    .0417475
    whlo |  -.0476866    .018308  -2.60   0.009   -.0835696   -.0118036
   _cons |  -1.048343    .280243  -3.74   0.000   -1.597609   -.4990767
----------------------------------------------------------------------
```

When we have an interaction with *los*, we are saying that the slope of *los* is different depending on the level of the risk factor, in this case the binary *white*. *White* has a different slope, and therefore, a different odds ratio, when *los* = 1 than when it is 10, 20, 30, 40, and so forth, up to 116 days. The coefficient, and therefore the odds ratio of the interaction, depends on the value of *los*. The formula to determine the log-odds ratio of the interaction is shown in Table 6.1.

TABLE 6.1

Formula to Calculate Binary Continuous Odds Ratio

B1 = slope or coefficient of binary predictor

B2 = slope of continuous predictor

B3 = slope of interaction of binary and continuous

x = value of continuous predictor

$$LOR = \beta_1(r_1-r_0) + \beta_3(r_1-r_0) \,^*x$$
$$= \beta_1 + \beta_3 \,^*x$$
$$OR = \exp(LOR)$$

To demonstrate the equality of the two formulae for LOR in the above table, use the following commands:

```
. gen lor1 = (.7703073*(white==1-white==0)) -
.0476866*(white==1 - white==0)*los
```

```
. gen lor = .7703073 -.0476866*los

. l lor1 lor in 1/5

     +---------------------+
     |      lor1          lor |
     |---------------------|
  1. | .5795609      .5795609 |
  2. | .3411279      .3411279 |
  3. | .6272475      .6272475 |
  4. | .3411279      .3411279 |
  5. | .7226207      .7226207 |
     +---------------------+
```

Using the above formula on the model, we have:

$$OR = \exp(.7703 + (-.0477*x))$$

When LOS = 20 we have:

$$\exp(.7703 + (-.0477*20)) = .8322$$

Other odds ratios can be calculated using the calculator built into most statistical software. Using Stata, we have, for example:

LOS 21:

```
. di exp(.7703 + (-.0477*21))
.79342203
```

Given the formula to calculate probability values based on odds ratios,

$$P = OR / (1 + OR) \qquad\qquad (6.10)$$

we have the following summary table:

LOS	OR	PROB	LOS	OR	PROB
1	2.06	.673			
10	1.34	.573	60	0.12	.110
20	0.83	.454	70	0.08	.071
30	0.52	.341	80	0.05	.045
40	0.32	.243	90	0.03	.029
50	0.20	.166	100	0.02	.018

6.4.4 Standard Errors and Confidence Intervals

Also important when calculating interaction terms is the modification that must also be given to the standard error and confidence intervals. We address interaction term standard errors and confidence intervals at this point but appreciate that the same logic applies to interactions of Binary × Binary, Binary × Categorical, and other interactions as well.

Recall that for interactions there is no one coefficient. For Binary × Continuous predictors, the coefficient varies with the values of the continuous predictor, based on the values of the coefficients of the binary and interaction terms, and by the values of the binary main effect. The standard error also requires the variance covariance matrix, displayed in Stata as:

```
. corr, _coef cov

             |    white       los      whlo     _cons
-------------+--------------------------------------------
       white |   .087415
         los |   .003223   .000262
        whlo |  -.003864  -.000262   .000335
       _cons |  -.078536  -.003223   .003223   .078536
```

The formula for the variance of the interaction coefficient is

$$\text{Variance} = (r_1 - r_0)^2 * V(\beta_1) + x(r_1 - r_0)^2 * V(\beta_3) + 2x(r_1 - r_0)^2 * CV(\beta_1, \beta_3) \quad (6.11)$$

The standard error is then:

$$SE = \text{sqrt}[(r_1 - r_0)^2 * V(\beta_1) + x(r_1 - r_0)^2 * V(\beta_3) + 2x(r_1 - r_0)^2 * CV(\beta_1, \beta_3)] \quad (6.12)$$

Where V is the variance and CV the covariance from the model variance-covariance table. When the interaction is Binary × Continuous, the above formula can be reduced to the simpler:

$$SE[B \times C] = \text{sqrt}[V(\beta_1) + x^2 V(\beta_3) + 2 * x * CV(\beta_1, \beta_3)] \quad (6.13)$$

The variance values of β_1 and β_3 are, respectively, .087415 and .000335. The covariance of (β_1, β_3) is −.003864. x is the value of the continuous main effect, in this case *los*.

```
. gen se = sqrt(.087415 + los^2 * .000335+ 2*los*(-.003864))
```

To determine the standard error of the interaction when LOS = 1,

```
. di sqrt(.087415 + 1^2 * .000335+ 2*1*(-.003864))
.2828816
```

The corresponding confidence intervals are given as:

$$CI[B \times C] = (\beta_1 + \beta_3 * x) +/- z_{1-\alpha/2} SE[B \times C] \qquad (6.14)$$

where $z_{1-\alpha/2}$ is the percent confidence interval based on a standard normal distribution. The traditional value for the latter is a 95 percent confidence interval, which is a z of 1.96. This value is sometimes expressed as $\alpha = 0.05$.

The lower and upper confidence intervals are then calculated as:

LOWER CI : 95%

```
. gen ci_lo1 = (.7709236 - .0477605*los) - 1.96* sqrt(.087415
+ los^2 * .000335+ 2*los*(-.003864))
```

HIGHER CI : 95%

```
. gen ci_hi1 = (.7709236 - .0477605*los) + 1.96* sqrt(.087415
+ los^2 * .000335+ 2*los*(-.003864))
```

FOR LOS = 1
COEFFICIENT

```
. di   (.7709236 - .0477605*1)
.7231631
```

95% CONFIDENCE INTERVALS AT LOS = 1

```
. di (.7709236 - .0477605*1) - 1.96* sqrt(.087415 + 1^2 *
.000335 + 2*1*(.003864))
.16871516
```

```
. di (.7709236 - .0477605*1) + 1.96* sqrt(.087415 + 1^2 *
.000335 + 2*1*(-.003864))
1.277611
```

Table 6.2 summarizes other coefficients and confidence intervals for the interaction *white* × *los*.

TABLE 6.2

Summary Table for Interaction *white* × *los*

LOS	Coef	SE	CI-Lo	CI-Hi
1	.72316	.28288	.16872	1.27761
5	.53212	.23906	.06356	1.00068
10	.29332	.20889	−.11611	.70274
15	.05452	.21649	−.36981	.47885
20	−.18429	.25856	−.69107	.32250
40	−1.13950	.56062	−2.23831	−.04068

We explained earlier that the odds ratio of various interaction effects are calculated as their exponentiation; in other words, $\exp[\beta_1(r_1-r_0) + \beta_3(r_1-r_0) *x]$, or for Binary × Continuous interactions, $\exp[\beta_1 + \beta_3*x]$. Likewise, the confidence intervals are exponentiated to obtain the appropriate values for the odds ratios of the interaction. The odds ratio of the example interaction at $los = 1$ is therefore 2.06094, with a lower 95 percent *CI* of 1.18378 and an upper 95 percent *CI* of 3.26671. Categorical × Continuous interactions are based on the same logic as Binary × Continuous interactions, with calculations for each level of the categorical effect.

6.4.5 Significance of Interaction

The significance of the exponentiated interaction can be determined if the confidence interval includes 1.0. If it does, the predictor is not significant. Likewise, if the interaction is left unexponentiated and the confidence interval includes 0, it is not significant. Since the confidence interval for the IRR of *mxa* at age 30 does not include 1.0, it is significant at the $p = 0.05$ level.

Another way to evaluate the significance of the interaction at a given value of the continuous predictor is to divide the coefficient by the *SE*. Doing so produces the z- or Wald statistic—how you refer to it is based on the statistic used to calculate the *p*-value. For the z-statistic, it is assumed that it is distributed normally. However, as a two-sided test, you divide the resultant value by two. The Wald statistic, used in SAS, is identical to *t* squared and is also employed to determine a predictor *p*-value. The probability of the interaction having a z-statistic as large as 1.404184 for the *whlo* interaction with $los = 10$ can be calculated as follows:

[Note: From the table of coefficients, *SEs* and *CIs* given on the previous page, we use the values of the coefficient and standard error at $los = 10$.]

```
. di .29332 / .20889
1.404184

. di normprob(1.404184)/2
.45993398
```

We find that the coefficient is not significant, as could be observed in the above table. When the confidence interval includes 0, the coefficient is not significant.

6.4.6 Graphical Analysis

We next address graphing of Binary × Continuous interactions. For this purpose, we shall use a simple scatter graph. More complex graphs may be developed. I should mention here that several libraries of graphical commands exist in Stata, and likewise for R. Also, there are excellent texts written using both Stata and R, especially if you wish to construct more complicated graphs than displayed here. With respect to Stata graphic libraries, the **postgr3** system developed by J. Scott Long (Univ. of Indiana)—which was used in the previous two sections—and the **viblmgraph** commands authored by Michael Mitchell (U.C.L.A.) are recommended. The book by Mitchell and Chen (2008), *A Visual Guide to Stata Graphics*, is also recommended. Paul Murrell's 2006 text, *R Graphics*, has already become a classic and is recommended for R users. New texts related to R are being published with considerable frequency; you are advised to search the sites of the foremost publishers of statistics texts for new books related to statistical graphing. This same advice holds for users of other software applications, e.g., SAS, SPSS.

When an interaction of *white* and *los* is constructed for the model, we wish to learn if length of hospital stay (LOS) is different for those who identify themselves as white than for those who identify themselves otherwise. And, of course, we want to see if the interaction significantly contributes to the model. To reiterate what has been expressed in previous sections, it is important to remember that when an interaction term is entered into a model, each main effect must also be entered. We have observed thus far in this chapter that the parameter estimates of the main effects terms cannot be interpreted the same as when there is no interaction—they are not independent of one another. They nevertheless must be entered into the model. A definition of the main effects terms when they are used in an interaction was given earlier in the chapter.

We again display the model with main effects and interaction.

```
. logit died white los whlo, nolog

Logistic regression                      Number of obs   =        1495
                                         LR chi2(3)      =       25.78
                                         Prob > chi2     =      0.0000
Log likelihood = -948.54483              Pseudo R2       =      0.0134
------------------------------------------------------------------------
    died |      Coef. Std. Err.      z  P>|z| [95% Conf. Interval]
---------+--------------------------------------------------------------
   white |   .7709236 .2955977    2.61 0.009    .1915627  1.350285
     los |   .0100189 .0161884    0.62 0.536   -.0217098  .0417475
➜   whlo |  -.0477605  .018294   -2.61 0.009 ➜ -.083616  -.011905
   _cons |  -1.048343  .280243   -3.74 0.000  -1.597609 -.4990767
------------------------------------------------------------------------
```

The interaction *appears* to significantly contribute to the model. However, as discussed earlier in this section, there are as many coefficients to the interaction as values of the continuous predictor, *los,* and levels of the binary predictor. The value that appears in the output is not a coefficient in the same sense as are coefficients in models without interactions.

The linear predictor of the above model, or $x\beta$, is calculated and then separated by *white* using the code displayed below. It is understood that each pattern of covariates has a distinct linear predictor. When there are continuous predictors, it is possible that there are as many distinct values of the linear predictor as observations in the model.

```
. predict xb, xb
. separate xb, by(white)
```

```
                  storage   display    value
variable name      type     format     label     variable label
------------------------------------------------------------------
xb0                float    %9.0g                 xb, white == 0
xb1                float    %9.0g                 xb, white == 1
```

```
. save medparint    /// save the data to keep for later analysis
```

We can use the results to create a scatter plot of the linear predictor value and *los* by race (*white*), as shown in Figure 6.4.

```
. scatter xb0 xb1 los, connect(1[-]1) symbol(ii)  xlabel(0 10
to 100) sort l1title(Predicted logit) title(Interaction of
White and LOS)
```

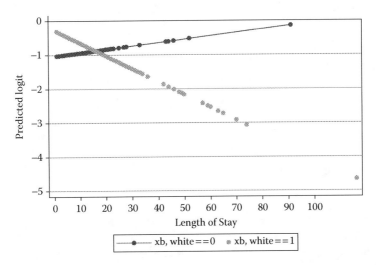

FIGURE 6.4
Interaction of white and LOS.

Notice the reverse effect of *white* versus non-*white*. There does appear to be a significant difference in the effect of the two levels of *white*.

We can also construct a similar scatter plot based on predicted or fitted values, that is, μ, as shown in Figure 6.5.

```
. drop xb xb1 xb0
. predict yhat, pr
. separate yhat, by(white)
```

```
                 storage  display   value
variable name    type     format    label      variable label
---------------------------------------------------------------
yhat0            float    %9.0g                 yhat, white == 0
yhat1            float    %9.0g                 yhat, white == 1
```

```
. scatter yhat0 yhat1 los, connect(1-[-]1) symbol(ii)
xlabel(0 10 to 100) l1title(Probability) title(Interaction of
White on LOS)
```

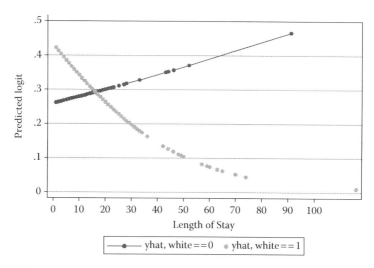

FIGURE 6.5
Interaction of white and LOS.

We see the same relationship. The two graphs above give a general visual representation of the differential relationship of *los* and *white*. This is the point of such graphs. To calculate more specific information, the researcher must employ quantitative methods.

6.5 Categorical × Continuous Interactions

We next extend discussion from Binary × Continuous interactions to Categorical × Continuous interactions. The logic is the same between the two, but the categorical aspect of the relationship can result in a bit messier model, with a correspondingly more tedious interpretation.

For this example interaction, we use the **heart01** data set, with patient killip level at admission employed as the categorical predictor and *age* as the continuous predictor. A tabulation of killip level is given as:

```
. tab killip

    Killip |
     level |      Freq.     Percent        Cum.
-----------+-----------------------------------
         1 |      3,768       73.12       73.12
         2 |      1,029       19.97       93.09
         3 |        292        5.67       98.76
         4 |         64        1.24      100.00
```

Higher killip levels indicate increasing cardiac impairment, with level 4 being a myocardial infarct.

There are comparatively few observations for Level 4—1¼ percent. This level represents patients who are admitted while experiencing a myocardial infarction or some other major cardiovascular trauma.

A logistic regression of killip level alone shows increasing odds of death for each level. The fact that there are comparatively few patients classified as killip level 4 is reflected in the wide confidence interval for its odds ratio.

```
. tab killip, gen(kk)
. logit death kk2 kk3 kk4, nolog or

Logistic regression                     Number of obs    =        5153
                                        LR chi2(3)       =      150.51
                                        Prob > chi2      =      0.0000
Log likelihood = -849.28983             Pseudo R2        =      0.0814

-----------------------------------------------------------------------
   death | Odds Ratio  Std. Err.    z   P>|z|  [95% Conf. Interval]
---------+-------------------------------------------------------------
     kk2 |   2.804823   .4497351  6.43  0.000   2.048429   3.840521
     kk3 |   4.333438   .9467527  6.71  0.000   2.824005   6.649666
     kk4 |   27.61688  7.545312  12.15  0.000   16.16645   47.17747
-----------------------------------------------------------------------
```

The table of parameter estimates can be displayed as:

```
. glm death kk2-kk4, nolog fam(bin) nohead

------------------------------------------------------------------
  death |     Coef.  Std. Err.     z  P>|z|  [95% Conf. Interval]
--------+---------------------------------------------------------
    kk2 |  1.031341  .1603435   6.43 0.000   .7170731   1.345608
    kk3 |  1.466361  .2184761   6.71 0.000   1.038156   1.894567
    kk4 |  3.318427  .2732137  12.15 0.000   2.782938   3.853916
   _cons | -3.633508 .1028673 -35.32 0.000  -3.835125  -3.431892
------------------------------------------------------------------
```

We add *age* to the model:

```
. logit death age kk2 kk3 kk4, nolog

Logistic regression                     Number of obs   =       5153
                                        LR chi2(4)      =     248.34
                                        Prob > chi2     =     0.0000
Log likelihood = -800.37241             Pseudo R2       =     0.1343


------------------------------------------------------------------
  death |     Coef.  Std. Err.     z  P>|z|  [95% Conf. Interval]
--------+---------------------------------------------------------
    age |  .0497254  .0051881   9.58 0.000   .0395569    .059894
    kk2 |   .874362  .1630111   5.36 0.000    .554866   1.193858
    kk3 |  1.098789  .2243805   4.90 0.000   .6590115   1.538567
    kk4 |  3.081978   .287323  10.73 0.000   2.518835   3.645121
   _cons | -7.080451 .3980213 -17.79 0.000  -7.860558  -6.300343
------------------------------------------------------------------
```

It appears that *age* does affect killip level, but only employing interactions into the model will let us know for certain. We create interactions of each level in case we need to change reference levels. Initially, we use the first level as the reference level. Note that *age* has values ranging from 40 to 100.

```
. gen byte agek1 = age * kk1

. gen byte agek2 = age * kk2

. gen byte agek3 = age * kk3

. gen byte agek4 = age * kk4


. logit death age kk2 kk3 kk4 agek2 agek3 agek4, nolog
```

```
Logistic regression                      Number of obs   =        5153
                                         LR chi2(7)      =      259.93
                                         Prob > chi2     =      0.0000
Log likelihood = -794.57564              Pseudo R2       =      0.1406
```

```
-----------------------------------------------------------------------
  death |     Coef.  Std. Err.     z    P>|z|   [95% Conf. Interval]
--------+--------------------------------------------------------------
    age |  .0635425  .0074949    8.48   0.000   .0488529   .0782322
    kk2 |    2.0918   .917526    2.28   0.023   .2934823   3.890118
    kk3 |   3.88023  1.294041    3.00   0.003   1.343956   6.416504
    kk4 |  7.320069  1.411028    5.19   0.000   4.554504   10.08563
  agek2 | -.0161336  .0118304   -1.36   0.173  -.0393207   .0070536
  agek3 | -.0359754   .016556   -2.17   0.030  -.0684246  -.0035262
  agek4 | -.0569094  .0187367   -3.04   0.002  -.0936326  -.0201861
   _cons | -8.120975  .5828596  -13.93   0.000  -9.263358  -6.978591
-----------------------------------------------------------------------
```

6.5.1 Interpretation

The same logic holds for Categorical × Continuous interactions as for Binary × Continuous interactions, except that there are levels of binary variables, each of which refers to a single reference level.

As discussed in the previous section, the formula used for a Binary × Continuous interaction is

$$IRR_{BxC} = \exp[\beta_1 + \beta_3 x] \qquad (6.15)$$

In a similar manner to Binary × Continuous interactions, the coefficient of each interaction is not the value used directly in interpretation. We calculate the interaction values for each nonreference level of the categorical predictor, as follows.

The coefficients for the above model are specified, for the purpose of interpreting interaction effects, as:

$\beta_1 = age$

$\beta_2 = kk2; \beta_3 = kk3; \beta_4 = kk4$

$\beta_5 = agek2; \beta_6 = agek3; \beta_7 = agek4$

$\beta_0 = intercept$

$agekk2 \quad : \quad \beta_{CatxCon2} = \beta_2 + \beta_5 x$

$\beta_{CatxCon2} = 2.0918 - 0.0161336 * age$

$agekk3 \quad : \quad \beta_{CatxCon3} = \beta_3 + \beta_6 x$

$$\beta_{\text{CatxCon3}} = 3.88023 - 0.0359754 * \text{age}$$

$$agekk4 \quad : \quad \beta_{\text{CatxCon3}} = \beta_4 + \beta_7 x$$

$$\beta_{\text{CatxCon3}} = 7.320069 - 0.0569094 * \text{age}$$

In order to obtain the odds ratios, simply exponentiate each of the above formulae.

For *agek2*, the OR may be calculated for *age* 65 by:

AGE 65

```
. di exp(2.0918-0.0161336*65)
2.8380466
```

OR values for *agek2*, *agek3*, and *agek4* can be calculated as shown in Table 6.3:

TABLE 6.3

Odds Ratios: Age × Killip Levels

AGE	agek2	agek3	agek4
40	4.25	11.49	155.04
50	3.62	8.02	87.76
60	3.08	5.59	49.68
70	2.62	3.90	28.12
80	2.23	2.72	15.92
90	1.90	1.90	9.01

The fact that killip level 4 has only 1.24 percent of the cases clearly bears on the odds ratios of both the main effect and any interaction of which it is a part. In a real study, an adjustment would need to be made to the levels. However, for our purpose in showing an example of how to construct and interpret interactions, it is fine.

The odds ratio of *agek2* at *age* 60 can be expressed as:

Patients *age* 60 who are rated as killip 2 increase the odds of death (within 48 hours of admission) by three times compared to similar patients who are rated at killip 1. That is, 60-year-old killip 2 patients have three times the odds of death compared to killip level 1 patients.

Other interactions are expressed in the same manner. If the reference level is altered, though, the data must be remodeled and new coefficients determined.

Note that the number of coefficients for a Categorical × Continuous interaction is M-1 × values of the continuous predictor, where M is the number of levels in the categorical variable. When reporting odds ratio for this type of interaction, it is preferable to categorize the results as displayed on the previous page. Of course, if we are interested in the odds ratio for a given level and value for the continuous predictor, then we may present a specific odds ratio. How we display results largely depends on the nature and purpose of the study.

6.5.2 Standard Errors and Confidence Intervals

The standard errors and confidence intervals for Categorical × Continuous interactions are calculated in a similar manner as they are for Binary × Continuous interactions.

6.5.3 Graphical Representation

Given the example model, and the Categorical × Continuous interaction we developed, it is instructive to graph the increasing probabilities of *death* for each killip level. Prior to looking at the interactive effects, we observe the main effects relations. To do this, we first need to calculate the probability of *death* for each patient.

```
. predict mu, pr      /// it may be necessary to drop previous
                          instances of mu*
```

Next, we separate the probabilities by killip level.

```
. separate mu, by(killip)
```

```
              storage  display   value
variable name   type    format   label        variable label
-------------------------------------------------------------
mu1            float    %9.0g                  mu, killip == 1
mu2            float    %9.0g                  mu, killip == 2
mu3            float    %9.0g                  mu, killip == 3
mu4            float    %9.0g                  mu, killip == 4
```

Now we can graph the relationships. See Figure 6.6.

```
. scatter mu1 mu2 mu3 mu4 age, title("Probability of death:
Killip Level by Age ") msymbol(Oh x o s) l1title("Prob of
Death")
```

The ranges of probability for each level match well with the calculated odds ratios:

2.4, 3.0, 21.8

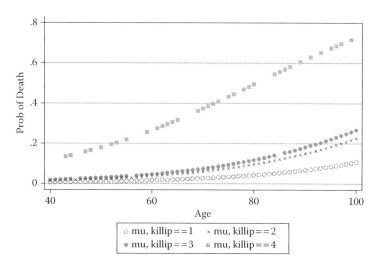

FIGURE 6.6
Probability of death, *killip* level by age.

```
. glm death age kk2 kk3 kk4, nolog eform fam(bin) nohead

-----------------------------------------------------------------
             |                  OIM
    death | Odds Ratio Std. Err.    z    P>|z|  [95% Conf. Interval]
-------+---------------------------------------------------------
     age |   1.050982 .0054528   9.58  0.000   1.040349    1.061724
     kk2 |   2.397345 .3908015   5.36  0.000   1.741697    3.299807
     kk3 |   3.000531 .6732703   4.90  0.000   1.932869     4.65794
     kk4 |   21.80148 6.264205  10.73  0.000   12.41397    38.28787
-----------------------------------------------------------------
```

We now estimate without *age* in the model, observing the effect on each killip level.

```
. glm death kk2 kk3 kk4, nolog eform fam(bin) nohead

-----------------------------------------------------------------
             |                  OIM
    death | Odds Ratio Std. Err.    z  P>|z|  [95% Conf. Interval]
-------+---------------------------------------------------------
     kk2 |   2.804823 .4497351   6.43 0.000   2.048429    3.840521
     kk3 |   4.333438 .9467527   6.71 0.000   2.824005    6.649666
     kk4 |   27.61688 7.545312  12.15 0.000   16.16645    47.17747
-----------------------------------------------------------------
```

Due to the fact that there appears to be a confounding effect between *age* and killip level, we create an interaction between them. We earlier created the requisite interactions by multiplying *age* by each nonreference level of killip. This time, however, we shall let Stata create the interaction terms.

```
xi: logistic death i.killip*age, nolog
i.killip        _Ikillip_1-4    (naturally coded; _Ikillip_1 omitted)
i.killip*age  _IkilXage_#  (coded as above)

Logistic regression                    Number of obs     =        5153
                                       LR chi2(7)        =      259.93
                                       Prob > chi2       =      0.0000
Log likelihood = -794.57564            Pseudo R2         =      0.1406
------------------------------------------------------------------------
      death | Odds Ratio Std. Err.     z  P>|z|  95% Conf. Interval]
------------+-----------------------------------------------------------
 _Ikillip_2 |   8.099483  7.431486  2.28 0.023  1.341089  48.91666
 _Ikillip_3 |   48.43537  62.67737  3.00 0.003  3.834182  611.8606
 _Ikillip_4 |   1510.308  2131.087  5.19 0.000  95.05963  23995.78
        age |   1.065605  .0079866  8.48 0.000  1.050066  1.081374
 _IkilXage_2 |  .9839959   .011641 -1.36 0.173  .9614424  1.007079
 _IkilXage_3 |   .964664   .015971 -2.17 0.030  .9338639   .99648
 _IkilXage_4 |  .9446797  .0177002 -3.04 0.002  .9106173  .9800163
------------------------------------------------------------------------
```

We drop the previous value of *mu*, and create values based on the model including *age*.

```
. drop mu*
. predict mu, pr
. separate mu, by(killip)
```

The scatter graph in Figure 6.7 clearly displays interactions between levels.

```
. scatter mu1 mu2 mu3 mu4 age, title("Probability of death:
Age by Killip Level, w Interactions") msymbol(Oh x o s)
```

We can clearly observe the interaction of levels 2 and 3, with the two crossing at approximately age 90. Moreover, level 3 approaches level 1 after the age of 80. Level 4 shows a substantially higher, and noninteractive, odds of death than any of the lower levels. For all four levels, there is a steady increase of risk of death with older age.

This type of graphic is truly valuable in *seeing* the relationships between levels. It is frequently used in research reports, with good effect.

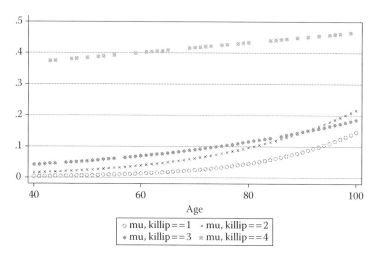

FIGURE 6.7
Probability of death, age by *killip* level, with interactions.

6.6 Thoughts about Interactions

Interactions are important to the modeling process. However, care must be taken when they are employed. Remember to keep the main effects of the interaction in the model. It is also advised that, if you are using more than one interaction in a model, to avoid overlap in the use of main effects terms. That is, it is preferred not to use the same main effect in more than one interaction. The confounding effect can be difficult to handle. If there are two interactions utilizing a common main effect, then consider a three-way interaction. For instance:

Interaction 1: $A \times B \Rightarrow AB$
Interaction 2: $A \times C \Rightarrow AC$

A three-way interaction would be between:

$$A \times B \times C$$

with A as the linking term in the interaction.

Some researchers check for all possible interactions in a model. This can be a tiring task when there are a large number of predictors. It is also difficult to make certain that there is no overlap in main effects terms.

There are two important points to remember when considering interactions in a logistic model. First, if you have prior evidence of interactive effects between two predictors from other research, then check for such an effect in your model. Second, keep in mind that models should be parsimonious—do not try to include everything possible, but rather only the most important significant terms. An important term in this context is determined by observing the results when taking it out of the model. If the deviance raises substantially, or other predictor coefficients change, then it is obvious that the predictor is important to that model.

I suggest that one should first create a table of univariable logistic regression estimates. Keep a list of the coefficient values for each predictor. When developing the full model, observe any substantial changes in coefficient values. If you find two statistically significant predictors with substantially changed values, particularly where one increases and the other decreases, there is a possible interactive effect between them. If this occurs between two continuous predictors, then categorize one before checking it as a Categorical × Continuous interaction. I have found that, in some interactions, only one of the main effects terms changes dramatically when modeled as a main effect with the other interactive term. But, in all cases, interaction effects very rarely occur between predictors whose full model coefficients differ little from the univariable model.

The terms *confounder* and *interaction* have been used in this chapter, as well as at other times in the text. We have also used the term *effect modifier* in previous chapters. These terms have been defined when first used, but it may help to clearly differentiate each of these model terms in one place.

> *Confounder*: a covariate that is associated with both the response and the risk factor. Confounding can be detected by observing the difference in, for example, the coefficient value of the risk factor when a covariate is added to the model.
>
> *Risk factor*: the predictor of foremost interest to the model. It is usually binary. Some authors employ the term for more than one predictor; I prefer to limit it to one primary predictor.
>
> *Effect modifier*: any predictor that interacts with a model risk factor.
>
> *Interaction*: occurs when a term significantly modifies the relationship of the risk factor and response. For a binary predictor, R, if $r == 1$ affects the relationship of another predictor to the response differently than $r == 0$. In general, any two or more predictors can enter into an interactive relationship based on their differential effect on the response.

Finally, it should be mentioned that other authors have interpreted interactions in a different manner than discussed in this chapter. I have followed the traditional manner of understanding logistic model interactions qua Hosmer and Lemeshow, Woodward, Long and Freese, Vittinghoff et al., Collett, and so forth. However, other authors from time to time have advanced different interpretations. For example, in 2003, Chunrong Ai and Edward Norton of the University of Florida proposed an alternative manner of interpreting coefficients and confidence intervals of logistic and probit interactions (Ai and

Norton, 2003; Norton, Wang, and Ai, 2004). They calculated the significance of the interaction based on the estimated cross derivative of the terms in the interaction, not on the coefficient of the interaction term. Of course, we have asserted that interactions have no single coefficient or odds ratio.

The formulae given by Norton, Wang, and Ai (2004) for various logistic model interactions include the following. Given:

$$F(u) = 1 / (1 + \exp(-(\beta_1 x_1 + \beta_2 x_2 + \beta_{12} x_{12} + x\beta))) \qquad (6.16)$$

where $\beta_1 x_1$ and $\beta_2 x_2$ are main effects coefficients and predictors, and $\beta_{12} x_{12}$ the interaction of the two. $x\beta$ refers to other predictors contributing to the overall logistic linear predictor.

6.6.1 Binary × Binary

If both predictors are binary, the interaction effect is given as the discrete double difference:

$$\Delta^2 F\left(u\right)/[\Delta x_1 x_2] = 1/(1 + \exp(-(\beta_1 + \beta_2 + \beta_{12} + x\beta))) - 1/(1 + \exp(-(\beta_1 + x\beta)))$$
$$- 1/(1 + \exp(-(\beta_2 + x\beta))) + 1/(1 + \exp(-x\beta)) \qquad (6.17)$$

Note that if there are only two binary predictors and their interaction in the model, the term $x\beta$ drops from the above equation.

6.6.2 Continuous × Binary

The interaction effect is the discrete difference wrt x_2 of the single derivative wrt x_1.

$$[\partial F\left(u\right)/\partial x_1]/\Delta x_2 = (\beta_1 + \beta_{12}) * [F\{(\beta_1 + \beta_{12})x_1 + \beta_2 + x\beta\}$$
$$(1 - F\{\{(\beta_1 + \beta_{12})x_1 + \beta_2 + x\beta)\}]$$
$$_1[F(\beta_1 x_1 + x\beta) * \{1 - F(\beta_1 x_1 + x\beta)\}] \qquad (6.18)$$

6.6.3 Continuous × Continuous

The interaction effect is the cross derivative wrt, both x_1 and x_2.

$$\partial^2 F\left(u\right)/[\partial x_1 \, \partial x_2] = \beta_{12}\{F * (1 - F)\} + (\beta_1 + \beta_{12}x_2)\,(\beta_2 + \beta_{12}x_1) *$$
$$\left[\{F(1 - F)\} * \{1 - 2F\}\right] \qquad (6.19)$$

An example should help clarify how the calculations are performed by a command the authors wrote to support their proposed interpretation. Titled **inteff**, the command requires that the response be made into a global variable. For Binary × Continuous interactions, the continuous main effect is placed on the command line directly following the command name, followed by the binary main effect, the interaction, and any other predictors in the model. $x\beta$ contains all predictors not involved in the interaction, including the intercept. If there are no terms other than those related to the interaction, then $x\beta$ is the value of the intercept.

Using the **medpar** data, the example we shall use is of *died* on binary predictors *white* and *hmo*. *wxh* is defined as the interaction of *white* and *hmo*.

```
. global y died

. gen byte wxh = white * hmo

. logit $y white hmo wxh, nolog

Logistic regression                    Number of obs   =        1495
                                       LR chi2(3)      =        2.45
                                       Prob > chi2     =      0.4853
Log likelihood = -960.21013            Pseudo R2       =      0.0013
------------------------------------------------------------------------
   died |     Coef.   Std. Err.      z    P>|z|   [95% Conf. Interval]
--------+---------------------------------------------------------------
  white |   .3327169   .2171455    1.53   0.125   -.0928805     .7583142
    hmo |   .2599575   .6467598    0.40   0.688  -1.007668     1.527583
    wxh |  -.2868072   .6645932   -0.43   0.666  -1.589386     1.015772
  _cons |  -.9531047   .2080822   -4.58   0.000  -1.360938    -.5452711
------------------------------------------------------------------------
```

Using the **inteff** command, the interaction effect, standard error, and z-value for each covariate pattern is given in a user-specified file named **bxb_inter.dta**. With Binary × Binary interactions, however, only one value is produced—when the interaction has a value of 1. It should read: Graphs are also produced. Interested readers should refer to Ai and Norton (2003) and Norton, Wang, and Ai (2004).

```
. inteff $y white hmo wxh, savedata(c:\ado\bxb_inter)
savegraph1(c:\ado\bxb1) savegraph2(c:\ado\bxb2)
Logit with two dummy variables interacted
file c:\ado\bxb_inter.dta saved
(file c:\ado\bxb1.gph saved)
(file c:\ado\bxb2.gph saved)
```

Variable	Obs	Mean	Std. Dev.	Min	Max
_logit_ie	1495	-.0611534	0	-.0611534	-.0611534
_logit_se	1495	.1464847	0	.1464847	.1464847
_logit_z	1495	-.4174728	0	-.4174728	-.4174728

The interaction effect may be calculated by hand using the formula for Binary × Binary interactions above.

```
. di 1/(1+exp(-(.3327169 +.2599575 -.2868072 - .9531047)))
- 1/(1+exp(-(.3327169 - .9531047))) - 1/(1+exp(-(.2599575-
.9531047))) + 1/(1+exp(.9531047))
-.06115339
```

The value of −0.06115339 is identical to that displayed in the **inteff** output. Note that this value differs from the one given for the interaction in the Stata output (0.2868072).

The authors express an interesting meaning for the exponentiation of the interaction term that is given in logistic model output. We have previously said that it is not the odds ratio, but what is it? Norton et al. show it to be a ratio of odds ratios.

Given:

$$\exp(\beta_1 + \beta_2 + \beta_{12} + x\beta) = \Pr(y = 1 \mid x_1 = 1; x_2 = 1)/[1 - \Pr(y = 1 \mid x_1 = 1; x_2 = 1)]$$

$$\exp(\beta_2 + x\beta) \qquad\qquad = \Pr(y = 1 \mid x_1 = 0; x_2 = 1)/[1 - \Pr(y = 1 \mid x_1 = 0; x_2 = 1)]$$

the odds ratio for $(x_1 \mid x_2 = 1)$:

$$\frac{\exp(\beta_1 + \beta_2 + \beta_{12} + x\beta)}{\exp(\beta_2 + x\beta)} \tag{6.20}$$

and given,

$$\exp(\beta_1 + x\beta) = \Pr(y = 1 \mid x_1 = 1; x_2 = 0)/[1 - \Pr(y = 1 \mid x_1 = 1; x_2 = 0)]$$

$$\exp(x\beta) \qquad = \Pr(y = 1 \mid x_1 = 0; x_2 = 0)/[1 - \Pr(y = 1 \mid x_1 = 0; x_2 = 0)]$$

the odds ratio for $(x_1 \mid x_2 = 0)$:

$$\frac{\exp(\beta_1 + x\beta)}{\exp(x\beta)} \tag{6.21}$$

Taking the ratio of odds ratios of $(x_1 \mid x_2 = 1)$ to $(x_1 \mid x_2 = 0)$, we have:

$$\frac{\exp(\beta_1 + \beta_2 + \beta_{12} + x\beta)}{\exp(\beta_2 + x\beta)} * \frac{\exp(x\beta)}{\exp(\beta_1 + x\beta)} = \exp(\beta_{12}) \tag{6.22}$$

The **inteff** command was developed by Norton et al. to calculate the predicted probability, and associated interaction, standard error, and z-statistic of the effect for each covariate pattern in the model. The command also displays what the authors call the traditional linear approach of determining the interaction effect. This traditional statistic assumes that there is a single coefficient and odds ratio for the interaction. However, as described in this chapter, most authors do not subscribe to this simplistic view of a logistic interaction. Unfortunately, though, the naïve view is frequently found in published research, and the authors are to be commended for pointing out the deficiency of this approach.

Exercises

6.1. The following logistic regression output displays a model using the **absent** data set found in Section 5.4.

```
. logit ovrmonth slow aborig girl grade2-grade4, nolog

Logistic regression                        Number of obs    =        140
                                           LR chi2(6)       =      22.25
                                           Prob > chi2      =     0.0011
Log likelihood = -54.567247                Pseudo R2        =     0.1693
----------------------------------------------------------------------
ovrmonth |     Coef.   Std. Err.      z    P>|z|   [95% Conf. Interval]
---------+------------------------------------------------------------
    slow |   .9606643   .6538048    1.47   0.142   -.3207696   2.242098
  aborig |   1.591901   .5246611    3.03   0.002    .563584    2.620218
    girl |  -.4236748   .5076044   -0.83   0.404   -1.418561   .5712115
  grade2 |  -.6426341   .9233743   -0.70   0.486   -2.452415   1.167146
  grade3 |   1.042728   .7633262    1.37   0.172   -.4533638   2.53882
  grade4 |   1.558996   .8516379    1.83   0.067   -.1101832   3.228176
   _cons |  -3.309335   .8176474   -4.05   0.000   -4.911894  -1.706775
----------------------------------------------------------------------
```

After creating a variable called *senior*, representing *grade4* versus the other levels of *grade*, create an interaction between *slow* and *senior*. called *slxsen*. Model *ovrmonth* on *slow, senior, aborig* and the interaction *slxsen*. What happens? Why?

6.2. Why is it important to have the main effects predictors together with the interaction in a model?

6.3. Given the **affairs** data set found in Exercise 5.2, model *affair* on the continuous predictors *age*, *yrsmarr* (years married), and the binary predictor, *kids* (1 = have kids; 0 = not have kids). Check for a significant interaction between years married and having children. Graph the interaction as discussed in the text.

6.4. Use the **environment** data to determine if having a literal belief in the Bible has a bearing on whether one considers oneself an environmentalist.

RESPONSE

    enviro (1/0)   1 = is environmentalist; 0 = not an environmentalist

PREDICTORS

    literal (1/0)   1 = literal belief in Bible; 0 = no literal belief
    gender (1/0)   1 = female; 0 = male
    married (1/0) 1 = married; 0 = single
    educate (1/0) 1 = college grad; 0 = not college grad
    age   Continuous: ages 15 to 91

    a. Model *enviro* on *literal* and other predictors as adjusters. Parameterize as odds ratios. Determine the significant predictors.

    b. Check for interaction effects. Model with appropriate interactions.

    c. Determine the optimal logistic model.

    d. Which combination of predictor levels has the highest probability of determining if someone is an environmentalist? The least probable?

6.5. The **genestudy** data set is a case-control study derived from the Gene-Environment Interaction Study. Gertig, Hunter, and Spiegelman (2000) provided the subset found in the LogXact Reference manual, which is used here.

RESPONSE

    case   1 = 144 colon cancer cases; 0 = 627 controls

PREDICTORS

gene	binary	1 = presence of mutation; 0 = mutation not present
exp	binary	1 = exposure to environmental agent; 0 = not exposed
agecat	categorical	5-level age group categories
geneXexp	interaction	interaction of gene and exp

a. Determine if the coefficients of *agecat* are statistically significant. Make combinations if necessary.

b. Evaluate the statistical worth of the interaction. Interpret its meaning.

c. Which combination of predictors and levels provides the highest probability of a subject having colon cancer?

6.6. Use the **medpar** data, creating a model of *died* on *los*, *hmo*, and an interaction of *los* and *hmo*.

a. For *los* = 25, compare the values of the coefficient of the interaction using the method described in the text and the method proposed by Norton et al.

b. Using the methods described in this chapter, calculate the standard error and confidence intervals for the interaction coefficient determined in (a).

c. Convert the interaction coefficient and confidence intervals from (b) to an odds ratio and odds ratio confidence intervals.

R Code

Section 6.0 <Introduction>

```
library('foreign')
medpar<- read.dta('medparx.dta')
summary(medpar)

with(medpar, table(age, type))   #show classification
library('gmodels') #for CrossTable
CrossTable(medpar$age, medpar$type, dnn=c('Age','Type'), prop.
   r=FALSE,
   prop.c=FALSE, prop.t=FALSE, prop.chisq=FALSE) #better table
      format
```

Section 6.1 Binary × Binary Interactions

```
titanic<- read.dta('titanicx.dta')  #get data from Stata file
summary(titanic)
tpass<- tpass<- titanic[-which(titanic$class=='crew'),]  #drop
  crew from dataset
with(tpass, table(survived))
with(tpass, table(sex))
with(tpass, table(age))
with(tpass, table(class))

fit6a<- glm(survived ~ sex + age, data=tpass,
  family=binomial(link=logit))
summary(fit6a)
exp(coef(fit6a))  #odds ratios

df<- data.frame(sex=c('women', 'women', 'man', 'man'),
  age=c('child', 'adults', 'child', 'adults'))
df

predict(fit6a, newdata=df, type='link') #show predictions
coef(fit6a)[1] + coef(fit6a)[2]*0 + coef(fit6a)[3]*0
  #prediction for sex='women', age='child'
coef(fit6a)[1] + coef(fit6a)[2]*0 + coef(fit6a)[3]*1
  #prediction for sex='women', age='adults'
coef(fit6a)[1] + coef(fit6a)[2]*1 + coef(fit6a)[3]*0
  #prediction for sex='man', age='child'
coef(fit6a)[1] + coef(fit6a)[2]*1 + coef(fit6a)[3]*1
  #prediction for sex='man', age='adults'

exp(predict(fit6a, newdata=df, type='link')) #predicted odds
  ratios
exp(coef(fit6a)[1] + coef(fit6a)[2]*0 + coef(fit6a)[3]*0)  #OR
  for sex='women', age='child'
exp(coef(fit6a)[1] + coef(fit6a)[2]*0 + coef(fit6a)[3]*1)  #OR
  for sex='women', age='adults'
exp(coef(fit6a)[1] + coef(fit6a)[2]*1 + coef(fit6a)[3]*0)  #OR
  for sex='man', age='child'
exp(coef(fit6a)[1] + coef(fit6a)[2]*1 + coef(fit6a)[3]*1)  #OR
  for sex='man', age='adults'

predict(fit6a, newdata=df, type='response') #show predictions
  in probabilities
1/(1+exp(1.4326))
0.1926889/(1 - 0.1926889)

fit6b<- glm(survived ~ sex + age + sex:age, data=tpass,
  family=binomial) #with interaction
summary(fit6b)
```

```
exp(coef(fit6b))  #odds ratios
exp(predict(fit6b, newdata=df, type='link'))  #predicted odds
  ratios
2.7924528/1.6470588 #OR for adults:child given sex=women
0.2215478/0.8285714 #OR for adults:child given sex=man
0.8285714/1.6470588 #OR for man:women given age=child
0.2215478/2.7924528 #OR for man:women given age=adults
0.07933806/0.5030612  #man:women|age=child /
  man:women|age=adults

#no plotgr3 in R

class2<- as.numeric(tpass$class=='2nd class') #indicator
  variable
class3<- as.numeric(tpass$class=='3rd class') #indicator
  variable
fit6c<- glm(survived ~ sex + age + sex:age + class2 + class3,
  data=tpass, family=binomial)
summary(fit6c)

fit6d<- glm(survived ~ sex*age + class, data=tpass,
  family=binomial)
summary(fit6d)
exp(coef(fit6d))   #ORs
exp(coef(fit6c))   #ORs
```

Section 6.2 Binary × Categorical Interactions

```
fit6e<- glm(survived ~ age + sex + class2 + class3,
  data=tpass, family=binomial)
summary(fit6e)
exp(coef(fit6e))   #ORs

agecl2<- as.numeric(tpass$age=='adults')*class2 #indicator
  variable
agecl3<- as.numeric(tpass$age=='adults')*class3 #indicator
  variable
fit6f<- glm(survived ~ age + sex + class2 + class3 + agecl2 +
  agecl3, data=tpass, family=binomial)
summary(fit6f)

sexcl2<- as.numeric(tpass$sex=='man')*class2 #indicator
  variable
sexcl3<- as.numeric(tpass$sex=='man')*class3 #indicator
  variable
fit6g<- glm(survived ~ age + sex + class2 + class3 + sexcl2 +
  sexcl3, data=tpass, family=binomial)
summary(fit6g)
```

```
fit6h<- glm(survived ~ age + sex*class, data=tpass,
   family=binomial)
summary(fit6h)

exp(coef(fit6g))   #ORs

#No plostgr3 in R

fit6i<- glm(survived ~ sex + class2 + class3, data=tpass,
   family=binomial)
summary(fit6i)
exp(coef(fit6i))   #ORs
fitcases<- matrix(c(1,0,0,0, 1,0,1,0, 1,0,0,1, 1,1,0,0,
   1,1,1,0, 1,1,0,1), ncol=4, byrow=TRUE) #cases for sex x
   class
adjust<- matrix(fitcases%*%coef(fit6i), ncol=3, byrow=TRUE)
   #link version
colnames(adjust)<- c('1st class', '2nd class', '3rd class')
rownames(adjust)<- c('women', 'man')
exp(adjust) #Odds version
0.136942/1.46764   #OR for sex | 3rd class
0.276349/0.71128   #OR for 2nd:1st | man
0.136942/0.71128   #OR for 3rd:1st | women

fit6j<- glm(survived ~ sex*class, data=tpass,
   family=binomial)
summary(fit6j)
exp(coef(fit6j))   #ORs
fitcases2<- matrix(c(1,0,0,0,0,0, 1,0,1,0,0,0, 1,0,0,1,0,0,
   1,1,0,0,0,0, 1,1,1,0,1,0, 1,1,0,1,0,1), ncol=6, byrow=TRUE)
   #cases for sex x class
adjust2<- matrix(fitcases2%*%coef(fit6j), ncol=3, byrow=TRUE)
   #link version
colnames(adjust2)<- c('1st class', '2nd class', '3rd class')
rownames(adjust2)<- c('women', 'man')
exp(adjust2) #Odds version
0.5254237/35.25 #OR for man:women | 1st class
7.1538462/35.25 #OR for 2nd:1st | women
0.8490566/35.25 #OR for 3rd:1st | women

0.162338/7.15385
0.525424/35.25
0.0226924/0.01490565

0.208531/0.849057
0.525424/35.25
0.24560306/0.01490565
```

Section 6.3 Binary × Continuous Interactions

```
heart<- read.dta('heart01.dta')
heart$anterior<- factor(heart$anterior) #fix up anterior,
  dropping unused levels
summary(heart$age)  #show age stats
cat('Non-missing:', sum(!is.na(heart$age)), 'Missing:',
  sum(is.na(heart$age)), '\n')

fit6k<- glm(death ~ anterior + age, data=heart,
  family=binomial)
summary(fit6k)
exp(coef(fit6k))  #ORs

axa<- as.numeric(heart$anterior=='Anterior')*heart$age
  #interaction
fit6l<- glm(death ~ anterior + age + axa, data=heart,
  family=binomial)
summary(fit6l)
exp(coef(fit6l))  #ORs

fit6m<- glm(death ~ anterior*age, data=heart,
  family=binomial)
summary(fit6m)

xage<- 40:100 #age abscissae
eta1<- coef(fit6m)[1] + coef(fit6m)[3]*xage #link values
p1<- 1/(1+exp(-eta1)) #response values
eta2<- coef(fit6m)[1] + coef(fit6m)[2] + coef(fit6m)[3]*xage +
  coef(fit6m)[4]*xage  #link values
p2<- 1/(1+exp(-eta2)) #response values

plot(xage, p1, type='l', ylim=c(0,0.25), xlab='Age (yr)',
  ylab='Probability', main='Relationship of Age and Anterior')
  #for Inferior
lines(xage, p2, col=2, pch=2) #for Anterior
legend('topleft', legend=c('Inferior', 'Anterior'),
  pch=c(1,2), col=c(1,2), bty='n')

plot(xage, eta1, type='l', ylim=c(-5,-1), xlab='Age (yr)',
  ylab='Logit', main='Logit Diff: Anterior vs Interior on
  Age') #for Inferior
lines(xage, eta2, col=2, pch=2) #for Anterior
legend('topleft', legend=c('Inferior', 'Anterior'),
  pch=c(1,2), col=c(1,2), bty='n')

fit6n<- glm(died ~ white + los, data=medpar, family=binomial)
summary(fit6n)
exp(coef(fit6n))  #ORs
```

```
whlo<- medpar$white*medpar$los  #interaction (both are
  numeric)
fit6o<- glm(died ~ white + los + whlo, data=medpar,
  family=binomial)
summary(fit6o)
exp(coef(fit6o))  #ORs

fit6p<- glm(died ~ white*los, data=medpar, family=binomial)
summary(fit6p)
plot(medpar$los,predict(fit6p, type='response'), ylim=c(0,.5),
  col=(medpar$white+1), pch=(medpar$white+1), xlab='Length of
  Stay', ylab='Logit', main='Interaction of White and LOS')
legend(locator(1), legend=c('White=0', 'White=1'), pch=c(1,2),
  col=c(1,2), bty='n')

exp(0.7703 + (-0.0477*21))

summary(medpar$los) #show some stats
quantile(medpar$los, probs=c(.01,.05,.10,.25,.5,.75,.9,.95,.99))
table(medpar$white) #classification

wlos<- medpar$white*medpar$los  #interaction (both numeric)
fit6q<- glm(died ~ white + los + wlos, data=medpar,
  family=binomial)
summary(fit6q)

plot(medpar$los,predict(fit6p, type='link'),col=(medpar$white+
  1), pch=(medpar$white+1), xlab='Length of Stay',
  ylab='Logit', main='Interaction of White and LOS')
legend(locator(1), legend=c('White=0', 'White=1'), pch=c(1,2),
  col=c(1,2), bty='n')

plot(medpar$los,predict(fit6p, type='response'), ylim=c(0,.5),
  col=(medpar$white+1), pch=(medpar$white+1), xlab='Length of
  Stay', ylab='Logit', main='Interaction of White and LOS')
legend(locator(1), legend=c('White=0', 'White=1'), pch=c(1,2),
  col=c(1,2), bty='n')
```

Section 6.4 Category × Continuous Interactions

```
table(heart$killip)
table(heart$killip)/sum(heart$killip>0, na.rm=TRUE)*100
#percents

fit6r<- glm(death ~ kk2 + kk3 + kk4, data=heart,
family=binomial)
summary(fit6r)
exp(coef(fit6r))  #ORs
```

```
fit6s<- glm(death ~ age + factor(killip), data=heart,
  family=binomial)
summary(fit6s)
exp(coef(fit6s))   #ORs
xheart<- na.omit(heart) #drop NA's
plot(xheart$age, predict(fit6s, xheart, type='response'),
  ylim=c(0,1), col=xheart$killip, pch=xheart$killip, xlab='Age
  (yr)', ylab='Probability of Death', main='Prob. of Death:
  Killip x Age')
legend(locator(1), legend=c('Killip=1', 'Killip=2',
  'Killip=3', 'Killip=4'), pch=1:4, col=1:4, bty='n')

fit6t<- glm(death ~ factor(killip)*age, data=heart,
  family=binomial)
summary(fit6t)
exp(coef(fit6t))   #ORs

plot(xheart$age, predict(fit6t, xheart, type='response'),
  ylim=c(0,1), col=xheart$killip, pch=xheart$killip, xlab='Age
  (yr)', ylab='Probability of Death', main='Prob. of Death:
  Killip x Age w/Interaction') legend(locator(1),
  legend=c('Killip=1', 'Killip=2', 'Killip=3', 'Killip=4'),
  pch=1:4, col=1:4, bty='n')

ageXk1<- heart$age*heart$kk1   #interaction
ageXk3<- heart$age*heart$kk3   #interaction
ageXk4<- heart$age*heart$kk4   #interaction
fit6u<- glm(death ~ age + kk1 + kk3 + kk4 + ageXk1 + ageXk3 +
  ageXk4, data=heart, family=binomial)   #fit relative to kk2
summary(fit6u)
exp(coef(fit6u))   #ORs
```

7

Analysis of Model Fit

Model fit has traditionally been assessed by means of summary test statistics or by residual analysis. Since the analysis of fit is so important to the modeling process, we shall discuss four categories of fit assessment:

- Traditional fit tests
- Hosmer–Lemeshow GOF test
- Information criteria tests
- Residual analysis

7.1 Traditional Fit Tests for Logistic Regression

Fit statistics were first developed for the **generalized linear models** (GLM) parameterization of logistic regression. They were first integrated into the GLIM statistical package, which was written in the early 1970s. Recall from earlier discussion that the GLM approach uses an **iteratively re-weighted least squares** (IRLS) algorithm to estimate parameters, standard errors, and other model statistics. The results are identical to the full maximum likelihood method of estimation. GLIM was the first statistical software to incorporate a full set of fit statistics and residual analysis for the various GLM family models.

The traditional GOF statistics that we shall discuss are

1. R^2 and Pseudo-R^2 statistics
2. Deviance statistic
3. Likelihood ratio statistic

7.1.1 R^2 and Pseudo-R^2 Statistics

The standard **goodness-of-fit** (GOF) statistic for ordinary least squares regression, or Gaussian regression, is the R^2 statistic, also called the **coefficient of determination.** The value represents the percentage variation in the data accounted for by the model. Ranging from 0 to 1, the higher the R^2 value, the better the fit of the model.

R^2 is equal to the ratio of the variance of the fitted values to the total variance. In Gaussian models, this is

$$RSS = \text{Residual SS} = \Sigma(y - \mu)^2 \tag{7.1}$$

$$TSS = \text{Total SS} = \Sigma(y - \text{mean}(y))^2 \tag{7.2}$$

where μ is the fitted value and y is the model response. The relationship of these two terms defines R^2 as:

$$R^2 = \frac{TSS - RSS}{TSS} = 1 - \frac{RSS}{TSS} \tag{7.3}$$

or in terms of the statistics:

$$R^2 = 1 - \frac{\Sigma(y - \mu)^2}{\Sigma(y - \text{mean}(y))^2} \tag{7.4}$$

where the sigma sign, Σ, represents summation over all observations in the model.

The R^2 statistic is not appropriate for use with the logistic model. It sometimes appears in the statistical output of various logistic regression routines, but it should be ignored. It does not inform us about the variability accounted for by the model, nor does it provide us with any information that can help decide between models. Statisticians have tried to devise other R^2 statistics that are more appropriate for logistic models, but the effort has been largely unsuccessful. The ones typically associated with logistic regression procedures are called Pseudo-R^2 statistics. Several have been created.

The most commonly used Pseudo-R^2 statistic is defined as:

$$1 - LL_F / LL_C \tag{7.5}$$

with

LL_C = the intercept-only model, and
LL_F = the full model

To demonstrate how the statistic can be calculated by hand, we model a constant only and full model, followed by a calculation of the pseudo-R^2.

CONSTANT-ONLY LOGIT MODEL

```
. logit death

Logistic regression                    Number of obs   =        5388
                                       LR chi2(0)      =       -0.00
                                       Prob > chi2     =           .
Log likelihood = -987.40611            Pseudo R2       =      -0.0000
-----------------------------------------------------------------------
  death |      Coef. Std. Err.     z  P>|z|  [95% Conf. Interval]
--------+--------------------------------------------------------------
  _cons | -3.057037 .0657766 -46.48 0.000 -3.185957    -2.928118
-----------------------------------------------------------------------
```

FULL LOGIT MODEL

```
. logit death age

Logistic regression                    Number of obs   =        5388
                                       LR chi2(1)      =      148.24
                                       Prob > chi2     =      0.0000
Log likelihood = -913.28575            Pseudo R2       =      0.0751
-----------------------------------------------------------------------
  death |      Coef. Std. Err.     z  P>|z|  [95% Conf. Interval]
--------+--------------------------------------------------------------
    age |   .0565292 .0048193  11.73 0.000  .0470834     .0659749
  _cons |  -7.076784 .3729006 -18.98 0.000 -7.807656    -6.345912
-----------------------------------------------------------------------
```

CALCULATION OF THE PSEUDO-R^2

```
. di 1-(-913.28575/-987.40611)
.07506573
```

Compare with the value of pseudo-R^2 displayed in the above model output. A list of alternative pseudo-R^2 statistics can be found in Hardin and Hilbe (2007), pages 59–63. The formal name of the version we have used here, and which is the one employed by Stata, is called **McFadden's likelihood-ratio index** (McFadden 1974a). This pseudo-R^2 statistic has been modified by subtracting the number of model predictors from the likelihood of the full model. This revised R^2 is called the **Ben-Akiva and Lerman** (1985) **adjusted likelihood-ratio index**.

Note that the constant log-likelihood for the binary logistic model, based on the Bernoulli PDF, may be calculated directly from the data; all one must know is the proportion of 1s in the response, y. Signifying the proportion of 1s as p, the formula for calculating the constant-only model is

$$LL_0 = n\{p * \ln(p) + (1-p) * \ln(1-p)\} \tag{7.6}$$

A good overview of various coefficients of determination, or R^2, for logistic regression, together with simulation studies, can be found in Liao and McGee (2003).

7.1.2 Deviance Statistic

Perhaps the most noteworthy fit statistic within the GLM tradition is the deviance statistic. It is defined from the model log-likelihood function as:

$$-2 \sum \{LL(y;y) - LL(y;\mu)\} \tag{7.7}$$

where the sigma sign represents summation over all observations in the data, y is the response, and μ (*mu*) is the fitted value—also signified as y hat ($\hat{y}$).

Deviance is the sum of the differences between the saturated and model log-likelihoods provided by each observation in the model. To calculate the saturated log-likelihood, you must substitute y for every μ in the logistic log-likelihood function.

The following equation is the logistic log-likelihood function:

$$LL = \sum \{y * \ln(\mu/(1-\mu)) + \ln(1-\mu)\} \tag{7.8}$$

The saturated log-likelihood function is therefore:

$$LLs = \sum \{y * \ln(y/(1-y)) + \ln(1-y)\} \tag{7.9}$$

The deviance is then calculated as:

$$-2 * \sum [\{y * \ln(y/(1-y)) + \ln(1-y)\} - \{y * \ln(\mu/(1-\mu)) + \ln(1-\mu)\}] \tag{7.10}$$

or simplified, as follows:

BERNOULLI (LOGISTIC) DEVIANCE FUNCTION

$$2 * \sum \{y * \ln(y/\mu) + (1-y) * \ln((1-y)/(1-\mu))\} \tag{7.11}$$

The deviance may be shortened based on whether $y = 0$ or $y = 1$.

$$D(y = 0) = -2 * \ln(1 - \mu) \qquad (7.12)$$

$$D(y = 1) = -2 * \ln(\mu) \qquad (7.13)$$

This is the preferred manner of formulation in a logistic regression algorithm. As an aside, the log-likelihood functions for $y = 0$ and $y = 1$ are, respectively,

$$LL(y = 0) = \Sigma\{\ln(1 - \mu)\}$$

$$LL(y = 1) = \Sigma\{\ln(\mu / (1 - \mu)) + \ln(1 - \mu)\}$$

The deviance statistic has primarily been used as a goodness-of-fit test for assessing the worth of model predictors. This is accomplished by setting up a deviance table where the deviance, the differences between sequential deviances, degrees of freedom, and mean deviance difference are given for each predictor being considered for the model. This method has proved particularly useful when assessing the worth of interaction terms to a model.

We shall use the **heart01** data to display how this method works. We begin by modeling *death*, the response, as a constant-only model. We then add predictors, checking the deviance statistic for each additional predictor.

Model	Deviance	Difference	DF	Mean Difference	
constant	1974.812				
Main Effects					
anterior	1543.845	430.967	1	430.967	
hcabg	1538.907	4.938	1	4.938	
killip	1372.575	166.332	3	55.444	
agegrp	1273.251	99.324	3	33.108	
Interaction Terms					
aXk	1265.631	7.620	4	1.905	///*anterior* X *killip*
hXa	1265.392	0.239	4	0.060	///*hcabg* X *agegrp*

Notice how each deviance value decreases as we add predictors and interactions to the model—until we reach *hcabg* X *agegrp*. The mean difference column informs us about the impact of added deviance values, with very low values indicating that the predictor or interactions fail to contribute significantly to the model.

We test the significance of the predictors, including interactions, as they enter the model by performing a *Chi2* test on the difference between deviances with the degrees of freedom as given in the table. Degrees of freedom reflect the number of added predictors to the model. These calculations are not shown. Rather, the table of parameter estimates corresponding to the deviance table is given as:

```
---------+------------------------------------------------------
         |               OIM
   death |    Coef.  Std. Err.     z   P>|z|  [95% Conf. Interval]
---------+------------------------------------------------------
 anterior| 1.541501 .3694139    4.17 0.000   .8174633   2.265539
    hcabg| 1.293652 1.147095    1.13 0.259  -.954613    3.541917
      kk2| 1.138065 .2172354    5.24 0.000   .7122915   1.563838
      kk3| 1.485787 .3623185    4.10 0.000   .7756561   2.195919
      kk4|  3.65986 .4910731    7.45 0.000   2.697375   4.622346
     age2| .4811162 .3113566    1.55 0.122  -.1291315   1.091364
     age3| 1.553419 .2689072    5.78 0.000   1.026371   2.080467
     age4| 2.239154 .2760787    8.11 0.000   1.69805    2.780258
      axk|-.5277404 .1890642   -2.79 0.005  -.8982994  -.1571814
      hxa|-.1948319  .393939   -0.49 0.621  -.9669382   .5772743
    _cons|-5.311577 .2843469  -18.68 0.000  -5.868887  -4.754267
---------+------------------------------------------------------
```

The deviance statistic has been commonly used in the manner shown above until the past few years. As we shall see, the use of the Akaike Information Criterion (AIC) and Bayesian Information Criterion (BIC) statistics have recently been employed as primary fit tests, but we shall delay a discussion of them until later.

7.1.3 Likelihood Ratio Test

We have seen the likelihood ratio test used before when deciding whether to include predictor(s) in a model. The test can also be formulated with the reduced log-likelihood function being the constant-only model:

$$-2[\text{LL reduced} - \text{LL full}] \tag{7.14}$$

The statistic has a *Chi2* distribution (χ^2), with the **degree of freedom** (DOF) equal to the difference in the number of predictors between the two models. Again, the reduced model is the constant-only model.

For an example, we shall use the model above showing how to calculate the pseudo-R^2 fit statistic. Recall that the constant-only log-likelihood was -987.40611. The model of death on *age* was -913.28575. Using the above formula, we calculate:

LIKELIHOOD RATIO

```
. di -2*(-987.40611 - (-913.28575))
148.24072
```

This matches the displayed output, shown again below. The *Chi2* p-value with 1 DOF is 0.000, indicating that the addition of *age* to the model is significantly better than the constant-only model.

```
. logit death age

Logistic regression                      Number of obs    =        5388
                                         LR chi2(1)       =      148.24
                                         Prob > chi2      =      0.0000
Log likelihood = -913.28575              Pseudo R2        =      0.0751
-----------------------------------------------------------------------
  death |      Coef. Std. Err.      z   P>|z|    [95% Conf. Interval]
--------+--------------------------------------------------------------
    age |  .0565292 .0048193   11.73 0.000    .0470834        .0659749
  _cons | -7.076784 .3729006  -18.98 0.000   -7.807656       -6.345912
-----------------------------------------------------------------------
```

Usually we are comparing a model with a number of predictors against the constant-only model. When the predictors are significant (at $p = 0.05$ level), the *Chi2* -value is nearly always 0.000. Except for the single predictor case, the statistic is not very useful.

We next address the goodness-of-fit test described in Hosmer and Lemeshow (2000). The statistics and its options have played a very important role in contemporary logistic fit analysis.

7.2 Hosmer–Lemeshow GOF Test

Hosmer and Lemeshow (2000) developed a number of residuals that can be used for the binary logistic model. They also developed a goodness-of-fit test that has been accepted by most statisticians as one of the best ways to assess the fit of logistic models.

Contrary to the standard GLM-based residuals used in commercial statistical software until the early 1990s, Hosmer and Lemeshow based logistic residuals, as well as types of fit analyses described in this section, on what they termed *m-asymptotics*. The term *m*-asymptotics refers to structuring observations in terms of covariate patterns. For instance, suppose that data are structured as:

	y	*x1*	*x2*	*x3*	covariate pattern
1:	1	1	0	1	1
2:	1	1	1	1	2
3:	0	1	0	1	1
4:	1	0	0	1	3
5:	0	0	1	1	4
6:	1	1	0	0	5
7:	0	0	1	0	6
8:	1	0	0	1	3
9:	0	1	0	1	1
10:	1	1	0	0	5

Considered as a normal data set, the above listing has 10 observations. The data are in *n-asymptotic* form. The data are structured having one line per observation. Residual and fit analyses are traditionally based on data having this form; *m*-asymptotics is based on covariate patterns where each is specified as a particular set of identical values, not counting the response. Therefore, observations 1, 3, and 9 all share the same pattern of predictor values, and are, therefore, a single *m*-asymptotic observation. The above data have six *m*-asymptotic observations.

Hosmer and Lemeshow argued that fit statistics for logistic regression are best evaluated using *m*-asymptotics. The post-estimation commands following Stata's **logistic** command are based on this type of format. On the other hand, post-estimation commands following Stata's **glm** command are based on *n*-asymptotics. Several other applications, e.g., S-Plus, employ *n*-asymptotics for post-estimation GLM fit analysis. Care must be taken to know which type of asymptotic structure is being used. When there are one or more continuous predictors in the model, *n*-asymptotics and *m*-asymptotics for the model will likely be identical, or closely so. The real difference occurs when the data consist of a few binary or categorical predictors, as in the above example list.

7.2.1 Hosmer–Lemeshow GOF Test

Perhaps the most important logistic regression fit statistic is the Hosmer–Lemeshow GOF statistic, together with a related table of deciles. The table divides the range of probability values, μ, into groups. Hosmer and Lemeshow recommended 10 groups for relatively large datasets, but the number can be amended to fit the data situation. The point is to provide a count, for each respective group, of actually observed 1s, fitted or expected 1s, observed 0s, and expected 0s. Theoretically, observed and expected counts should be close. Diagnostically, it is easy to identify groups where deviations from the expected occur. When this is the case, we can go back to the data to see the covariate pattern(s) giving rise to the discrepancy.

A fit statistic is provided that measures the overall correspondence of counts. Based on the χ^2 distribution, a Hosmer–Lemeshow (H–L) statistic with a *p*-value greater than 0.05 is considered a good fit. The lower the H–L statistic, the less variance in fit, and the greater the *p*-value.

A problem arises when there are ties (multiple equal outcomes). When ties occur, selecting a decile partition induces randomness to the statistic. This is particularly the case when the independent variables are different over observations producing the ties. We need to ensure that this randomness does not affect a particular application of the test. The Hosmer–Lemeshow test is appropriate only if the number of observations that are tied at a given covariate pattern is small when compared with the total of all observations. Binary and categorical predictors tend to result in a large number of tied observations. Models with more predictors generally have fewer tied observations. When dealing with a model having few predictors, at least one must be a continuous variable with few equal values. Models with more continuous predictors will help guarantee fewer tied observations. However, having more continuous predictors and fewer ties results in data where m-asymptotically based statistics cannot be applied. Moreover, as we have discussed, continuous predictors may need to be categorized in the modeling process. Continuous predictors may assist in producing a Hosmer–Lemeshow test with fewer ties, but they may not be wanted for the most appropriate model fit test. As a result of these difficulties, statisticians use the Hosmer–Lemeshow test with caution.

A method commonly used to ameliorate the difficulty with ties entails employing the test with a different number of groupings. We can request a table with 10 groups, as well as tables with 8 and 12 groups, looking at the respective tables to see if any substantial differences occur. If they do not, the test statistic may more likely be trusted; if they do, the test is not reliable. For smaller data situations, the number of groups will likely need to be reduced.

Consider the full model on *death*. The data are modeled as follows:

```
. logistic death anterior hcabg kk2-kk4 age3 age4, nolog

Logistic regression                     Number of obs   =         4503
                                        LR chi2(7)      =       209.90
                                        Prob > chi2     =       0.0000
Log likelihood = -638.15957             Pseudo R2       =       0.1412
--------+------------------------------------------------------------
   death| Odds Ratio Std. Err.    z   P>|z|  [95% Conf. Interval]
--------+------------------------------------------------------------
anterior|    1.894096 .3173418  3.81 0.000 1.363922     2.630356
   hcabg|    2.195519 .7744623  2.23 0.026 1.099711     4.383246
     kk2|    2.281692 .4117012  4.57 0.000 1.602024     3.249714
     kk3|    2.218199 .5971764  2.96 0.003 1.308708     3.759743
     kk4|   14.63984 5.218374   7.53 0.000 7.279897    29.44064
    age3|    3.549577 .7119235  6.32 0.000 2.395823     5.258942
    age4|    6.964847 1.44901   9.33 0.000 4.632573    10.47131
--------+------------------------------------------------------------
```

The **estat** or **lfit** commands are issued, which produce identical results.

```
estat gof, table group(10) /// (or) lfit, table group(10)

Logistic model for death, goodness-of-fit test
```

```
(Table collapsed on quantiles of estimated probabilities)
(There are only 8 distinct quantiles because of ties)
+---------------------------------------------------------------------+
| Group |   Prob | Obs_1 | Exp_1 | Obs_0 |   Exp_0 | Total |
|-------+--------+-------+-------+-------+---------+-------|
|     2 | 0.0080 |    10 |   9.9 |  1220 |  1220.1 |  1230 |
|     4 | 0.0151 |     9 |  13.4 |   878 |   873.6 |   887 |
|     5 | 0.0181 |     5 |   5.9 |   325 |   324.1 |   330 |
|     6 | 0.0279 |     8 |  12.9 |   453 |   448.1 |   461 |
|     7 | 0.0338 |    11 |  10.5 |   302 |   302.5 |   313 |
|-------+--------+-------+-------+-------+---------+-------|
|     8 | 0.0534 |    30 |  26.7 |   484 |   487.3 |   514 |
|     9 | 0.0965 |    35 |  25.6 |   291 |   300.4 |   326 |
|    10 | 0.6100 |    68 |  71.1 |   374 |   370.9 |   442 |
+---------------------------------------------------------------------+

        number of observations =       4503
               number of groups =          8
      Hosmer-Lemeshow chi2(6) =       7.91
                  Prob > chi2 =       0.2449
```

The Hosmer–Lemeshow statistic is rather low (7.91), resulting in a *p*-value of 0.2449. This value indicates an excellent fit. The low H–L statistic indicates little variability. Observe the table above, paying attention to the differences in values between the observed and predicted values at each level, and in both 1 and 0 response groups. The 0 columns appear to be close at each level; the 1 columns are more sporadic in their association. Levels 4, 6, and 9 in particular have considerable separation in values—especially level 9 (35 to 25.6). Again, the closer the overall fit, the lower the Hosmer–Lemeshow *Chi2* statistic, and the greater the *p*-value. Values over 0.05 do not provide statistical evidence of a poorly fitted model, that is, they may be considered as well fitted.

It should be noted that the decile tables of various software applications often differ—and at times differ greatly. The primary reason is that each handles ties differently; classification into groups is based on slightly different criteria, with sometimes significant results. Compare Stata's table with a like table on the same model using SAS or SPSS. Any difference is usually due to differing grouping criteria as well as criteria on how ties are handled.

Notice that the program automatically reduced the number of levels to 8. To confirm consistency of the model, we use six levels.

```
. estat gof, table group(6)

Logistic model for death, goodness-of-fit test
```

```
(Table collapsed on quantiles of estimated probabilities)
+----------------------------------------------------------------+
| Group |    Prob | Obs_1 | Exp_1 | Obs_0 |   Exp_0 | Total |
|-------+---------+-------+-------+-------+---------+-------|
|     1 | 0.0080 |    10 |   9.9 |  1220 | 1220.1 |  1230 |
|     2 | 0.0151 |     9 |  13.4 |   878 |  873.6 |   887 |
|     3 | 0.0181 |     5 |   5.9 |   325 |  324.1 |   330 |
|     4 | 0.0338 |    19 |  23.4 |   755 |  750.6 |   774 |
|     5 | 0.0594 |    31 |  27.8 |   501 |  504.2 |   532 |
|-------+---------+-------+-------+-------+---------+-------|
|     6 | 0.6100 |   102 |  95.6 |   648 |  654.4 |   750 |
+----------------------------------------------------------------+

              number of observations =      4503
                    number of groups =         6
          Hosmer-Lemeshow chi2(4)    =      3.36
                       Prob > chi2   =      0.4992
```

The results are consistent.

Calling for 12 levels results in a table with 9 levels, and with statistics appearing as:

```
Hosmer-Lemeshow chi2(7)  =        6.69
            Prob > chi2  =        0.4614
```

Therefore, with 6, 8, and 9 levels, the results all show a well-fitted model. Testing for consistency is very important, and if neglected, can result in an incorrect assessment.

In Stata, the Hosmer–Lemeshow GOF test is available as a post-estimation command following **logit, logistic,** and **glm.** SAS's GENMOD and SPSS's GENLIN procedures automatically produce the test. The results of all three tests are typically different since each defines how ties are dealt with in a different manner.

A Pearson χ^2 goodness-of-fit test may also be given as a test of the observed versus expected number of responses based on levels defined by the covariate patterns. Values of p greater than 0.05 indicate a well-fitted model:

```
. lfit

Logistic model for death, goodness-of-fit test

              number of observations =      4503
         number of covariate patterns =        42
                  Pearson chi2(34)    =     46.87
                       Prob > chi2   =     0.0699
```

The Pearson χ^2 GOF test is part of most logistic regression software, and is more consistent in its results than is the Hosmer–Lemeshow test. On the other hand, if the number of covariate patterns is close to the number of observations, the test statistic has little value. In this case, though, the test is appropriate, and significant.

7.2.2 Classification Matrix

The classification or confusion matrix is one of the original fit tests used by statisticians with logistic regression. It was originally used in Discriminant Analysis, a multivariate classification and prediction tool we discuss in Chapter 14. In fact, because logistic distribution constrains predicted values to the range of probability values, 0–1, Discriminant Analysis has used the logistic distribution for its table matrix.

In Stata, the classification matrix can be obtained as a post-estimation command following **logit, logistic**, and **glm**. Other software typically displays it automatically, or it can easily be selected.

A classification matrix is based on a cutoff point, which answers the question, "What is the optimal probability value to separate predicted versus observed successes and failures?" Since the definitions of sensitivity and specificity are based on these relationships, we use them to determine the appropriate cutoff point.

The default cutoff is .5, which is the mean of the logistic distribution. However, this is not the appropriate cutoff point for the majority of logistic models. The problem is in determining the optimal cutoff point so that the classifications are correct for the data and model.

A traditional way of determining the cutoff point was to take the mean value of the response. Since the binary response is 1/0, and the fitted value is a probability, the idea was to simply use a cutoff point as the proportion of the response. A slightly more sophisticated procedure that has been used to determine the optimal cutoff point is to model the data, predict the fitted value, μ, which is the probability of success (1), and use the mean of μ as the cutoff point. For the present model, these two values are

MEAN OF RESPONSE

```
. mean death

Mean estimation                          Number of obs    =     5388

-----------------------------------------------------------------
            |        Mean    Std. Err.      [95% Conf. Interval]
------------+----------------------------------------------------
      death |     .0449146    .0028219        .0393826    .0504467
-----------------------------------------------------------------
```

MEAN OF PREDICTED FIT

The model must be re-estimated since the **mean** command erased the post-estimation statistics saved by the previous modeling of the data. We then use the **predict** command to obtain the predicted probability of success, or μ.

```
. qui logistic death anterior hcabg kk2-kk4 age3 age4

. predict mu

. mean mu

Mean estimation                      Number of obs   =    4503

-----------------------------------------------------------------
             |        Mean    Std. Err.    [95% Conf. Interval]
-------------+---------------------------------------------------
         mu  |    .0390851    .0008121     .0374929     .0406772
-----------------------------------------------------------------
```

Note that the number of observations used in the calculation of the mean of *death* (5388) is different from the number of observations used to calculate μ, or *mu*, above. Some 1285 observations were dropped from the model due to missing values, which originated with the killip levels. *Death* has no missing values. We may recalculate the mean of *death* based on the observations used in the model by quietly re-estimating the model, and then using only model data with the **mean** command.

```
. qui  logistic death anterior hcabg kk2-kk4 age3 age4

. mean death if e(sample)

Mean estimation                      Number of obs   =    4503

-----------------------------------------------------------------
             |        Mean    Std. Err.    [95% Conf. Interval]
-------------+---------------------------------------------------
      death  |    .0390851    .0028883     .0334225     .0447476
-----------------------------------------------------------------
```

Note that the values of the mean of *death* and *mu* are identical. However, the standard errors differ appreciably.

Using this method, the optimal cutoff probability point is .039, or approximately 0.04. The concepts of sensitivity and specificity need to be defined. They are commonly used terms in epidemiology, and are employed in other areas of research as well. Using the epidemiological paradigm, suppose that we have patients who have a disease, D, and those who do not, ~D. The probability of testing positive for having the disease, given that the patient actually has the disease, is called the *sensitivity* of a test. The probability of testing negative for having a disease, given the fact that the person does not have the disease, is called the *specificity* of a test. Therefore,

$$Sensitivity = Pr(+test \mid D)$$
$$Specificity = Pr(-test \mid \sim D)$$

These concepts are closely related to false negative and false positive tests. A false negative test occurs when a test is negative for having a disease, when in fact the patient has the disease. A false positive occurs when a patient tests positive for having a disease when in fact the person does not have it. Therefore,

$$False\ negative = Pr(-test \mid D)$$
$$False\ positive = Pr(+test \mid \sim D)$$

All of these relationships are given with a typical classification table.

Following estimation of the logistic regression model, we graph the values of sensitivity and specificity against a cutoff probability (Figure 7.1). Where they cross is the ideal cutoff probability point. We graph this relationship using the **lsens** postestimation command. Also created are values for the sensitivity and specificity, together with the cutoff probabilities. In effect, the graph is the range of sensitivity and specificity values that would be observed if we calculated a classification table for cutoff values from 0 to 1. Mramor and Valentincic (2003) have a nice discussion on appropriate cutoff points.

```
. lsens, gense(se) gensp(sp) genp(cutp)
```

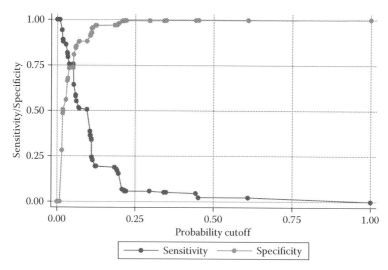

FIGURE 7.1
Sensitivity-specificity.

Sort the values on sensitivity, then on specificity and probability. We list those values where the sensitivity and specificity are closest.

```
. sort se sp cutp

. list se sp cutp in 31/35

   31. | 0.642045    0.808643    0.053400 |
   32. | 0.755682    0.734458    0.037949 |
   33. | 0.755682    0.734689    0.038993 |
   34. | 0.755682    0.737000    0.051643 |  <-
   35. | 0.795455    0.678299    0.033820 |
```

Using .05 for the cutoff point, a classification matrix is created that appears as:

```
. estat class, cut(.05)

Logistic model for death

              -------- True --------
Classified |         D            ~D  |      Total
-----------+--------------------------+-----------
     +     |        133          1138  |       1271
     -     |         43          3189  |       3232
-----------+--------------------------+-----------
   Total   |        176          4327  |       4503

Classified + if predicted Pr(D) >= .05
True D defined as death != 0
--------------------------------------------------
Sensitivity                     Pr( +| D)    75.57%
Specificity                     Pr( -|~D)    73.70%
Positive predictive value       Pr( D| +)    10.46%
Negative predictive value       Pr(~D| -)    98.67%
--------------------------------------------------
False + rate for true ~D        Pr( +|~D)    26.30%
False - rate for true D         Pr( -| D)    24.43%
False + rate for classified +   Pr(~D| +)    89.54%
False - rate for classified -   Pr( D| -)     1.33%
--------------------------------------------------
Correctly classified                         73.77%
--------------------------------------------------
```

We selected the line where the sensitivity and specificity were closest. The associated probability value is the optimal cutoff point for the classification matrix. Note that the optimal cutoff point is very close to the values we determined using more basic statistics, that is, mean response = .045, mean response based on the model observations = .039, and the mean predicted fit = .039.

The classification matrix shows that the model correctly classifies 73.77 percent of the time. This is an acceptable value, lending further credence that the model is well fitted.

Note that sensitivity and specificity are not only important in their own right, but are used for constructing ROC curves.

7.2.3 ROC Analysis

ROC, or Receiver Operator Characteristic, curves are generally used when statisticians wish to use the logistic model to classify cases. The ROC curve is defined as the sensitivity (vertical axis) by the inverse specificity (horizontal axis). A model with no predictive value will have a slope of 1, resulting in an ROC of 0.5. Higher predictive values arch in a manner shown in Figure 7.2, which displays an ROC curve *p*-value of 0.8. Values of 0.95 and greater are highly unlikely. The typical values found for this test range from about 0.6 to 0.9. Again, we have confirmation of a well-fitted model.

Note that SAS calls this value a (Harrell's) C statistic. In either case, higher values indicate a better fit.

```
. lroc

Logistic model for death


number of observations =      4503
area under ROC curve    =    0.7968
```

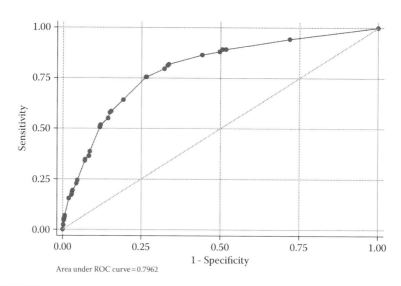

Area under ROC curve = 0.7962

FIGURE 7.2
ROC curve = .7968.

A number of other ROC-related plots exist to give visual assistance to model assessment. They are of particular importance when the model is ill-fitted. Graphical tests many times, like residual analysis, allow us to identify problem areas in the data.

7.3 Information Criteria Tests

We now come to the fit statistics that are currently regarded as valuable tools for the comparison of model fit. All of the Information Criterion tests are comparative in nature, with a lower value regarded as related to the preferred model—the model that fits the data better than others.

There are two general varieties of Information Criterion tests—the Akaike Information Criterion (AIC) and Bayesian Information Criterion (BIC). Each has several versions, which can cause some confusion unless one is clear about the version being used for a particular model.

7.3.1 Akaike Information Criterion—AIC

The original version of the Akaike Information Criterion was developed by Akaike in 1974. However, its widespread use did not occur until the beginning of the 21st century. Defined in terms of the model log-likelihood statistic, its original formulation appears as:

$$AIC = \frac{-2*LL + 2*k}{n} \tag{7.15}$$

or

$$AIC = \frac{-2(LL - k)}{n} \tag{7.16}$$

where LL represents the model log-likelihood, k is the number of predictors in the model, and n the number of observations. The second term, $2k$, is referred to as the penalty term, and adjusts for the size and complexity of the model. k also includes the constant.

As the number of parameters increase, $-2LL$ becomes smaller since more parameters make what is observed more likely. To adjust for this bias, AIC adds the term $2k$ to $-2LL$ as a penalty for increasing the number of parameters in a model.

Larger n also affects $-2LL$. AIC divides $(-2LL + 2k)$ by n to obtain a per observation contribution to the adjusted $-2LL$. Everything being equal, smaller values of AIC indicate a better-fitting model.

Many statistical software applications drop division by n, which usually results in a much larger value than the traditional AIC statistic. When one observes a small AIC value, it is more likely that it is associated with the original version. Stata and several other major statistical software applications use the without-n version. The use of n when comparing models where one is observation based and the other is panel or covariate pattern based gives biased weight to n. In such cases in particular, use of the non-n AIC is preferred.

ALTERNATE AIC STATISTIC

$$AIC = -2 * LL + 2 * k \qquad (7.17)$$

$$AIC = -2(LL - k) \qquad (7.18)$$

Again, when comparing models, whether they are nested models or non-nested models associated with entirely different probability functions, the model having an associated AIC statistic that is lower in value to others with which it is being compared is the preferred model.

Unfortunately, there is no criterion that can be used to determine the statistical significance between AIC statistics, that is, of two models, when do we decide if a lower value of the AIC is significantly lower than the AIC statistic with which it is compared? At present, no such p-value exists. However, I have developed guidelines for Equations 7.17 and 7.18 that can be used to select the preferred model.

Difference between Models A and B	Result if A < B
>0.0 & < = 2.5	No difference in models
>2.5 & < = 6.0	Prefer A if $n > 256$
>6.0 & < = 9.9	Prefer A if $n > 64$
10+	Prefer A

We first demonstrate the AIC statistic for use with nested models.

```
. logistic death anterior hcabg kk2-kk4 age3 age4

Logistic regression                      Number of obs   =       4503
                                         LR chi2(7)      =     209.90
                                         Prob > chi2     =     0.0000
Log likelihood = -638.15957              Pseudo R2       =     0.1412
```

```
-----------------------------------------------------------------
 death | Odds Ratio Std. Err.  z  P>|z| [95% Conf. Interval]
-------+---------------------------------------------------------
anterior |  1.894096 .3173418 3.81 0.000 1.363922    2.630356
   hcabg |  2.195519 .7744623 2.23 0.026 1.099711    4.383246
     kk2 |  2.281692 .4117012 4.57 0.000 1.602024    3.249714
     kk3 |  2.218199 .5971764 2.96 0.003 1.308708    3.759743
     kk4 | 14.63984  5.218374 7.53 0.000 7.279897   29.44064
    age3 |  3.549577 .7119235 6.32 0.000 2.395823    5.258942
    age4 |  6.964847 1.44901  9.33 0.000 4.632573   10.47131
-----------------------------------------------------------------
```

Stata's AIC (and BIC) function is a member of the **estat** series of commands—specifically the **estat** *ic*. The version of AIC used here does not divide the main terms by n.

```
. estat ic
```

```
-----------------------------------------------------------------
 Model |  Obs  ll(null) ll(model)   df      AIC       BIC
-------+---------------------------------------------------------
     . | 4503 -743.1095 -638.1596    8  1292.319  1343.619
-----------------------------------------------------------------
```

```
    Note:  N=Obs used in calculating BIC; see [R] BIC note
. di -2*(-638.15957 -8) /// 8 being the number of predictors
plus constant
1292.3191
```

Now we drop the *age* predictors from the model, comparing the resultant AIC statistic with that of the full model.

```
. logistic death anterior hcabg kk2-kk4
```

```
Logistic regression                 Number of obs   =       4503
                                    LR chi2(5)      =     113.64
                                    Prob > chi2     =     0.0000
Log likelihood = -686.28751         Pseudo R2       =     0.0765
-----------------------------------------------------------------
 death | Odds Ratio Std. Err.  z  P>|z| [95% Conf. Interval]
-------+---------------------------------------------------------
anterior |  2.05989  .3391907 4.39 0.000   1.4917    2.844505
   hcabg |  2.051369 .7098963 2.08 0.038 1.041063    4.042131
     kk2 |  2.573692 .4558429 5.34 0.000 1.818841    3.641818
     kk3 |  3.244251 .8504555 4.49 0.000  1.94079    5.423135
     kk4 | 18.55298  6.12317  8.85 0.000 9.715963   35.4276
-----------------------------------------------------------------
```

```
. estat ic

------------------------------------------------------------------
   Model |   Obs    ll(null)   ll(model)    df        AIC        BIC
---------+--------------------------------------------------------
       . |   4503   -743.1095   -686.2875    6   1384.575   1423.05
------------------------------------------------------------------
        Note:   N=Obs used in calculating BIC; see [R] BIC note
. di -2*(-686.28751 -6)
1384.575
```

A comparison of the nested-model AIC statistics give us:

    AIC (full): 1292.319 <= preferred

    AIC (reduced): 1384.575

```
. di 1-(1292.3191/1384.575)
.0666312
```

The full model is some 7 percent less than the reduced model, indicating that it is the preferred model. The AIC statistic is therefore consistent with the likelihood ratio test for this model comparison.

    Again, the important feature of the AIC statistic rests in its ability to compare models that are not nested. The likelihood ratio test is not appropriate for non-nested models.

7.3.2 Finite Sample AIC Statistic

The Finite Sample AIC is defined as:

$$AIC_{FS} = -2\{LL - k - k(k+1)/(n-k-1)\}/n \tag{7.19}$$

and has a value that is usually close to that of AIC with n. Calculating the statistic from the values given in the set of models used in the previous subsection, we have:

```
. di -2*(-638.1596-8-8*(8+1)/(4503-8-1))/4503
.28699783

. di -2*(-686.28751 -6-6*(6+1)/(4503-6-1))/4503
.3074825
```

Multiplying the Finite Sample AIC statistics by their respective n gives us:

```
. di .28699783*4503
1292.3512
```

```
. di .3074825*4503        /// AIC = 1384.575
1384.5937
```

The AIC and AIC_{FS} statistics are nearly identical.

7.3.3 LIMDEP AIC

William Greene, author of LIMDEP, uses the following formula for another unnamed version of AIC:

$$AIC_{LIMDEP} = (LL - k)/(n/2) - (1 + \ln(2\pi)) \qquad (7.20)$$

```
. di (-638.15957 -8)/(4503/2) - (1+ln(2*_pi)) <= preferred
-3.1248678
```

```
. di (-686.28751 -6)/(4503/2) - (1+ln(2*_pi))
-3.1453554
```

The preferred model is the one having the lowest absolute value for the respective AIC statistic.

7.3.4 SWARTZ AIC

The Swartz AIC statistic, designed by Joel Swartz of Harvard University in 1978, is defined as:

$$AIC_s = (-2LL + k * \ln(n))/n \qquad (7.21)$$

with the AIC statistic having the lowest absolute value as the preferred model.

```
. di (-638.15957 + 8*ln(4503))/4503  <= preferred
-.12677317
```

```
.di (-686.28751 + 6*ln(4503))/4503
-.14119754
```

7.3.5 Bayesian Information Criterion (BIC)

Designed by the University of Washington's Adrian Raftery in 1986, the Bayesian Information Criterion (BIC) is one of the most popular goodness-of-fit tests currently used in commercial statistical software. Typically, both the AIC and BIC statistics are displayed as output for a given modeling situation. Like the AIC statistic, the BIC can compare non-nested models across different samples.

The BIC statistic is defined using either the GLM deviance function or the log-likelihood. The original version described by Raftery is based on the deviance:

$$\text{BIC} = D - df * \ln(n) \tag{7.22}$$

where D is the model deviance statistic, n is the number of model observations, and df is the model degrees of freedom.

We model the same predictors as before but use the GLM algorithm to have a deviance statistic. I also apply a Stata AIC/BIC command that I wrote called **abic** after the model output for comparison purposes. The right-hand column has the AIC and BIC statistics produced by **estat** *ic*, the left-hand side has the traditional AIC statistic and the BIC statistic used in LIMDEP software (to be discussed shortly).

```
. glm death anterior hcabg kk2-kk4 age3 age4, nolog fam(bin) eform

Generalized linear models                No. of obs      =        4503
Optimization       : ML                  Residual df     =        4495
                                         Scale parameter =           1
Deviance         =   1276.319134         (1/df) Deviance =     .283942
Pearson          =   4212.631591         (1/df) Pearson  =    .9371817

Variance function: V(u) = u*(1-u)
Link function     : g(u) = ln(u/(1-u))

                                         AIC             =    .2869907
Log likelihood   = -638.1595669          BIC             =   -36537.86
-------------------------------------------------------------------------
             |                 OIM
     death | Odds Ratio Std. Err.    z  P>|z| [95% Conf. Interval]
---------+---------------------------------------------------------------
  anterior |   1.894096 .3173418  3.81 0.000 1.363922     2.630356
     hcabg |   2.195519 .7744623  2.23 0.026 1.099711     4.383246
       kk2 |   2.281692 .4117012  4.57 0.000 1.602024     3.249714
       kk3 |   2.218199 .5971764  2.96 0.003 1.308708     3.759743
       kk4 |  14.63984 5.218374   7.53 0.000 7.279897    29.44064
      age3 |   3.549577 .7119235  6.32 0.000 2.395823     5.258942
      age4 |   6.964847  1.44901  9.33 0.000 4.632573    10.47131
-------------------------------------------------------------------------

. abic
AIC Statistic =     .2869907              AIC*n       = 1292.3191
BIC Statistic =     .2908262              BIC(Stata)  = 1343.6191
```

Notice that the model output includes both an AIC and a BIC statistic. The displayed BIC statistic is −36537.86, which is different from Stata's BIC—with a value of 1343.6191. The model BIC is based on the deviance definition above. It may be calculated as:

```
. di 1276.319134  -  4495*ln(4503)
-36537.864
```

which gives the same value as shown in the model output. Stata's (non-GLM) version of the BIC statistic is based on the log-likelihood, with no reference to the deviance.

$$BIC_{LL} = -2 * LL + \ln(n) * k \qquad (7.23)$$

```
. di -2*(-638.15957)  +  ln(4503)*8
1343.6191
```

The calculated value based on this definition produces the value shown in **abic** output, as well as **estat** *ic*.

```
. estat ic
```

Model	Obs	ll(null)	ll(model)	df	AIC	BIC
.	4503	.	-638.1596	8	1292.319	1343.619

```
         Note:  N=Obs used in calculating BIC; see [R] BIC note
```

Following the work of Raftery, the degree of model preference can be based on the absolute difference between the BIC statistics of two models.

difference	Degree of preference
0–2	Weak
2–8	Positive
6–10	Strong
> 10	Very Strong

Models A and B:

If $BIC_A - BIC_B < 0$, then A preferred

If $BIC_A - BIC_B > 0$, then B preferred

or

Model with lower BIC value preferred.

For the example models we have worked with in this chapter, the reduced model has the following partial output:

```
                                   No. of obs    =       4503
                                   Residual df   =       4497
                                   AIC           =   .3074784
Log likelihood    = -686.2875063   BIC           = -36458.43


. abic
AIC Statistic =    .3074783      AIC*n       = 1384.5751
BIC Statistic =    .3095883      BIC(Stata)  = 1423.05
```

The deviance-based BIC statistics are −36537.86 to −36458.43.
The log-likelihood-based BIC statistics are 1343.62 to 1423.05.

DEVIANCE

```
. di -36537.86 - (-36458.43)
-79.43
```

LOG-LIKELIHOOD

```
. di 1343.62  -  1423.05
-79.43
```

Both differences are identical, but this is just happenstance. In either case, the absolute difference between the full and reduced models is substantially greater than 10, indicating a very strong preference for the full model.

The AIC and BIC statistics give us consistent advice. Both the AIC and BIC tests tell us that the full model is preferred.

Recall that the true value of the AIC and BIC statistics rests in the fact that they can both compare non-nested models. For example, modeling the same data using the full model with a Bernoulli *loglog* link, the likelihood BIC statistic is 1335.4843. Compare this to the same statistic *logit* link value of 1343.6191. The difference is 8.13, in favor of the *loglog* link. This indicates that the *loglog* link is strongly preferred over the *logit* link.

Note that the deviance statistic is used only for GLM-based statistical procedures. The log-likelihood is the normal way to estimate all other maximum likelihood models. Most contemporary GLM algorithms have a calculated log-likelihood function as part of the output; therefore, the deviance-based formula is rarely used now.

The log-likelihood version of the BIC test is also referred to as the **Swartz Binary Goodness of Fit test**. It is commonly used for binary response data, such as logistic regression.

7.3.6 HQIC Goodness-of-Fit Statistic

The HQIC, or Hannan and Quinn Information Criterion (Hannan and Quinn, 1979), is defined as:

$$HQIC = -2\{LL - k * \ln(k))/n \tag{7.24}$$

The calculated values for the nested models we have been discussing in this chapter are

```
. di -2*(-638.15957 -8*ln(8))/4503
.29082616

. di -2*(-686.28751 -6*ln(6))/4503
.3095883
```

The HQIC test is an alternate version of BIC, used in LIMDEP. If the values of the BIC and HQIC differ greatly, it is wise to check the models.

It is vital to be certain you know which version of AIC and BIC is being used. Whatever the version, be consistent throughout the comparative evaluation of models.

7.3.7 A Unified AIC Fit Statistic

A useful procedure to evaluate the comparative statistical contribution of each predictor to a logistic model using the AIC test statistic was authored by Zhiqiang Wang of the Menzies School of Health Research in Australia (Wang, 2000a). Written as a Stata command called **lrdrop1**, the test is performed following the estimation of the full model. Typing the command name results in a table displaying each predictor and the AIC statistic for the model excluding that predictor. Other statistics are provided as well, but for the purposes of this section, we are interested only in the AIC test results.

The best way to evaluate the comparative AIC statistics is to determine which values are closest to the original full model. The closer the value, the less the predictor contributes to the model. We observe below that *age2* (1292.32) is nearly the same as the original model (1291.65), and is not a significant contribution to the model. Although *hcagb* and *kk3* are significant contributors, they model *p*-values of .035 and .004, respectively. These values are confirmed by the AIC values displayed in the table. The *p > Chi2* test, the results of which are shown in the table is based on the likelihood ratio test, is a preferred method of evaluating the contribution of a predictor. The test indicates that *hcabg* does not in fact significantly contribute to the model.

```
. qui logit death anterior hcabg kk2-kk4 age2-age4

. lrdrop1
Likelihood Ratio Tests: drop 1 term
logit regression
number of obs = 4503
----------------------------------------------------------------
     death     Df     Chi2    P>Chi2   -2*log ll  Res. Df    AIC
----------------------------------------------------------------
Original Model                          1273.65     4494    1291.65
-anterior      1    15.22    0.0001    1288.87     4493    1304.87
  -hcabg       1     3.78    0.0519    1277.43     4493    1293.43
   -kk2        1    19.12    0.0000    1292.78     4493    1308.78
   -kk3        1     7.35    0.0067    1281.00     4493    1297.00
   -kk4        1    43.22    0.0000    1316.88     4493    1332.88
   -age2       1     2.67    0.1025    1276.32     4493    1292.32
   -age3       1    39.59    0.0000    1313.25     4493    1329.25
   -age4       1    79.62    0.0000    1353.27     4493    1369.27
----------------------------------------------------------------
Terms dropped one at a time in turn.
```

Care should be taken though when using **lrdrop1** with interaction terms in the model. The command does not appear to be reliable in these cases.

Wang designed other methods of assessing predictor worth, including a stepwise AIC deletion procedure (Wang, 2000b). The reader is referred to the original source for the various commands Wang prepared that are useful adjuncts to assessing predictor worth.

7.4 Residual Analysis

Residual analysis is a vital aspect of model fit analysis and is the oldest manner statisticians have used to evaluate the worth of a model. Residuals originally were developed for ordinary least squares regression models, but were adapted with adjustment into the GLM framework in the early 1970s. Residuals used for logistic regression models until 1990 were nearly all based on standard GLM residuals. New residuals were developed by statisticians during this time, but they were not nearly as well used as those associated with the standard GLM methods.

With the publication of Hosmer and Lemeshow's 1989 text, *Applied Logistic Regression*, a new cadre of residuals was popularized, based on the m-asymptotics described earlier. The GLM residuals, as I shall refer to them, were traditionally n-asymptotic, or observation based. Recall that m-asymptotic residuals are constructed from the covariate patterns, which means that many times fewer "cases" are used to develop a residual graph. We shall begin by enumerating GLM residuals.

7.4.1 GLM-Based Residuals

A listing of the traditional GLM residuals is shown in Table 7.1.

Each of these residuals has been employed frequently in published research for the past 25 to 35 years. Except for the raw residual each has a unique formulation designed for a particular GLM family. GLM families include those shown in Table 7.2.

We will define each residual in terms of the binomial/binary family, of which the logit link is canonical, meaning that it is directly related to the underlying binomial probability function. The logit-linked Bernoulli model is called logistic regression. The logit-linked binomial model is typically referred to as grouped logistic regression. However, many statisticians refer to the grouped, or what we call binomial or proportional, logistic model as simply logistic regression, giving no difference in name. Recall that the Bernoulli distribution handles binary (1/0) responses. It is a version of the greater binomial distribution, which relates to the number of successes (numerator) out of a specified number of covariate patterns (denominator). The Bernoulli model has a denominator of 1 and is the manner in which most researchers understand logistic regression.

TABLE 7.1

GLM Residuals

RAW RESIDUAL: $y-\mu$

PEARSON RESIDUAL: $r^p = (y-\mu)/\text{sqrt}(V)$

   Note: $\Sigma(r^p)^2 = X^2 = Chi2$

   May be modified as $(y-\mu)/\text{sqrt}(V*k/w)$ where

   k = scale. Used to ensure that the denominator is a reasonable estimator of $V(y)$.

DEVIANCE RESIDUAL: $r^d = \text{sgn}(y-\mu)*\text{sqrt(deviance)}$

   Note: $\Sigma(r^d)^2 = \text{DEVIANCE}$

STANDARDIZED PEARSON RESIDUAL : $r^p/\text{sqrt}(1-h)$

STANDARDIZED DEVIANCE RESIDUAL: $r^d/\text{sqrt}(1-h)$

LIKELIHOOD RESIDUAL:

   $\text{sgn}(y-\mu)*\text{sqrt}\{(h(r^p)^2+(1-h)(r^d)^2\}$

ANSCOMBE RESIDUAL:

   $\int_y^\mu d\mu*V^{-1/3}(\mu)$

Bernoulli: $\{A(y)-A(\mu)\}/\{\mu(1-\mu)^{1/6}\}$

   $A=2.05339*\{\text{Incomplete Beta}(2/3,2/3,z)\}$

Binomial : $\{A(y)-A(\mu)\}/\{\mu(1-\mu)^{1/6} \text{ sqrt}((1-h)/m))\}$

TABLE 7.2

GLM Families and Main Binomial Links

Continuous response
Gaussian (normal)
Gamma
Inverse Gaussian
Binomial response
Bernoulli (binary)
Logit link
Probit link
Complementary loglog link
Loglog link
Binomial (proportional)
<same as Bernoulli>
Count response
Poisson
Negative Binomial
Geometric (a variety of negative binomial)

7.4.1.1 Raw Residual

$$y - \mu \tag{7.25}$$

The raw residual, for each observation in the model, is the response minus the fitted value. It is a measure of discrepancy, or variation. The closer the value of the individual response, y, is to its fitted value, μ, or $\hat{y}$, the better the fit. The method of least squares analysis is based on the total sum of squared residuals, $\Sigma(y - \mu)^2$. Most of the other residuals we will discuss are based on the raw residual. The raw residual is not itself used in residual analysis.

```
. qui glm death anterior hcabg kk2-kk4 age3 age4, fam(bin)
. predict mu, mu
. predict raw, response
. l death mu raw in 1/5
```

```
     +-------------------------------+
     | death       mu          raw  |
     |-------------------------------|
  1. |     0   .0151095    -.0151095 |
  2. |     1   .1140369     .8859631 |
  3. |     0   .0279465    -.0279465 |
  4. |     1   .1105182     .8894818 |
  5. |     0   .0338204    -.0338204 |
     +-------------------------------+
```

7.4.1.2 Pearson Residual

$$r^p = (y - \mu)/\text{sqrt}(V) \tag{7.26}$$

The Pearson residual is simply the raw residual adjusted by the square root of the variance. It is a per-observation statistic. The Bernoulli variance function is $\mu(1 - \mu)$, and the binomial is $\mu(1 - \mu/m)$, where m is the binomial denominator. Recall that for the binary or Bernoulli models, $m = 1$.

$$r^p = (y - \mu)/\text{sqrt}(\mu(1 - \mu/m)) \tag{7.27}$$

The Pearson χ^2, or *Chi2*, goodness-of-fit statistic is defined in terms of the Pearson residual. Summing the square of all observation Pearson residuals yields the Pearson χ^2 statistic. It was at one time a popular GOF test, but has lost favor in recent years due to its bias.

```
. predict pearson, pearson
. gen variance=mu*(1-mu)
. l death mu raw variance pearson in 1/5
```

	death	mu	raw	variance	pearson
1.	0	.0151095	-.0151095	.0148812	-.1238602
2.	1	.1140369	.8859631	.1010325	2.787308
3.	0	.0279465	-.0279465	.0271655	-.1695583
4.	1	.1105182	.8894818	.098304	2.836949
5.	0	.0338204	-.0338204	.0326765	-.1870941

We can verify the calculation of the Pearson statistic by using the values from the first observation:

```
.di -.0151095/sqrt(.0148812)
-.12386001
```

PEARSON *Chi2* STATISTIC

$$\text{Pearson } \chi^2 = \Sigma(r^p)^2 \tag{7.28}$$

```
. egen pear=sum(pearson*pearson)
. su pear if e(sample)
```

Variable	Obs	Mean	Std. Dev.	Min	Max
pear	4503	4212.631	0	4212.631	4212.631

which is identical to the value of the model Pearson statistic, displayed in the full model output. We used the *e(sample)* option to be consistent with the model observations.

The Pearson residual can be thought of as the contribution of the individual residual to the Pearson *Chi2* GOF statistic. When calculated using *m*-asymptotics, it is the contribution of each covariate pattern residual to the Pearson statistic.

The Pearson residual may itself be modified by introducing a scale constant, k, into the formula. Used with binomial logistic models, the value of k is used to ensure that the denominator is a reasonable estimator of $V(y)$. Most GLM software provides for its calculation, but it is rarely used in research.

$$(y - \mu)/\text{sqrt}(V * k/w) \tag{7.29}$$

We have also shown where w, or prior weights, enter into the formula.

The Pearson residual statistic is used in **generalized estimating equations** (GEE), a statistical technique to model longitudinal and clustered data. Each iteration of a GLM estimating algorithm recalculates the Pearson residuals and uses them to calculate an adjustment to the model inverse Hessian matrix, from which standard errors are derived. GEE models are discussed at length in Chapter 13.

7.4.1.3 Deviance Residual

$$r^d = \text{sgn}(y - \mu) * \text{sqrt(deviance)} \tag{7.30}$$

Like the Pearson *Chi2* residual, the deviance residual is a per-observation statistic. However, in this case, the residual is based on the individual deviance statistics. For the binomial family, the deviance statistic is

$$d = +/- \text{sqrt}[2 \sum y * \ln(y/m * \mu) - (m - y) * \ln((m - y)/(m * (1 - \mu)))] \tag{7.31}$$

where $m = 1$ for the binary logistic deviance function. This produces:

$$d = +/- \text{sqrt}[2 \sum y * \ln(y/\mu) - (1 - y) * \ln((1 - y)/(1 - \mu))] \tag{7.32}$$

or

$$d = \text{sqrt}\left[2 \sum \{\ln(1/\mu)\}\right] \quad \text{if } y = 1 \tag{7.33}$$

$$d = \text{sqrt}\left[2 \sum \{\ln(1/(1 - \mu)\}\right] \quad \text{if } y = 0 \tag{7.34}$$

For binomial models with *m* as the binomial denominator, we have:

$$d = -\text{sqrt}(2 * m^* \,|\, \ln(1-\mu)\,|) \text{ if } y = 1 \tag{7.35}$$

$$d = \text{sqrt}(2 * m^* \,|\, \ln(\mu)\,|) \text{ if } y = 0 \tag{7.36}$$

```
. predict deviance, deviance

. 1 death mu deviance in 1/5

    +-------------------------------+
    | death        mu     deviance |
    |-------------------------------|
1.  |     0   .0151095   -.1744984 |
2.  |     1   .1140369    2.083858 |
3.  |     0   .0279465   -.2380944 |
4.  |     1   .1105182    2.098845 |
5.  |     0   .0338204   -.2623185 |
    +-------------------------------+
```

Again, like the overall Pearson statistic, the sum of squared deviance residuals produces the model deviance goodness-of-fit statistic:

$$\text{Deviance} = \Sigma(r^d)^2 \tag{7.37}$$

```
. egen dev=sum(deviance*deviance)

. su dev if e(sample)

    Variable |    Obs      Mean   Std. Dev.       Min        Max
-------------+------------------------------------------------------
         dev |   4503  1276.319          0   1276.319   1276.319
```

The model deviance goodness-of-fit statistic is identical to the value of the full model deviance.

In general, the deviance residual is preferred over the Pearson residual for evaluating the distributional properties of the fitted model. The primary reason is that its distribution is closer to residuals produced in normal or OLS regression, that is, it is generally less skewed than Pearson residuals. An adjustment to the deviance residual makes approximation to normality even closer for binomial models.

$$\text{Adj } r^d = r^d \, 1/6 * ((1-2\mu/m)/(\text{sqrt}(m * \mu * (1-\mu))) \tag{7.38}$$

7.4.1.4 Standardized Pearson Residual

Standardized Pearson residuals are normalized to a standard deviation of 1.0. Normalization is approximated by dividing the statistic by sqrt(1–*hat*).

$$r_{\text{Std}}{}^p = r^p/\text{sqrt}(1-h) \tag{7.39}$$

The *hat* statistic is a *hat* matrix diagonal statistic, a measure of the influence of a predictor to the model. Since it is an important regression statistic, we need to spend a little time defining it more carefully.

7.4.1.4.1 Hat Matrix Diagonal

The *hat* matrix diagonal, h, is defined in terms of matrix algebra as:

$$h = W^{1/2}\, X(X'WX)^{-1}\, X'W^{1/2} \tag{7.40}$$

where

$$W = \text{diag}\{1/\{V(\mu)\} * (\partial\mu/\partial\eta)^2\} \tag{7.41}$$

For the binomial family of distributions, $\partial\mu/\partial\eta$ is the inverse partial derivative of the binomial link, which is

$$\text{BINOMIAL LINK } \partial\mu/\partial\eta : ln(\mu/(m-\mu)) \tag{7.42}$$

$$\text{BERNOULLI LINK } \partial\mu/\partial\eta : ln(\mu/(1-\mu)) \tag{7.43}$$

and, as earlier defined:

$$\text{BINOMIAL VARIANCE } V(\mu) = \mu(m-\mu) \tag{7.44}$$

$$\text{BERNOULLI VARIANCE } V(\mu) = \mu(1-\mu) \tag{7.45}$$

The *hat* matrix statistic can easily be calculated using standard maximum likelihood and GLM derived statistics. It is a shortcut calculation, defined as:

$$h = V(\mu) * \text{std}p\verb|^|2 \tag{7.46}$$

where *stdp* = standard error of prediction, that is, the standard error of μ.

Using Stata's programming language, the Bernoulli *hat* statistic can be calculated as:

```
* mu and variance previously calculated; inserted to show
logic or calculation

. qui glm death anterior hcabg kk2-kk4 age3 age4, fam(bin)
                                /// no display
. predict mu                    /// predicted y; ie mu
. predict stdp, stdp            /// stand. error of mu
. gen variance = mu*(1-mu)      /// calc binary logistic
                                    variance
. gen hat = stdp*stdp*variance /// calc hat statistic
```

To compare with Stata's hard-coded calculation:

```
. predict hmd, hat              /// hat matrix diagonal
```

Now list the results:

```
. l hmd hat in 1/5             /// compare values

     +---------------------+
     |      hmd        hat |
     |---------------------|
  1. | .0005095   .0005095 |
  2. | .0046063   .0046063 |
  3. | .0008865   .0008865 |
  4. | .0032823   .0032823 |
  5. | .0013681   .0013681 |
     +---------------------+
```

High values for *hat* indicate abnormal covariate patterns.

7.4.1.4.2 Standardized Pearson Residual

We can now return to the calculation of the standardized Pearson residual. The term *standardized* may be taken to mean that the Pearson residual has been adjusted to account for the correlation between y and μ, or $\hat{y}$. The standardized Pearson residual has been frequently found in graphics of logistic regression models. Typically, the graph is of the standardized Pearson by the fitted value, μ, or by the standardized Pearson by the linear predictor, also called η.

```
. gen pearson, p
. gen stdpear = pearson/sqrt(1-hmd)
. predict spear, pearson standard
. l pearson spear stdpear in 1/5
```

```
+---------------------------------+
|   pearson       spear    stdpear |
|---------------------------------|
1. | -.1238602    -.1238917   -.1238917 |
2. |  2.787308     2.79375     2.79375  |
3. | -.1695583    -.1696335   -.1696335 |
4. |  2.836949     2.841617    2.841616 |
5. | -.1870941    -.1872222   -.1872222 |
+---------------------------------+
```

Notice that, because the *hat* values associated with the first five observations are very small, with the largest value .004 and the smallest .0005, the difference in values between the Pearson and standardized Pearson is miniscule. A table showing the moments of the *hat* statistic and its range can be produced by:

. su hmd, detail

```
                              hat diagonal
-----------------------------------------------------------------
          Percentiles      Smallest
  1%        .0002982        .0002982
  5%        .0002982        .0002982
 10%        .0002982        .0002982      Obs                 4503
 25%        .0002982        .0002982      Sum of Wgt.         4503

 50%        .0008815                      Mean            .0017766
                            Largest       Std. Dev.       .0035467
 75%        .0013992        .0344714
 90%        .0032823        .0344714      Variance        .0000126
 95%        .0056677        .0414185      Skewness        5.922438
 99%        .019962         .0419859      Kurtosis        46.24533
```

The range of *hat* values is from .0003 to .04199, with a median value of .0009.

7.4.1.5 *Standardized Deviance Residual*

The standardized deviance residual is one of the central residuals used for graphing GLM-based regression models, including logistic regression. It comes closer to achieving normality than any of the previous residuals. We have found that producing a scatterplot of the standardized deviance (vertical axis) by the fitted value, μ (horizontal axis), is the best graphical method of assessing the internal shape of the modeled data from among the standard GLM-based residuals.

. predict sdeviance, deviance standard

. l deviance sdeviance in 1/5

```
         +----------------------+
         |  deviance    sdeviance |
         |----------------------|
    1.  | -.1744984    -.1745429 |
    2.  |  2.083858     2.088674 |
    3.  | -.2380944     -.2382   |
    4.  |  2.098845     2.102298 |
    5.  | -.2623185    -.2624981 |
         +----------------------+
```

Again, the difference between the deviance and standardized deviance is negligible. For graphing purposes, it seems to be most instructive to square the standardized deviance statistic and graph by μ. Values above 4.0 are considered outliers. The two streams of values represent $y = 1$ and $y = 0$ values. Off-stream values are influential, and not necessarily contributory to the model. See Figure 7.3.

```
. gen sdeviance2 = sdeviance*sdeviance
. scatter sdeviance2 mu, title(square standardized deviance by fit)
. su sdeviance2 if death==1, detail
```

```
                              sdeviance2
-------------------------------------------------------------
         Percentiles      Smallest
   1%      1.019326       1.019326
   5%      2.180668       1.019326
  10%      3.277268       1.019326     Obs               176
  25%      4.419656       1.019326     Sum of Wgt.       176

  50%      4.688557                    Mean         5.368926
                         Largest       Std. Dev.    1.923443
  75%      5.935105       9.650903
  90%      8.389132       9.650903     Variance     3.699633
  95%      9.650903       9.650903     Skewness      .312205
  99%      9.650903       9.650903     Kurtosis     3.065412
```

```
. su sdeviance2 if death==0, detail
```

```
                              sdeviance2
-------------------------------------------------------------
         Percentiles      Smallest
   1%      .0161387       .0161387
   5%      .0161387       .0161387
  10%      .0161387       .0161387     Obs              4327
  25%      .0161387       .0161387     Sum of Wgt.      4327

  50%      .0356742                    Mean          .0778522
                         Largest       Std. Dev.     .1253747
  75%      .1061972       1.942057
```

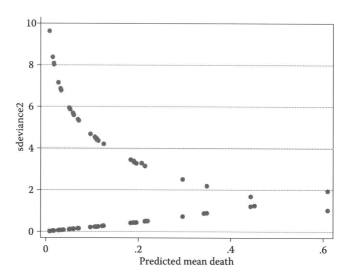

FIGURE 7.3
Squared standardized deviance by fit.

```
90%        .203595        1.942057        Variance        .0157188
95%        .2350039       1.942057        Skewness        7.641566
99%        .4387647       1.942057        Kurtosis        95.17601
```

There are many outliers for those patients who have died within 48 hours of admission. No patients who survived until 48 hours are outliers, at the $\alpha = 0.05$ level (approximately 2 standard deviations, squared –4).

The covariate pattern(s) having the greatest outlier value, that is, squared standardized deviances, are

```
. 1 death anterior hcabg kk1-kk4 age1-age4 if
sdeviance2>9.637270 & anterior!=. & kk1!=. & age2!=.
```

	death	anterior	hcabg	kk1	kk2	kk3	kk4	age1	age2	age3	age4
23.	1	Inferior	0	1	0	0	0	1	0	0	0
135.	1	Inferior	0	1	0	0	0	1	0	0	0
460.	1	Inferior	0	1	0	0	0	0	1	0	0
587.	1	Inferior	0	1	0	0	0	0	1	0	0
895.	1	Inferior	0	1	0	0	0	1	0	0	0
1003.	1	Inferior	0	1	0	0	0	0	1	0	0
1961.	1	Inferior	0	1	0	0	0	1	0	0	0
3399.	1	Inferior	0	1	0	0	0	0	1	0	0
3642.	1	Inferior	0	1	0	0	0	0	1	0	0
5350.	1	Inferior	0	1	0	0	0	1	0	0	0

7.4.1.6 Likelihood Residuals

The likelihood statistic was specifically designed for logistic regression models. It is a weighted average of standardized Pearson and standardized deviance residuals, and is defined as:

$$r^L = \text{sgn}(y - \mu) * \text{sqrt}\{h(r^p)^2 + (1 - h)(r^d)^2\} \tag{7.47}$$

```
. predict like, likelihood

. l sdeviance like in 1/5

     +---------------------------+
     |  sdeviance          like  |
     |---------------------------|
  1. |  -.1745429     -.1745208  |
  2. |   2.088674      2.092467  |
  3. |     -.2382      -.238148  |
  4. |   2.102298      2.105149  |
  5. |  -.2624981     -.2624099  |
     +---------------------------+
```

The likelihood and standardized deviance statistics are normally close in value. Because of this, the likelihood residual is now rarely used in research.

7.4.1.7 Anscombe Residuals

The Anscombe residual is the closest to normal of any other residual. It, too, is usually close to the standardized deviance, but is rarely found in commercial statistical software. At present, only Stata, SPSS (beginning with version 15), and XploRe software incorporate Anscombe residuals in their GLM residual offerings. The Anscombe residual is defined as:

$$r^L = \int_y^\mu \frac{d\mu}{V^{1/3}(\mu)} \tag{7.48}$$

Each GLM family has its own unique solution of the above equation. For the Bernoulli and binomial families, we have:

$$\text{Bernoulli: } \{A(y) - A(\mu)\}/\{\mu(1-\mu)\}^{1/6} \tag{7.49}$$

with
$$A(z) = \text{Beta}(2/3, 2/3) * [\text{Incomplete Beta}(2/3, 2/3, z)]$$
$$A(z) = 2.05339 * [\text{Incomplete Beta}(2/3, 2/3, z)]$$

where z takes the value of y or μ as appropriate. The constant value of the two-term *beta* function with both parameters at 2/3, is 2.05339.

```
. di exp(lngamma(.666667)+lngamma(.666667)-lngamma(.666667+.666667))
2.0533894
```

The code for calculating the residual can be given as:

```
((2.0533902*ibeta(2/3,2/3,y))-(2.0533902*ibeta(2/3,2/3,mu)))/
(mu*(1-mu))^(1/6)
```

The Anscombe residuals for the binomial family appear as

$$\frac{\left(A\left(\frac{2}{3}, \frac{2}{3}, \frac{y}{m}\right) - A\left(\frac{2}{3}, \frac{2}{3}, \mu\right)\right)}{\mu^{1/6}\left(1-\mu\right)^{1/6}} \tag{7.50}$$

```
((2.0533902*ibeta(2/3,2/3,(y/m)))-(2.0533902*ibeta(2/3,2/3,(mu))))/
((mu*(1-mu))^(1/6))
```

Some statisticians multiply $\sqrt{1-h}/m$ to the denominator, where h is the hat matrix diagonal.

Anscombe residuals for the current example model are calculated as:

```
. predict anscombe, anscombe

. l sdeviance like anscombe in 1/5    /// Note: "l" is short
                                              for "list"
```

```
        +----------------------------------------+
        | sdeviance         like      anscombe |
        |----------------------------------------|
   1.   | -.1745429     -.1745208    -.1852245 |
   2.   |  2.088674      2.092467     2.483685 |
   3.   |   -.2382       -.238148    -.2528956 |
   4.   |  2.102298      2.105149     2.506215 |
   5.   | -.2624981     -.2624099    -.2787101 |
        +----------------------------------------+
```

Values of Anscombe greater than 2.0 are usually considered outliers.

7.4.2 *m*-Asymptotic Residuals

M-asymptotic residuals are those that are based on covariate patterns rather than individual observations. In Stata, post-estimation residuals following the logistic command are based on *m*-asymptotics. Traditionally, GLM post-estimation residuals are *n*-asymptotic, but recently applications such as Stata and SAS have incorporated *m*-asymptotics into post estimation algorithms. We will discuss the primary *m*-asymptotic residuals detailed in Hosmer and Lemeshow (2000).

7.4.2.1 Hat Matrix Diagonal Revisited

We discussed observation-based *hat* matrix diagonal statistics in the previous section. We now expand the discussion to the context of *m*-asymptotics.

Large *hat* values indicate covariate patterns far from the average covariate pattern, regardless of the residual value. When *hat* statistics are graphed by the standardized Pearson residuals, values on the horizontal extremes are high residuals (Figure 7.4). Covariate patterns with high leverage (*hat*) and low residuals are interesting because they are not easily detected by usual analysis—they do not fit the model, but do not appear as residual outliers. Graphing leverage or *hat* statistics by *m*-asymptotic standardized Pearson residuals can be performed in Stata using the following commands:

```
. qui logistic death anterior hcabg kk2-kk4 age3 age4
. predict mhat, hat
. predict mspear, rstandard
. scatter mhat mspear, xline(0)
```

Values with high leverage and outside the range of +/− 2 are covariate patterns that act against a well-fitted model.

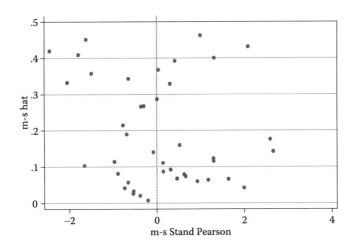

FIGURE 7.4
Standardized Pearson by hat *m*-asymptotic.

7.4.2.2 Other Influence Residuals

DeltaX (*dx2*) residual values indicate the decrease in the Pearson *Chi2* statistic due to deletion of the covariate pattern associated with the respective *deltaX* residual. It is defined as the square of the standardized Pearson residual.

deltaD (*ddeviance* or *dd*) is the change in deviance created by deletion of the covariate pattern associated with the respective *deltaD* residual. It is defined as $(d^2 + (r^2*h)/(1 - h))$, but may be approximated by normalization of the deviance; that is, by dividing *d* by sqrt$(1 - h)$. *h* above is the hat matrix diagonal, or *hat*, and *r* is the Pearson residual, or *pear*. Lastly, *dbeta* or *db*, evaluates the effect that the deletion of a covariate pattern has on the value of the estimated coefficients. We calculate these by:

```
. predict dx2, dx2
```

```
. predict dd, ddeviance
```

```
. predict db, dbeta
```

I shall now show a graph of *dx2* by μ (Figure 7.5). Values above the line at *dx2^2* = 4 are considered likely outliers (upper 95 percentile of *Chi2* = 3.84). The points extending from the upper left to the lower right are singular patterns with low = 1. Points from the upper right to the lower left are singular patterns with low = 0. Other off-curve points are multi-case covariate patterns. Most covariate patterns in this model are multi-case.

```
. predict mu, pr
```

```
. scatter dx2 mu,   yline(4)
```

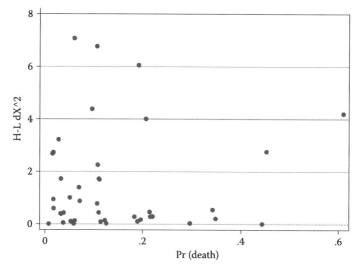

FIGURE 7.5
*Delta*X-squared by probability of death.

Five covariate patterns may be considered outliers.

Pregibon's *db* is a measure of the change in the coefficient vector that would be caused by deleting a covariate pattern. *dx2* measures the decrease in *mspear* caused by deleting a covariate pattern. These two measures may be combined in such a manner that information regarding both leverage and model GOF are provided. We do this by graphing *dx2* by *mu*, but weighted by *db*. Poorly fit patterns are vertically high, and patterns with high leverage are formed as large circles. Large circles represent more observations in the covariate pattern. Those points that are dually suspect are generally identified as being large high circles, as well as those in the middle of the graph or in the cup. Although not clearly seen in Figure 7.6, many scatter graphs of *dx2* and *mu* have one line of covariates starting high on the left side, swooping down to the lower right of the graph. A second line starts high on the right and swoops down to the left. The image forms a cup, with the two lines crossing near the low center. We see this image in Figure 7.7, but not for our example model as displayed in Figure 7.6. Here, the four circles above the line are influential outliers, likely altering the more standard image we see in Figure 7.7. The smaller circle on the line is also influential. Together, they are the patterns that highly influence the model coefficients but do not fit well with the rest of the model.

PARADIGM FORM OF GRAPH

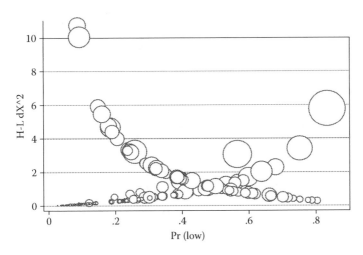

FIGURE 7.6
Paradigm form of graph.

```
. scatter dx2 mu [w=db], yline(4) mfcolor(white) t1(Symbol
size proportional to dBeta)
```

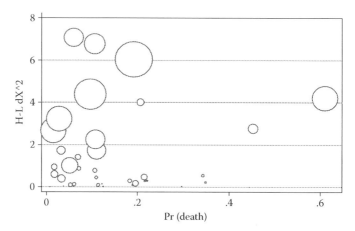

FIGURE 7.7
Symbol size proportional to dBeta.

7.4.3 Conditional Effects Plot

Perhaps one of the more interesting features of logistic regression is the ability to use a model to compare probabilities of predictor levels. For example, we can analyze the relationship of infarction site (*anterior* or *inferior*) and *age* on the odds for dying within 48 hours of hospital admission. To do this, we may construct a conditional effects plot of *death* on *age* controlled for infarct site. Figure 7.8 demonstrates this relationship.

```
. logit death anterior age

Logistic regression                      Number of obs   =        4696
                                         LR chi2(2)      =      144.62
                                         Prob > chi2     =      0.0000
Log likelihood = -713.68569             Pseudo R2       =      0.0920
------------------------------------------------------------------------
    death |    Coef. Std. Err.     z   P>|z|  [95% Conf. Interval]
--------+---------------------------------------------------------------
 anterior | .7266114 .1578098    4.60 0.000 .4173099     1.035913
      age | .0567263    .00545  10.41 0.000 .0460445      .0674081
    _cons |-7.603426 .4293466 -17.71 0.000 -8.44493     -6.761922
------------------------------------------------------------------------

. gen L1 = _b[_cons]+_b[age]*age+_b[anterior]*1
. gen Y1 = 1/(1+exp(-L1))
. gen L2 = _b[_cons]+_b[age]*age+_b[anterior]*0
. gen Y2 = 1/(1+exp(-L2))
. scatter Y1 Y2 age, title("Probability of Death w/i 48 hrs
admission") subtitle("Myocardial Infarct Site")
```

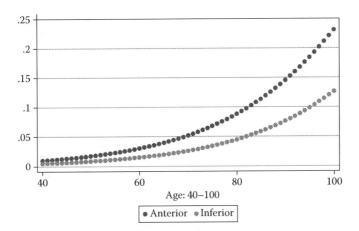

FIGURE 7.8
Myocardial infarct site: Probability of death within 48 hours admission.

Note the higher and steadily increasing probability of death for patients having had an *anterior* myocardial infarction versus an *inferior* site, which is under *anterior* in the above graph. The older the patient, the greater the risk— and more so for *anterior* site patients.

Construct a conditional effects plot of *death* on *age*, controlled by *killip* level. Figure 7.9 demonstrates this relationship.

```
. logit death age killip
. gen K1 =  _b[_cons]+_b[age]*age+_b[killip]*1
. gen R1 = 1/(1+exp(-K1))
. gen K2 =  _b[_cons]+_b[age]*age+_b[killip]*2
. gen R2 = 1/(1+exp(-K2))
. gen K3 =  _b[_cons]+_b[age]*age+_b[killip]*3
. gen R3 = 1/(1+exp(-K3))
. gen K4 =  _b[_cons]+_b[age]*age+_b[killip]*4
. gen R4 = 1/(1+exp(-K4))
. scatter R1 R2 R3 R4 age, title("Probability of Death w/i
48 hrs admission") subtitle("Killip Level")
```

Although it may not be clear from the lines, which come out in color in Stata, the higher killip level is above the others. That is, killip level 1 is the lowest line on the graph. Even if a patient is 100 years of age, their probability of *death*, based on killip level alone, is less than 0.1. For killip level 4 patients, those at the highest risk, they have a near 0.6 probability of *death* if 100 years of age. Actually, the graph is as we would expect, based on prior clinical background.

If we wish to add other predictors to the model, they must be entered into the calculation set at their mean value. The median value may also be specified. Conditional effects plots are valuable visual tools in research, but

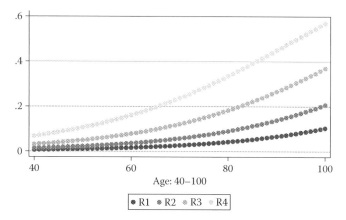

.6
.4
.2
0
        40   60   80   100
                   Age: 40–100

                ● R1   ● R2   ● R3   ● R4

FIGURE 7.9
Killip level: Probability of death within 48 hours admission.

necessitate having a continuous predictor. Some statisticians use a continuous predictor for constructing a conditional effects plot, but when actually constructing the model, they categorize the continuous predictor into three or four levels. So doing allows one to see any differential odds at various points along the continuum of the continuous predictor.

7.5 Validation Models

Statistical models are developed, in general, to understand the contributing factors giving rise to a particular distribution of the response. We seek the most parsimonious model that passes the criteria for a well-fitted model. Inherent to this understanding are two additional features of the model—the ability to predict values of the response both in and outside the fitted model, and the ability to classify. Care must be taken, however, when extrapolating to outside-model predictions. Usually there is little difficulty if the predictions are based on values that are within the range actually fitted. If we predict based on covariate or predictor values that exceed the range used in constructing the model, we may find that predictions turn out to be mistaken.

Many researchers use a validation data set to help confirm the use of the model to a greater population. This is particularly the case when the data come from a sample of observations within a population.

Validation data can be taken from the model itself or can be based on data that were not used in constructing the model. Usually the validation data consist of 20 percent of the observations being modeled—but there is no hard rule as to the actual percentage used. If taken from the model itself, the procedure is to take a sample from the fitted model and remodel it. The parameter estimates, or coefficients, should be similar to those produced by the fitted model. A Hausman test (1978) on the equality of coefficients can be used to access the statistical difference in the two groups of observations.

A researcher may also withhold a percentage of data from the model fitting process. Once the selected data have been modeled, the withheld data—the validation data—are modeled for comparison purposes. If the coefficients are similar, the two sets of data are combined and remodeled for a final fitted model. These methods help prevent us from creating overfitted models— models that cannot be used for classification or prediction of non-model data from a greater population.

It is easy to overfit a model. We tend to do this when we employ many predictors in the model and fine-tune it with highly specific transformations and statistical manipulations. That is, when we try every statistical trick to fit the data to the model, it is more likely that we cannot use that model to evaluate outside data. Testing a model against validation data helps eliminate the possibility of overfitting. At least it is a good check for it.

Several of the residuals that we have considered in the previous section actually mimic the notion of subsample testing, although the subsample is either an individual observation or a covariate pattern. In the section on *m*-asymptotic residuals we discussed three residuals in particular—the *deltaX* (Pearson *Chi2*), *deltaD* (change in deviance, or *dd*), and *delta*β. All of these residuals are based on the exclusion of the associated covariate pattern, calculating the change in the Pearson *Chi2*, the change in deviance, or the change in the coefficient vector, respectively. This gives us the impact on the summary statistic, or parameters, if the covariate pattern was excluded from the model. Taking a 20 percent sample from the model and remodeling the remaining data inform us if the model is stable. We may jackknife the model using percentages of data from 1 percent to 20 percent, testing for the overall model consistency. This tactic is also a version of validation, albeit internal validation of the model. Bringing in additional data from the population to evaluate the stability of the parameter estimates is an external validation technique. Whichever method we prefer to evaluate the stability of the data, it is clear that employing some type of validation test provides us with means to access the goodness-of-fit of the model to the data.

Two points should be remembered when accessing fit. First, if the model is of a sample from a greater population of observations, we expect that the fitted model can be used to predict response values in the population. If the model only tells us about the sample and cannot be extrapolated to a greater population, as well as to future observations as the population grows, it is of little use. Of course, it is dangerous to predict values of the response beyond the range specified by the fitted model.

I shall give a simple example of a validation sample randomly taken from the model. We model the **heart01** data as before, then take a 20 percent random sample from the data, and remodel. It is fully expected that the coefficients will differ, but we look to see if major differences occur. We also look at the consistency of the deviance and Pearson dispersion statistics, which are adjusted by the degrees of freedom, and which in turn are based on the number of model observations. To obtain the relevant statistics and parameter values, we use the GLM command.

```
. glm death anterior hcabg kk2-kk4 age3-age4, nolog fam(bin)

Generalized linear models                    No. of obs      =        4503
Optimization       : ML                      Residual df     =        4495
                                             Scale parameter =           1
Deviance        =   1276.319134              (1/df) Deviance =     .283942
Pearson         =   4212.631591              (1/df) Pearson  =    .9371817

Variance function: V(u) = u*(1-u)        [Bernoulli]
Link function     : g(u) = ln(u/(1-u))   [Logit]

                                             AIC             =    .2869907
Log likelihood   = -638.1595669              BIC             =   -36537.86
-----------------------------------------------------------------------------
             |                 OIM
       death |     Coef.  Std. Err.      z    P>|z|    [95% Conf. Interval]
-------------+---------------------------------------------------------------
    anterior |   .6387416  .1675426    3.81   0.000    .3103642     .9671191
       hcabg |   .7864185  .3527468    2.23   0.026    .0950475      1.47779
         kk2 |   .8249174  .1804368    4.57   0.000    .4712678     1.178567
         kk3 |   .7966954  .2692168    2.96   0.003    .2690401     1.324351
         kk4 |   2.683746  .3564503    7.53   0.000    1.985117     3.382376
        age3 |   1.266828  .2005658    6.32   0.000    .8737267      1.65993
        age4 |   1.940876  .2080462    9.33   0.000    1.533112     2.348639
       _cons |  -4.815946  .1934408  -24.90   0.000   -5.195083    -4.436809
-----------------------------------------------------------------------------
```

Model data can be preserved in memory for later use by issuing the *preserve* command.

```
. preserve
```

Take a random 20 percent sample:

```
. sample 20
```

Now remodel on the remaining approximate 1/5 of the data.

```
. glm death anterior hcabg kk2-kk4 age3-age4, nolog fam(bin)
```

```
Generalized linear models          No. of obs       =        896
Optimization       : ML            Residual df      =        888
                                   Scale parameter  =          1
Deviance         =   298.7737834   (1/df) Deviance  =    .336457
Pearson          =   849.3522476   (1/df) Pearson   =   .9564778

Variance function: V(u) = u*(1-u)       [Bernoulli]
Link function    : g(u) = ln(u/(1-u))   [Logit]

                                   AIC              =     .35131
Log likelihood   = -149.3868917    BIC              =  -5737.797
---------------------------------------------------------------------
           |             OIM
    death  |   Coef.   Std. Err.    z   P>|z|  [95% Conf. Interval]
-----------+---------------------------------------------------------
  anterior | .4698369  .3412397   1.38  0.169  -.1989806   1.138654
     hcabg | .4327004  .7874227   0.55  0.583  -1.11062    1.976021
       kk2 | .9164658  .3756461   2.44  0.015   .1802131   1.652719
       kk3 | .9092588  .5489248   1.66  0.098  -.1666139   1.985132
       kk4 |  2.95479  .7732182   3.82  0.000   1.43931    4.47027
      age3 | 1.030879  .3836705   2.69  0.007   .2788987   1.782859
      age4 | 1.498619  .4590615   3.26  0.001   .5988748   2.398363
     _cons |-4.290099  .3608704 -11.89  0.000  -4.997392  -3.582806
---------------------------------------------------------------------
```

First, check the two dispersion statistics. The deviance dispersion changes from 0.284 to 0.336, and the Pearson dispersion from 0.94 to 0.96. The consistency is quite good. Other similar samples will likely not be as close to what we see here. With respect to the coefficients, each value is well within the confidence interval of the other. Evidence of overfitting, or of a poorly fit model, exists when this coherence fails to be the case.

Finally, to restore the original data back into memory, erase the 20 percent random sample data from active RAM and issue the *restore* command. Once submitted, the preserved data are brought back into memory as if no intermediate analyses had been executed.

Exercises

7.1. Model the same **absent** data as found in Exercise 6.1, then evaluate the model using the following fit statistics.

 a. Evaluate the Hosmer–Lemeshow test, grouping the data by decile. What are the test statistic and *p*-value? What other tests related to the Hosmer–Lemeshow test are important to access?

  b. Calculate the Pearson *Chi2* test. What do the statistic and *p*-value tell us?

  c. Calculate the AIC and BIC statistics. What is their importance?

  d. Construct a classification test. What is the percent correctly classified? The specificity? The sensitivity?

  e. Develop and graph the ROC curve.

  f. Graph standardized deviance residuals by the predicted value. How are outliers assessed using this type of graph?

7.2. What is the *hat* matrix diagonal? How is it best used to assess influential observations?

7.3. How do *m*-asymptotics differ from *n*-asymptotics? When is it appropriate to use *m*-asymptotics?

7.4. On January 28, 1986, NASA's space shuttle, *Challenger*, exploded only 1 minute and 13 seconds after takeoff, killing all seven astronauts aboard. A commission found that O-rings failed to seal due to a number of factors. The cold temperature appears to have been the major cause of the crash, although other factors may have been contributory. Rubber O-rings sealing the joints of the booster rocket tend to become rigid in colder weather. Use the **shuttle** data provided on the book's Web site to demonstrate that a proper evaluation of the data would have provided evidence that the O-rings would fail, resulting in a crash of the shuttle. Use a binary logistic model to calculate the probabilities of O-ring failure. Note that the ill-fated *Challenger* flight was STS 51-L. Determine if the model used to determine the appropriate probabilities is well fitted.

```
Contains data from shuttle.dta
obs: 25                      First 25 space shuttle flights
-------------------------------------------------------------
flight      byte   flbl    Flight  Challenger: STS 51-L
month       byte           Month of launch
day         byte           Day of launch
year        int            Year of launch
distress    byte   dlbl    Thermal distress incidents
temp        byte           Joint temperature, degrees F
damage      byte           Damage severity index (Tufte 1997)
comments    str55          Comments (Tufte 1997)
any         float          1= distress>0
date        float
-------------------------------------------------------------
```

7.5. Using the **heartprocedure** data, determine the comparative risks of having to have both CABG and PTCA heart procedures compared to only one based on age and a prior medical history of having had a myocardial infarction or a TIA.

a. Calculate the probabilities of having both a PTCA and CABG compared with only one of the procedures.

b. Construct a conditional effects plot of the binary response, *cabgptca2*, and *age*.

c. The $\Delta\beta$ statistic is defined *m*-asymptotically as,

$$\Delta\beta = h * (R_P)^2/(1 - h),$$

where R_P is the standardized Pearson residual and h the *hat* matrix diagonal statistic. With the model having both *age* and the predictor created in 7.5a as predictors, graph $\Delta\beta$ by μ and interpret the results.

d. Perform the various fit analyses, that is, classification table, Hosmer–Lemeshow test with deciles of risk, and ROC test.

7.6. The following clinical trial data come from Zelterman (1999), page 127.

	Men		Women	
	Treated	Control	Treated	Control
Success	8000	4000	12000	2000
Failure	5000	3000	15000	3000

a. Determine the probabilities of success for patients based on whether they were in the treatment or control groups. Control for the *sex* of the patients.

b. Is gender a significant predictor of success status?

c. Determine the probabilities of success for patients based on both their treatment and gender status. What combination of predictor values provide the greatest likelihood of success? The greatest likelihood of failure?

d. What is unique about the odds ratios of both predictors?

R Code

Section 7.1 Traditional Fit Tests

```
library('foreign')
heart<- read.dta('heart01.dta') #read Stata format file
heart$anterior <- heart$anterior[,drop=TRUE] #drop empty
   levels that complicate tabulations
```

```
heart$center<- factor(heart$center) #convert to factor from
  numeric levels
heart$killip<- factor(heart$killip) #convert to factor from
  numeric levels

fit7_1a<- glm(death ~ 1, data=heart,family=binomial(link=logit))
  #intercept only
summary(fit7_1a)

fit7_1b<- glm (death ~ age, data=heart, family=binomial)
summary(fit7_1b)

1- 1826.6/1974.8 #pseudo-R^2 from deviances

1974.8-1826.6 #Likelihood ratio from deviances
anova(fit7_1a, fit7_1b, test='Chisq') #LRT
```

Section 7.2 Hosmer–Lemeshow GOF Test

```
fit7_2a<- glm(death ~ anterior + hcabg + kk2 + kk3 + kk4 +
  age3 + age4, data=heart, family=binomial)
summary(fit7_2a)
exp(coef(fit7_2a)) #ORs

source('ralhlGOFtest.r') #macro for H-L GOF test
heart2<- na.omit(heart) #drop rows with missing data
hlGOF.test(heart2$death, predict(fit7_2a, heart2,
  type='response'), breaks=10)
hlGOF.test(heart2$death, predict(fit7_2a, heart2,
  type='response'), breaks=6)

mean(heart$death)
sqrt(mean(heart$death)*(1-mean(heart$death))/
  length(heart$death))
library('binGroup') #binomial package
binBlaker(length(heart$death), sum(heart$death)) #Blaker C.I.
  for proportion

mu<- predict(fit7_2a, heart2, type='response') #get predicted
  values again
muSE<- sd(mu)/sqrt(length(mu)) #S.E. of mean
cat('Mean:', mean(mu), 'S.E.:', muSE, '95% C.I.:',
  mean(mu)-1.96*muSE, mean(mu)+1.96*muSE, '\n')
length(mu)

mean(heart2$death) #new mean for rows without missing data
sqrt(mean(heart2$death)*(1-mean(heart2$death))/
  length(heart2$death))
```

```
binBlaker(length(heart2$death), sum(heart2$death)) #Blaker
  C.I. for proportion

library('epicalc') #package for cmx() and lroc()
cmxdf<- data.frame(id=1:nrow(heart2), death=heart2$death,
  pred=mu)
cmx(cmxdf, threshold=0.05) #confusion matrix

lroc(fit7_2a) #ROC plot
```

Section 7.3 Information Criteria Tests

```
fit7_3a<- glm(death ~ anterior + hcabg + kk2 + kk3 + kk4 +
  age3 + age4, data=heart, family=binomial)
summary(fit7_3a)
exp(coef(fit7_3a)) #ORs
cat('AIC:',AIC(fit7_3a, k=2), 'BIC:',
  AIC(fit7_3a, k=log(nrow(!is.na(heart)))), '\n')
1276.3+8*2 #calc. AIC from deviance

fit7_3b<- glm(death ~ anterior + hcabg + kk2 + kk3 + kk4,
  data=heart, family=binomial)
summary(fit7_3b)
exp(coef(fit7_3b)) #ORs
cat('AIC:',AIC(fit7_3b, k=2), 'BIC:',
  AIC(fit7_3b, k=log(nrow(!is.na(heart)))), '\n')
1372.6 + 6*2 #calc. AIC from deviance

1 - 1292.319/1384.575 #change in AIC between 2 models

-2*(as.numeric(logLik(fit7_3a)) - 8 - 8*(8+1)/(4503-8-1))/4503
  #AICfs
-2*(as.numeric(logLik(fit7_3b)) - 6 - 6*(6+1)/(4503-6-1))/4503
  #AICfs
0.2869978*4503 #compare to AIC=1292.3
0.3074825*4503 #compare to AIC=1384.6

(as.numeric(logLik(fit7_3a)) - 8)/(4503/2) - (1+log(2*pi))
  #AIClimdep
(as.numeric(logLik(fit7_3b)) - 6)/(4503/2) - (1+log(2*pi))
  #AIClimdep

(as.numeric(logLik(fit7_3a)) + 8*log(4503))/4503 #AICs
(as.numeric(logLik(fit7_3b)) + 6*log(4503))/4503 #AICs

cat('AIC:',AIC(fit7_3a, k=2), 'BIC:',
  AIC(fit7_3a, k=log(nrow(!is.na(heart)))), '\n')
deviance(fit7_3a) - 4495*log(4503)
-2*as.numeric(logLik(fit7_3a)) -4495*log(4503) #BIC
```

```
AIC(fit7_3a, k=log(nrow(!is.na(heart)))) - AIC(fit7_3b,
  k=log(nrow(!is.na(heart))))

-2*(as.numeric(logLik(fit7_3a)) - 8*log(8))/4503 #HQIC
-2*(as.numeric(logLik(fit7_3b)) - 6*log(6))/4503 #HQIC

step(fit7_3a, direction='backward', k=2) #backward selection
  via AIC
```

Section 7.4 Residual Analysis

```
fit7_4a<- glm(death ~ anterior + hcabg + kk2 + kk3 + kk4 +
  age3 + age4, data=heart, family=binomial)
mu<- predict(fit7_4a, type='response')
raw<- heart2$death - mu #use heart2 because dropped missing
data rows
head(cbind(mu, raw)) #predicted and raw residuals

pearson<- residuals(fit7_4a, type='pearson')
variance<- mu*(1-mu)
head(cbind(death=heart2$death, mu=mu, raw=raw,
  variance=variance, pearson=pearson))
-0.015109539/sqrt(0.01488124) #calc. pearson residual
pear<- sum(pearson*pearson) #chi-sq.
pear
length(pearson)

devianze<- residuals(fit7_4a, type='deviance')
head(cbind(death=heart2$death, mu=mu, deviance=devianze))
dev<- sum(devianze*devianze) #chi-sq.
dev

pred<- predict(fit7_4a, se.fit=TRUE, type='response')
mu<- pred$fit #predictions of probs
stdp<- pred$se.fit #S.E. of fit
variance<- mu*(1-mu)
h<- stdp*stdp*variance #hat statistics
hmd<- hatvalues(fit7_4a)
head(cbind(hmd=hmd, hat=h))

stdpear<- pearson/sqrt(1-hmd) #std. pearson resid.
head(cbind(pearson, stdpear))
summary(hmd)

sdeviance<- rstandard(fit7_4a) #standardized deviance
  residuals
head(cbind(devianze, sdeviance))
```

```
sdeviance2<- sdeviance*sdeviance
summary(sdeviance2[heart2$death==1])
summary(sdeviance2[heart2$death==0])

heart2[sdeviance2>9.63727 & heart2$anterior=='Inferior',c(1:7,
  13:16)]

mhat<- hatvalues(fit7_4a)
msdev<- rstandard(fit7_4a) #use std. deviance, std. pearson
  not available
plot(msdev, mhat, main='Std. Deviance by Hat', xlab='Std. Dev.
  Residual', ylab='Hat value', col='blue', xlim=c(-3,4))
abline(v=0, col='red')

rstud<- rstudent(fit7_4a)
dbeta<- dfbeta(fit7_4a)
dbetas<- dfbetas(fit7_4a)

layout(matrix(1:4, ncol=2))
plot(fit7_4a)

fit7_4b<- glm(death ~ anterior + age, data=heart,
  family=binomial)
summary(fit7_4b)
L1<- coef(fit7_4b)[1] + coef(fit7_4b)[3]*heart$age +
  coef(fit7_4b)[2]*(heart$anterior=='Anterior')
Y1<- 1/(1+exp(-L1))
L2<- coef(fit7_4b)[1] + coef(fit7_4b)[3]*heart$age +
  coef(fit7_4b)[2]*(heart$anterior!='Anterior')
Y2<- 1/(1+exp(-L2))
layout(1)
plot(heart$age, Y1, col=1, main='P[Death] within 48 hr
  admission', xlab='Age')
lines(heart$age, Y2, col=2, type='p')

knum<- as.numeric(heart$killip) #convert factor to level values
fit7_4c<- glm(death ~ age + knum, data=heart, family=binomial)
K1<- coef(fit7_4c)[1] + coef(fit7_4c)[2]*heart$age +
  coef(fit7_4c)[3]*1
R1<- 1/(1+exp(-K1))
K2<- coef(fit7_4c)[1] + coef(fit7_4c)[2]*heart$age +
  coef(fit7_4c)[3]*2
R2<- 1/(1+exp(-K2))
K3<- coef(fit7_4c)[1] + coef(fit7_4c)[2]*heart$age +
  coef(fit7_4c)[3]*3
R3<- 1/(1+exp(-K3))
K4<- coef(fit7_4c)[1] + coef(fit7_4c)[2]*heart$age +
  coef(fit7_4c)[3]*4
R4<- 1/(1+exp(-K4))
```

```
layout(1)
plot(heart$age, R1, col=1, main='P[Death] within 48 hr
  admission', sub='Killip Level', xlab='Age', ylim=c(0,0.6))
lines(heart$age, R2, col=2, type='p')
lines(heart$age, R3, col=3, type='p')
lines(heart$age, R4, col=4, type='p')
```

Section 7.5 Validation Models

```
fit7_5a<- glm(death ~ anterior + hcabg + kk2 + kk3 + kk4 +
  age3 + age4, data=heart, family=binomial)
summary(fit7_5a)
deviance(fit7_5a)/fit7_5a$df.residual #dispersion

i20<- sample(1:nrow(heart), round(0.2*nrow(heart)),
  replace=FALSE)
heart20<- heart[i20,]
fit7_5b<- glm(death ~ anterior + hcabg + kk2 + kk3 + kk4 +
  age3 + age4, data=heart20, family=binomial)
summary(fit7_5b)
deviance(fit7_5b)/fit7_5b$df.residual #dispersion
```

8

Binomial Logistic Regression

In Chapter 4, we provided an overview of the binomial distribution, upon which logistic regression is based. We also differentiated between two varieties of logistic regression: the binary response (1/0) logistic model and the proportional (n/m) response, or binomial logistic model. The proportional response model derives from the full binomial probability distribution (PDF); the binary response logistic model is derived from the Bernoulli distribution, which is the binomial model with each observation having a response denominator of one (1). That is, the Bernoulli is the most elementary form of binomial model.

The derivation of the Bernoulli algorithm—including all of the statistics required in the binary logistic regression estimating algorithm—was detailed in Chapter 4. We shall do likewise for the full binomial algorithm in this chapter.

Basic to all binomial/Bernoulli distributions and the derived logistic regression algorithms is the notion of p as the probability of success. GLM theorists have typically referred to p as μ, which ranges from near 0 to near 1, and y as the binary response 1/0. A value of $y = 1$ indicates a success. $y = 0$ represents a lack of success, which may be thought of as a failure. The binomial PDF, incorporating a binomial denominator, may be expressed as:

BINOMIAL PDF

$$f(y_i; p_i, m) = \Pi_{i=1}^{n} (m \ y) \ p_i^{y_i} (1 - p_i)^{m - y_i} \tag{8.1}$$

Expressed in terms of the exponential family form, without subscripts:

$$f(y; \theta, \phi) = \Pi \exp\{(y\theta - b(\theta)) / \alpha(\phi) + c(y; \phi)\} \tag{8.2}$$

we have:

$$f(y; p, m) = \Pi \exp\left\{ y * \ln(p) + m * \ln(1 - p) - y * \ln(1 - p) + \ln\binom{m}{y} \right\} \tag{8.3}$$

or

$$f(y;p,m) = \Pi \exp\left\{ y * \ln(p/(1-p)) + m * \ln(1-p) + \ln\binom{m}{y} \right\} \qquad (8.4)$$

The probability function may be reparameterized as a likelihood, expressed in the binomial case as:

$$L(p;y,m) = \Pi \exp\left\{ (y\ln(p/(1-p)) + m * \ln(1-p) + \ln\binom{m}{y} \right\} \qquad (8.5)$$

As discussed with respect to the Bernoulli distribution in Chapter 4, the value of the exponential family of distribution is such that if a distribution is parameterized to reflect the above structure, the link, mean, and variance of the distribution can be calculated from the following relationships.

LINK

$$\theta = \ln(p/(m-p)) = \eta = \ln(\mu/(m-\mu)) \qquad (8.6)$$

DERIVATIVE OF THE LINK

$$g'(\mu) = m/(\mu * (m-\mu)) \qquad (8.7)$$

MEAN

The mean is calculated as the first derivative of $b(\theta)$ wrt θ:

$$b(\theta) = m * \ln(1-p) \qquad (8.8)$$

$$b'(\theta) = mp = \mu \qquad (8.9)$$

INVERSE LINK

$$\mu = g^{-1}(\mu) = m/(1+\exp(-\eta)) \qquad (8.10)$$

VARIANCE

The variance is calculated as the second derivative of $b(\theta)$ wrt θ:

$$b''(\theta) = p(1-p) = \mu\left(1-\mu/m\right) \qquad (8.11)$$

therefore,

$$p = \mu/m \qquad (8.12)$$

$$V = \mu\left(1-\mu/m\right) \qquad (8.13)$$

LOG-LIKELIHOOD

The log-likelihood is determined by logging (natural or Naperian log) the likelihood function (Equation 8.5), appearing as:

$$L(\mu;y,m) = \Sigma\left\{y\ln(\mu/(1-\mu)) + m*\ln(1-\mu) + \binom{m}{y}\right\} \qquad (8.14)$$

DEVIANCE

The deviance function, which is also used as a goodness-of-fit statistic for logistic models, is defined as:

$$D = 2\Sigma\{LL(y;y) - LL(\mu;y)\} \qquad (8.15)$$

which calculates to:

$$D = 2\Sigma\{y\ln(y/\mu) + (m-y)\ln((m-y)/(m-\mu))\} \qquad (8.16)$$

or:

$$D(m > 1; y = 0) = 2m*\ln(m/(m-\mu)) \qquad (8.17)$$

$$D(m > 1; y = m) = 2m*\ln(m/\mu) \qquad (8.18)$$

$$D(m > 1; 0 < y < m) = 2y\ln(y/\mu) + 2(m-y)\ln((m-y)/(m-\mu)) \qquad (8.19)$$

For completeness, the first and second derivatives of the log-likelihood function are also provided. These values are necessary for maximum likelihood implementations.

$$\frac{\partial(L)}{\partial\mu} = \frac{y}{\mu} - (m-y)(1-\mu)^{-1} \tag{8.20}$$

$$\frac{\partial^2(L)}{\partial\mu^2} = \frac{y}{\mu^2} - \frac{m-y}{(1-\mu)^2} \tag{8.21}$$

We now can construct the GLM-based binomial logistic algorithm.

GLM BINOMIAL LOGISTIC REGRESSION ALGORITHM

```
Dev=0
µ = (y+.5)/(m+1)
η = ln(µ/(m-µ))
WHILE (abs(ΔDev) > tolerance level)   {
    u = m*(y-µ)/(µ*(m-µ))
    w⁻¹ = m/(µ*(m-µ))
    w = µ*(m-µ)/m
    z = η + u - offset
    β = (X'WX)-1 X'Wz
    η = Xβ  + offset
    µ = m/(1+exp(-η))
    OldDev = Dev
    Dev = 2Σ{y*ln(y/µ)  - (m-y)*ln((m-y)/(m-µ))}
    ΔDev = Dev - OldDev
}
χ2 = Σ (y-µ)2 / (µ*(1-µ/m))
LL = Σ{y*ln(m/m)+(m-y)*ln(1-(µ/m))}
```

The basic formulae and binomial logistic algorithms have now been covered. We now turn to a fully worked-out example.

Historically, the initial implementations of logistic regression were parameterized as grouped or binomial models. In binomial or grouped logistic models, the response is formulated in terms of a proportion with the numerator indicating how many successes (1s) there are for a given pattern of covariates. The number of observations sharing the same covariate pattern is the response denominator. We dealt with this notion when discussing m-asymptotics, but here the data are structured m-asymptotically from the outset.

Consider the example we used for discussing covariate patterns. The data are structured as:

	y	$x1$	$x2$	$x3$	covariate pattern
1:	1	1	0	1	1
2:	1	1	1	1	2
3:	0	1	0	1	1

	y	*x*1	*x*2	*x*3	covariate pattern
4:	1	0	0	1	3
5:	0	0	1	1	4
6:	1	1	0	0	5
7:	0	0	1	0	6
8:	1	0	0	1	3
9:	0	1	0	1	1
10:	1	1	0	0	5

We don't have to, but it makes it easier to see the patterns we are discussing by combining like covariate patterns.

	y	*x*1	*x*2	*x*3	covariate pattern
1:	1	1	0	1	1
3:	0	1	0	1	1
9:	0	1	0	1	1
2:	1	1	1	1	2
4:	1	0	0	1	3
8:	1	0	0	1	3
5:	0	0	1	1	4
6:	1	1	0	0	5
10:	1	1	0	0	5
7:	0	0	1	0	6

To be the same covariate pattern, the values of *x*1, *x*2, and *x*3 must be identical—therefore, the first three cases directly above share the same pattern of 1-0-1. The value of *y* indicates the count of successes for a given covariate pattern. Restructuring the above gives:

y	cases	*x*1	*x*2	*x*3
1	3	1	0	1
1	1	1	1	1
2	2	0	0	1
0	1	0	1	1
2	2	1	0	0
0	1	0	1	0

The grouped logistic model is generally estimated as a GLM since it is the natural or canonical form of the binomial distribution. We use Stata's **glm** command to model the data above.

```
. glm y x1 x2 x3, fam(bin cases) eform nolog

Generalized linear models                     No. of obs      =        6
Optimization         : ML                     Residual df     =        2
                                              Scale parameter =        1
Deviance         =    8.231021724             (1/df) Deviance = 4.115511
Pearson          =    6.630002874 =>          (1/df) Pearson  = 3.315001

Variance function: V(u) = u*(1-u/cases)
Link function     : g(u) = ln(u/(cases-u))

                                              AIC             =   2.97548
Log likelihood   = -4.926441078               BIC             =   4.647503
------------------------------------------------------------------------
             |                OIM
    y | Odds Ratio  Std. Err.     z    P>|z|    [95% Conf. Interval]
---+--------------------------------------------------------------------
x1 |    1.186947   1.769584   0.11   0.908    .0638853    22.05271
x2 |    .2024631   .3241584  -1.00   0.318    .0087803    4.668551
x3 |    .5770337   .9126937  -0.35   0.728     .025993     12.8099
---+--------------------------------------------------------------------
```

Given the fact that all predictors have *p*-values substantially greater than 0.05, this is not a well-fitted model. Moreover, the Pearson dispersion, defined as "(1/df) Pearson," is considerably greater than 1.0, indicating that the model is overdispersed. The term *Pearson* refers to the Pearson χ^2 goodness-of-fit statistic; degree of freedom is the number of model predictors less one.

$$\text{Dispersion Statistic} = \text{Pearson}/(\text{residual df})$$

Here we have a dispersion statistic of 3.315, indicating that the model is not appropriate for the data. There is an excess of correlation in the data, which can be a result of a variety of different situations. We will devote the entire next chapter to the problem of overdispersion; in this chapter, we simply detail the basic nature of the binomial logistic model. Overdispersion primarily occurs in logistic models when they are binomially parameterized. Binary response models can only be overdispersed if the data are clustered, and the value of the dispersion statistic is irrelevant to its status as overdispersed. The criterion for binary response model overdispersion differs from the standard meaning of the term, as we shall elaborate on in the next chapter. Although we will not discuss it further at this point, keep the problem of overdispersion in mind when observing a binomial model dispersion statistic greater than one.

We may convert a binary response logistic model to a grouped model if there are not too many predictors, and if the predictors are binary or categorical with few levels. We return to the **heart01** data since it allows rather easy modeling of both individual and grouped data.

The following Stata code will convert the observation-based model to a grouped or proportional response model. We begin, though, by producing a binary response model of *death* on the explanatory predictors previously found to be significantly predictive.

```
. glm death anterior hcabg kk2-kk4 age3-age4, nolog fam(bin)

Generalized linear models              No. of obs       =        4503
Optimization       : ML                Residual df      =        4495
                                       Scale parameter  =           1
Deviance        =    1276.319134       (1/df) Deviance  =     .283942
Pearson         =    4212.631591       (1/df) Pearson   =    .9371817

Variance  function: V(u) = u*(1-u)
Link function      : g(u) = ln(u/(1-u))

                                       AIC              =    .2869907
Log likelihood  =  -638.1595669        BIC              =   -36537.86
```

```
-------------------------------------------------------------------------
             |                OIM
      death  |     Coef.  Std. Err.      z   P>|z|   [95% Conf. Interval]
-------------+-----------------------------------------------------------
   anterior  |  .6387416  .1675426    3.81  0.000   .3103642    .9671191
      hcabg  |  .7864185  .3527468    2.23  0.026   .0950475    1.47779
        kk2  |  .8249174  .1804368    4.57  0.000   .4712678   1.178567
        kk3  |  .7966954  .2692168    2.96  0.003   .2690401   1.324351
        kk4  |  2.683746  .3564503    7.53  0.000   1.985117   3.382376
       age3  |  1.266828  .2005658    6.32  0.000   .8737267    1.65993
       age4  |  1.940876  .2080462    9.33  0.000   1.533112   2.348639
      _cons  | -4.815946  .1934408  -24.90  0.000  -5.195083  -4.436809
-------------------------------------------------------------------------
```

To parameterize the above table in terms of odds ratios, we need simply type:

```
. glm, eform nohead    /// nohead option suppresses display of
                           header statistics
```

```
-------------------------------------------------------------------------
             |                OIM
      death  | Odds Ratio  Std. Err.      z   P>|z|   [95% Conf. Interval]
-------------+-----------------------------------------------------------
   anterior  |  1.894096  .3173418    3.81  0.000   1.363922   2.630356
      hcabg  |  2.195519  .7744623    2.23  0.026   1.099711   4.383246
        kk2  |  2.281692  .4117012    4.57  0.000   1.602024   3.249714
        kk3  |  2.218199  .5971764    2.96  0.003   1.308708   3.759743
        kk4  |  14.63984  5.218374    7.53  0.000   7.279897   29.44064
```

```
    age3 |   3.549577   .7119235   6.32 0.000   2.395823    5.258942
    age4 |   6.964847    1.44901   9.33 0.000   4.632573   10.47131
---------------------------------------------------------------------
```

All statistics above the table of odds ratios are identical to the model estimating the "parameter estimates." Therefore, we chose not to repeat their display. We now convert the binary response model to a binomial or proportional response model. Comments following three foreslashes after each command provide a rationale for the terms used in the respective commands.

```
. keep death anterior hcabg kk1 kk2 kk3 kk4 age1 age2 age3
age4
```

```
. egen grp = group(anterior hcabg kk1 kk2 kk3 kk4 age1 age2
age3 age4)
```

/// Assign a unique number, beginning with 1, for each pattern of covariates from *anterior* to *age4*. The group or covariate pattern number is stored in the newly created variable, *grp*.

```
. egen cases = count(grp), by(grp)
```

/// The number of cases having the same covariate pattern is calculated and stored in the new variable, *cases*. *Cases* is the binomial denominator.

```
. egen dead=total(death), by(grp)
```

/// For each covariate pattern, the sum of the number of times the variable *death* is equal to 1 is calculated and stored in the new variable, *dead*. *Dead* is the binomial numerator.

```
. sort grp
```

/// Covariate pattern group numbers are sorted from 1 to n, with all *grp* numbers grouped together.

```
. by grp: keep if _n ==1
```

/// For each covariate pattern, only the first observation is kept; all others are discarded.

We now construct the binomial logistic model, noting that the number of observations in the data has been reduced from 4483 to 895. There are 895 distinct covariate patterns in the data. Given that there are 10 binary predictors involved (includes reference variables), there are $2^{10} = 1024$ maximum possible covariate pattern configurations. Some 87 percent (895/1024) of the possible configurations were expressed in the model.

```
. glm dead anterior hcabg kk2 kk3 kk4 age3 age4, nolog fam(bin
cases) eform

Generalized linear models           No. of obs      =        53
Optimization      : ML              Residual df     =        45
                                    Scale parameter =         1
Deviance        =  57.39847115      (1/df) Deviance =  1.275522
Pearson         =  56.74320607      (1/df) Pearson  =   1.26096

Variance function: V(u) = u*(1-u/cases)
Link function    : g(u) = ln(u/(cases-u))

                                    AIC             =  3.387615
Log likelihood    = -81.77178466    BIC             = -121.2647
```

		OIM				
dead	Odds Ratio	Std. Err.	z	P>|z|	[95% Conf. Interval]	
anterior	1.894096	.3173418	3.81	0.000	1.363922	2.630356
hcabg	2.195519	.7744623	2.23	0.026	1.099711	4.383246
kk2	2.281692	.4117012	4.57	0.000	1.602024	3.249714
kk3	2.218199	.5971764	2.96	0.003	1.308708	3.759743
kk4	14.63984	5.218374	7.53	0.000	7.279897	29.44064
age3	3.549577	.7119235	6.32	0.000	2.395823	5.258942
age4	6.964847	1.44901	9.33	0.000	4.632573	10.47131

We should mention that Stata has a command that makes conversion from a binary to a grouped binomial model easier than using the above six-step algorithm. In fact, only two lines of code are required to convert the data. The operations and associated options of Stata's **collapse** command are currently not available as a command or procedure in other commercial statistical software. Therefore, those using a statistical application other than Stata will not be able to employ the conversion method shown here. On the other hand, the method detailed above is easily replicable using the command-line facilities of the majority of commercial software packages.

```
. use heart01, clear

. drop if anterior==. | kk2==.

. gen count=1
```

```
. collapse (sum) count (sum)  death, by(anterior hcabg kk2 kk3
kk4 age3 age2 age4)

. glm death anterior hcabg kk2-kk4 age3-age4, nolog fam(bin
count) eform

Generalized linear models              No. of obs      =        53
Optimization       : ML                Residual df     =        45
                                       Scale parameter =         1
Deviance       =   57.39847115         (1/df) Deviance =  1.275522
Pearson        =   56.74320607         (1/df) Pearson  =   1.26096

Variance function: V(u) = u*(1-u/count)
Link function    : g(u) = ln(u/(count-u))

                                       AIC             =  3.387615
Log likelihood   = -81.77178466        BIC             = -121.2647
```

```
-----------------------------------------------------------------
            |                  OIM
      death | Odds Ratio  Std. Err.    z   P>|z|  [95% Conf. Interval]
------------+----------------------------------------------------
   anterior |   1.894096  .3173418  3.81 0.000   1.363922   2.630356
      hcabg |   2.195519  .7744623  2.23 0.026   1.099711   4.383246
        kk2 |   2.281692  .4117012  4.57 0.000   1.602024   3.249714
        kk3 |   2.218199  .5971764  2.96 0.003   1.308708   3.759743
        kk4 |  14.63984   5.218374  7.53 0.000   7.279897  29.44064
       age3 |   3.549577  .7119235  6.32 0.000   2.395823   5.258942
       age4 |   6.964847  1.44901   9.33 0.000   4.632573  10.47131
-----------------------------------------------------------------
```

Note that the two methods of converting observation to aggregated data result in identical outcomes.

As an aside, the observation-based binary response model may be estimated as a grouped binomial model by using frequency weights. The response term is brought into the *by()* option of the **collapse** function, unlike the above method, which creates binomial denominators. This method will be used in Chapter 10 to convert observation-based models, such as the proportional odds model, into an aggregated format.

For those who are using their computers to follow this discussion, the data in current memory must be preserved so that we can return to it when finished with this example.

We converted observation-based data to grouped format, which is the form of the data now in memory. Stata's **preserve** command allows the data in current memory to be stored while the data—or even new data files—are loaded and/or operated on in any manner we desire. The original data can subsequently be restored to active memory by using the **restore** command. Therefore, we preserve the current data and re-load the **heart01** data into active memory.

```
. preserve

. use heart01,clear

. gen count=1

. drop if anterior==. | kk2==.          /// optional

. collapse (sum) count , by(death anterior hcabg kk2 kk3 kk4
age3 age2 age4)

. glm death anterior hcabg kk2-kk4 age3-age4 [fw=count] ,
nolog fam(bin) eform
```

```
Generalized linear models          No. of obs      =        4503
Optimization      : ML             Residual df     =        4495
                                   Scale parameter =           1
Deviance        =  1276.319134     (1/df) Deviance =     .283942
Pearson         =  4212.631591     (1/df) Pearson  =    .9371817

Variance function: V(u) = u*(1-u)
Link function    : g(u) = ln(u/(1-u))

                                   AIC             =    .2869907
Log likelihood   = -638.1595669    BIC             =   -36537.86
```

```
-----------------------------------------------------------------
             |              OIM
      death | Odds Ratio  Std. Err.    z   P>|z|  [95% Conf. Interval]
-----------+-----------------------------------------------------
    anterior |  1.894096   .3173418  3.81  0.000  1.363922   2.630356
       hcabg |  2.195519   .7744623  2.23  0.026  1.099711   4.383246
         kk2 |  2.281692   .4117012  4.57  0.000  1.602024   3.249714
         kk3 |  2.218199   .5971764  2.96  0.003  1.308708   3.759743
         kk4 |  14.63984   5.218374  7.53  0.000  7.279897   29.44064
        age3 |  3.549577   .7119235  6.32  0.000  2.395823   5.258942
        age4 |  6.964847    1.44901  9.33  0.000  4.632573   10.47131
-----------------------------------------------------------------
```

```
. count
91

. restore
```

Note that the AIC and Pearson dispersion statistics for the six-step algorithm and the collapse method are identical. The dispersion is 1.26 and represents the amount of overdispersion in the binomial logistic model. The method using frequency weights results in the same AIC and Pearson dispersion statistics as we found with the binary observation-based model. This indicates that the frequency-weighted model is essentially an observation-based model, not a binomial model. The count of observations in the data set is 91, compared with 53 in the two truly converted binomial data sets.

We now return to the discussion related to the binomial format created by the earlier conversions. The above **restore** command brought the previously stored data into memory.

Recall that we combined age group levels 1 and 2 for a unified reference level. We found that *age2*, when used in the model, was not statistically significant. Note that the odds ratios, standard errors, and derived table statistics are all identical to the observation-based model. The binomial deviance statistic, however, is half the value of the Bernoulli deviance. The Pearson dispersion, unlike the binary response model, now informs the statistician of possible overdispersion in the data, that is, model variability not accounted for by the model. The dispersion is 1.26. Given the large size of the data set, the model should ideally have a dispersion approximating 1.0. Admittedly though, it is possible that much of the evident overdispersion can be accounted for by random variation, and it may not be related to extra correlation in the data.

The simplest method to adjust for the 26 percent overdispersion in the above model is to scale the standard errors by the Pearson dispersion. As we discussed earlier, the method is a *post hoc* procedure, applied after the parameters and standard errors have been estimated. Applying it to our model gives us:

```
. glm death anterior hcabg kk2 kk3 kk4 age3 age4, nolog
fam(bin count) irls scale(x2) eform
```

```
Generalized linear models              No. of obs      =        53
Optimization       : MQL Fisher scoring  Residual df     =        45
                     (IRLS EIM)           Scale parameter =         1
Deviance           = 57.39847115         (1/df) Deviance =  1.275522
Pearson            = 56.74320353         (1/df) Pearson  =   1.26096

Variance function: V(u) = u*(1-u/count)
Link function     : g(u) = ln(u/(count-u))

                                         BIC             = -121.2647
```

```
         |                 EIM
  death  | Odds Ratio  Std. Err.    z   P>|z|  [95% Conf. Interval]
---------+---------------------------------------------------------
anterior |   1.894096  .3563509   3.40  0.001  1.309962   2.738704
   hcabg |   2.195519   .869663   1.99  0.047  1.010112   4.772051
     kk2 |   2.281692  .4623095   4.07  0.000  1.533872   3.394102
     kk3 |   2.218199  .6705841   2.64  0.008  1.226517    4.01169
     kk4 |  14.63984  5.859841   6.70  0.000    6.6808   32.08071
    age3 |   3.549577  .7994365   5.62  0.000  2.282804   5.519305
    age4 |   6.964847  1.627129   8.31  0.000  4.406093   11.00955
```

(Standard errors scaled using square root of Pearson X2-based dispersion)

Scaling only minimally adjusted the standard errors. The interpretation of the scaled standard errors is that they represent the standard errors of the model if the dispersion had a value of 1.0, or scaling attempts to eliminate the effect of the dispersion. Given the relatively small percentage of overdispersion in the model, we did not expect the standard errors to be changed more than we observe.

Scaling adjusts weights, W, in:

$$\beta = (X'WX)^{-1}\, X'Wz \tag{8.22}$$

which produces the parameter estimates for a generalized linear model. It replaces W by the inverse square root of the Pearson dispersion statistic. Doing this in effect adjusts the model standard errors to the value that would have been calculated if the dispersion statistic had originally been 1.0. Fortunately, the parameter p-values are still all significant.

We may also employ a robust or sandwich variance estimator to account for the extra correlation in the data due to the clustering effect of provider site. As with the binary response model, we can presume that the standard of care is more similar within a provider or hospital than between providers. This violates the assumption upon which likelihood, or the derived deviance, theory is based—that is, that observations are independent of one another. Since the explanatory impact of observations is collectively stored in the model covariate patterns, it must be understood that covariate patterns are themselves assumed to be independent of one another.

We model the data, basing the clustering effect on *center* (or hospital), as we did when we considered the binary response model. The data, however, must be reconverted. We may convert the observation-based data to binomial format with *center* included in *group*(), or, if the **collapse** method is employed, *center* is included in its *by*() option.

```
. glm dead anterior hcabg kk2-kk4 age3 age4, fam(bin cases)
vce(cluster center) eform nolog

Generalized linear models              No. of obs      =        895
Optimization      : ML                 Residual df     =        887
                                       Scale parameter =          1
Deviance       = 616.200182            (1/df) Deviance =  .6947014
Pearson        = 936.5039284           (1/df) Pearson  =  1.055811

Variance function: V(u) = u*(1-u/cases)
Link function    : g(u) = ln(u/(cases-u))

                                       AIC             =  .9345461
Log pseudolikelihood = -410.209401  BIC             = -5412.582

                (Std. Err. adjusted for 47 clusters in center)
-----------------------------------------------------------------
             |              Robust
        dead | Odds Ratio  Std. Err.   z    P>|z|  [95% Conf. Interval]
---------+-------------------------------------------------------
    anterior |   1.894096  .3866809  3.13  0.002   1.269488    2.82602
       hcabg |   2.195519  .6870412  2.51  0.012   1.188972   4.054177
         kk2 |   2.281692  .4066383  4.63  0.000   1.609006   3.235611
         kk3 |   2.218199  .7249049  2.44  0.015   1.169038   4.208934
         kk4 |   14.63984  4.396843  8.94  0.000   8.126271   26.37431
        age3 |   3.549577  .5975619  7.53  0.000    2.55199   4.937125
        age4 |   6.964847  1.668158  8.10  0.000   4.355513    11.1374
-----------------------------------------------------------------

. abic
AIC Statistic =    .9345462                AIC*n      = 836.41882
BIC Statistic =    .9538434                BIC(Stata) = 874.7934
```

Generally, such a method is used when one knows a grouping or clustering factor. In this case, we do, and it does appear to make a difference to the model.

Parameter coefficients and odds ratios for binomial logistic models are interpreted in exactly the same manner as for binary response logistic regression. We would also likely want to merge predictors *kk2* and *kk3* since their parameter estimates are close. There appears to be little differential effect in the two predictors. To show this relationship for binomial models, we repeat the creation of *kk23*, the merger of *kk2* and *kk3*. We do not need to check for missing values this time, though, since the data set is now parameterized in terms of covariate patterns. There are no more missing values.

```
. gen byte kk23=kk2==1 | kk3==1
```

We model the data, employing a robust variance estimator based on *center.*
The data appears to have a slight clustering effect now that it is formatted in
terms of covariate patterns.

```
. glm dead anterior hcabg kk23 kk4 age3 age4,fam(bin cases)
vce(cluster center) eform

Generalized linear models              No. of obs       =        895
Optimization       : ML                Residual df      =        888
                                       Scale parameter  =          1
Deviance         =    616.210437       (1/df) Deviance  =   .6939307
Pearson          =    932.5089613      (1/df) Pearson   =   1.050123

Variance function: V(u) = u*(1-u/cases)
Link function    : g(u) = ln(u/(cases-u))

                                       AIC              =    .932323
Log pseudolikelihood = -410.2145285   BIC              =  -5419.369

                   (Std. Err. adjusted for 47 clusters in center)
------------------------------------------------------------------
             |               Robust
        dead | Odds Ratio   Std. Err.    z   P>|z|  [95% Conf. Interval]
-------------+----------------------------------------------------
    anterior |   1.892059   .3778025   3.19  0.001   1.279288   2.798342
       hcabg |   2.196883   .6855977   2.52  0.012   1.191698   4.049932
        kk23 |   2.265916   .4131094   4.49  0.000   1.585103   3.239144
         kk4 |   14.64305   4.405796   8.92  0.000   8.119366   26.40832
        age3 |   3.546141   .5849571   7.67  0.000   2.566524    4.89967
        age4 |   6.952853   1.621527   8.31  0.000   4.401978   10.98192
------------------------------------------------------------------

. abic
AIC Statistic   =   .9345462          AIC*n        = 836.41882
BIC Statistic   =   .9538434          BIC(Stata)   = 874.7934
```

Again, as with the binary response model, given smaller values of both the
AIC and BIC statistics, the model using the merged *kk23* rather than *kk2*
and *kk3* appears to be better fitted. In addition, the dispersion statistic has
dropped a bit, but not substantially. Moreover, comparing the binomial logistic AIC and BIC statistics with the binary logistic values gives a preference to
the binomial model, but with a qualification. A comparison is found below:

```
BINARY LOGISTIC
AIC Statistic =    .2869907           AIC*n        = 1292.3191
BIC Statistic =    .2908262           BIC(Stata)   = 1343.6191
```

```
BINOMIAL LOGISTIC
. abic
AIC Statistic    =    .9345462         AIC*n       = 836.41882
BIC Statistic    =    .9538434         BIC(Stata)  = 874.7934
```

The qualification is based on the fact that the right column statistics favor the binomial, whereas the left column values prefer the binary. With respect to the AIC, the conflict apparently relates to the use of n in the formula. This subject was discussed in Section 6.3.1, where we covered the use of AIC when comparing models having different definitions of n. In this case, the binary model is observation-based or n-asymptotic, with n equal to the number of observations in the model. The binomial model parameterizes observations as being based on covariate patterns or panels. An observation represents a panel, not the constituent observations within the panel. When comparing models with differing observation formats, n can seriously bias the comparative value of AIC with n.

In such situations one should use AIC without adjustment by n.

All definitions of BIC use n for an adjustment. Recalling that n can refer to two vastly different numbers—based on observations or based on panels—how a particular definition of BIC weights the main terms by n will have a bearing on results.

The BIC used in LIMDEP, displayed in the left column of *abic* (an author-written command posted to this book's Web site), is defined as:

$$BICL = -2 * (LL - k * \ln(k)) / n \qquad (8.23)$$

whereas the BIC used by Stata (in the *estat ic* command), is

$$BICS = -2 * LL + k * \ln(n) \qquad (8.24)$$

The comparative weight given to n in BICL has much more impact with observation data than panel data. When using it to compare observation and panel models, n biases the results to rather substantially favor the observation-based data as preferred. When comparing models with differing meanings of n, BIC methods are preferred.

As a result of the above discussion, we favor the binomial over the Bernoulli logistic regression model. Specifically, the preferred model for the **heart01** data, given *death* (binary) or *dead* (binomial) as the response, appears to be the binomial model with *kk23* included. It offers a slight improvement over having both *kk2* and *kk3* included in the model. Either model is acceptable, but the *kk23* model is better fitted when evaluated by the preferred versions of AIC and BIC.

There are a number of reasons why a binomial logistic model may be overdispersed, aside from a clustering effect. We can see from the above dispersion statistic that there still exists residual correlation in the model (1.05). To determine if we can further reduce dispersion in the model, we need to explore in more detail the nature and scope of overdispersion. This takes us to the next chapter.

Exercises

8.1. Use the **medpar** data set. The response is *died*, and the predictors are *white, hmo, los, age80,* and *type*. *Type* is a three-level categorical variable, which will need to be made into three indicator variables, with *type1* (elective) as the reference.

  a. Model *died* on *white, hmo, los, age80, type2 type3* using a binary logistic regression. Parameterize as odds ratios.

  b. Assess the goodness of fit of the model given in (a). Also calculate the AIC and BIC statistics.

  c. Determine if the model data are correlated by virtue of being collected from different providers or hospitals.

  d. Using only the variables from the model in (c), convert the casewise format of the data to grouped format. Model the data, assuring that the estimates and standard errors are identical in both models.

  e. Assess the goodness-of-fit of the model given in (c). Calculate the AIC and BIC statistics, comparing them to results in (b).

  f. Graph the standardized deviance residuals on the fitted value. What information do we get about the model?

8.2.

Age Group	Housing	Coronary	Person Years
40–44	rent	2	1107.447
	own	3	619.338
45–49	rent	24	3058.986
	own	19	4550.166
50–54	rent	31	3506.53
	own	25	4857.904
55–59	rent	28	3756.65
	own	27	4536.832
60–64	rent	28	2419.622
	own	26	2680.843
65–69	rent	2	351.71
	own	4	356.394

The above data come from the Scottish Health Study of coronary heart disease. The data in the table above are appropriate for modeling the number of coronary events, occurring in a given number of person years, on age group and housing status. Are renters or owners more likely to have a coronary event, such as myocardial infarction, CHF, and so forth? Are older patients more likely to have such events?

a. Convert the table above into electronic form using your favored statistical package.

b. Model coronary event on housing status and age group, adjusted by person years.

c. Assess the odds of sustaining a coronary event if one is a renter compared to an owner. Is being a renter a higher risk?

d. Assess the odds of sustaining a coronary event by age group. Are older age groups more likely to have such an event than younger?

e. Evaluate model fit. Is the model over- or underdispersed? If so, what needs to be done to adjust for any extradispersion? If needed, does such an adjustment make a difference?

8.3. The following table provides data to determine if smoking adds to the risk of developing coronary heart disease among physicians throughout the U.K. [The **doll** data set is from Doll and Hill (1966) and is analyzed in Hardin and Hilbe (2007).]

	Person Years		Coronary Deaths	
Age	Non-Smokers	Smokers	Non-Smokers	Smokers
35–44	18790	52407	2	32
45–54	10673	43248	12	104
55–64	5710	28612	28	206
65–74	2585	12663	28	186
75–84	1462	5317	31	102

a. Model *deaths* on *smokers* and *age* (categorized) as a grouped or binomial logistic regression. *Person years* is the binomial denominator.

b. Evaluate the model in (a). Is it well fitted? If not, what are some of the ways the model can be improved?

c. Model the same data using both rate parameterized Poisson and negative binomial models. Evaluate the fit.

d. Is the data fit better using a binomial, Poisson, or negative binomial (NB-2) model? Why?

8.4. Using the **drugprob** data, model *drabuse* on *drugdep, alcabuse, gender,* and *race* using a binary logistic regression. The data come from a 7-year survey of 750 youths between the ages of 10 and 22 related to predictors of alcohol and drug abuse. The covariates are defined as:

RESPONSE

drabuse (1/0)   drug abuse 1=is drug abuser; 0=not drug abuser

PREDICTORS

drugdep (1/0)	dependent on drugs	1=is dependent; 0=not dependent
alcabuse (1/0)	abuse of alcohol	1=does abuse alcohol; 0=not abuse
gender (1/0)	gender	1=male; 0=female
race (1/0)	race	1=white; 0=other

a. Parameterize the model for odds ratios. Determine which predictors are statistically significant.

b. Test the goodness of fit of the model using a Hosmer–Lemeshow test with a table of deciles, classification matrix, ROC test, and so forth.

c. Graph the standardized deviance residuals by the fitted value and *delta* deviance by fit weighted by *delta beta*. What covariate patterns are outliers?

d. Convert the data into proportional or aggregated format, making certain that the parameter estimates and standard errors are identical in both models.

e. Is the binomial logistic model overdispersed? How can you tell?

f. Construct a goodness-of-link test to determine if another binomial model might better fit the aggregated data.

g. Which combination of predictors yields the highest probability of a youth abusing drugs?

8.5. For this exercise, use the **hiv** data set. It is formatted in grouped or aggregated form.

a. Model *infec* on *cd4* and *cd8*, with cases as the population adjustment, using a logistic, rate Poisson, and rate negative binomial model. Which of the three models best fits the data?

b. Convert the aggregated data to individual-level format. Model the individual level data as a binary logistic regression to be assured that the conversion yields identical estimates and standard errors.

8.6. Use the **oak Ridge** data set from a study on the effects of radiation
levels and *age* for workers at the Oak Ridge, Tennessee nuclear facility. The data comes from Selvin (1995), page 474. Categorize the two
categorical predictors.

RESPONSE

    deaths   deaths of workers at the facility

    pyears   person years

PREDICTORS

age	age group	1: < 45	2: 45–49	3: 50–54	
		4: 55–59	5: 60–64	6: 65–69	7: > 69
rad	radiation level	Levels 1 through 8			

  a. Model the data both as a grouped logistic and rate Poisson model.
Compare the results. Explain the difference, or similarity, of the
models.

  b. Combine levels if necessary to effect a better fitted logistic
model.

  c. What combination of *age* and radiation level results in the highest probability of death?

R Code

Binomial Logistic Regression

```
eg<- data.frame(y=c(1,1,2,0,2,0), cases=c(3,1,2,1,2,1),
  x1=c(1,1,0,0,1,0), x2=c(0,1,0,1,0,1), x3=c(1,1,1,1,0,0))
eg$noty<- eg$cases - eg$y    #get failures
fit8a<- glm(cbind(y, noty) ~ x1 + x2 + x3, data=eg,
  family=binomial(link=logit))
summary(fit8a)
exp(coef(fit8a))    #ORs

binomialX2<- function(y, m, e) {   #Compute Pearson chi-square
for binomial
  #y: number successes
  #m: number trials
  #e: expected probability
  return(sum(m*(y/m-e)^2/(e*(1-e))))
}
```

```
fit8aX2<- binomialX2(eg$y, eg$cases, fitted(fit8a,
  type='response'))
cat('Pearson X2:', fit8aX2, 'Dispersion:', fit8aX2/fit8a$df.
  residual, '\n')

library('foreign')
heart<- read.dta('heart01.dta')  #read Stata format file
heart$anterior <- heart$anterior[,drop=TRUE] #drop empty
  levels that complicate  tabulations
heart$center<- factor(heart$center) #convert to factor from
  numeric levels
heart$killip<- factor(heart$killip) #convert to factor from
  numeric levels

fit8b<- glm(death ~ anterior + hcabg + kk2 + kk3 + kk4 + age3
  + age4, data=heart, family=binomial)
summary(fit8b)
exp(coef(fit8b))  #ORs

library(reshape)  #for cast(), melt()
heart2 <- na.omit(data.frame(cast(melt(heart, measure="death"),
  anterior + hcabg + kk1 + kk2 + kk3 + kk4 + age1 + age2 +
  age3 + age4 + center ~ .,
  function(x) { c(alive=sum(x == 0), dead=sum(x == 1)) })))
fit8c<- glm(cbind(dead, alive) ~ anterior + hcabg + kk2 + kk3
  + kk4 + age3 + age4, data=heart2, family=binomial)
summary(fit8c)
exp(coef(fit8c))  #ORs
exp(confint(fit8c)) #CI for ORs
fit8cX2<- binomialX2(heart2$dead, (heart2$dead+heart2$alive),
  fitted(fit8c, type='response'))
cat('Pearson X2:', fit8cX2, 'Dispersion:', fit8cX2/fit8c$df.
  residual, '\n')

fit8d<- glm(cbind(dead, alive) ~ anterior + hcabg + kk2 +
  kk3 + kk4 + age3 + age4, data=heart2, family=quasibinomial)
summary(fit8d)
exp(coef(fit8d))  #ORs
fit8dX2<- binomialX2(heart2$dead, (heart2$dead+heart2$alive),
  fitted(fit8d, type='response'))
cat('Pearson X2:', fit8dX2, 'Dispersion:', fit8dX2/fit8d$df.
  residual, '\n')
```

9

Overdispersion

9.1 Introduction

In the last chapter, we mentioned the problem of overdispersion and said that an overdispersed model is not a well-fitted model, regardless of the status of the parameter estimate p-values. An effect of overdispersion is to underestimate the standard errors of the parameter estimates. A predictor may appear to contribute significantly to the model when it in fact does not. We previously mentioned that overdispersion can be identified in binomial logistic models, also called grouped or proportional logistic models, by referring to the Pearson dispersion statistic. Binary response models are not overdispersed unless they entail a significant clustering effect.

In this chapter, we will discuss the nature and scope of overdispersion, differentiating between apparent and real overdispersion. We have already shown methods of dealing with overdispersed logistic model; for example, scaling, robust variance estimators, bootstrapping, jackknifing, and so forth. In this chapter, we hone in on what causes it and how it may be ameliorated.

9.2 The Nature and Scope of Overdispersion

Overdispersion has traditionally been applied to discrete response models either without a scale parameter or where the scale parameter is set at a constant value of 1.0. Discrete response models include members of the binomial (binary and binomial), count (Poisson and negative binomial), and categorical (ordered and multinomial models) response family of models. However, overdispersion takes two different forms, based on the type of response involved. Binary and categorical response models have one type of overdispersion, while binomial (by which we mean grouped binomial, such as grouped logistic regression) and count response models have another.

Traditionally, discussion of overdispersion has focused on the latter, since there are specific ways of identifying and handling overdispersion in such models.

It should be mentioned here that models may also be underdispersed. Together, underdispersed and overdispersed models are termed extradispersed. Few real data situations are underdispersed, but they do occur. Usually, most examples of underdispersion occur when simulated data are generated. When dealing with binomial models, we handle any type of extradispersion in the same manner. Count models, however, are different. Underdispersed and overdispersed models are generally dealt with in quite different manners, but we do not deal with them in this text. Just remember that binomial and count overdispersion have common causes, and that they run up against one another in certain circumstances, which we discuss later in this chapter. For a thorough evaluation of count models and the problem of overdispersion, see Hilbe (2007a).

In this chapter, we shall separate the discussions of binary/categorical response overdispersion and binomial/count model overdispersion. Moreover, since we specifically address categorical response logistic models in Chapters 10, 11, and 12, we shall delay addressing overdispersion for these models until that time. Additionally, we only provide a brief look at count overdispersion, concentrating instead on overdispersion as it pertains to logistic models. The discussion in this chapter is therefore centered on detailing the nature of overdispersion for binary logistic models and for binomial logistic models, respectively. We first turn to binomial overdispersion.

9.3 Binomial Overdispersion

We first define overdispersion as having more correlation in the data than is allowed by model distributional assumptions. The problem when model building is to identify overdispersion. Not all overdispersion is real; many times a model may give the *prima facie* indication of being overdispersed, when in fact it is not. We shall discuss this problem at length in the sections to follow.

To get a clearer notion about what is involved when dealing with model overdispersion, Table 9.1 outlines major points that should be considered when dealing with this issue.

As mentioned at the outset of the chapter, we have already dealt with various methods of handling overdispersion. We do this when attempting to reduce the value of the dispersion statistic in binomial logistic models or in count models. Our emphasis here will be on identifying and handling apparent overdispersion.

TABLE 9.1

Binomial Overdispersion

---

1. What is overdispersion?

   Overdispersion exists when there is more correlation in the data than allowed by model distributional assumptions.

2. What causes overdispersion?

   Overdispersion generally arises when there are violations in the distributional assumptions of the model due to excessive correlation in the data.

3. Why is overdispersion a problem?

   Overdispersion may cause estimated standard errors to be underestimated, e.g., a variable may appear to be a significant predictor when it is in fact not.

4. How is overdispersion recognized?

   Possible overdispersion is likely when the value of the model dispersion statistic, Pearson χ^2/DOF, is greater than 1.0. The quotient is called the dispersion. For count models, overdispersion is indicated when the variance exceeds the mean.

5. How much is too much?

   For moderate-sized models, when the dispersion is greater than 1.25.

   For large sized models, when the dispersion is greater than 1.05.

6. When is apparent overdispersion likely?

   a. When the model omits important explanatory predictors.

   b. When the model fails to include needed interaction terms.

   c. When a predictor needs to be transformed to another scale.

   d. When the assumed linear relationship between the response and the link function and predictors is mistaken, in other words, when the link is misspecified.

   e. When the data include outliers.

As mentioned at the outset of the chapter, we have already dealt with various methods of handling overdispersion. We do this when attempting to reduce the value of the dispersion statistic in binomial logistic models or in count models. Our emphasis here will be on identifying and handling apparent overdispersion.

9.3.1 Apparent Overdispersion

Apparent overdispersion exists in a binomial logistic model when the model appears to be overdispersed, that is, when the model has extra correlation as indicated by having a model dispersion value greater than 1.0. We call the model apparently overdispersed if we can make simple transformations to eradicate the correlation, thereby reducing the dispersion to approximately 1.0. If such transformations do not reduce the dispersion, we can conclude that the model is most likely truly overdispersed.

We develop simulated data sets showing how each of the five indicators of apparent overdispersion can be identified and amended. We list them again

TABLE 9.2

Criteria Apparent Overdispersion

---

1. Add appropriate predictor.
2. Construct required interactions.
3. Transform predictor(s).
4. Use correct link function.
5. Check and accommodate existing outliers.

---

here in Table 9.2 and will demonstrate how each situation arises, as well as how it can be eliminated.

9.3.1.1 Simulated Model Setup

First, create a randomly generated 10,000 observation grouped logistic data set with a binomial denominator having ten different values.

```
. set seed0 70785
. set seed 123044
. set obs 10000
. gen x1=invnorm(uniform())
. gen x2=invnorm(uniform())
. gen x3=invnorm(uniform())
. gen d = 50+5*int((_n-1)/1000)
. tab d
```

d	Freq.	Percent	Cum.
50	1,000	10.00	10.00
55	1,000	10.00	20.00
60	1,000	10.00	30.00
65	1,000	10.00	40.00
70	1,000	10.00	50.00
75	1,000	10.00	60.00
80	1,000	10.00	70.00
85	1,000	10.00	80.00
90	1,000	10.00	90.00
95	1,000	10.00	100.00
Total	10,000	100.00	

Save the data so that they may be used in the future.

```
. save overex                /// save data
```

Now create a binomial logistic model with the following parameters:

$$\text{Constant}: \quad 0.5$$
$$X1: \quad 1.0$$

```
                        X2:  -1.25
                        X3:   0.25

. gen y =rbinomial(d, 1/(1+exp(-(.5+1*x1-1.25*x2+.25*x3))))
                                /// a user authored command
```

AQ: We have retain the commends as in Previous edition Please confirm

```
. glm y x1 x2 x3, fam(bin d)

Generalized linear models            No. of obs      =    10000
Optimization     : ML                Residual df     =     9996
                                     Scale parameter =        1
Deviance       =  10097.05196        (1/df) Deviance = 1.010109
Pearson        =  9906.109333      → (1/df) Pearson  = .9910073

Variance function: V(u) = u*(1-u/d)  [Binomial]
Link function    : g(u) = ln(u/(d-u))  [Logit]

                                   → AIC           =  5.117194
Log likelihood   = -25581.9681       BIC           = -81969.51
-----------------------------------------------------------------
             |            OIM
          y  |    Coef.  Std. Err.     z   P>|z|  [95% Conf. Interval]
-------------+---------------------------------------------------
         x1  |  .9995431 .0033734  296.30 0.000   .9929312  1.006155
         x2  | -1.245868 .0035972 -346.34 0.000  -1.252918 -1.238818
         x3  |  .2491513 .002952   84.40  0.000   .2433655  .2549371
       _cons |  .4985296 .0029547  168.73 0.000   .4927386  .5043207
-----------------------------------------------------------------

. abic
AIC Statistic =   5.117194    AIC*n       = 51171.938
BIC Statistic =   5.117503    BIC(Stata)  = 51200.777
```

The predictors have nearly identical values to the values we specified. Moreover, the Pearson dispersion statistic is .991, only 9 one-thousandths from the theoretically ideal 1.0. The AIC statistic has a value of 5.11, or, without adjustment by n, 51171.9.

9.3.1.2 Missing Predictor

Exclude the $x1$ predictor, observing the values of the coefficients, dispersion, and AIC statistic.

```
. glm y x2 x3, fam(bin d)

Generalized linear models            No. of obs      =    10000
Optimization     : ML                Residual df     =     9997
                                     Scale parameter =        1
Deviance       =  121318.8128        (1/df) Deviance = 12.13552
Pearson        =  117527.2422      → (1/df) Pearson  = 11.75625
```

```
Variance  function: V(u) = u*(1-u/d)     [Binomial]
Link  function    : g(u) = ln(u/(d-u))   [Logit]

                                    → AIC          = 16.23917
Log likelihood     = -81192.84853    BIC          = 29243.04
-------------------------------------------------------------
           |              OIM
     y |    Coef.   Std. Err.     z   P>|z|  [95% Conf. Interval]
-------+-----------------------------------------------------
    x2 |-1.046665   .0031577 -331.47 0.000 -1.052854   -1.040476
    x3 |  .1989551  .0026972   73.76 0.000   .1936687    .2042415
 _cons |  .4234062  .0026815  157.90 0.000   .4181506    .4286617
-------------------------------------------------------------
```

The dispersion statistic rose from .99 to 11.75, clearly indicating overdispersion. Also, AIC rose threefold in value from 5.12 to 16.24. The model coefficients were altered as well:

	Correct Model	Model with Missing Predictor
X2	−1.25	−1.05
X3	0.25	0.20
Con	0.50	0.42

What this indicates is that if a model fails to have a predictor that happens to be essential to the model, it will appear to be overdispersed. By including the appropriate predictor, the apparent overdispersion vanishes. The problem, of course, is that the data may not have the needed predictor. It may not have been collected, or if originally collected, not kept with the data. When designing a research project, it is therefore important to include the collection of all relevant clinical data, especially data that have proved to be important in similar studies.

9.3.1.3 Needed Interaction

We next demonstrate the effect of excluding an interaction, when one is required for a well-fit model. An interaction of $x2$ and $x3$ is created, giving it a parameter of .2.

```
. gen x23 = x2*x3
. gen yi =rbinomial(d, 1/(1+exp(-(.5+1*x1-1.25*x2+.25*x3+.2*x23))))
. glm yi x1 x2 x3 x23, fam(bin d)

Generalized linear models          No. of obs     =      10000
Optimization     : ML              Residual df    =       9995
                                   Scale parameter =         1
Deviance        = 10380.43339      (1/df) Deviance = 1.038563
Pearson         = 10105.68926    → (1/df) Pearson = 1.011074
```

```
Variance function: V(u) = u*(1-u/d)      [Binomial]
Link function     : g(u) = ln(u/(d-u))   [Logit]

                                       → AIC               =    5.125379
Log likelihood    = -25621.89295         BIC               =  -81676.92
-----------------------------------------------------------------------
            |                 OIM
        yi  |      Coef.  Std. Err.      z   P>|z|  [95% Conf. Interval]
--------+--------------------------------------------------------------
        x1  |   1.002341  .0033892  295.75  0.000   .9956988   1.008984
        x2  |  -1.255015  .0036378 -345.00  0.000  -1.262145  -1.247886
        x3  |   .2564922  .0030142   85.09  0.000   .2505844      .2624
       x23  |   .1998892  .0035116   56.92  0.000   .1930065   .2067719
      _cons |   .5043715  .0029755  169.51  0.000   .4985397   .5102033
-----------------------------------------------------------------------
```

The coefficients, including the interaction with a specified parameter of .2, are well represented by the model. The dispersion is 1.01, with a one one-hundredth difference from the theoretically correct value due to randomness of the random number generator—as expected. The AIC statistic is 5.125.

Now model the data without the interaction, *x23*.

```
. glm yi x1 x2 x3, fam(bin d)

Generalized linear models              No. of obs        =      10000
Optimization      : ML                 Residual df       =       9996
                                       Scale parameter   =          1
Deviance          =   13678.84537      (1/df) Deviance   =   1.368432
Pearson           =   13408.43927    → (1/df) Pearson    =    1.34138
Variance function: V(u) = u*(1-u/d)      [Binomial]
Link function     : g(u) = ln(u/(d-u))   [Logit]

                                       AIC               =    5.45502
Log likelihood    = -27271.09894         BIC               =  -78387.72
-----------------------------------------------------------------------
            |                 OIM
        yi  |      Coef.  Std. Err.      z   P>|z|  [95% Conf. Interval]
--------+--------------------------------------------------------------
        x1  |   1.000164  .0033806  295.86  0.000   .9935381    1.00679
        x2  |  -1.249446  .0036071 -346.38  0.000  -1.256516  -1.242376
        x3  |   .2810576  .0029665   94.74  0.000   .2752434   .2868719
      _cons |    .517251  .0029659  174.40  0.000    .511438   .5230641
-----------------------------------------------------------------------
```

The coefficients differ little from the true values, with *x3* varying the most (0.28 to 0.25). The only indication that a problem might exist rests with the dispersion statistic having a value of 1.34. Given an observation base of 10,000, 1.34 clearly indicates overdispersion. In this case, it was a result of a missing interaction. When the dispersion statistic is high, as in this case, consider checking for interactions in the data.

9.3.1.4 Predictor Transformation

Another problem exists when a predictor needs to be transformed. For this example, create a model where the true model incorporates the squared value of $x1$ rather than simply $x1$. We call squared $x1$, $x1sq$, and assign it a parameter value of 0.5:

```
. gen x1sq = x1*x1      /* square of x1 */
. gen ysq =rbinomial(d, 1/(1+exp(-(.5+.5*x1sq-1.25*x2+.25*x3))))
                                                    /* ysq */

. glm ysq x1sq x2 x3, nolog fam(bin d)
```

AQ: We have retain the commends as in Previous edition Please confirm

```
Generalized linear models              No. of obs      =      10000
Optimization       : ML                Residual df     =       9996
                                       Scale parameter =          1
Deviance       =    10340.03543        (1/df) Deviance =   1.034417
Pearson        =    10177.21956        (1/df) Pearson  =   1.018129

Variance function: V(u) = u*(1-u/d)    [Binomial]
Link function: g(u) = ln(u/(d-u))      [Logit]

                                       AIC             =   5.103278
Log likelihood     = -25512.39102      BIC             =  -81726.53
-------------------------------------------------------------------
            |                OIM
     ysq |       Coef. Std. Err.      z   P>|z| [95% Conf. Interval]
-------------+-----------------------------------------------------
    x1sq |   .4956539 .0030406  163.01 0.000   .4896943    .5016134
      x2 |  -1.246583 .0036182 -344.53 0.000  -1.253674   -1.239491
      x3 |   .2472632 .0029582   83.59 0.000   .2414653    .2530611
   _cons |   .5002603 .0037002  135.20 0.000   .4930079    .5075126
-------------------------------------------------------------------
```

Again, the parameter estimates appear to be consistently the same as how they were defined, less random error. The dispersion is 1.018, only 18 thousandths from unity. The AIC statistic remains at approximately 5.1, the same as each of the above "true" models.

Next, model the data with $x1$, as if we have no prior knowledge that $x1$ is really $x1$-squared.

```
. glm ysq x1 x2 x3, nolog fam(bin d)

Generalized linear models              No. of obs      =      10000
Optimization       : ML                Residual df     =       9996
                                       Scale parameter =          1
Deviance       =    46879.64294        (1/df) Deviance =    4.68984
Pearson        =    44215.14913    →   (1/df) Pearson  =   4.423284
```

```
Variance function: V(u)  = u*(1-u/d)    [Binomial]
Link function:  g(u) = ln(u/(d-u))      [Logit]

                                    → AIC          =   8.757239
Log likelihood     = -43782.19478      BIC         = -45186.92
-----------------------------------------------------------------
         |                 OIM
   ysq   |     Coef.  Std. Err.    z  P>|z| [95% Conf. Interval]
---------+-------------------------------------------------------
     x1  |  .0023782 .0028272   0.84 0.400  -.003163    .0079194
     x2  | -1.190336 .0034652 -343.51 0.000 -1.197128  -1.183545
     x3  |  .2322037 .0028751  80.76 0.000   .2265685    .2378388
   _cons |  .9103651 .0029962 303.84 0.000   .9044927    .9162375
-----------------------------------------------------------------
```

The parameter estimates vary considerably in their closeness to specified values. The true constant is 0.5; it is now 0.91. The non-squared $x1$ has a parameter estimate of 0.002, far from the assigned value of 0.50. The dispersion statistic rose from near unity to 4.42—a huge increase considering the number of observations in the data. Lastly, the AIC is 8.76, far above the true value of 5.1. If we were given the data, and created a model from it using main effects only, we would know that the model was overdispersed because of the high dispersion. However, we would not know if it were truly overdispersed, or only apparently. As it turns out, a simple squaring of $x1$ eliminates the apparent overdispersion.

9.3.1.5 Misspecified Link Function

Suppose that we wish to construct a binomial or grouped model. Naturally, we first try employing a binomial logistic model. However, we notice that the dispersion statistic indicates that the model is overdispersed. We check for missing predictors, possible interactions, and transformations of the various predictors. Recall that we have no knowledge of the "true" model parameters. We suspect a problem because of the high dispersion. What should we do?

We demonstrate this type of situation by creating a simulated binomial probit model. We then subject the data to a binomial logistic model, observing the parameter estimates and fit statistics.

Random probit data is best generated using a pseudo-random uniform generator. Keeping the same seed, we create three pseudo-random variates, and a linear predictor with the same values used for the logit models above.

```
. gen xx1 = runiform()   // create xx2 and xx3 in the same manner
. gen yp = rbinomial(d, normprob(.5+1*xx1-1.25*xx2+.25*xx3))
. glm yp xx1 xx2 xx3, nolog fam(bin d) link(probit)
```

```
Generalized linear models          No. of obs       =      10000
Optimization       : ML            Residual df      =       9996
                                   Scale parameter  =          1
```

```
Deviance          =   9920.985699        (1/df) Deviance =   .9924956
Pearson           =   9803.934329        (1/df) Pearson  =   .9807857

Variance function: V(u) = u*(1-u/d)      [Binomial]
Link function    : g(u) = invnorm(u/d) [Probit]

                                          AIC             =   5.392163
Log likelihood   =  -26956.81434          BIC             = -82145.58
------------------------------------------------------------------
          |                 OIM
    yp |      Coef.  Std. Err.      z   P>|z|  [95% Conf. Interval]
------+-----------------------------------------------------------
   xx1 |   .9981278  .0055718  179.14  0.000   .9872073  1.009048
   xx2 | -1.247718   .005676  -219.82  0.000  -1.258843 -1.236594
   xx3 |   .2470518  .0055181   44.77  0.000   .2362365   .2578671
 _cons |   .5017432  .0049456  101.45  0.000   .4920499   .5114365
------------------------------------------------------------------

. abic
AIC Statistic   =     5.392163        AIC*n      =  53921.629
BIC Statistic   =     5.392472        BIC(Stata) =  53950.469
```

The simulated binomial probit model has parameter estimates very close to those assigned, the dispersion is 0.98, only two one-thousandths from unity, and the AIC statistic is 53922, higher than the 51172 value for the logit model.

We have no knowledge that the true model for these data is a binomial probit. First, attempt a binomial logistic model on the same data.

MODEL PROBIT DATA WITH A BINOMIAL LOGISTIC MODEL

```
. glm yp xx1 xx2 xx3, nolog fam(bin d)

Generalized linear models             No. of obs       =      10000
Optimization     : ML                 Residual df      =       9996
                                      Scale parameter =          1
Deviance          =   9989.42777        (1/df) Deviance =   .9993425
Pearson           =   9838.864611       (1/df) Pearson  =   .9842802

Variance function: V(u) = u*(1-u/d)      [Binomial]
Link function    : g(u) = ln(u/(d-u)) [Logit]

                                          AIC             =   5.399007
Log likelihood   =  -26991.03538          BIC             = -82077.13
------------------------------------------------------------------
          |                 OIM
    yp |      Coef.  Std. Err.      z   P>|z|  [95% Conf. Interval]
------+-----------------------------------------------------------
   xx1 |  1.665472  .0094186  176.83  0.000  1.647012   1.683932
   xx2 | -2.083331  .0096725 -215.39  0.000 -2.102288  -2.064373
```

```
  xx3 |   .4147304 .0092279    44.94 0.000    .396644     .4328168
 _cons |   .8288387 .0082547   100.41 0.000    .8126598    .8450176
-----------------------------------------------------------------------

. abic
AIC Statistic    =   5.399007        AIC*n       =   53990.07
BIC Statistic    =   5.399316        BIC(Stata)  =   54018.914
```

The logistic model is different, with an AIC of 53990, nearly 70 higher than the probit model on the same data. The logistic model is not overdispersed, but the AIC and BIC values are substantially higher than the probit.

Table 9.3 displays a comparison of the parameter estimates. The binomial logistic parameter estimates are nearly two times greater than the true probit estimates. Without prior knowledge of the true model, the only indication of mis-specification are the AIC and BIC statistic. Monte Carlo simulation, however, shows that with repeated estimation without seeds, a logistic model of true probit data will be overdispersed.

TABLE 9.3

Probit Data: True, Probit, Logistic Models

	True	Probit	Logistic
XX1	1.000	0.998	1.665
XX2	−1.250	−1.248	−2.083
XX3	0.250	0.247	0.415
Cons	0.500	0.502	0.829
AIC		53922	53990
DISP	1.00	53950	54019

9.3.1.6 Existing Outlier(s)

The final criterion of apparent overdispersion, or perhaps better termed as misspecification, relates to outliers in the data. Of course, an outlier is dependent on the distribution of which it is a part. What may be an outlier in one circumstance may not be an outlier in another. For our purposes, outliers relate to the Bernoulli and binomial distributions. For this section, however, we emphasize binomial outliers.

Because the binomial logistic and rate parameterized Poisson distributions are related, and because creating outliers with binomial models can be trickier to interpret, we shall demonstrate how an outlier or outliers can affect overdispersion by using a Poisson model.

For this example, use the **cancer** data set, with the following variables:

response = *time*   /// days in study
predictor = *drug1-drug3* (*drug1* = referent)   /// dummy (1/0) predictors

[Note: Some varieties of the **cancer** data sets name the time variable as *studytim*. If this is the case, simply rename *studytim* to *time*. If *studytim* is to be used with subsequent models, then create *time* as identical to *studytim*. In Stata: *gen time = studytim*.]

First, model *time* on *drug*, with *drug1* as the referent. It is the placebo, with *drugs* 2 and 3 being certain types of cancer-retarding drugs.

```
. tab drug, gen(drug)

. glm time drug2 drug3, fam(poi)
```

```
Generalized linear models                No. of obs         =          48
Optimization         : ML                Residual df        =          45
                                         Scale parameter  =           1
Deviance           =   192.396951        (1/df) Deviance  =    4.275488
Pearson            =   185.1355512    →  (1/df) Pearson   =    4.114123
Variance function: V(u) = u              [Poisson]
Link function    : g(u) = ln(u)          [Log]
                                         AIC                =    8.444248
Log likelihood     = -199.6619534        BIC                =    18.19291
-------------------------------------------------------------------------
             |                 OIM
        time |    Coef.  Std. Err.     z  P>|z|    [95% Conf. Interval]
-------------+-----------------------------------------------------------
       drug2 | .5060523   .101687   4.98 0.000    .3067495    .7053552
       drug3 | 1.035836  .0915011  11.32 0.000     .856497    1.215175
       _cons | 2.197225  .0745356  29.48 0.000    2.051137    2.343312
-------------------------------------------------------------------------
```

The dispersion statistic is 4.11, indicating that the model is overdispersed. Let's see what happens when we create a typo—multiplying the largest number of days by 10.

```
. sum time, detail
```

```
                     Months to death or end of exp.
-------------------------------------------------------------------------
          Percentiles      Smallest
   1%           1              1
   5%           2              1
  10%           4              2        Obs                      48
  25%         7.5              3        Sum of Wgt.              48

  50%        12.5                       Mean                   15.5
```

		Largest	Std. Dev.	10.25629
75%	23	33		
90%	32	34	Variance	105.1915
95%	34	35	Skewness	.5412136
99%	39	39	Kurtosis	2.243608

The largest value of *time* is 39, hence:

```
. replace time = 390 if time==39

. glm time drug2 drug3, fam(poi) nolog
```

```
Generalized linear models          No. of obs       =       48
Optimization     : ML              Residual df      =       45
                                   Scale parameter  =        1
Deviance         =   1319.87876    (1/df) Deviance  = 29.33064
Pearson          =   2620.232188   (1/df) Pearson   = 58.22738

Variance function: V(u) = u        [Poisson]
Link function    : g(u) = ln(u)    [Log]

                                   AIC              = 31.98134
Log likelihood   = -764.5522272    BIC              = 1145.675
```

time	Coef.	OIM Std. Err.	z	P>|z|	[95% Conf.	Interval]
drug2	.5060523	.101687	4.98	0.000	.3067495	.7053552
drug3	1.723333	.0834984	20.64	0.000	1.559679	1.886987
_cons	2.197225	.0745356	29.48	0.000	2.051137	2.343312

The Pearson dispersion statistic has increased from 4.11 to 58.23—on the basis of one altered value of the response. In addition, the parameter estimate of *drug3* has changed from 1.036 to 1.723, a substantial increase. We would suspect that the changed value of *time* is associated with *drug3*, that is, that *drug3* = 1. We tabulate on the outlier to find out.

```
. l if time==390          /// "l" is an abbreviation for list.
```

	time	drug	drug1	drug2	drug3
48.	390	3	0	0	1

Indeed, the outlier value of *time* is associated with *drug3* being equal to 1.

We can show the effect of an outlier in a predictor by estimating a binary logistic model. Using the same **cancer** data set, we model *died* (1/0) on *age*, which has a range of 47 to 67, and a mean value of 56.

```
. mean age

Mean estimation                          Number of obs    =       48

------------------------------------------------------------------
           |       Mean    Std. Err.     [95% Conf. Interval]
-----------+------------------------------------------------------
       age |     55.875    .8168358      54.23174     57.51826
------------------------------------------------------------------
```

The logit results are

```
. glm died age, fam(bin) nolog

Generalized linear models              No. of obs       =           48
Optimization      : ML                 Residual df      =           46
                                       Scale parameter  =            1
Deviance       =    59.89075733        (1/df) Deviance  =     1.301973
Pearson        =    48.05722403        (1/df) Pearson   =     1.044722

Variance function: V(u) = u*(1-u)      [Bernoulli]
Link function: g(u) = ln(u/(1-u))      [Logit]

                                       AIC              =     1.331057
Log likelihood  =   -29.94537866       BIC              =    -118.1845
------------------------------------------------------------------
           |               OIM
     died  |     Coef. Std. Err.     z    P>|z| [95% Conf. Interval]
-----------+------------------------------------------------------
       age |  .0893535 .0585925   1.52   0.127 -.0254857    .2041928
     _cons | -4.353928 3.238757  -1.34   0.179 -10.70178    1.993919
------------------------------------------------------------------
```

As we did for the Poisson model above, multiply a value of age by 10. This time, determine *age* at the 75th percentile, multiplying it by 10.

```
. centile age, centile(75)

                                             -- Binom. Interp. --
 Variable | Obs   Percentile    Centile      [95% Conf. Interval]
----------+-------------------------------------------------------
      age |  48           75         60            58     62.1327
```

The 75 percent percentile of *age* is 60. However, we only wish to create one typo, so it is important to know how many patients in the data are *age* 60.

```
. count if age==60
    2
```

To determine where in the data each instance of *age* = 60 is located, type:

```
. l age if age==60      /// or . l age[] if age==60; "l" = list

        +-----+
        | age |
        |-----|
  37. |   60 |
  45. |   60 |
        +-----+

. replace age=600 in 37

. glm died age, fam(bin)

Generalized linear models              No. of obs      =          48
Optimization        : ML               Residual df     =          46
                                       Scale parameter =           1
Deviance          =   60.81662735      (1/df) Deviance =    1.322101
Pearson           =   47.48936134      (1/df) Pearson  =    1.032377

Variance function: V(u) = u*(1-u)      [Bernoulli]
Link function: g(u) = ln(u/(1-u))      [Logit]

                                       AIC             =    1.350346
Log likelihood    =  -30.40831367      BIC             =   -117.2586
------------------------------------------------------------------------
           |               OIM
     died  |     Coef. Std. Err.     z    P>|z|  [95% Conf. Interval]
---------+--------------------------------------------------------------
     age  | -.0062006 .0075743  -0.82  0.413  -.021046      .0086449
    _cons |  1.001645  .540915   1.85  0.064 -.0585286      2.061819
------------------------------------------------------------------------
```

The dispersion statistic did not change, but the parameter estimates changed dramatically from:

age	.089	to	−.006	and
constant	−4.354	to	1.002	

Because the dispersion does not change in this situation (even though it has no bearing on assessment of overdispersion), it is more difficult to tell if an error exists in the data. If we did not check and left the incorrect value in the model, the parameter estimates and resulting predictions made on their basis would be mistaken. The differences in odds ratio would also lead to mistaken conclusions. We can view the odds ratios of the two models by typing:

```
. glm died age, fam(bin)  nohead eform

-----------------------------------------------------------------
         |                OIM
    died | Odds Ratio Std. Err.    z  P>|z|  [95% Conf. Interval]
---------+-------------------------------------------------------
     age |   .9938186 .0075275 -0.82 0.413   .9791739    1.008682
-----------------------------------------------------------------

. webuse cancer, clear       /// or, if file is in working directory: use
                                 cancer, clear


. glm died age, fam(bin) nolog nohead eform

-----------------------------------------------------------------
         |                OIM
    died | Odds Ratio Std. Err.    z  P>|z| [95% Conf. Interval]
---------+-------------------------------------------------------
     age |  1.093467   .064069  1.52 0.127 .9748363    1.226535
-----------------------------------------------------------------
```

The single typo produces a 9 percent drop in odds ratio, with a substantial change in *p*-value. In fact, many typos and incorrectly entered data elements result in a predictor appearing to significantly contribute to a model when in fact is does not, and likewise, a predictor may appear not to contribute to a model when it in fact does.

The point is that an outlier, or outliers, whether typos or true values, can produce an overdispersed model or substantially alter the parameter estimates. If a model is highly overdispersed, check for typos and make the correction. If the outlier is a true value in the sense that it differs greatly from other values of the variable, then decide whether to keep it in the model. This is a separate problem. In any event, it is important to evaluate a model for outliers. A simple correction may result in apparent overdispersion being eliminated from the model.

9.3.2 Relationship: Binomial and Poisson

A binomial logistic model is comparable to a rate parameterized Poisson regression model. The binomial numerator is the number of successes; the denominator is the number of cases or observations with the same pattern of covariates. The Poisson numerator is the count of events; the denominator is the area or time frame in which the counts take place. Refer to Section 5.5 for an earlier discussion of the relationship of logistic odds ratios and their interpretation as risk ratios, as well as an overview of Poisson regression. However, the previous discussion was based on observation-level data rather than aggregated and time/area-based models, which we address in the coming discussion.

A binomial logistic model is sometimes better modeled by using a Poisson model. This particularly occurs when the mean values of the numerator are very small compared with the values of the denominator. When numerator values are closer on average to the denominator, the binomial model will be a better fit. The reason for this comes from the theoretical derivation of the Poisson as a binomial with a low ratio of numerator to denominator. In general, proportional response data are better fit as Poisson if the ratio is below 0.1. Of course, there is considerable variability in this ratio, based on the structure of the data. The ratio of numerator to denominator for the **overex** data we have used in this chapter is

```
[ . clear
    . restore overex
    . gen y =rbinomial(d, 1/(1+exp(-(.5+1*x1-1.25*x2+.25*x3))))]


. gen yd = y/d

. mean yd
```

```
Mean estimation                      Number of obs   =    10000
-------------------------------------------------------------------
             |       Mean    Std. Err.     [95% Conf. Interval]
-------------+-----------------------------------------------------
         yd  |    .5823412    .0028244      .5768049     .5878775
-------------------------------------------------------------------
```

The median value may be determined by:

```
. centile yd, centile(50)

                                         -- Binom. Interp. --
    Variable |    Obs  Percentile  Centile  [95% Conf. Interval]
-------------+-----------------------------------------------------
         yd  |  10000          50  .6210526  .6133333     .6324575
```

In either case, the ratio is quite high, indicating that the binomial logistic would fit the data better than a Poisson. A comparison is given below, with the middle of the header statistics deleted as unnecessary.

BINOMIAL LOGISTIC

```
. glm y x1 x2 x3, fam(bin d) nolog          /// Header statistics
                                                modified
```

```
Generalized linear models          No. of obs      =      10000
Optimization     : ML              Residual df     =       9996
                                   Scale parameter =          1
Deviance         =  10097.05196    (1/df) Deviance =   1.010109
Pearson          =  9906.109333    (1/df) Pearson  =   .9910073
                                   AIC             =   5.117194
Log likelihood   =  -25581.9681    BIC             =  -81969.51
```

| OIM
        y | Coef. Std. Err.   z   P>|z| [95% Conf. Interval]
--------+---
      x1 | .9995431 .0033734   296.30 0.000   .9929312   1.006155
      x2 | -1.245868 .0035972 -346.34 0.000 -1.252918 -1.238818
      x3 | .2491513   .002952   84.40 0.000   .2433655   .2549371
   _cons | .4985296 .0029547   168.73 0.000   .4927386   .5043207

RATE POISSON

```
. glm y x1 x2 x3, fam(poisson) lnoffset(d) nolog
                                    /// or use exposure(d)
```

```
Generalized linear models          No. of obs      =      10000
Optimization     : ML              Residual df     =       9996
                                   Scale parameter =          1
Deviance         =  25461.59852    (1/df) Deviance =   2.547179
Pearson          =  22293.73717    (1/df) Pearson  =   2.230266
                                   AIC             =   7.904474
Log likelihood   =  -39518.36939   BIC             =  -66604.96
```

| OIM
        y | Coef. Std. Err.   z   P>|z| [95% Conf. Interval]
--------+---
      x1 | .289489 .0015408   187.88 0.000   .2864691   .292509
      x2 | -.3668839 .0015483 -236.95 0.000 -.3699186 -.3638492
      x3 | .076376   .001569   48.68 0.000   .0733009   .0794512
   _cons | -.6466301 .0016961 -381.25 0.000 -.6499543 -.6433058
       d | (exposure)

The Poisson clearly fits worse than the binomial logistic. Since the data were created as binomial logistic, this is no surprise. Be aware, however, that not all cases will be like this, and it may be preferable to employ a Poisson, or more likely, as in this model, a negative binomial.

The model below obtains the negative binomial heterogeneity or ancillary parameter value of 0.0357547 from a maximum likelihood negative binomial algorithm called from within the **glm** program. It supplies the value and employs it as a constant to the **glm** estimating equations. See Hilbe (2007a) for a comprehensive discussion of this subject.

RATE NEGATIVE BINOMIAL

```
. glm y x1 x2 x3, fam(nb ml) lnoffset(d) nolog
```

```
Generalized linear models            No. of obs        =      10000
Optimization        : ML             Residual df       =       9996
                                     Scale parameter   =          1
Deviance       =   11349.87047       (1/df) Deviance   =   1.135441
Pearson        =   9375.329411       (1/df) Pearson    =   .9379081

Variance function: V(u)  = u+(.0357547)u^2
Link function     : g(u) = ln(u)
                                     AIC               =   7.360041
Log likelihood   = -36796.20398      BIC               = -80716.69
```

y	Coef.	OIM Std. Err.	z	P>|z|	[95% Conf. Interval]	
x1	.3453102	.0026425	130.68	0.000	.340131	.3504894
x2	-.4367985	.0027094	-161.22	0.000	-.4421089	-.4314882
x3	.0919314	.0025739	35.72	0.000	.0868867	.0969761
_cons	-.6679007	.0025827	-258.60	0.000	-.6729628	-.6628387
d	(exposure)					

Note that the negative binomial has overcompensated for the Poisson, displaying a dispersion of 0.94. However, the AIC statistic is substantially higher than the binomial, although lower than the Poisson.

We have discussed the comparison of binomial and count models to show that at times one may consider using a count model for proportional data. Poisson and negative binomial coefficients can be parameterized as risk ratios in a similar manner as logistic coefficients can be parameterized as odds ratios.

9.4 Binary Overdispersion

9.4.1 The Meaning of Binary Model Overdispersion

Binary response logistic regression cannot be overdispersed. That is, if the likelihood assumptions of the logistic model are upheld, and every observation is independent from one another, then overdispersion will not occur. The error in such a model is also binary, taking only the form of 1 and 0. However, as we have already discussed, binary response models may be clustered, in which case the assumption of the independence of observations—essential to maximum likelihood theory—is biased. When we model clustered data adjusting by the variable causing the cluster effect, the model is no longer a likelihood, or a deviance-based model, but rather a log-pseudo-likelihood or pseudo-deviance model. Likelihood ratio tests are theoretically not applicable, nor are other tests that assume the independence of observations.

Clustered data were traditionally modeled by adjusting the standard errors by scaling. We discussed scaling in Section 5.6. Although scaling makes theoretical sense, that is, it adjusts the standard errors to calibrate them to the values they would be if the Pearson dispersion statistic were 1.0, it is generally only calculated for models estimated using Fisher Scoring and it does not indicate any source of extradispersion. The currently most popular way statisticians handle clustering is by employing a robust variance estimator, with clustering based on a specific variable. We discussed robust estimators in Section 5.7.

The point I am making is that a clustered binary response logistic model is not a true logistic model. Therefore, it is the case that binary logistic models are not overdispersed, but if we enlarge the scope of binary models to include clustered models, then they can indeed be overdispersed. It is largely a matter of semantics. However, the underlying theoretical distinction between the two should be kept in mind.

The **heart01** data will again be used for an example.

```
. glm death anterior hcabg kk2-kk4 age3-age4, nolog fam(bin)

Generalized linear models          No. of obs      =        4503
Optimization     : ML              Residual df     =        4495
                                   Scale parameter =           1
Deviance         =   1276.319134   (1/df) Deviance =     .283942
Pearson          =   4212.631591   (1/df) Pearson  =    .9371817

Variance function: V(u) = u*(1-u)   [Bernoulli]
Link function: g(u) = ln(u/(1-u))   [Logit]
```

```
                                   AIC              =   .2869907
Log likelihood    = -638.1595669   BIC              = -36537.86
------------------------------------------------------------------
           |              OIM
   death   |   Coef.   Std. Err.    z   P>|z|  [95% Conf. Interval]
-----------+------------------------------------------------------
 anterior  |  .6387416  .1675426   3.81  0.000   .3103642   .9671191
    hcabg  |  .7864185  .3527468   2.23  0.026   .0950475   1.47779
      kk2  |  .8249174  .1804368   4.57  0.000   .4712678   1.178567
      kk3  |  .7966954  .2692168   2.96  0.003   .2690401   1.324351
      kk4  | 2.683746   .3564503   7.53  0.000   1.985117   3.382376
     age3  | 1.266828   .2005658   6.32  0.000   .8737267   1.65993
     age4  | 1.940876   .2080462   9.33  0.000   1.533112   2.348639
    _cons  |-4.815946   .1934408 -24.90  0.000  -5.195083  -4.436809
------------------------------------------------------------------
```

Note that the Pearson dispersion statistic of 0.937 has no bearing on over-dispersion. In fact, when we earlier converted it to grouped format, where overdispersion does make sense, the dispersion statistic was 1.04. Unlike the binomial form of the model, no statistic indicates overdispersion for a binary logistic model.

We know from the data set that the patients come from a number of different hospitals, which are specified in the variable *center*. It may be the case that the relationship between *death* and the explanatory predictors is more highly correlated within than between hospital providers. If this is the case, then the data are clustered, and can be better modeled as a quasilikelihood—or pseudolikelihood—logistic model than by using standard methods. We model this as:

```
. glm death anterior hcabg kk2-kk4 age3-age4, nolog fam(bin)
  cluster(center)

Generalized linear models             No. of obs      =       4503
Optimization     : ML                 Residual df     =       4495
                                      Scale parameter =          1
Deviance         =  1276.319134       (1/df) Deviance =    .283942
Pearson          =  4212.631591       (1/df) Pearson  =
.9371817

Variance function: V(u) = u*(1-u)     [Bernoulli]
Link function: g(u) = ln(u/(1-u))     [Logit]

                                      AIC             =   .2869907
Log pseudolikelihood = -638.1595669   BIC             = -36537.86
```

```
                      (Std. Err. adjusted for 47 clusters in center)
      --------------------------------------------------------------------
            |                Robust
    death   |    Coef.  Std. Err.      z   P>|z|  [95% Conf. Interval]
  ----------+---------------------------------------------------------------
  anterior  |  .6387416  .2041506    3.13  0.002   .2386138   1.038869
     hcabg  |  .7864185  .3129288    2.51  0.012   .1730893   1.399748
       kk2  |  .8249174  .1782178    4.63  0.000   .4756168   1.174218
       kk3  |  .7966954  .3267989    2.44  0.015   .1561813    1.43721
       kk4  |  2.683746  .3003342    8.94  0.000   2.095102   3.272391
      age3  |  1.266828  .1683474    7.53  0.000   .9368735   1.596783
      age4  |  1.940876  .2395111    8.10  0.000   1.471442   2.410309
     _cons  | -4.815946  .1944683  -24.76  0.000  -5.197097  -4.434795
  --------------------------------------------------------------------------
```

Again, we cannot pay attention to the dispersion statistic, which is inapplicable for analysis on this model. Overdispersion is checked by comparing the standard errors of the model with the robust or empirical standard errors of the model adjusted by the provider (See Table 9.4).

TABLE 9.4

Differences in Model and Robust Standard Errors

Predictor	Model	Center Adj Model	
anterior	.168	.204	+
hcabg	.353	.313	−
kk2	.180	.178	0
kk3	.269	.327	+
kk4	.356	.300	−
age3	.201	.168	−
age4	.208	.240	+
_cons v	.193	.194	0

The standard errors of the two columns differ, but note that there is not a uniform inflation or deflation of standard errors when employing the adjustment. Moreover, *hcabg* and *kk3* were the only predictors with non-zero *p*-values, and they were significant in both models. There is no appreciable difference in the two models. That is, the fact that data were collected for a number of different providers did not result is a clustering effect. The observations in the data are still independent.

Suppose that there was a significant difference in the standard errors—that adjustment of the model greatly inflated standard errors. In what sense is the unadjusted model overdispersed? It is overdispersed in the sense that the data are being modeled and the first model did not take into consideration the clustering effect of the various hospitals treating the individual patients. The data are then better modeled using a cluster effect. And, insofar as there is a significant difference in the values of the standard errors, the first model is overdispersed. Again, this is indeed a different sense of overdispersion than we understood for binomial models.

Not all statisticians have agreed that binary models are overdispersed. However, the majority of authors on the subject favor the idea, even though the guidelines have not been delineated previously.

9.4.2 Implicit Overdispersion

There is another sense in which a binary model can be overdispersed. Recall in the previous chapter when we converted a binary model to a binomial format. We could only do this if the model predictors were binary or categorical. Nevertheless, we discovered that the parameter estimates, standard errors, p-values, and confidence intervals were all identical. The individual case model and aggregated or grouped model were the same, with the exception of various summary statistics. This is largely due to the differences in the number of model observations between the two models.

Given that the parameter estimates and standard errors are identical between the binary and converted binomial response models, if the binomial model is then proved to be overdispersed, the structure of observations in the binary format must have been such that it gave rise to the overdispersed binomial model. The binary model is not itself overdispersed, but is implicitly overdispersed in that conversion to binomial format results in an overdispersed model. There is nothing known that can be identified in the binary model to indicate binomial overdispersion. However, it would make an interesting study.

Calling a binary model implicitly overdispersed only tells you that its binomial conversion is overdispersed. It does not have a bearing on the goodness of fit of the binary model, which we discuss in Chapter 7.

9.5 Real Overdispersion

9.5.1 Methods of Handling Real Overdispersion

For binomial models, once we have evaluated a model to determine if it can be adjusted to eliminate apparent overdispersion and found that we cannot, it may be concluded that the model is most likely truly or really overdispersed. Table 9.5 provides alternative logistic models that have been used to adjust for binomial overdispersion. Some we have considered in earlier chapters. Williams' procedure will be examined in the following subsection, with a brief overview of the generalized binomial model in the subsequent subsection.

TABLE 9.5

Methods of Handling Real Binomial Overdispersion

1. Scale SEs by Chi^2 dispersion.
2. Scale iteratively; Williams' procedure.
3. Robust variance estimators.
4. Bootstrap or jackknife SE.
5. Generalized binomial.
6. Parameterize as a rate-count response model.
7. Parameterize as a panel model.
8. Nested logistic regression.

9.5.2 Williams' Procedure

Williams' procedure was developed by Williams (1982) as a method to adjust clustered binomial or grouped logistic models. The procedure is a built-in adjustment to the generalized linear models algorithm, employing an extra parameter, called ϕ, which iteratively scales the GLM variance function. Essentially, the Williams' algorithm continuously dampens the χ^2-dispersion statistic until it reaches 1.0. At that point, the statistic is used to weight a grouped or binomial logistic regression. The formula for the weight function is given as:

$$\frac{1}{1+(m-1)\phi} \tag{9.1}$$

where m is the binomial denominator.

The Williams' procedure algorithm can be schematized as shown in Table 9.6.

TABLE 9.6

Williams' Procedure

```
φ == user input of initial value
else (φ = 1)
scale = 1 + (m − 1) φ
scale = 1  if φ = 1
μ = m(y + 0.5)/(m − 1)
η = ln(μ/(m − μ))
WHILE (abs(ΔDev) > tolerance) {
    w = (scale*μ*(m − μ))/m
    z = η + (m*(y − μ))/(scale*μ*(m − μ))
    β = (xᵀwx)⁻¹xᵀwz
```

$$\eta = x\beta$$
$$\text{OldDev} = \text{Dev}$$
$$\text{Dev} = m\ln(m-y)/(m-\mu)) \text{ if } (y=0)$$
$$\text{Dev} = y\ln(y/\mu) \qquad\qquad \text{if } (y=m)$$
$$\text{Dev} = y\ln(y/\mu) + (m-y)\ln((m-y)/(m-\mu)) \text{ if } (y! = 0; y! = m)$$
$$\text{Dev} = 2\Sigma(\text{Dev})/\text{scale}$$
$$\}$$
$$\chi^2 = \Sigma(y-\mu)^2/(\text{scale}*\mu*(1-\mu/m))$$
$$\text{dof} = \#\text{obs} - \#\text{predictors}$$
$$\text{dispersion} = \chi^2/\text{dof}$$
/// Note: Adjust ϕ so dispersion=1, then with final ϕ value:
$$w = 1 + (m-1)\phi$$
/// Rerun algorithm with prior weight equal to w and $\phi = 1$

The problem with the above algorithm is that the user must supply the initial value of ϕ. An IRLS algorithm that can be used to estimate a value of ϕ for the Williams' algorithm may be presented as shown in Table 9.7.

TABLE 9.7

Williams' Estimating Algorithm

Let k = scale /// any value
wt = 1
Williams y m predictors /// using the above algorithm
$$\chi^2 = \Sigma(y-\mu)^2/(\mu(1-\mu/m))$$
$$\text{dof} = \#\text{obs} - \#\text{predictors}$$
$$\text{disp} = \chi^2/\text{dof}$$
$$\phi = 1/\text{disp}$$
WHILE (abs(ΔDev) > tolerance) {
  OldDisp = Disp
  Williams y m predictors, $k = \phi$
  $$\chi^2 = \Sigma(y-\mu)^2/\mu(1-\mu/m))$$
  Disp = χ^2/dof
  ΔDisp = Disp − OldDisp
}
wt = $1 + (m-1)\phi$
Williams y m predictors [w = wt]

A Stata Williams' procedure program may be downloaded from the Web as:

```
. net from http://www.stata.com/users/jhilbe
. net install Williams
```

Using the same **heart01** data we have used for most of this section, a Williams' procedure may be use to estimate odds ratios.

```
. williams dead cases anterior hcabg kk2-kk4 age3 age4, eform
  nolog

Resid DF    =        887              No obs.     =        895
Chi2        =   870.4991              Deviance    =   572.4562
Dispersion  =    .981397              Prob>chi2   =          1
                                      Dispersion  =   .6453847
CHI2(7)     =   183.1971
Prob>CHI2   =   4.11e-36              Pseudo R2   =   .1391893

logistic regression: Williams' procedure
-------------------------------------------------------------------
    dead |      OR Std. Err.    z  P>|z|   [95% Conf. Interval]
---------+---------------------------------------------------------
anterior | 1.873635  .3315315  3.55 0.000  1.324551    2.650339
   hcabg | 2.157318  .7657057  2.17 0.030  1.075954    4.325483
     kk2 | 2.240909  .4248089  4.26 0.000  1.545475    3.249275
     kk3 | 2.202764  .6044194  2.88 0.004  1.286488    3.771641
     kk4 | 14.32207  5.142361  7.41 0.000  7.085665   28.94883
    age3 | 3.481389  .7451306  5.83 0.000  2.288588    5.295871
    age4 | 6.738025  1.476571  8.71 0.000   4.38531    10.35297
-------------------------------------------------------------------
Log Likelihood = -566.48779
Phi            =   .0196564
```

Comparing the Williams result with an estimation of the data using an unadjusted binomial logistic regression gives us:

```
Log likelihood    =  -410.209401
-------------------------------------------------------------------
         |              OIM
    dead | Odds Ratio Std. Err.   z  P>|z| [95% Conf. Interval]
---------+---------------------------------------------------------
anterior |   1.894096 .3173418  3.81 0.000 1.363922    2.630356
   hcabg |   2.195519 .7744623  2.23 0.026 1.099711    4.383246
     kk2 |   2.281692 .4117012  4.57 0.000 1.602024    3.249714
     kk3 |   2.218199 .5971764  2.96 0.003 1.308708    3.759743
     kk4 |   14.63984 5.218374  7.53 0.000 7.279897   29.44064
    age3 |   3.549577 .7119235  6.32 0.000 2.395823    5.258942
    age4 |   6.964847  1.44901  9.33 0.000 4.632573    10.47131
-------------------------------------------------------------------
```

The results do not appreciably differ. However, there is a significant difference in the log-likelihood functions, favoring the standard model. Comparisons of AIC statistics (not shown) also indicate a clear preference for the standard model. It may be concluded, therefore, that employing the Williams' procedure with grouped data does not always result in a satisfactory adjustment of clustered or overdispersed data.

It should be emphasized that the Williams' procedure, commonly used throughout the 1980s and 1990s, is rarely used in current research. Given the above example, it appears that it will likely not regain widespread use.

9.5.3 Generalized Binomial Regression

A generalized binomial model was constructed by Famoye (1995) from a generalized negative binomial model, which was first introduced by Jain and Consul (1971). The generalized binomial model has been discussed in some detail in Hardin and Hilbe (2007); those who are interested in this model are referred to that source.

The log-likelihood of the generalized binomial may be expressed, without subscripts, as:

$$L(\beta,\phi) = \Sigma\{\ln m + y \ln\pi + (m + \phi y - y)\ln(1 + \phi\pi - \pi) - (m + \phi y)\ln(1 + \phi\pi) +$$
$$\ln \Gamma(1 + \phi y + 1) - \ln \Gamma(y + 1) - \ln \Gamma(m + \phi y - y + 1)\} \tag{9.2}$$

The mean and variance are given as, respectively:

$$E(Y) = \mu = m\pi \tag{9.3}$$

and

$$V(Y) = m\pi(1 + \phi\pi)(1 + \phi\pi)(1 + \phi\pi - \pi) \tag{9.4}$$

where π is the probability of success, m is the binomial denominator, and ϕ is the ancillary parameter to be estimated.

During estimation, π is initialized using the inverse logistic link function, $1/(1 + \exp(-\eta))$. Modeling alternative binomial links only take a substitution of their inverse link functions; for example, probit $[\phi(\eta)]$, complementary loglog $[1 - \exp(-\exp(\eta)]$.

The generalized binomial model has not been implemented in any commercial software. It experiences rather severe problems with convergence and for the most part does not appear to accommodate overdispersion better than other models we have discussed. It may well be the case that other parameterizations of the generalized binomial may result in handling overdispersed—and underdispersed—data better than the method presented here. However, this discussion goes beyond the scope of this text.

9.6 Concluding Remarks

Overdispersion is a rather complex notion. In this chapter, we have attempted to identify and categorize the various types and meanings of the term and how it has been applied to models. It is important to keep in mind that an overdispersed model is not well fitted. It represents a violation of the assumptions upon which the models themselves are based. The means that have been invented to adjust for, or accommodate, overdispersion have taken us to quasi- and pseudolikelihood models.

We next discuss extensions to the basic logistic model. Included in this discussion will be models having an ordered discrete or categorical response, models having an unordered discrete response, and longitudinal and clustered logistic models. Models in this last group are many times collectively referred to as panel models, particularly in econometrics. We thereupon discuss various alternative logistic models, as well as the method largely used before binary and categorical response logistic models became popular—discriminant analysis. Finally, we shall touch on a new area involving computer intensive permutations, called exact statistics, and its alternatives. Only Stata, SAS, and Cytel's LogXact provide statisticians with the ability to estimate logistic models using exact methods. As we shall observe, exact statistics are appropriate when there are relatively few observations, or when the data are highly unbalanced.

Exercises

9.1. This example comes from a famous data set quoted in Finney (1971).

Dose of Rotenone (mg/l)	Number of Insects	Number Dead
0	49	0
2.6	50	6
3.8	48	16
5.1	46	24
7.7	49	42
10.2	50	44

   a. Model the number of killed insects (*dead*) from the total number identified on dose using a logistic model. Is *dose* a significant contributor to understanding the number of insects killed?

   b. Take the *natural log of dose* and model using it rather than *dose*. What is the difference, if any?

c. Construct an empirical proportion of dead insects over number of insects. Graph the proportion by both the *dose* and *lndose*.

d. Calculate the predicted probability of death, modeling on both *dose* and *lndose*. Are there any differences in the graphs of predicted proportions and empirical proportions?

e. Is the *lndose* model overdispersed? How do you tell? What can be done to ameliorate the overdispersion?

f. If there is evidence of overdispersion, is it real or only apparent?

9.2. These data come from a study involving respiratory disease in infants. The study tested how many infants developed either bronchitis or pneumonia during the first year of life. Risk factors included gender and type of feeding, which included bottle only, breast plus a supplement, or breast only. Data are from the GLIM 4 manual (Francis, Green, and Payne, 1993), page 451.

	Bottle Only	Breast + Supplement	Breast Only
Boys	77/458	19/147	47/494
Girls	48/384	16/127	31/464

a. Is there a significant risk of developing respiratory disease (bronchitis of pneumonia) based on the type of feeding?

b. Is the risk the same for both boys and girls?

c. Is there an interactive effect between gender and type of feeding? Graph the main effects in order to observe if there is an interactive effect.

d. Is the model overdispersed? If so, is it real or apparent?

e. Interpret the model results.

f. Perform any related goodness-of-fit tests to be assured that the model is well fit. Would another binomial link result in a statistically superior model?

9.3. Use the same data and model as given in Exercise 8.4. Answer the following questions.

a. If the grouped or binomial model appears to be overdispersed, is it real or apparently overdispersed?

b. Model the grouped binomial model as a rate Poisson and rate negative binomial model. Which model best fits the data?

c. Are all overdispersed binomial models implicitly overdispersed Bernoulli models? Why or why not?

9.4. Based on the modeling results of Exercise 8.6, how can the logistic model be best adjusted for the extradispersion existing in the data? Does the adjustment result in a statistically superior model?

9.5. Use the same grouped logistic model of Exercise 9.4 and remodel it using Williams' procedure. Compare the results of this model to the logistic and Poisson models performed earlier.

9.6. The table below incorporates data gathered in the Framingham study on the risk factors for coronary heart disease (CHD), as analyzed by Truett, Cornfield, and Kannel (1967) and used in Collett (1991), page 224. The response values consist of the number of subjects developing CHD given the number of subjects tested at the pre-established levels of gender, age group, and serum cholesterol level.

Sex	Age Group	Serum Cholesterol Level			
		<190	190–219	220–249	≥250
Male	30–49	13/340	18/408	40/421	57/362
	50–62	13/123	33/176	35/174	49/183
Female	30–49	6/542	5/552	10/412	18/357
	50–62	9/58	12/135	21/218	48/395

a. Model the number of subjects diagnosed with CHD from within each group with similar characteristics of gender, age group, and serum cholesterol level. Which estimates contribute significantly to the model?

b. Interpret the odds ratios for the various estimates.

c. Is the model overdispersed? If so, make appropriate adjustments to accommodate the standard errors.

d. Compare the logistic model with a similar model using Poisson and negative binomial regression. Which is the best-fitted model?

e. Which combination of predictor levels results in the highest probability of being diagnosed with CHD? Which combination results in the lowest probability?

R Code

Section 9.2 The Nature and Scope of Overdispersion

No examples

Section 9.3 Binomial Overdispersion

```
library('foreign')
simul<- read.dta('overexx.dta') #read Stata format file
simul$noty<- simul$d - simul$y #numbers of failures
```

```
simul$notyi<- simul$d - simul$yi #numbers of failures
simul$notysq<- simul$d - simul$ysq #numbers of failures
simul$notyp<- simul$d - simul$yp #numbers of failures
summary(simul)
head(simul)

source('pearsonX2.r') #reinstall binomialX2()

fit9_2a<- glm(cbind(y, noty) ~ x1 + x2 + x3, data=simul,
    family=binomial)
summary(fit9_2a)
fit9_2aX2<- binomialX2(simul$y, simul$d, fitted(fit9_2a,
    type='response'))
cat('Pearson X2:', fit9_2aX2, 'Dispersion:', fit9_2aX2/
    fit9_2a$df.residual, 'AIC/d.f.:', fit9_2a$aic/fit9_2a$df.
    residual, '\n')

fit9_2b<- glm(cbind(y, noty) ~ x2 + x3, data=simul,
    family=binomial)
summary(fit9_2b)
fit9_2bX2<- binomialX2(simul$y, simul$d, fitted(fit9_2b,
    type='response'))
cat('Pearson X2:', fit9_2bX2, 'Dispersion:', fit9_2bX2/
    fit9_2b$df.residual, 'AIC/d.f.:', fit9_2b$aic/fit9_2b$df.
    residual, '\n')

fit9_2c<- glm(cbind(yi, notyi) ~ x1 + x2 + x3 + x23,
    data=simul, family=binomial)
summary(fit9_2c)
fit9_2cX2<- binomialX2(simul$yi, simul$d, fitted(fit9_2c,
    type='response'))
cat('Pearson X2:', fit9_2cX2, 'Dispersion:', fit9_2cX2/
    fit9_2c$df.residual, 'AIC/d.f.:', fit9_2c$aic/fit9_2c$df.
    residual, '\n')

fit9_2d<- glm(cbind(yi, notyi) ~ x1 + x2 + x3, data=simul,
    family=binomial)
summary(fit9_2d)
fit9_2dX2<- binomialX2(simul$yi, simul$d, fitted(fit9_2d,
    type='response'))
cat('Pearson X2:', fit9_2dX2, 'Dispersion:', fit9_2dX2/
    fit9_2d$df.residual, 'AIC/d.f.:', fit9_2d$aic/fit9_2d$df.
    residual, '\n')

fit9_2e<- glm(cbind(ysq, notysq) ~ x1sq + x2 + x3, data=simul,
    family=binomial)
summary(fit9_2e)
fit9_2eX2<- binomialX2(simul$ysq, simul$d, fitted(fit9_2e,
    type='response'))
```

```
cat('Pearson X2:', fit9_2eX2, 'Dispersion:', fit9_2eX2/
    fit9_2e$df.residual, 'AIC/d.f.:', fit9_2e$aic/fit9_2e$df.
    residual, '\n')

fit9_2f<- glm(cbind(ysq, notysq) ~ x1 + x2 + x3, data=simul,
    family=binomial)
summary(fit9_2f)
fit9_2fX2<- binomialX2(simul$ysq, simul$d, fitted(fit9_2f,
    type='response'))
cat('Pearson X2:', fit9_2fX2, 'Dispersion:', fit9_2fX2/
    fit9_2f$df.residual, 'AIC/d.f.:', fit9_2f$aic/fit9_2f$df.
    residual, '\n')

fit9_2g<- glm(cbind(yp, notyp) ~ x1 + x2 + x3, data=simul,
    family=binomial(link=probit))
summary(fit9_2g)
fit9_2gX2<- binomialX2(simul$yp, simul$d, fitted(fit9_2g,
    type='response'))
cat('Pearson X2:', fit9_2gX2, 'Dispersion:', fit9_2gX2/
    fit9_2g$df.residual, 'AIC/d.f.:', fit9_2g$aic/fit9_2g$df.
    residual, '\n')

fit9_2h<- glm(cbind(yp, notyp) ~ x1 + x2 + x3, data=simul,
    family=binomial(link=logit))
summary(fit9_2h)
fit9_2hX2<- binomialX2(simul$yp, simul$d, fitted(fit9_2h,
    type='response'))
cat('Pearson X2:', fit9_2hX2, 'Dispersion:', fit9_2hX2/
    fit9_2h$df.residual, 'AIC/d.f.:', fit9_2h$aic/fit9_2h$df.
    residual, '\n')

cancer<- read.dta('cancerx.dta') #read Stata format file
cancer$drug<- factor(cancer$drug) #convert to factor
summary(cancer)
head(cancer)
nrow(cancer)

fit9_2i<- glm(studytim ~ drug, data=cancer, family=poisson)
summary(fit9_2i)
fit9_2iX2<- poissonX2(cancer$studytim, fitted(fit9_2i,
    type='response'))
cat('Pearson X2:', fit9_2iX2, 'Dispersion:', fit9_2iX2/
    fit9_2i$df.residual, 'AIC/d.f.:', fit9_2i$aic/fit9_2i$df.
    residual, '\n')

cancer$time<- cancer$studytim #make copy
cancer$time[cancer$studytim==39]<- 390 #create outliers
fit9_2j<- glm(time ~ drug, data=cancer, family=poisson)
summary(fit9_2j)
```

```
fit9_2jX2<- poissonX2(cancer$time, fitted(fit9_2j,
    type='response'))
cat('Pearson X2:', fit9_2jX2, 'Dispersion:', fit9_2jX2/
    fit9_2j$df.residual, 'AIC/d.f.:', fit9_2j$aic/fit9_2j$df.
    residual, '\n')
cancer[cancer$time==390,] #show rows for time=390

summary(cancer$age) #show stats
cat('S.D.:', sd(cancer$age), 'S.E. mean:', sd(cancer$age)/
    sqrt(length(cancer$age)), '\n')

fit9_2k<- glm(died ~ age, data=cancer, family=binomial)
summary(fit9_2k)
fit9_2kX2<- binomialX2(cancer$died, rep(1, nrow(cancer)),
    fitted(fit9_2k, type='response'))
cat('Pearson X2:', fit9_2kX2, 'Dispersion:', fit9_2kX2/
    fit9_2k$df.residual, 'AIC/d.f.:', fit9_2k$aic/fit9_2k$df.
    residual, '\n')

quantile(cancer$age, probs=0.75)
sum(cancer$age==60)
which(cancer$age==60)
cancer$age[37]<- 600 #make outlier

fit9_2l<- glm(died ~ age, data=cancer, family=binomial)
summary(fit9_2l)
fit9_2lX2<- binomialX2(cancer$died, rep(1, nrow(cancer)),
    fitted(fit9_2l, type='response'))
cat('Pearson X2:', fit9_2lX2, 'Dispersion:', fit9_2lX2/
    fit9_2l$df.residual, 'AIC/d.f.:', fit9_2l$aic/fit9_2l$df.
    residual, '\n')
exp(coef(fit9_2l)) #ORs
exp(coef(fit9_2k)) #ORs

yd<- simul$y/simul$d #proportions
summary(yd) #stats
cat('S.D.:', sd(yd), 'S.E. mean:', sd(yd)/sqrt(length(yd)), '\n')

summary(fit9_2a) #show previous fit again
cat('Pearson X2:', fit9_2aX2, 'Dispersion:', fit9_2aX2/
    fit9_2a$df.residual, 'AIC/d.f.:', fit9_2a$aic/fit9_2a$df.
    residual, '\n')

ld <- ln(d) # natural log of offset
fit9_2m<- glm(y ~ x1 + x2 + x3, data=simul, offset=ld,
    family=poisson)
summary(fit9_2m)
fit9_2mX2<- poissonX2(simul$y, fitted(fit9_2m,
    type='response'))
```

```
cat('Pearson X2:', fit9_2mX2, 'Dispersion:', fit9_2mX2/
   fit9_2m$df.residual, 'AIC/d.f.:', fit9_2m$aic/fit9_2m$df.
   residual, '\n')


require('MASS')
fit9_2n<- glm.nb(y ~ x1 + x2 + x3 + offset(ld), data=simul)
summary(fit9_2n)
fit9_2nX2<- binomialX2(simul$y, simul$d, fitted(fit9_2n,
   type='response'))
cat('Pearson X2:', fit9_2nX2, 'Dispersion:', fit9_2nX2/
   fit9_2n$df.residual, 'AIC/d.f.:', fit9_2n$aic/fit9_2n$df.
   residual, '\n')
```

Section 9.4 Binary Overdispersion

```
heart<- read.dta('heart01.dta') #read Stata format file
heart$anterior <- heart$anterior[,drop=TRUE] #drop empty
   levels that complicate
tabulations
heart$center<- factor(heart$center) #convert to factor from
   numeric levels
heart$killip<- factor(heart$killip) #convert to factor from
   numeric levels


heart2<- na.omit(heart) #drop NAs


fit9_3a<- glm(death ~ anterior + hcabg + kk2 + kk3 + kk4 +
   age3 + age4, data=heart2, family=binomial)
summary(fit9_3a)
fit9_3aX2<- binomialX2(heart2$death, rep(1,
   length(heart2$death)), fitted(fit9_3a, type='response'))
cat('Pearson X2:', fit9_3aX2, 'Dispersion:', fit9_3aX2/
   fit9_3a$df.residual, 'AIC/d.f.:', fit9_3a$aic/fit9_3a$df.
   residual, '\n')


fit9_3b<- glm(death ~ anterior + hcabg + kk2 + kk3 + kk4 +
   age3 + age4 + center, data=heart2, family=binomial)
summary(fit9_3b)
fit9_3bX2<- binomialX2(heart2$death, rep(1,
   length(heart2$death)), fitted(fit9_3b, type='response'))
cat('Pearson X2:', fit9_3bX2, 'Dispersion:', fit9_3bX2/
   fit9_3b$df.residual, 'AIC/d.f.:', fit9_3b$aic/fit9_3b$df.
   residual, '\n')
```

Section 9.5 Real Overdispersion

<Williams procedure not implemented in R>

10

Ordered Logistic Regression

10.1 Introduction

The ordered logistic model is an extension of the binary response model discussed in Chapters 2 through 7, where the response has two or more levels. When a two-level response is estimated using an ordered logistic algorithm, the results are the same as those produced using standard logistic regression software. The model is nearly always employed, therefore, on responses having three or more ordered levels. For example, school grades of A, B, C, D, and F represent such an ordered response—with five levels. The ordered levels, Poor, Fair, Average, Good, Better, and Best are also appropriate for modeling with an ordered logistic procedure.

Statisticians have devised a variety of ordered or ordinal response models. The foremost ordered logistic model is the proportional odds model, which will be the emphasis of our discussion in this chapter. The proportional odds model assumes that the model coefficients for each level or response are equal. When this assumption is challenged, or found not to be the case, a partial proportional model may be developed for the data. Additionally, a generalized ordered logistic model may be constructed; the model assumes no proportionality in the level coefficients. This chapter will address each of these models, with an examination of how to test when proportionality criteria have been violated. Moreover, we will pay particular attention to the interpretation of parameter estimates for each of these models.

Other rather underutilized ordinal response models have also been developed. We shall discuss these in Chapter 12, following an intermediate chapter on the multinomial model, which assumes no order to the levels of a categorical response model. Models discussed in Chapter 12 include the continuation-ratio model, stereotype logistic model, heterogeneous choice or the location scale logistic model, adjacent category model, and other nonlogistic-ordered binomial models. McFadden's choice model and rank-ordered logistic regression are discussed in Section 13.3, the section related to conditional logistic models.

Categorical response models have been classified by some as a member of the family of generalized linear models. However, they are not of the traditional variety employing Fisher Scoring, or the IRLS algorithm, as a means for estimation. Although statisticians working with GLIM software, which was described in the Introduction, were able to manipulate the software so that it could estimate the proportional odds and the multinomial logit models, the applications were limited in terms of the number of levels and predictors that could be estimated. Today no commercial application uses a GLM approach to the estimation of categorical response models. Rather, some variety of full maximum likelihood estimation is normally employed for the modeling process. Quadrature and simulation methods are used many times when the model is complex, for example, with random coefficient ordered logistic models. These types of categorical response models will be addressed in Chapter 13.

Ordered models assuming the proportionality of coefficients or slopes can be parameterized for any of the binomial links found in generalized linear models procedures. These include the logit or logistic, the probit, complementary loglog, loglog, and Cauchit. These same parameterizations maintain for the generalized and partial proportional slopes models as well. Several of the models we discuss in Chapter 12 can also be parameterized in a similar manner, but generally are not. I shall defer this discussion to Chapter 12, but keep it in mind as we discuss the models of this chapter.

Although Zavonia and McElvey (1975) are generally credited with developing this class of models, creating what they called an ordinal regression model, Walker and Duncan (1967) were in fact the first to actually propose an ordered response logistic model, which they termed a cumulative logit model. However, Zavonia and McElvey's presentation resulted in an increased interest in the models, which is likely the result of the superior computing capabilities from 1967 to 1975. It was during the 1970s that computers became more available, and the first personal computers appeared in the marketplace. Of course, 1981 initiated a new era in computing, with the release of the first IBM Personal Computer (PC) in August of that year. Shortly thereafter, the major statistical software companies, and new ones that evolved as a result, developed PC-based statistical applications, which allowed statisticians and researchers to perform complex statistical analysis on their own, without having to submit batch programs to mainframe facilities.

Zavonia and McElvey's ordered response model was based on a normally distributed underlying latent response variable. As an extension to the previously developed probit model, it was later renamed an ordinal or ordered probit. The underlying continuous response, typically referred to as y^*, was construed to be divided into a number of categories or levels, in a similar manner to how we categorize a continuous predictor. This manner of envisioning how categories arise is complemented by the alternative of viewing categories as natural divisions. However, if you carefully consider the matter, it is difficult to think of an ordered set of categories that do not have an underlying continuous range. For example, the grades A, B, C, D, and F are categories we assign to percentages, with traditional grades cut at A = 100–90 percent, B = 89–80 percent, C = 79–70 percent, D = 69–60 percent, and F = below 60 percent. Other cut points can, and have, been assigned. Likewise, the killip levels used for some of our previous examples are based on conditions of severity, with an underlying range of risk. Political categories, such as extremely liberal, liberal moderate, conservative, and extremely conservative, also imply an underlying range of attitudes. If there is no latent continuous variable underlying the categories we find, or construct for ourselves, then there is no implicit order to the categories, and if there is no order, then a multinomial model should be used for estimation, not an ordered model.

We will admit, though, that a latent variable need not be in mind when we are considering the levels of categories of the response, but it does help us understand the cumulative nature of how parameter estimates are interpreted.

Peter McCullagh (1980) of the University of Chicago constructed an ordered logit model based on the same logic as Zavonia and McElvey's ordered probit model; however, he employed a logistic function for the underlying latent response rather than a normal distribution. The model, which he termed a proportional odds model, was structured so that the model coefficients or slopes are parallel across categories. We now turn to this model.

10.2 The Proportional Odds Model

Viewed from the point of view of an underlying latent variable, as described in the previous section, the relationship of y^* to cut points and levels or categories can be graphically displayed as:

```
LEVELS            1         2         3         4
y*        -∞< --------|----------|----------|--------->+∞
CUTS      <τ₀>       τ₁         τ₂        τ₃         <τⱼ>
```

Notice that for a four-level categorical response, there are three cut points. The cuts can be anywhere along the y^* line.

The ordered relationship can be described in terms of probabilities along the y^* line. For a four-level response, y, we have:

LEVEL 1: $\tau_0 \leq y^* < \tau_1$

LEVEL 2: $\tau_1 \leq y^* < \tau_2$

LEVEL 3: $\tau_2 \leq y^* < \tau_3$

LEVEL 4: $\tau_3 \leq y^* < \tau_4 \, (\tau_j)$

In a concise form, the logistic form of the model is defined as the log of the ratio of probability of levels under or equal to a given cut to the probability of those over the cut, for each ascending level of cuts.

$$\ln[\Pr(Y \le y_j | x) / \Pr(Y > y_j | x)] \tag{10.1}$$

where

$$\Pr(Y \le y_j | x) = \frac{\exp(\alpha_j - x\beta)}{1 + \exp(\alpha_j - x\beta)} \tag{10.2}$$

or

$$\Pr(Y \le y_j | x) = \frac{1}{1 + \exp(-(\alpha_j - x\beta))} \tag{10.3}$$

α_js are the cuts or thresholds, also referred to as the intercepts or constants. In some parameterizations, however, such as the Stata parameterization being discussed here, the cut points are the negatives of the intercepts. The values of the cuts will, therefore, have an opposite sign to those algorithms that directly relate cuts and intercepts. In any case, subscript j reflects the number of model intercepts. The intercepts always meet the following criterion:

$$\alpha_1 \le \alpha_2 \le \alpha_3 \le \ldots \le \alpha_J \tag{10.4}$$

An example of this criterion can be shown best with a response having many categories. We shall use the **gss2002_10** data, modeling political views, a seven-level categorical variable, on the binary predictor *sex*. Notice the ascending values in the cutpoints, reflecting the cumulative nature of the underlying response.

```
. ologit polviews sex, nolog

Ordered logistic regression              Number of obs   =        1331
                                         LR chi2(1)      =        0.05
                                         Prob > chi2     =      0.8152
Log likelihood = -2219.9696              Pseudo R2       =      0.0000
------------------------------------------------------------------------------
polviews |      Coef. Std. Err.     z    P>|z|    [95% Conf. Interval]
---------+--------------------------------------------------------------------
     sex |   .0230673 .0986795  0.23  0.815    -.170341     .2164755
---------+--------------------------------------------------------------------
   /cut1 |  -3.272389 .2114207                 -3.686766   -2.858012
   /cut2 |  -1.757354   .169988                 -2.090525   -1.424184
```

```
/cut3 |  -.9992154   .163324          -1.319325   -.6791062
/cut4 |   .6736439  .1613901            .357325    .9899628
/cut5 |  1.494413   .1659575           1.169143   1.819684
/cut6 |  3.483993   .2187003           3.055349   3.912638
```

The proportional odds model may be expressed symbolically in terms of probability levels, as follows, supposing a response having four levels or categories:

CATEGORY OR LEVEL 1

$$\text{Logit} = \ln(p_1/(1 - p_1))$$

CATEGORY OR LEVEL 2

$$\text{Logit} = \ln[(p_1 + p_2)/(1 - p_1 - p_2)]$$

CATEGORY OR LEVEL 3

$$\text{Logit} = \ln[(p_1 + p_2 + p_3)/(1 - p_1 + p_2 + p_3)]$$

Response categories or levels for any number of levels may be symbolized as:

$$\text{Logit} = \ln[(p_1 + p_2 + \cdots + p_{n-1})/(1 - p_1 - p_2 - \cdots - p_{n-1})]$$

The probabilities of all levels must sum to 1:

$$p_1 + p_1 + \cdots + p_{n+1} = 1$$

Each of the four log-odds relationships takes the form of:

$$\alpha_n - x\beta \quad (x\beta \text{ is also represented as } x'\beta \text{ and sometimes as } \beta x)$$

Note, however, that SAS parameterizes the relationship as:

$$\alpha_n + x\beta.$$

This example has four constants associated with the main model parameter estimates. These are also called cutpoints, and they define how probabilities are assigned to each post-estimation level. It is important to understand that the proportional odds model is in fact a multiple equation model related to each nonreference level of the response. The parameter estimates for each level (equation) are identical. The only difference rests with the equation constants, which differ. The output of a proportional odds model is therefore tradition-ally formatted so that only one table of parameter estimates is displayed, together with an ancillary table of cut points, which are the multiple-equation constants. Again, we can do this because of the assumption that the slopes, or coefficients, are the same.

An example should help clarify the above relationships. I will show two methods that can be used to calculate probabilities. Remember, using the formulae we examined earlier in this book, probabilities can be converted to odds and odds ratios.

Use the **medpar_olr** data set, which can be downloaded from the text's Web site. **Medpar** refers to U.S. Medicare data. At the age of 65, U.S. citizens may opt to join a nationalized health-care system that provides substantially reduced costs for hospitalization, physician visits, and other specified health-related services.

During a given year, each hospitalized Medicare patient becomes an observation or case in the Medicare data. The year we are using for this example comes from the 1991 Arizona **Medpar** data, for a specific type of disease. The diagnostic related group (DRG) for this data is known, but for privacy, it is left unspecified.

Regardless, the response variable in this 1,495 observation data set is *type*, indicating the type of hospital admission for each patient in the data. For our initial example, we will model *type* on a single binary predictor, *age80*, representing patients over the age of 80 (1 = 80+; 0 < 80).

A tabulation of the response, *type*, can be presented as:

```
. tab type

    Type of |
  Admission |      Freq.      Percent        Cum.
------------+-----------------------------------
   Elective |      1,134        75.85       75.85
     Urgent |        265        17.73       93.58
  Emergency |         96         6.42      100.00
------------+-----------------------------------
      Total |      1,495       100.00
```

There are relatively few Emergency admissions (6.4 percent) compared with other types of admissions to the hospital, as would be expected. With such a disparity in level numbers, problems with modeling may occur. One solution, of course, is to combine Urgent and Emergency admits, but this puts us back to using a straightforward logistic regression for modeling. Information is also lost. The levels will therefore be retained as found in the data.

Notice that *type* is ordered according to severity of admission. Ordered logistic regression does not specify units or quantities of order; for example, level 2 is twice the odds of level 1, level 3 is thrice the level of level 1, or twice the odds of level 2, or whatever specific relationship. All that is required is that there is an order to the levels.

```
. ologit type age80, nolog
```

```
Ordered logistic regression          Number of obs    =        1495
                                     LR chi2(1)       =        1.35
                                     Prob > chi2      =      0.2456
Log likelihood = -1034.7966          Pseudo R2        =      0.0007
--------------------------------------------------------------------
  type |      Coef.  Std. Err.     z   P>|z|   [95% Conf. Interval]
-------+------------------------------------------------------------
 age80 | -.1708724   .1485981  -1.15   0.250  -.4621193   .1203746
-------+------------------------------------------------------------
 /cut1 |  1.108138   .0677373                  .9753753   1.2409
 /cut2 |  2.643507   .1096271                  2.428642   2.858372
--------------------------------------------------------------------
```

We can let the software calculate the predicted values for each category and level of *age80* by using the *predict* command, followed by the variable names that the user wishes to use for defining each level. *pr1, pr2* and *pr3* will be used for each probability level.

. predict pr1 pr2 pr3

Next summarize each probability level, first for *age80* $= 0$ (<80) followed by *age80* $= 1$ (80+). Stata's **summarize** command is used since all probability values are equal within a level and predictor value. Following an examination of the basic logistic regression model, we can regard *age80* as the exposure, or risk factor.

. su pr1 pr2 pr3 if age80==0

```
Variable |      Obs       Mean   Std. Dev.        Min        Max
---------+----------------------------------------------------------
    pr1  |     1165  .7517818          0   .7517818   .7517818
    pr2  |     1165  .1818279          0   .1818279   .1818279
    pr3  |     1165  .0663903          0   .0663903   .0663903
```

. su pr1 pr2 pr3 if age80==1

```
Variable |      Obs       Mean   Std. Dev.        Min        Max
---------+----------------------------------------------------------
    pr1  |      330  .7822812          0   .7822812   .7822812
    pr2  |      330  .1611667          0   .1611667   .1611667
    pr3  |      330  .0565521          0   .0565521   .0565521
```

To demonstrate the calculations, it will be helpful to provide a base estimates table, including the coefficient value for *age80* and the values of the two cuts or intercepts. Recall from our earlier discussion that the model presented below is actually three models, each having the same parameter estimate, but differing intercepts. The results are displayed in this manner to save space, and to reinforce the notion that the three slopes are equal.

```
--------------------------------------------------------------------
 type |      Coef.  Std. Err.     z  P>|z|  [95% Conf. Interval]
------+-------------------------------------------------------------
age80 |  -.1708724  .1485981  -1.15 0.250  -.4621193   .1203746
------+-------------------------------------------------------------
/cut1 |   1.108138  .0677373                .9753753     1.2409
/cut2 |   2.643507  .1096271               2.428642   2.858372
--------------------------------------------------------------------
```

AGE80 == 0

Recall that when the predictor has a value of 0, it is effectively excluded from the calculation of the probability. In the case of a single predictor with a value of 0, we must rely entirely on the cut values.

For the first level, we employ the inverse logit link, $1/(1+\exp(-xb))$, which may also be formulated as $\exp(xb)/(1+\exp(xb))$. The first formula is used for our calculations. In Stata, one may also use the saved post-estimation values for the cuts, accessed by $\_b[/cut_j]$, where j takes the cut number.

METHOD 1

This method of calculating the probability values for each level stems directly from the formula described in the paragraph above.

PR1

```
. di -_b[/cut1]
-1.1081379

. di 1/(1+exp(-_b[/cut1]))    or  di 1/(1+exp(-1.1081379))
.75178179
```

PR2

```
. di -_b[/cut2]
-2.6435072

. di  1/(1+exp(-_b[/cut2])) - .75178179
.18182789
```

PR3

```
. di 1- 1/(1+exp(-_b[/cut2]))
.06639032
```

These three probability values are identical to those calculated by the software.

METHOD 2

This method exploits the ratio nature of the relationship. Notice that at the second step we divide the previously calculated value by one plus the value. For example, if the calculated value is 0.05, the denominator is 1.05. The probabilities calculated using this method are identical to those of the first method.

PR1

```
. di exp(-_b[/cut1])
.33017322

. di .33017322/1.33017322              ///  (0.*)/(1.*)
.24821821

. di 1 - .24821821
.75178179
```

PR2

```
. di exp(-_b[/cut2])
.07111143

. di .07111143/1.07111143
.06639032

. di .24821821 - .06639032
.18182789
```

```
or
```

```
Calculate PR1 and PR3, subtract sum of both from 1
```

```
. di 1 - (.75178179 + .06639032)
.18182789
```

PR3

```
. di exp(-_b[/cut2])
.07111143

. di .07111143/1.07111143
.06639032
```

AGE80 == 1

METHOD 1

With the predictor value of 1, we now utilize the coefficient value, subtracting the cut value from it. Since the value of *age80* is 1, we need only use the coefficient without any multiplication. Recall that the logistic probability is determined from the inverse link function, expressed here as $1/(1+\exp(-x\beta))$.

PR1

```
. di -.1708724 - _b[/cut1]          /// linear predictor
-1.2790103

. di 1/(1+exp(-1.2790103))          /// probability
.78228126
```

PR2

```
. di -.1708724 - _b[/cut2]
-2.8143796

. di 1/(1+exp(-2.8143796)) - 1/(1+exp(-1.2790103))
.16116668
```

PR3

```
. di 1-1/(1+exp(-2.8143796))
.05655206
```

METHOD 2

PR1

```
. di -.1708724 - _b[/cut1]
-1.2790103

. di exp(-1.2790103)
.27831261

. di .27831261/1.27831261
.21771874

. di 1-.21771874
.78228126
```

PR2

```
. di exp(-.1708724 - _b[/cut1])
.27831262
```

```
. di .27831261/1.27831261
.21771874

. di exp(-.1708724 - _b[/cut2])
.0599419

. di .0599419/1.0599419
.05655206

. di .21771874   -.05655206
.16116668
```

or

Calculate PR1 and PR3, subtract sum of both from 1

```
. di  1-(.78228126 + .05655206)
`.16116668
```

PR3

```
. di -.1708724 - _b[/cut2]
-2.8143796

. di exp(-2.8143796)
.0599419

. di .0599419/1.0599419
.05655206
```

What we see from the above is the varied ways that the probabilities can be calculated. In fact, statistical software companies differ in how they calculate the values, and in some cases, the ordering of the cuts is entirely different.

MULTIVARIABLE MODEL

We shall next expand the model to include four predictors. We do not include *age80*, and *provnum* is only used when we test the model for excessive correlation. An overview of the variables included in the model can be seen using the *describe* command, shortened to a simple *d*.

```
. d

Contains data from medpar_olr.dta
 obs: 1,495                    1991 AZ Medicare: Ordered LR
vars:     6                    28 Apr 2007 08:34
size: 32,890 (99.7% of memory free)
```

```
----------------------------------------------------------------
                storage   display   value
variable name    type     format    label     variable label
----------------------------------------------------------------
type             byte      %9.0g     type      Type of Admission
died             float     %9.0g               1=Died; 0=Alive
white            float     %9.0g               1=White; 0=Other
hmo              byte      %9.0g               1=HMO member
los              int       %9.0g               Length of Stay
provnum          str6      %9s                 Provider number
----------------------------------------------------------------
Sorted by:  provnum
```

We check the values of each variable and look for missing values. The number of observations listed as *Obs* represent non-missing values. If a variable shows less than 1,495, we know that it has missing values. We can tabulate the identified variable to obtain the percentage of values.

```
. su

Variable |     Obs        Mean     Std. Dev.       Min         Max
---------+--------------------------------------------------------
    type |    1495    1.305686     .5838647          1           3
    died |    1495     .3431438     .4749179          0           1
   white |    1495     .9150502     .2789003          0           1
     hmo |    1495     .1598662     .3666046          0           1
     los |    1495    9.854181     8.832906          1         116
---------+--------------------------------------------------------
 provnum |       0
```

Provnum represents the hospital providing hospital services. Of the 1,495 patients in the data, there are 54 providers. It may be the case that there is a clustering effect in the data due to observations being taken from different providers. There may be a higher correlation of values related to the relationship of type of admission to patient profile within providers than between providers. If this is the case, the independence-of-observations criterion that is essential to models based on a likelihood function is violated. The result is biased standard errors. Adjustments for the extra correlation, which we identified as overdispersion in Chapter 9, may be made by applying robust standard errors to the model, or bootstrapping or jackknifing standard errors. We shall test the multivariable model estimated below to determine if robust errors are required. The **distinct** command below is from Longton and Cox (http://fmwww.bc.edu/repec/bocode/d/distinct.ado).

```
. distinct provnum

                 |        Observations
        Variable |      total     distinct
-----------------+----------------------
         provnum |       1495           54
```

The other model predictors are fairly straightforward. For instance, *died* refers to the patient dying within 48 hours of hospitalization; *white* refers to the patient's self-identified ethnicity; *hmo* refers to whether the patient belongs to a Health Maintenance Organization (HMO); and *los* (Length Of Stay) refers to the total number of days that the patient was hospitalized for his or her adverse health incident. It is the only continuous predictor in the model.

We subject *type* to an ordered logistic, or proportional odds, model. As before, the *nolog* option is used to suppress the iteration log.

```
. ologit type died white hmo los, nolog

Ordered logistic regression               Number of obs   =        1495
                                          LR chi2(4)      =      115.61
                                          Prob > chi2     =      0.0000
Log likelihood = -977.66647               Pseudo R2       =      0.0558
------------------------------------------------------------------------
  type |      Coef.  Std. Err.      z    P>|z|     [95% Conf. Interval]
-------+----------------------------------------------------------------
  died |   .5486314   .1278436    4.29   0.000     .2980626    .7992002
 white |  -.5051675   .1963398   -2.57   0.010    -.8899864   -.1203486
   hmo |  -.7107835   .1985937   -3.58   0.000    -1.10002    -.321547
   los |   .0577942   .0068349    8.46   0.000     .0443979    .0711904
-------+----------------------------------------------------------------
 /cut1 |   1.393847   .2104434                      .981385    1.806308
 /cut2 |   3.046913   .2321391                     2.591928    3.501897
------------------------------------------------------------------------
```

The model output provides the usual table of parameter estimates, standard errors, and so forth, plus the above-described constants or cut points. The probability of group membership for each patient in the data may be determined using the parameter estimates together with the cut points. Stata lets the user circumvent hand calculation of these probabilities by simply employing the post-estimation *predict* command. Remember, it must be used directly following the model.

Predicted group probabilities may be calculated for our example by using the following command:

```
. predict elective urgent emergency
(option pr assumed; predicted probabilities)
```

The first five cases are listed, including the response, *type,* and each probability level.

```
. l type elective-emergency in 1/5

    +---------------------------------------------+
    |       type    elective      urgent   emerge~y |
    |---------------------------------------------|
 1. | Elective    .8412873    .1238517    .034861 |
 2. | Elective    .8898931    .0869654   .0231415 |
 3. | Elective    .8685026    .1033254    .028172 |
 4. | Elective    .7988089    .1551872   .0460039 |
 5. | Elective     .784585    .1654727   .0499423 |
    +---------------------------------------------+
```

The proportional odds assumption of the ordered logistic model is evaluated by using Brant's test.

```
. brant, detail

Estimated coefficients from j-1 binary regressions

                y>1             y>2
 died    .53481304      .74345738
white   -.60541135      .00834463
  hmo   -.66807378     -1.6938295
  los    .05276231      .06607137
_cons   -1.2556358     -3.6528028


Brant Test of Parallel Regression Assumption

    Variable |     chi2    p>chi2     df
-------------+-------------------------------
         All |     8.83     0.065      4
-------------+-------------------------------
        died |     1.04     0.308      1
       white |     2.84     0.092      1
         hmo |     3.22     0.073      1
         los |     2.34     0.126      1
-----------------------------------------

A significant test statistic provides evidence that the
parallel regression assumption has been violated.
```

Since all of the predictors, as well as the overall model, have *p*-values greater then 0.05, the proportional-odds assumption appears not to have been violated. However, we need to evaluate the effect of *prov-num*, which is tested by adjusting the standard errors by robust standard errors, clustered by provider. The unadjusted and robust adjusted models are compared, evaluating the difference in the two arrays of standard errors.

```
. ologit type died white hmo los, nolog
                              [HEADER STATISTICS NOT DISPLAYED]

-----------------------------------------------------------------
   type |     Coef.  Std. Err.     z   P>|z|   [95% Conf. Interval]
--------+--------------------------------------------------------
   died | .5486314  .1278436    4.29  0.000    .2980626   .7992002
  white | -.5051675  .1963398   -2.57  0.010   -.8899864  -.1203486
    hmo | -.7107835  .1985937   -3.58  0.000   -1.10002   -.321547
    los | .0577942  .0068349    8.46  0.000    .0443979   .0711904
--------+--------------------------------------------------------
  /cut1 | 1.393847  .2104434                    .981385   1.806308
  /cut2 | 3.046913  .2321391                   2.591928   3.501897
-----------------------------------------------------------------

. ologit type died white hmo los, nolog cluster(provnum)
                              [HEADER STATISTICS NOT DISPLAYED]
-----------------------------------------------------------------
        |            Robust
   type |     Coef.  Std. Err.     z   P>|z|   [95% Conf. Interval]
--------+--------------------------------------------------------
   died | .5486314  .1662668    3.30  0.001    .2227544   .8745084
  white | -.5051675  .1959724   -2.58  0.010   -.8892664  -.1210686
    hmo | -.7107835  .2263512   -3.14  0.002   -1.154424  -.2671432
    los | .0577942  .0120382    4.80  0.000    .0341997   .0813887
--------+--------------------------------------------------------
  /cut1 | 1.393847  .2879427                    .8294892   1.958204
  /cut2 | 3.046913  .3509648                   2.359034   3.734791
-----------------------------------------------------------------
```

There is no appreciable difference in the respective standard errors. We seem to have developed a well-fitted ordered logistic model.

We may convert the estimated coefficients to odds ratios in a similar manner as we did with standard binary logistic regression, in other words, by exponentiating the coefficients. Most software has a selection button that allows the display of odds ratios rather than the default parameter estimates or coefficients. In Stata, the option, *or*, is appended to the command line.

```
. ologit type died white hmo los, nolog or

Ordered logistic regression          Number of obs   =      1495
                                     LR chi2(4)      =    115.61
                                     Prob > chi2     =    0.0000
Log likelihood = -977.66647          Pseudo R2       =    0.0558
-----------------------------------------------------------------
   type | Odds Ratio  Std. Err.    z   P>|z|   [95% Conf. Interval]
--------+--------------------------------------------------------
   died | 1.730882   .2212822    4.29  0.000   1.347246   2.223762
  white | .6034045   .1184723   -2.57  0.010   .4106613   .8866113
```

```
hmo |   .4912592   .097561   -3.58  0.000   .3328644   .7250266
los |  1.059497   .0072416    8.46  0.000  1.045398   1.073786
------+---------------------------------------------------------------
/cut1 |  1.393847   .2104434                .981385   1.806308
/cut2 |  3.046913   .2321391               2.591928   3.501897
---------------------------------------------------------------------
```

The odds ratios of the above estimates may be interpreted as follows in Box 10.1:

Box 10.1

died: The expected odds of being admitted to the hospital as an emergency patient compared to an urgent patient, or an urgent patient compared to an elective patient, is some 1.73 times greater among those who died within 48 hours of admission than those who did not die, holding the other predictors constant.

white: The expected odds of being admitted to the hospital as an emergency patient compared to an urgent patient, or an urgent patient compared to an elective patient, is some 40 percent less among those who died within 48 hours of admission than those who did not die, holding the other predictors constant.

hmo: Patients not belonging to an HMO (Health Maintenance Organization) have roughly half the odds of being admitted to the hospital as an emergency patient compared to an urgent patient, or an urgent patient compared to an elective patient, than are patients having some other insurance of payment plan.

los: For each one-unit increase in *los*, there is a 6% increase in the odds of being admitted to the hospital as an emergency patient compared to an urgent patient, or an urgent patient compared to an elective patient.

The proportional odds model, assuming the equality of slopes among response levels or categories, allows one to interpret the model in the same manner for the other categories. For example, the same odds apply for admission to the hospital as an urgent patient as was the case for admission to the hospital as an emergency patient.

A graphical display of the probabilities may be constructed to provide the user with an overview of how the range of probabilities relate with one another with respect to a continuous predictor. Such a graph is shown in Figure 10.1. Our model has one such predictor, *los*. The model can employ all predictors such that *los* is being graphed, but adjusted by the remaining predictors.

```
. qui ologit type los      /// qui requests that the model not be displayed.

. predict Elective Urgent Emergency

. scatter Elective Urgent Emergency los, c(lll) s(Oh S Dh)
sort title("Hospital Admission Types on LOS")
```

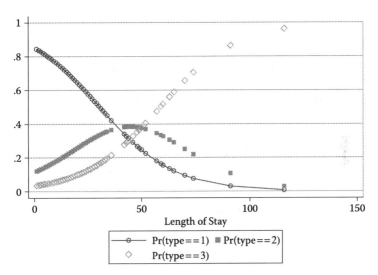

FIGURE 10.1
Hospital admission types on LOS.

As may be observed in Figure 10.1, the probabilities for each of the three levels differ considerably in their distribution. The left vertical axis represents probability values; the horizontal axis represents days in the hospital.

SIMULATION

We have previously developed a number of simulated data sets that can be used to evaluate discrepancies from the distributional assumptions upon which a model is based. The following batch, or *do*, file, called **syn_ologit.do**, generates random numbers that are structured in such a manner that they are most appropriately modeled using an ordered logistic, or proportional odds, model. The line beginning as *"gen y = …"* defines the linear predictor, with an attached logit function. Two coefficients are specified as $b1 = 0.75$ and $b2 = 1.25$. A four-level categorical response, *ys*, is defined on the next lines. They designate the three cut points, at 2, 3, and 4. It is quite simple to amend the code for any desired coefficient and cut values, if appropriate to the underlying distributional assumptions.

The *do* file adds additional lines that model the structured data using Stata's **ologit** command, and then uses **predict** to generate predicted values for each defined level. The predicted values are termed *olpr1* through *olpr4*. Note that the seed value allows remodeling to result in the identical estimates and calculated statistics. Placing a * in front of the seed line tells the software to ignore the command. This provides variation in the results. Bootstrapping the procedure results in values closely identical to the user-specified estimates and cuts.

```
. do syn_ologit

* DO FILE FOR CREATING AN ORDERED LOGIT MODEL WITH DEFINED
COEFFICIENTS/CUTS
* Hilbe, Joseph 18Feb 2008
qui {
drop _all
set obs 50000
set seed 13444
gen double x1 = 3*uniform()+1
gen double x2 = 2*uniform()-1
gen err = uniform()
gen y = .75*x1 + 1.25*x2 + log(err/(1-err))
gen int ys = 1 if y<=2
replace ys=2 if y<=3 & y>2
replace ys=3 if y<=4 & y>3
replace ys=4 if     y>4
noi ologit ys x1 x2, nolog
predict double (olpr1 olpr2 olpr3 olpr4), pr
}
```

```
Ordered logistic regression            Number of obs    =       50000
                                       LR chi2(2)       =    10929.44
                                       Prob > chi2      =      0.0000
Log likelihood = -54852.219            Pseudo R2        =      0.0906
------------------------------------------------------------------------
     ys |      Coef.   Std. Err.     z    P>|z|    [95% Conf. Interval]
--------+---------------------------------------------------------------
     x1 |   .7529516   .0107529   70.02   0.000    .7318764   .7740269
     x2 |    1.26078   .0163425   77.15   0.000    1.228749   1.292811
--------+---------------------------------------------------------------
  /cut1 |   2.004859   .0292476                    1.947535   2.062184
  /cut2 |   2.994769   .0309955                    2.934019   3.055519
  /cut3 |   4.018492   .0333522                    3.953123   4.083861
------------------------------------------------------------------------
```

A simulated ordered probit may be generated by amending the linear predictor line, and deleting the previous line defining *err*. The ordered probit linear predictor is defined as:

$$\text{gen double } y = .75\text{*}x1 + 1.25\text{*}x2 + \text{invnorm(uniform())}$$

```
Ordered probit regression          Number of obs    =      50000
                                   LR chi2(2)       =   24276.71
                                   Prob > chi2      =     0.0000
Log likelihood = -44938.779        Pseudo R2        =     0.2127
------------------------------------------------------------------
     ys |     Coef.  Std. Err.      z   P>|z|  [95% Conf. Interval]
--------+---------------------------------------------------------
     x1 |   .7461112    .006961  107.18  0.000  .7324679   .7597544
     x2 |   1.254821   .0107035  117.23  0.000  1.233842   1.275799
--------+---------------------------------------------------------
  /cut1 |   1.994369   .0191205                 1.956894   2.031845
  /cut2 |   2.998502   .0210979                 2.957151   3.039853
  /cut3 |   3.996582   .0239883                 3.949566   4.043599
------------------------------------------------------------------
```

We discuss the ordered probit and other ordered binomial-linked models in Chapter 12. A nice overview of the proportional odds model and small sample size can be found in Murad et al. (2003).

Recall when we converted a binary response model to a grouped or binomial model. We did this so that we could evaluate binomial overdispersion. However, statisticians more often use this method to save space. For example, an observation-based binary response model may have 10,000 observations, yet due to the profile of the covariate pattern, the associated binomial or grouped form may have 20 or fewer observations.

We have also shown how to model tables of data using frequency weights. The logic of the two methods is the same. It is also possible to convert observation-based proportional odds models to grouped format. We do this by collapsing the data so that there is only one set of identical covariate patterns for each level of the response. A detailed explanation of how to employ this method with binary response models was given in Chapter 8. For a three or more level response model, whether it be an ordered or unordered response, it is easier to use the method described here.

A grouped proportional odds model may be created in Stata by use of the **collapse** command. Users of other software will likely find it easier to use the method described in Chapter 8. We shall use the **medpar** data with *type*, a three level variable, as the response and binary variables *died* and *white* as predictors. The model is initially shown in its standard observation-based form.

```
. use medpar,clear

. ologit type died white, nolog

Ordered logistic regression        Number of obs    =       1495
                                   LR chi2(2)       =      22.07
                                   Prob > chi2      =     0.0000
Log likelihood =   -1024.435       Pseudo R2        =     0.0107
```

```
-------------------------------------------------------------------
 type |      Coef.  Std. Err.     z    P>|z|   [95% Conf. Interval]
------+------------------------------------------------------------
 died |  .4293199  .1238931    3.47   0.001    .1864939   .6721459
white | -.6564387  .1904411   -3.45   0.001   -1.029696  -.283181
------+------------------------------------------------------------
/cut1 |  .7102926  .1836698                     .3503063  1.070279
/cut2 | 2.260941   .2007527                    1.867473   2.654409
-------------------------------------------------------------------
```

The grouped data format is created by using the following code.

```
. gen byte count = 1

. collapse (sum) count, by(type died white)
```

We now re-estimate the model using *count* as a frequency weight. *Count* indicates the number of observations sharing the same pattern of covariates for a given level of *type*.

```
. ologit type died white [fw=count], nolog

Ordered logistic regression              Number of obs   =        1495
                                         LR chi2(2)      =       22.07
                                         Prob > chi2     =      0.0000
Log likelihood =  -1024.435              Pseudo R2       =      0.0107


-------------------------------------------------------------------
 type |      Coef.  Std. Err.     z    P>|z|   [95% Conf. Interval]
------+------------------------------------------------------------
 died |  .4293199  .1238931    3.47   0.001    .1864939   .6721459
white | -.6564387  .1904411   -3.45   0.001   -1.029696  -.283181
------+------------------------------------------------------------
/cut1 |  .7102926  .1836698                     .3503063  1.070279
/cut2 | 2.260941   .2007527                    1.867473   2.654409
-------------------------------------------------------------------
```

Note that the results are identical. The number of observations is listed as 1,495, as was the observation-based model. However, this value reflects the frequency weight, *count*, which, when all of its values are summed, equals the value of the observation-based model. In fact, the number of observations is reduced from 1495 to 12.

```
. count
      12
```

A listing of the entire data set is given as:

```
. list

     +-------------------------------------+
     | died   white       type    count |
     |-------------------------------------|
 1.  |    0       0    Elective       62 |
 2.  |    0       1    Elective      708 |
 3.  |    1       0    Elective       18 |
 4.  |    1       1    Elective      346 |
 5.  |    0       0      Urgent       23 |
     |-------------------------------------|
 6.  |    0       1      Urgent      138 |
 7.  |    1       0      Urgent       14 |
 8.  |    1       1      Urgent       90 |
 9.  |    0       0   Emergency        6 |
10.  |    0       1   Emergency       45 |
     |-------------------------------------|
11.  |    1       0   Emergency        4 |
12.  |    1       1   Emergency       41 |
     +-------------------------------------+
```

It is also possible to use a clustering effect with grouped data. However, it does add to the number of observations in the grouped data. The reason it increases the data is because the cluster effect must be included as a variable in the covariate pattern.

We wish to use *provnum* as the cluster variable, upon which a robust variance estimator is based. However, *provnum* is stored as a string, and **ologit** does not accept string variables. **Logit, logistic,** and **glm** have no such restriction. *Provnum* can be converted to a real number by using the *real()* function. We create a new variable called *hospital*, which has the same values as *provnum*, but is stored in numeric format. The data are estimated first in observation-based format, followed by the generation of grouped data and then re-estimated. Note the addition of *hospital* to the list of variables, which constitute the terms of the covariate patterns.

```
. use medpar, clear

. gen hospital = real(provnum)

. ologit type died white, nolog vce(cluster hospital)
```

```
Ordered logistic regression              Number of obs   =        1495
                                         Wald chi2(2)    =       15.82
                                         Prob > chi2     =      0.0004
Log pseudolikelihood =  -1024.435        Pseudo R2       =      0.0107
```

```
                    (Std. Err. adjusted for 54 clusters in hospital)
        -----------------------------------------------------------------
              |              Robust
      type |     Coef.   Std. Err.      z    P>|z|   [95% Conf. Interval]
      ------+----------------------------------------------------------
       died |  .4293199   .1386094    3.10   0.002    .1576506    .7009892
      white | -.6564387   .2138771   -3.07   0.002   -1.07563   -.2372474
      ------+----------------------------------------------------------
      /cut1 |  .7102926   .2647388                     .1914141   1.229171
      /cut2 |  2.260941   .4793465                    1.321439   3.200443
        -----------------------------------------------------------------

. gen  count = 1

. collapse (sum) count, by(type died white hospital)

. ologit type died white [fw=count], nolog vce(cluster
hospital)

Ordered logistic regression            Number of obs   =      1495
                                       Wald chi2(2)    =     15.82
                                       Prob > chi2     =    0.0004
Log pseudolikelihood = -1024.435       Pseudo R2       =    0.0107

                    (Std. Err. adjusted for 54 clusters in hospital)
        -----------------------------------------------------------------
              |              Robust
      type |     Coef.   Std. Err.      z    P>|z|   [95% Conf. Interval]
      ------+----------------------------------------------------------
       died |  .4293199   .1386094    3.10   0.002    .1576506    .7009892
      white | -.6564387   .2138771   -3.07   0.002   -1.07563   -.2372474
      ------+----------------------------------------------------------
      /cut1 |  .7102926   .2647388                     .1914141   1.229171
      /cut2 |  2.260941   .4793465                    1.321439   3.200443
        -----------------------------------------------------------------
```

Because *hospital* was added as a term in the covariate pattern, the number of observations in the grouped data increases from 12 to 249. It is still a substantial reduction from the original 1,495.

```
. count
  249
```

We now turn to generalizing the restriction of the proportionality of coefficients, or slopes. Note, however, that the majority of models we discuss in this and the next two chapters can be grouped and estimated in the manner shown in this section.

10.3 Generalized Ordinal Logistic Regression

The generalized ordered logit model differs from the standard ordinal or proportional odds model in that it relaxes the proportional-odds assumption. It allows that the predictors may have different effects on the odds that the response is above a cutpoint, depending on how the outcomes are dichotomized. In this respect, there is a similarity to the unordered categorical response model, called the multinomial logistic model, in that there are $r - 1$ estimated coefficient vectors that correspond to the effect of changing from one set of responses to a higher response not in the set.

The sets of coefficient vectors are defined for the cutpoints between the r outcomes. These cutpoints partition the responses into two groups. The first (lowest) coefficient vector corresponds to partitioning the responses into sets $\{1\}$ and $\{2, \ldots, r\}$. The second coefficient vector corresponds to partitioning the responses into the sets $\{1, 2\}$ and $\{3, \ldots, r\}$. The $(r - 1)$th coefficient vector corresponds to partitioning the responses into the sets $\{1, \ldots, r - 1\}$ and $\{r\}$. The generalized ordered logit model fits $r - 1$ simultaneous logistic regression models, where the responses for these models are defined by collapsing the response variable into new binary dependent variables defined by the partitions described above.

We shall next model the **medpar_olr** data using a generalized ordered logit. We found that the traditional model did not violate basic model assumptions, but it will be instructive to assume that such was the case.

The Stata user-created **gologit2** command (Williams, 2006) provides generalized ordered logit parameter estimates. The proportional odds and generalized model outputs will both be displayed.

```
. ologit type died white hmo los, nolog
```

```
Ordered logistic regression            Number of obs   =        1495
                                        LR chi2(4)      =      115.61
                                        Prob > chi2     =      0.0000
Log likelihood = -977.66647             Pseudo R2       =      0.0558
```

type	Coef.	Std. Err.	z	P>|z|	[95% Conf.	Interval]
died	.5486314	.1278436	4.29	0.000	.2980626	.7992002
white	-.5051675	.1963398	-2.57	0.010	-.8899864	-.1203486
hmo	-.7107835	.1985937	-3.58	0.000	-1.10002	-.321547
los	.0577942	.0068349	8.46	0.000	.0443979	.0711904
/cut1	1.393847	.2104434			.981385	1.806308
/cut2	3.046913	.2321391			2.591928	3.501897

```
. gologit2 type died white hmo los, nolog

Generalized Ordered Logit Estimates          Number of obs =    1495
                                             LR chi2(8)     =  126.32
                                             Prob > chi2    =  0.0000
Log likelihood = -972.30884                  Pseudo R2      =  0.0610
-----------------------------------------------------------------------
        type |      Coef.  Std. Err.     z   P>|z|   [95% Conf. Interval]
-------------+---------------------------------------------------------
Elective     |
        died |  .5235821  .1293354   4.05  0.000   .2700894    .7770748
       white | -.5928813  .2029161  -2.92  0.003  -.9905895    -.195173
         hmo | -.6717372  .1993216  -3.37  0.001    -1.0624   -.2810741
         los |  .0532246  .0076009   7.00  0.000   .0383272    .0681221
       _cons | -1.267827  .2173879  -5.83  0.000  -1.693899   -.8417541
-------------+---------------------------------------------------------
Urgent       |
        died |  .7207729  .2211275   3.26  0.001   .2873709   1.154175
       white |  .0521476  .3758461   0.14  0.890  -.6844972    .7887925
         hmo | -1.660555  .5924799  -2.80  0.005  -2.821794   -.4993154
         los |   .066057    .00891   7.41  0.000   .0485936    .0835203
       _cons | -3.686187  .4071814  -9.05  0.000  -4.484248   -2.888126
-----------------------------------------------------------------------
```

Again, the generalized ordered logit model takes into account the different effects of (1) moving from *Elective* to a higher level, and (2) moving from a lower level (*Elective* and/or *Urgent*) to *Emergency*.

The above results tell us that the probability of having a more severe admission if the patient is *white* is important only when moving from *Elective* to a higher level. All other effects are as the model assumes.

10.4 Partial Proportional Odds

As we have seen, the generalized ordered logistic model is a viable alternative to the ordered logistic model when the assumption of proportional odds is untenable. In fact, the generalized model eliminates the restriction that the regression coefficients are the same for each response for all of the predictors. This may be a rather drastic solution when the proportional-odds assumption is violated by only one or a few of the regressors.

The **gologit2** command allows estimation of the partial proportional-odds models. Such a model is somewhere between the ordered logistic model and the generalized ordered logistic model.

When the *brant* test was used earlier with the ordered logistic model, both a global and per-predictor test was obtained of the proportional-odds assumption. No violations were found in the model. However, for the sake of having an example, a partial proportional ordered logistic model will be modeled and the results displayed.

We use the *autofit* option with the **gologit2** command to effect the partial model:

```
. gologit2 type died white hmo los, nolog autofit


----------------------------------------------------------------
Testing parallel lines assumption using the .05 level of
significance...

Step   1:   Constraints for parallel lines imposed for died
            (P Value = 0.3265)
Step   2:   Constraints for parallel lines imposed for los
            (P Value = 0.1684)
Step   3:   Constraints for parallel lines imposed for white
            (P Value = 0.0931)
Step   4:   Constraints for parallel lines imposed for hmo
            (P Value = 0.0761)
Step   5:   All explanatory variables meet the pl assumption

Wald test of parallel lines assumption for the final model:

 ( 1)   [Elective]died - [Urgent]died = 0
 ( 2)   [Elective]los - [Urgent]los = 0
 ( 3)   [Elective]white - [Urgent]white = 0
 ( 4)   [Elective]hmo - [Urgent]hmo = 0

        chi2(  4) =     8.86
    Prob > chi2 =     0.0648

An insignificant test statistic indicates that the final model
does not violate the proportional odds/ parallel lines
assumption

If you re-estimate this exact same model with gologit2,
instead of autofit you can save time by using the parameter

pl(died los white hmo)
----------------------------------------------------------------

Generalized Ordered Logit Estimates      Number of obs =      1495
                                         LR chi2(4)    =    115.61
                                         Prob > chi2   =    0.0000
Log likelihood = -977.66647              Pseudo R2     =    0.0558

 ( 1)   [Elective]died - [Urgent]died = 0
 ( 2)   [Elective]los - [Urgent]los = 0
```

```
 ( 3)   [Elective]white - [Urgent]white = 0
 ( 4)   [Elective]hmo  - [Urgent]hmo  = 0
--------------------------------------------------------------------
        type |     Coef. Std. Err.    z  P>|z|  [95% Conf. Interval]
-------------+------------------------------------------------------
Elective     |
        died |  .5486314 .1278436   4.29 0.000  .2980626   .7992002
       white | -.5051675 .1963398  -2.57 0.010 -.8899864 -.1203486
         hmo | -.7107835 .1985937  -3.58 0.000  -1.10002  -.321547
         los |  .0577942 .0068349   8.46 0.000  .0443979  .0711904
       _cons | -1.393847 .2104434  -6.62 0.000 -1.806308  -.981385
-------------+------------------------------------------------------
Urgent       |
        died |  .5486314 .1278436   4.29 0.000  .2980626   .7992002
       white | -.5051675 .1963398  -2.57 0.010 -.8899864 -.1203486
         hmo | -.7107835 .1985937  -3.58 0.000  -1.10002  -.321547
         los |  .0577942 .0068349   8.46 0.000  .0443979  .0711904
       _cons | -3.046913 .2321391 -13.13 0.000 -3.501897 -2.591928
--------------------------------------------------------------------
```

With this model, the top of the output indicates that the proportional odds assumptions are satisfied (showing *p*-values above 0.05). The table of parameter estimates for the partial proportional ordered logistic model is constrained to display a single parameter estimate for the various levels, with the exception of the constant. The parameter estimates are the same as the traditional ordered logistic parameter estimates. The constants are the same as the cutpoints, except that they have opposite signs.

A side note: The ordered and generalized ordered models may be fit with links other than the logit. All of the generalized linear models binomial links are available in Stata, but not in other statistical packages. Richard Williams wrote a general ordered binomial model command called **oglm,** which can be used to model logit, probit, loglog, complementary loglog, and Cauchit linked models. He also wrote the **gologit2** command we used for modeling partial proportional odds. If it is not necessary to interpret the model results in terms of odds ratios, then the model of choice is the one having the lowest AIC and/or BIC goodness-of-fit statistic; see Hardin and Hilbe (2007). Interested readers should refer to Williams (2006) for a discussion of gologit2. For categorical response panel and mixed models, refer to Rabe-Hesketh and Skrondal (2008).

Exercises

10.1. The following table is from Forthofer and Lehnen (1981), in which the authors studied the effects of *smoking* and *age* on the development of breathing difficulties throughout Texas industrial plants. The three-level response is ordered as *normal, borderline,* and *abnormal.* The data were recorded in table format.

Age	Smoking Status	Breathing Test Results		
		Normal	Borderline	Abnormal
<40	Never smoked	577	27	7
	Former smoker	192	20	3
	Current smoker	682	46	11
40–59	Never smoked	164	4	0
	Former smoker	145	15	7
	Current smoker	245	47	27

a. Model breathing test results using a proportional odds model on *age* and *smoking* status. Parameterize the results so that odds ratios are displayed.

b. Interpret the odds ratios from (a).

c. Assess the model for goodness of fit.

d. Can the slopes be evaluated as to their proportionality? If so, engage the test.

10.2. Use the **affairs** data set that is on this book's Web site. Using a proportional odds procedure, model the five-level *ratemarr* variable on *notrel, naffairs, educ, kids,* and *age*. The covariates are defined as:

RESPONSE

ratemarr   1 = very unhappy;   2 = somewhat unhappy;   3 = average unhappy; 4 = happier than average;   5 = very happy

PREDICTORS

notrel	not religious (1/0) relig == 2
naffairs	number of affairs (continuous)
educ	7 levels; model as categorical or continuous
kids	Have kids (1/0)
age	9 levels; model as categorical or continuous

a. Model as a proportional odds model; parameterize as odds ratios.

b. Test the proportionality of slopes. If violated, model as partial proportional and generalized ordered logistic model. Which is preferred?

10.3. Using the **autoorder** data set, model the three-level *rep78* on the three continuous predictors, *price*, *weight*, and *length*.

    a. Parameterize the results in terms of odds ratios.

    b. Are the slopes proportional?

    c. Interpret the odds ratio of each predictor.

    d. How are the intercepts, or cutpoints, to be interpreted?

    e. Calculate probabilities for each level. Which level has the highest probability value associated with it?

    f. If the slopes are not proportional, construct a partial proportional odds model, interpreting the resulting odds ratios.

10.4. The table below comes from the 1991 General Social Survey and tabulates *income* on *job satisfaction*. The tabulation is stratified by *gender*, but has been modified by the author.

		Job Satisfaction			
Gender	In 1000's Income$	Not	Somewhat	Moderate	Very
Male	<15	300	400	1100	200
	15–25	200	500	1700	300
	25–50	50	200	800	500
	>50	10	300	1000	300
Female	<15	150	200	300	50
	15–25	100	300	600	100
	25–50	50	50	800	800
	>50	25	50	1000	1500

    a. Create the data set in electronic format, and suitable for modeling as a proportional odds model.

    b. Model the four-level *satisfaction* response on levels *income* and the binary covariate, *gender*.

    c. Evaluate the equality of slopes using the Brant test. If any level fails, remodel the data using a partial proportional odds model.

    d. What are the most important factors in determining job satisfaction for men? For women?

    e. Model using a proportional slopes probit model. Statistically compare the logistic and probit models; for instance, which is the preferred model? Why?

10.5. Create a synthetic proportional odds model using the **syn_ologit.do** procedure that is given in Chapter 10. Amend the program so that there are five levels having the following user-defined parameter estimates and cutpoints (intercepts). Users of packages other than Stata may create a similar macro appropriate for their own statistical application.

$$\beta 1 = 0.5; \beta 2 = 1.75; \beta 3 = -1.25$$
$$\text{Cut1} = .8; \text{Cut2} = 1.6; \text{Cut3} = 2.4; \text{Cut4} = 3.2$$

a. Run the model and confirm that a 50,000 observation proportional odds model with the assigned parameters and cutpoints has been correctly estimated.

b. Calculate the five vectors of probabilities associated with each level. Find the mean value of each probability vector, checking if the mean of each level is less than the one prior to it (except for the first level).

R Code

Section 10.2 Proportional Odds

```
library('foreign')
gss2002<- read.dta('gss2002_10x.dta')  #read Stata format file
gss2002$sex<- factor(gss2002$sex) #convert to factor
gss2002$polviews<- factor(gss2002$polviews, ordered=TRUE)
  #convert to ordered factor
summary(gss2002)
head(gss2002)

library('MASS') #for polr()
fit10_1a<- polr(polviews ~ sex, data=gss2002,
  method='logistic')
summary(fit10_1a)

medpar<- read.dta('medparx.dta')  #read Stata format file
summary(medpar)
head(medpar)
table(medpar$type)
fit10_1b<- polr(type ~ age80, data=medpar, method='logistic')
summary(fit10_1b)

pfit10_1b<- predict(fit10_1b, type='p')
summary(pfit10_1b)

fit10_1b$zeta[1]  #first intercept
1/(1+exp(-fit10_1b$zeta[1]))
fit10_1b$zeta[2]  #2nd intercept
1/(1+exp(-fit10_1b$zeta[2])) - 0.7517818
1-1/(1+exp(-fit10_1b$zeta[2]))
```

```
exp(-fit10_1b$zeta[1])
0.3301732/1.3301732
1-0.2482182

exp(-fit10_1b$zeta[2])
0.07111143/1.07111143
0.2482182-0.06639032

1-(0.7517818 + 0.06639032)

exp(-fit10_1b$zeta[2])
0.07111143/1.07111143

-0.1708724 - fit10_1b$zeta[1]
1/(1+exp(-1.2790103))
-0.1708724 - fit10_1b$zeta[2]
1/(1+exp(-2.8143796)) - 1/(1+exp(-1.2790103))
1 - 1/(1+exp(-2.8143796))

-0.1708724 - fit10_1b$zeta[1]
exp(-1.2790103)
0.27831261/1.27831261
1-0.21771874

exp(-0.1708724 - fit10_1b$zeta[1])
0.27831261/1.27831261
exp(-0.1708724 - fit10_1b$zeta[2])
0.0599419/1.0599419
1-(0.78228126 + 0.05655206)

-0.1780724 - fit10_1b$zeta[2]
exp(-2.8143796)
0.0599419/1.0599419

str(medpar)
summary(medpar)
unique(medpar$provnum)
sum(is.na(medpar))

fit10_1c<- polr(type ~ died + white + hmo + los, data=medpar,
  method='logistic')
summary(fit10_1c)
pfit10_1c<- predict(fit10_1c, type='p')
head(cbind(medpar$type, pfit10_1c))

#No Brant's test in R available
confint(fit10_1c)
exp(coef(fit10_1c))
exp(fit10_1c$zeta)
```

```
fit10_1d<- polr(type ~ los, data=medpar, method='logistic')
pfit10_1d<- predict(fit10_1d, type='p')
plot(medpar$los, pfit10_1d[,1], pch=1, col=1, xlab='LOS',
  ylab='Probability', main='Hospital Admission Types on LOS')
points(medpar$los, pfit10_1d[,2], pch=2, col=2)
points(medpar$los, pfit10_1d[,3], pch=3, col=3)

fit10_1e<- polr(type ~ died + white, data=medpar,
  method='logistic')
summary(fit10_1e)

medpar2<- medpar[,c('died', 'white', 'type')]
ftable(medpar2)
```

Note: Generalized Ordinal (10.3) and Partial Proportional Odds (10.4) Models do not have R support at this time. When it becomes available, the code will be placed on the book's Web site or made part of the subsequent printings if possible.

11

Multinomial Logistic Regression

11.1 Unordered Logistic Regression

The multinomial logistic model is an extension of generalized linear models allowing for an estimation of an unordered categorical response. Unlike the ordinal model, unordered categorical response values have no order. An example of an unordered response can be given as White, Black, Hispanic, or other. One may use ordered levels for a multinomial model, but the information entailed in the order is lost to the understanding of the response.

11.1.1 The Multinomial Distribution

The multinomial probability distribution is an extension, or can be regarded as a more general form, of the binomial distribution. However, whereas the binomial model assesses the odds of one category or level compared to another, the multinomial model tests the probability or risk of being in a given category or level compared to other categories. Instead of the relationship of levels being construed in terms of odds ratios, for multinomial models, the relationships are thought of as relative risk ratios. Therefore, we may use probability language when referring to the exponentiated multinomial coefficients. We assess the probability of choosing one category over the others. The probability mass function may be expressed as:

$$\Pr(n_1, \ldots, n_k) = \frac{n!}{n_1!\, n_2! \ldots n_k!}\, p_1^{n1} p_2^{n2} \cdots p_k^{nk} \tag{11.1}$$

where n is the frequency with which a choice or level is selected, k is the number of possible choices, and p represents the probability of each choice being selected.

A general formulation of the log-likelihood function, parameterized as $x\beta$ rather than p or μ, can be expressed as:

$$L(\beta) = \sum_{i=1}^{n} \{y(x'\beta_1) + \ldots + y(x'\beta_k) - \ln(1 + \exp(x'\beta_1) \ldots + \exp(x'\beta_k))\} \qquad (11.2)$$

where

$$(x'\beta_j) = \ln(\Pr(y=j|x)/\Pr(y=1|x)), \quad j=1, 2, \ldots k \qquad (11.3)$$

There are k levels of j, where each level is indicated as $j_1, \ldots, j_k$. Each level j has associated observations, X.

The relationship between choices or levels 2 and 1, and 3 and 1, are expressed as:

$$(x'\beta_1) = g_1(x) = \ln(\Pr(y=2|x)/\Pr(y=1|x)) = \beta_{01} + \beta_{11}x_{11} + \ldots \qquad (11.4)$$

$$(x'\beta_2) = g_2(x) = \ln(\Pr(y=3|x)/\Pr(y=1|x)) = \beta_{02} + \beta_{12}x_{12} + \ldots \qquad (11.5)$$

We may use the same formulation through $g_k(x)$ to represent levels 1 through k. A separate table of parameter estimates is presented for each non-reference level, with estimates at each nonreference level interpreted with respect to the reference level.

The conditional probabilities at each of the levels 1 (reference), 2 and 3 may be expressed as:

$$\Pr(y=1|x) = \frac{1}{1 + \exp(x'\beta_2) + \exp(x'\beta_3)} \qquad (11.6)$$

$$\Pr(y=2|x) = \frac{\exp(x'\beta_2)}{1 + \exp(x'\beta_2) + \exp(x'\beta_3)} \qquad (11.7)$$

$$\Pr(y=3|x) = \frac{\exp(x'\beta_3)}{1 + \exp(x'\beta_2) + \exp(x'\beta_3)} \qquad (11.8)$$

These probabilities may be placed into a generalized form for k levels of y.

$$\Pr(y=j|x) = \frac{\exp(x'\beta_j)}{\sum_{j=0..k} \exp(x'\beta_j)}$$ (11.9)

The important to remember that the above parameterization is based on setting $\beta_i = 0$, with the other coefficients being evaluated as to their change with respect to $y = 1$. The reference level may be changed, with the other coefficients representing the measure of change with respect to the new reference level.

11.1.2 Interpretation of the Multinomial Model

We shall continue to use the same **medpar_olr** data as in the last section. The three levels of response do appear to be ordered, but we shall ignore this order for the sake of our example. It takes nothing away from the multinomial model.

Before proceeding to the full model, I would like to show the multinomial relationship between the response and a single predictor. We commence by looking at the data again.

```
. d                     /// Note: "d" is an abbreviation for describe

Contains data from medpar_olr.dta
  obs:    1,495                      1991 AZ Medicare: Ordered LR
  vars:        6                     28 Apr 2007 09:08
  size:  32,890 (99.7% of memory free)
-----------------------------------------------------------------
              storage  display  value
variable name    type   format  label   variable label
-----------------------------------------------------------------
type            byte    %9.0g    type    Type of Admission
died            float   %9.0g            1=Died; 0=Alive
white           float   %9.0g            1=White; 0=Other
hmo             byte    %9.0g            1=HMO member
los             int     %9.0g            Length of Stay
provnum         str6    %9s             Provider number
-----------------------------------------------------------------
Sorted by:  provnum
```

First, observe the basic odds ratio that may be calculated from tabulating *white* on *type*.

```
. tab white type

  1=white; |          Admission type
0=non-white |  Elective    Urgent  Emergency |      Total
------------+------------------------------------+----------
        0 |         80        37        10 |        127
        1 |      1,054       228        86 |      1,368
------------+------------------------------------+----------
    Total |      1,134       265        96 |      1,495
```

Cross multiply to calculate the odds ratio. The method used is based on the table:

		Response	
		1	0
Predictor	1	A	B
	0	C	D

so that OR = (A*D)/(B*C).

With *Elective* the referent, the relative risk ratio of *urgent* is

```
. di (80*228)/(37*1054)
.46771629
```

and the relative risk ratio of *emergency* to *elective* is

```
. di (80*86)/(1054*10)
.65275142
```

Modeling *type*, given the predictor *white*, produces:

```
. mlogit type white, rrr nolog

Multinomial logistic regression           Number of obs   =       1495
                                          LR chi2(2)      =      12.30
                                          Prob > chi2     =     0.0021
Log likelihood = -1029.3204               Pseudo R2       =     0.0059
------------------------------------------------------------------
        type|      RRR  Std. Err.    z   P>|z|  [95% Conf. Interval]
------------+-----------------------------------------------------
Urgent      |
       white|.4677163 .0990651  -3.59 0.000  .3088112      .7083893
------------+-----------------------------------------------------
Emergency   |
       white|.6527514 .2308533  -1.21 0.228   .32637      1.305526
------------------------------------------------------------------
(type==Elective is the base outcome)
```

which was correctly calculated by hand.

Next model *white* on *type*, using *urgent*, or level 2, as the base. First, using *urgent* as the referent, obtain the relative risk ratio of *elective*:

```
. di (37*1054)/(80*228)
2.1380482
```

Next, calculate the risk ratio of *emergency* (with respect to *urgent*):

```
. di (37*86)/(10*228)
1.395614
```

The statistical model appears as:

```
. mlogit type white, rrr nolog base(2)

Multinomial logistic regression        Number of obs   =       1495
                                        LR chi2(2)      =      12.30
                                        Prob > chi2     =     0.0021
Log likelihood = -1029.3204             Pseudo R2       =     0.0059
------------------------------------------------------------------
       type |     RRR  Std. Err.    z   P>|z|  [95% Conf. Interval]
------------+-----------------------------------------------------
Elective    |
      white | 2.138048 .4528512 3.59 0.000 1.411653     3.238225
------------+-----------------------------------------------------
Emergency   |
      white | 1.395614 .5278317 0.88 0.378 .6650197     2.928843
------------------------------------------------------------------
(type==Urgent is the base outcome)
```

This type of univariable relationship can be calculated for all levels. Of course, once we have more than one predictor, this straightforward relationship breaks down.

A note should be made about why the multinomial logit employs a relative risk ratio rather than an odds ratio as in standard logistic regression. The reason is that the values of the binary response logistic model are assumed to be ordered, that is, that the values of 1 and 0 are not independent. When we consider that 1 typically represents success, even if the success is *death*, and 0 is lack of success or failure, compared with 1, then this assumption makes sense. There is no question that the two values are not independent. For this reason, the ordered logistic models and continuation ratio models (to be discussed) retain the use of odds ratio. Multinomial models assume that the categories of the response are not ordered and are independent of one another. Because of this, risk ratios are used rather than odds ratios. Note that the logistic model produces the same coefficients as both the multinomial and proportional odds models, but the intercept of the latter is reversed in many parameterizations.

The full multinomial model can be displayed as:

```
mlogit type died white hmo los, nolog

Multinomial logistic regression        Number of obs   =      1495
                                        LR chi2(8)      =    125.36
                                        Prob > chi2     =    0.0000
Log likelihood = -972.79279            Pseudo R2       =    0.0605
-----------------------------------------------------------------
       type|    Coef. Std. Err.    z  P>|z|   [95% Conf. Interval]
-----------+-----------------------------------------------------
Urgent     |
       died| .424642  .144354   2.94 0.003   .1417134     .7075705
      white|-.7117305 .2160032 -3.30 0.001  -1.135089    -.2883721
        hmo|-.4524631 .2083697 -2.17 0.030  -.8608602     -.044066
        los| .0400675 .0085351  4.69 0.000    .023339      .056796
      _cons|-1.293815 .2315819 -5.59 0.000  -1.747707    -.8399226
-----------+-----------------------------------------------------
Emergency  |
       died| .8513779 .2271924  3.75 0.000   .4060891     1.296667
      white|-.2159545 .3822292 -0.56 0.572  -.9651099     .5332009
        hmo|-1.782408  .595745 -2.99 0.003  -2.950047    -.6147691
        los| .0802779 .0100075  8.02 0.000   .0606635     .0998922
      _cons|-3.399962 .4107698 -8.28 0.000  -4.205056    -2.594868
-----------------------------------------------------------------
(type==Elective is the base outcome)
```

Interpretation of coefficients is much easier if they are parameterized to relative risk ratios. Moreover, for research purposes, you will most likely always want to use relative risk ratios, which have an acronym of RRR. We reparameterize them using the *rrr* option.

```
. mlogit type died white hmo los, nolog rrr

Multinomial logistic regression        Number of obs   =      1495
                                        LR chi2(8)      =    125.36
                                        Prob > chi2     =    0.0000
Log likelihood = -972.79279            Pseudo R2       =    0.0605
-----------------------------------------------------------------
       type|     RRR Std. Err.    z  P>|z|   [95% Conf. Interval]
-----------+-----------------------------------------------------
Urgent     |
       died|1.529043 .2207234  2.94 0.003  1.152246     2.029056
      white| .4907942 .1060131 -3.30 0.001   .3213936     .7494827
        hmo| .6360596 .1325355 -2.17 0.030   .4227982     .9568908
        los|1.040881  .008884  4.69 0.000  1.023614      1.05844
-----------+-----------------------------------------------------
```

```
Emergency   |
      died | 2.342873   .5322828    3.75  0.000  1.500936      3.657086
     white | .8057719   .3079895   -0.56  0.572  .3809413      1.704379
       hmo | .1682326   .1002237   -2.99  0.003  .0523373       .5407657
       los | 1.083588    .010844    8.02  0.000  1.062541      1.105052
-----------------------------------------------------------------------
```

(type==Elective is the base outcome)

As before, we have level 1, *elective*, as the reference level. Relative risk ratios are interpreted for each level with respect to the reference. We may conclude, on the basis of the model, that *white* patients are half as likely to be admitted as an *urgent* admission compared to an *elective* admission. *Whites* are also some 20 percent less likely to be admitted as an *emergency* admission than an *elective*. We may also conclude that for a one-unit increase in the predictor, *los*, the relative risk of having an urgent admission rather than an elective admission increases by some 4 percent. Likewise, for a one-unit change in the predictor, *los*, the relative risk of having an emergency admission rather than an elective admission increases by about 8 percent. The other predictor values are held at their mean values.

If you wish to look at the relationship of *urgent* to *emergency*, you must change the base or reference level using the *base(#)* option.

We may also express the results of the above multinomial logistic regression as probabilities. If *died*, *hmo*, and *los* are all set at their mean values for each level, we may calculate probabilities for each level of both *white* and *nonwhite* categories. We use the **prtab** command (spost9_ado, from http://www.indiana.edu/~jslsoc/stata) to do our work (Long 1997).

```
. prtab white

mlogit: Predicted probabilities for type

Predicted probability of outcome 2 (Urgent)

--------------------
1=White;   |
0=Other    | Prediction
---------+----------
      0  |     0.2896
      1  |     0.1686
--------------------


Predicted probability of outcome 3 (Emergency)

--------------------
1=White;   |
0=Other    | Prediction
---------+----------
      0  |     0.0490
      1  |     0.0469
--------------------
```

```
Predicted probability of outcome 1 (Elective)

----------------------
1=White;   |
0=Other    | Prediction
----------+-----------
        0 |     0.6613
        1 |     0.7845
----------------------


         died       white          hmo          los
x=  .34314381  .91505017   .15986622   9.8541806
```

Interpretation of the probabilities is straightforward. We may perform a variety of tests of both the predictors as well as omnibus tests on the model by using the *all* option.

```
. mlogtest, all

**** Likelihood-ratio tests for independent variables (N=1495)
 Ho: All coefficients associated with given variable(s) are 0.

             |      chi2    df    P>chi2
------------+------------------------
      died |    19.464     2     0.000
     white |    10.091     2     0.006
       hmo |    18.820     2     0.000
       los |    77.275     2     0.000
------------------------------------


**** Wald tests for independent variables (N=1495)
 Ho: All coefficients associated with given variable(s) are 0.

             |      chi2    df    P>chi2
------------+------------------------
      died |    19.604     2     0.000
     white |    10.875     2     0.004
       hmo |    12.945     2     0.002
       los |    65.928     2     0.000
------------------------------------


**** Hausman tests of IIA assumption (N=1495)
 Ho: Odds(Outcome-J vs Outcome-K) are independent of other
     alternatives.

Omitted  |    chi2    df    P>chi2    evidence
---------+----------------------------------
  Urgent |   1.097     5    0.954    for Ho
Emergenc |   2.042     5    0.843    for Ho
----------------------------------------
```

```
**** Small-Hsiao tests of IIA assumption (N=1495)
  Ho: Odds(Outcome-J vs Outcome-K) are independent of other
      alternatives.

Omitted | lnL(full) lnL(omit)  chi2  df  P>chi2  evidence
--------+-------------------------------------------------
 Urgent |  -151.223  -145.260 11.927   5   0.036  against Ho
Emergenc |  -333.458  -329.755  7.406   5   0.192  for Ho
---------------------------------------------------------
```

```
**** Wald tests for combining alternatives (N=1495)
  Ho: All coefficients except intercepts associated with a
      given pair of alternatives are 0 (i.e., alternatives can
      be combined).

Alternatives tested|   chi2   df   P>chi2
-------------------+----------------------
  Urgent-Emergenc  |  24.136   4    0.000
  Urgent-Elective  |  44.678   4    0.000
Emergenc-Elective  |  82.346   4    0.000
------------------------------------------
```

```
**** LR tests for combining alternatives (N=1495)
  Ho: All coefficients except intercepts associated with a
      given pair of alternatives are 0 (i.e., alternatives can
      be collapsed).

Alternatives tested|   chi2   df   P>chi2
-------------------+----------------------
  Urgent-Emergenc  |  29.565   4    0.000
  Urgent-Elective  |  44.501   4    0.000
Emergenc-Elective  | 100.513   4    0.000
------------------------------------------
```

Post-estimation probabilities may be obtained in a manner similar to ordered logistic models:

```
. predict ProbElective ProbUrgent ProbEmergency
```

We can evaluate the ordinality of the response. The ordered and unordered models may be compared using the likelihood ratio test.

```
. qui mlogit type died white hmo los
. estimates store mlogit
. qui ologit type died white hmo los
. lrtest mlogit, force

Likelihood-ratio test                 LR chi2(4)   =      9.75
(Assumption: . nested in mlogit)      Prob > chi2  =    0.0449
```

Since the models are not recognized as being nested due to the naming conventions of equations, we used the *force* option. Because the hypothesis is rejected, we conclude that the response is significantly ordered, that is, the response is best modeled using the ordered logistic regression command. But, since we earlier found that the coefficients were proportional, we suspected that we would have this conclusion.

SYNTHETIC MODEL

As done in the previous chapter, we create a synthetic multinomial logit model with user-defined coefficients. Here we define constants for each separate nonreference level. The code, in **syn_mlogit1.do**, is expressed as:

```
* SYNTHETIC MULTINOMIAL LOGISTIC REGRESSION
* Joseph Hilbe 20 Feb, 2008
qui {
  clear
  set mem 50m
  set seed 111322
  set obs 100000
  gen x1 =  uniform()
  gen x2 = uniform()
  gen denom = 1+exp(.4*x1 - .5*x2 +1 ) + exp(-.3*x1+.25*x2 +2)
  gen p1 = 1/denom
  gen p2 = exp(.4*x1-.5*x2 + 1) / denom
  gen p3 = exp(-.3*x1+.25*x2 + 2) / denom

  gen u = uniform()
  gen y = 1 if u <= p1
  gen p12 = p1 + p2
  replace y = 2 if y==. & u<=p12
  replace y = 3 if y==.
}
  /* coefficients for y=2 are .4,  -.5 and 1
     coefficients for y = 3 are -.3, .25 and 2*/

* mlogit y x1 x2,  baseoutcome(1) nolog
                              ///delete * to model results
```

The response, *y*, is created by the simulation, and may be tabulated as:

```
. tab y

        y |      Freq.     Percent       Cum.
----------+---------------------------------
        1 |      9,176        9.18        9.18
        2 |     24,228       24.23       33.40
        3 |     66,596       66.60      100.00
----------+---------------------------------
    Total |    100,000      100.00
```

With defined coefficient values for level 2 [.4, −.5, 1] and level 3 [−.3, .25, 2], and having level 1 as the base of reference level, we have:

```
.   mlogit y x1 x2,  baseoutcome(1) nolog

Multinomial logistic regression     Number of obs   =       100000
                                     LR chi2(4)      =      1652.17
                                     Prob > chi2     =       0.0000
Log likelihood = -82511.593          Pseudo R2       =       0.0099
---------------------------------------------------------------------
      y |    Coef.  Std. Err.     z    P>|z|   [95% Conf. Interval]
--------+------------------------------------------------------------
2       |
     x1 |  .4245588  .0427772    9.92  0.000   .3407171    .5084005
     x2 | -.5387675  .0426714  -12.63  0.000  -.6224019   -.455133
  _cons |  1.002834  .0325909   30.77  0.000   .9389565   1.066711
--------+------------------------------------------------------------
3       |
     x1 | -.2953721  .038767    -7.62  0.000  -.371354   -.2193902
     x2 |  .2470191  .0386521    6.39  0.000   .1712625    .3227757
  _cons |  2.003673  .0295736   67.75  0.000   1.94571    2.061637
---------------------------------------------------------------------
(y==1 is the base outcome)
```

It is important to emphasize that no matter which response level is assigned as the reference, the probabilities for each level will be the same. The coefficients, of course, differ, but the underlying model is the same. If the probabilities across levels differed with differing reference levels, then this would not be the case.

```
. qui mlogit y x1 x2, baseoutcome(1)
. predict pr1_1 pr1_2 pr1_3              /// reference level
                                             indicated after pr

. qui mlogit y x1 x2, baseoutcome(2)
. predict pr2_1 pr2_2 pr2_3

. qui mlogit y x1 x2, baseoutcome(3)
. predict pr3_1 pr3_2 pr3_3

. l pr1_1 pr1_2 pr1_3  pr2_1 pr2_2 pr2_3 pr3_1 pr3_2 pr3_3 in 1/5

+----------------------------------------------------------------------+
|     pr1_1      pr1_2      pr1_3      pr2_1      pr2_2      pr2_3 |
|----------------------------------------------------------------------|
1. | .0901045  .0901045  .0901045  .2876121  .2876121  .2876121 |
2. | .0914752  .0914752  .0914752  .2281509  .2281509  .2281509 |
3. | .0890073  .0890073  .0890073  .2021004  .2021004  .2021004 |
4. |  .093086   .093086   .093086  .2575265  .2575265  .2575265 |
5. | .0916807  .0916807  .0916807  .2989847  .2989847  .2989847 |
+----------------------------------------------------------------------+
```

```
+--------------------------------+
|      pr3_1      pr3_2      pr3_3 |
|--------------------------------|
1. | .6222834 .6222834 .6222834 |
2. | .6803738 .6803738 .6803738 |
3. | .7088923 .7088923 .7088923 |
4. | .6493874 .6493874 .6493874 |
5. | .6093346 .6093346 .6093346 |
+--------------------------------+
```

11.2 Independence of Irrelevant Alternatives

A foremost problem facing those who are using multinomial models is the so-called **independence of irrelevant alternatives** assumption, many times simply referred to as IIA. The assumption is that a preference for a given choice of response level is unaffected by the presence of the other choices or levels—the ones not involved in the relative risk ratio. Another way of expressing the assumption is to say that the choice of one alternative level over another is unaffected by the existence of other choices, or levels. Unfortunately, this assumption is violated in many research endeavors—and needs to be checked.

We may check for the independence of irrelevant alternatives by estimating the full model, then fitting a model reduced by a level and employing the Hausman–McFadden test of IIA (Hausman and McFadden, 1984). The test is run as follows, where the model is reduced by excluding the *urgent* level. The null hypothesis is understood as there being no systematic difference in the parameter estimates for either of the models.

```
. quietly mlogit type died white hmo los
. estimates store all
. quietly mlogit type died white hmo los if type!=2
. estimates store partial
. hausman partial all, alleqs constant
```

```
           ---- Coefficients ----
        |      (b)         (B)            (b-B)     sqrt(diag(V_b-V_B))
        |    partial       all         Difference         S.E.
--------+---------------------------------------------------------------
  died  |  .8688752    .8513779        .0174973         .0404346
 white  |  -.272366   -.2159545       -.0564115         .0908705
  hmo   | -1.772366   -1.782408        .0100416         .0355235
  los   |  .0760452    .0802779       -.0042327         .0036925
 _cons  | -3.303338   -3.399962        .096624          .0760595
--------------------------------------------------------------------------
```

```
           b = consistent under Ho and Ha; obtained from mlogit
   B = inconsistent under Ha, efficient under Ho; obtained from mlogit
       Test:  Ho:  difference in coefficients not systematic
                 chi2(5) = (b-B)'[(V_b-V_B)^(-1)](b-B)
                         =        1.10
                 Prob>chi2 =      0.9543
                 (V_b-V_B is not positive definite)
```

The Hausman test indicates that the IIA assumption is not violated for this model.

Long and Freese (2006b) designed the Hausman test in such a manner that the displayed output shows the *Chi2* test results of independently excluding each level. The null hypotheses is that the IIA holds. *p>chi2* values less than 0.05 indicate that the IIA assumption has been violated at the 95% significance level.

Note that for the previous version of the test, level 2 (*urgent*) was excluded, producing a *Chi2* test of 1.10 and *p*-value of 0.9543. This same set of values is given to the *urgent* level in the test results below; the results of excluding the third level (*emergency*) are also provided.

```
. mlogtest, hausman

**** Hausman tests of IIA assumption (N=1495)

Ho: Odds(Outcome-J vs Outcome-K) are independent of other
    alternatives.

Omitted  |     chi2    df    P>chi2    evidence
---------+------------------------------------------
  Urgent |    1.097     5    0.954     for Ho
Emergenc |    2.042     5    0.843     for Ho
---------------------------------------------------
```

The Small–Hsiao (1985) test is also used as a test for a violation of the IIA assumption. It is commonly used in conjunction with the Hausman test, with the aim of having both tests yield consistent results. The test divides the data into two sub-samples. Separate multinomial logit models are estimated on both subsamples, with a weighted average taken of the two sets of parameter estimates using the following formula:

$$\beta_a{}^{\text{wmean}} = \beta_1{}^{-.5} + (\beta * \beta_2{}^{-.5}) \qquad (11.10)$$

A new estimation is performed on the second sub-sample, eliminating a level of the response. The estimates are given as β_b. The resulting likelihood of the restricted model, that is, the model with a level eliminated, is subtracted from the likelihood based on the first averaged set of estimates.

$$SM = -2\{LL(\beta_a{}^{\text{wmean}}) - LL(\beta_b)\tag{11.11}$$

The Small–Hsiao test may be calculated as an option to **mlogtest**. The option is taken from an earlier software implementation developed by Nick Winter (1999). The Small–Hsiao test results may be produced using the *smhsiao* option of **mlogtest**, after estimating the same multinomial model we employed for the Hausman example.

```
. mlogtest, smhsiao

**** Small-Hsiao tests of IIA assumption (N=1495)

 Ho: Odds(Outcome-J vs Outcome-K) are independent of other
     alternatives.

 Omitted | lnL(full) lnL(omit)   chi2   df   P>chi2   evidence
---------+----------------------------------------------------------
  Urgent |  -151.223  -145.260 11.927    5   0.036    against Ho
Emergenc |  -333.458  -329.755  7.406    5`  0.192    for Ho
------------------------------------------------------------------
```

Unlike the Hausman results, we have conflicting evidence for and against IIA. Unfortunately, this is not uncommon. Running the model again randomly creates new sub-samples, which can result in different *p*-values. One may be more confident that the Hausman and Small–Hsiao tests are providing worthwhile advice if they both consistently yield the same results, even with repeatedly re-estimating the Small–Hsiao test. Otherwise, the tests are not reliable.

Concerning the question of when to model data using a multinomial or variety of ordered response model, Cheng and Long (2007) conducted simulations that found that neither the Hausman nor Small–Hsaio test by itself is conclusive for determining if the IIA assumption has been violated. Both tests appear to have poor sample size conditions.

The authors concluded that tests of the IIA assumption that are based on eliminating a level or levels, in other words, that are based on the estimation of a restricted set of choices, are not useful for research purposes.

Although it did not occur with the above example, it often happens that the *Chi2* statistics are negative. This situation may appear strange, but it often can be used as evidence that the IIA assumption has not been violated. See Long and Freese (2006) for more details. In any case, it is not a situation that immediately validates a conclusion regarding the violation of IIA.

For years, there was a paucity of fit tests associated with the multinomial logit model. Various *ad hoc* methods would appear from time to time, but there was no single test that became well accepted in the statistical community. A new goodness-of-fit test has been developed for the multinomial logit model that is based on the same logic as the Hosmer–Lemeshow goodness-of-fit test for logistic models. Developed by Fagerland, Hosmer, and Bofin (2008), the test employs a deciles-of-risk approach, which sorts the observations by the complement of the estimated reference level probabilities, then places observations across levels into equal-sized groups corresponding to the probability levels of the reference level. As with the standard Hosmer–Lemeshow test, a Pearson χ^2 test is used to assess the goodness-of-fit of the multinomial model. At the time of this writing, only MATLAB® code is available for executing this test. I suspect, however, that R and Stata code will soon be available.

One final point. We may compare the multinomial logit model with like models of alternative links, just as we could with the ordered logit and generalized ordered logit. Interested readers should consult Hardin and Hilbe (2007), as cited at the conclusion of the previous section on ordinal logit models. I also suggest Long and Freese (2006). Non-Stata users may instead want to access Long (1997).

11.3 Comparison to Multinomial Probit

The multinomial probit model is at times used in econometric research, where it is many times referred to as a random utility model, or RUM. The idea here is that the choice a person makes from among r alternatives is the choice having maximum utility. It was first described by Albright, Lerman, and Manski (1977) and was subsequently developed to its current form by Daganzo (1979). It is a difficult model to compute since it entails multivariate Gaussian or normal probabilities, which are highly complex functions. As a result, the multinomial probit model has only recently appeared in commercial software.

One of the foremost reasons that researchers have given for using the multinomial probit with unordered categorical response data relates to the fact that it is not subject to the assumption of irrelevant alternatives, or IIA. Violation of this assumption plagues many multinomial logit modeling endeavors. Without this assumption, a researcher does not have to subject the model to related tests, such as the Hausman test, nor does the model need to be amended in light of such violations. For this reason, it has recently been a fairly popular alternative to the multinomial logistic model.

Recall that the multinomial logit model could be considered as an expansion of the logit model. To show this relationship, we use the **gss_polviews2002** data, taken from the 2002 General Social Survey data. Earlier, we manipulated the data, creating new variables by categorizing many-valued discrete variables. There are three variables, with the following structure.

   politics3 is the response. Another response variable was created from *politics3*, called *party*, which consists of the liberal and conservative levels, with the latter level given the value of 1.

```
. use gss_polviews2002

. tab politics3     /// or party

   politics3 |       Freq.      Percent         Cum.
-------------+-----------------------------------------
     Liberal |         322        26.72        26.72
    Moderate |         467        38.76        65.48
Conservative |         416        34.52       100.00
-------------+-----------------------------------------
       Total |       1,205       100.00

. tab agegrp

      agegrp |       Freq.      Percent         Cum.
-------------+-----------------------------------------
         <32 |         578        23.24        23.24
       32-59 |       1,320        53.08        76.32
         >59 |         589        23.68       100.00
-------------+-----------------------------------------
       Total |       2,487       100.00

. tab gradschool

1=post grad |
     school |       Freq.      Percent         Cum.
-------------+-----------------------------------------
          0 |       2,175        87.42        87.42
          1 |         313        12.58       100.00
-------------+-----------------------------------------
       Total |       2,488       100.00
```

Again, when we considered the multinomial logistic model, it was mentioned that the multinomial is an extension of logit such that the former can be used to model the binary response, but not as risk ratios.

```
. logit party agegrp2 agegrp3 gradschool, nolog

Logistic regression                       Number of obs   =        1196
                                          LR chi2(3)      =       14.93
                                          Prob > chi2     =      0.0019
Log likelihood = -764.63102               Pseudo R2       =      0.0097
```

```
---------------------------------------------------------------------
      party|    Coef. Std. Err.    z   P>|z|   [95% Conf. Interval]
----------+----------------------------------------------------------
    agegrp2|  .4279824 .1619694  2.64 0.008   .1105283    .7454365
    agegrp3|  .6548711 .1830566  3.58 0.000   .2960867   1.013655
 gradschool|-.2744726  .191063  -1.44 0.151  -.6489492    .100004
     _cons|-.9951536 .1392079 -7.15 0.000  -1.267996   -.7223111
---------------------------------------------------------------------

. mlogit party agegrp2 agegrp3 gradschool, nolog

Multinomial logistic regression      Number of obs   =      1196
                                     LR chi2(3)      =     14.93
                                     Prob > chi2     =    0.0019
Log likelihood = -764.63102          Pseudo R2       =    0.0097
---------------------------------------------------------------------
      party|    Coef. Std. Err.    z   P>|z|   [95% Conf. Interval]
----------+----------------------------------------------------------
1         |
    agegrp2|  .4279824 .1619694  2.64 0.008   .1105283    .7454365
    agegrp3|  .6548711 .1830566  3.58 0.000   .2960867   1.013655
 gradschool|-.2744726  .191063  -1.44 0.151  -.6489492    .100004
     _cons|-.9951536 .1392079 -7.15 0.000  -1.267996   -.7223111
---------------------------------------------------------------------
(party==0 is the base outcome)
```

Unlike the proportional logit model, which has identical predictors as the logit, but with intercepts having opposite signs (dependent on the algorithm used), the multinomial and logit results are identical throughout.

The above relationship is not the case for the multinomial probit and probit.

```
. probit party agegrp2 agegrp3 gradschool, nolog

Probit regression                    Number of obs   =      1196
                                     LR chi2(3)      =     14.90
                                     Prob > chi2     =    0.0019
Log likelihood = -764.64231          Pseudo R2       =    0.0097
---------------------------------------------------------------------
      party|    Coef. Std. Err.    z   P>|z|   [95% Conf. Interval]
----------+----------------------------------------------------------
    agegrp2|  .2585123 .0971542  2.66 0.008  .0680936    .448931
    agegrp3|  .3991735 .1109625  3.60 0.000  .181691    .616656
 gradschool|-.1663334 .1155034  -1.44 0.150 -.392716    .0600492
     _cons|-.6122916 .0828696 -7.39 0.000 -.774713   -.4498702
---------------------------------------------------------------------

. mprobit party agegrp2 agegrp3 gradschool, nolog

Multinomial probit regression        Number of obs   =      1196
                                     Wald chi2(3)    =     14.72
```

```
Log likelihood = -764.64231          Prob > chi2     =     0.0021
-----------------------------------------------------------------
       party|    Coef. Std. Err.   z  P>|z|  [95% Conf. Interval]
-----------+-----------------------------------------------------
_outcome_2  |
     agegrp2|  .3655915 .1373969  2.66 0.008   .0962986    .6348844
     agegrp3|  .5645165 .1569248  3.60 0.000   .2569496    .8720834
  gradschool|-.2352309 .1633466 -1.44 0.150  -.5553843    .0849225
       _cons| -.865911 .1171954 -7.39 0.000   -1.09561   -.6362122
-----------------------------------------------------------------
```

(party=0 is the base outcome)

The fact, however, that the log-likelihoods of the two models are the same, and the *z*- and *p*-values are the same, tells us that there is likely a ratio between the two sets of coefficients.

```
. di   .3655915/.2585123          . di .5645165/.3991735
1.4142132                         1.4142132

. di -.2352309/ -.1663334         . di -.865911/ -.6122916
1.4142132                         1.4142132
```

The ratio is identical for them all to within the nearest one-millionth. The ratio of the standard errors also have the same value—$2^{.5}$, or the square root of 2.

This ratio exists for any multinomial probit versus probit relationship, and reflects the difference in the variance of the errors, $var(\varepsilon)$, for which the probit has a constant value of 1 (and the logit has a constant value of $\pi^2/3$). The error variance of the multinomial probit is 2, with the standard deviation being the square root of the variance, or sqrt(2). We therefore find this ratio between both the respective model coefficients and standard errors. But, because both coefficient and standard errors differ by the same amount, the ratio of coefficient to standard error between the probit and multinomial probit is identical. This results in identical *z*-values and therefore *p*-values.

We now compare the two multinomial models using a three-level response. The multinomial probit assumes that the errors associated with each response level are distributed as multivariate normal.

$$[\varepsilon_{1i}, \varepsilon_{2i}, \varepsilon_{3i}, ..., \varepsilon_{ki}] \sim N(0_{kx1}, \Sigma_{kxk}) \tag{11.12}$$

In addition, the multinomial probit algorithm normalizes the variances (resulting in the above discussed discrepancy in coefficients and SEs) so that the diagonal elements of Σ_{kxk} are all 1, and off diagonal covariances are 0. The probabilities that are generated from the linear predictor are based on the cumulative bivariate normal distribution. Again, the computations are tedious, and were not available in commercial software until computers had sufficient Hz and RAM to make estimation feasible.

The three-level *politics3*, with liberal as the reference category,

```
. mprobit politics3 agegrp2 agegrp3 gradschool, nolog base(1)

Multinomial probit regression      Number of obs   =        1196
                                    Wald chi2(6)    =       41.32
Log likelihood = -1280.1157         Prob > chi2     =      0.0000
---------------------------------------------------------------------
  politics3|    Coef.  Std. Err.    z  P>|z| [95% Conf. Interval]
-----------+---------------------------------------------------------
Moderate   |
    agegrp2| .1323305 .1402914  0.94 0.346 -.1426356   .4072967
    agegrp3|  .331427 .1674431  1.98 0.048  .0032445   .6596095
 gradschool|-.8133637 .1710697 -4.75 0.000 -1.148654  -.4780734
      _cons| .2470274  .116529  2.12 0.034  .0186348    .47542
-----------+---------------------------------------------------------
Conservative|
    agegrp2| .4128574 .1462157  2.82 0.005  .1262799   .6994349
    agegrp3| .7119887 .1715873  4.15 0.000  .3756838  1.048294
 gradschool|-.6244265 .1684097 -3.71 0.000 -.9545034  -.2943496
      _cons|-.1035662 .1233009 -0.84 0.401 -.3452316   .1380992
---------------------------------------------------------------------
(politics3=Liberal is the base outcome)

. estat ic

---------------------------------------------------------------------
Model |  Obs  ll(null) ll(model)   df        AIC         BIC
------+--------------------------------------------------------------
    . | 1196        .  -1280.116    8    2576.231    2616.925
---------------------------------------------------------------------
```

We may compare the multinomial probit with the multinomial logit using the AIC and BIC fit statistics. Lower values of these statistics indicate a better-fitted, and therefore preferred, model. Estimating the multinomial logit, without displaying results, and determining the values of AIC and BIC will allow us to observe if major differences occur between values.

```
. qui mlogit politics3 agegrp2 agegrp3 gradschool, base(1)
nolog

. estat ic

---------------------------------------------------------------------
Model |  Obs  ll(null) ll(model)   df        AIC         BIC
------+--------------------------------------------------------------
    . | 1196 -1301.006 -1280.205    8    2576.409    2617.103
---------------------------------------------------------------------
```

The multinomial probit produces lower AIC and BIC statistics, but the difference in their value is statistically inconsequential. Neither model is preferred on the basis of these statistics.

It should be noted that the multinomial probit may be parameterized so that estimates and standard errors take values as if they were based on probit regression. That is, coefficients and standard errors are divided by the square root of 2, resulting in the following output.

[Stata's *probitparam* option calls for the conversion.]

```
. mprobit politics3 agegrp2 agegrp3 gradschool, nolog
probitparam base(1)

------------------------------------------------------------------
   politics3 |    Coef. Std. Err.     z  P>|z|  [95% Conf. Interval]
-------------+----------------------------------------------------
Moderate     |
     agegrp2 | .0935718 .1179706   0.79 0.428 -.1376462   .3247899
     agegrp3 | .2343543 .1408023   1.66 0.096 -.0416132   .5103218
  gradschool | -.575135 .1438519  -4.00 0.000 -.8570795  -.2931905
       _cons | .1746748 .0979888   1.78 0.075 -.0173797   .3667293
-------------+----------------------------------------------------
Conservative |
     agegrp2 | .2919343 .1229522   2.37 0.018  .0509523   .5329163
     agegrp3 | .5034521 .1442871   3.49 0.000  .2206545   .7862496
  gradschool |-.4415362 .1416151  -3.12 0.002 -.7190967  -.1639758
       _cons |-.0732323 .1036833  -0.71 0.480 -.2764479   .1299832
------------------------------------------------------------------

(politics3=Liberal is the base outcome)
```

It appears that there is no appreciable difference between the logit and probit multinomial models; therefore, we might evaluate the multinomial logit to determine if the model violates IIA assumptions.

The IIA assumption can be assessed for the multinomial logit model of the **gss_polviews2002** data by using the following code.

```
. qui mlogit politics3 agegrp2 agegrp3 gradschool, base(1) nolog
. est store all
. qui mlogit politics3 agegrp2 agegrp3 gradschool if politics3 !=
      "Moderate":politics3, base(1)
. est store reduce
. hausman reduce all, alleqs constant

              ---- Coefficients ----
          |      (b)         (B)        (b-B)     sqrt(diag(V_b-V_B))
          |    reduce       all      Difference        S.E.
----------+-------------------------------------------------------
   agegrp2 |  .5494224    .5303685    .0190539        .0180871
   agegrp3 |  .9420392    .9270442    .0149951        .0200287
```

```
gradschool| -.8146784 -.7972547  -.0174237         .021525
    _cons| -.1412037 -.1301227   -.011081              .
-----------------------------------------------------------------
b = consistent under Ho and Ha; obtained from mlogit
B = inconsistent under Ha, efficient under Ho; obtained from
mlogit

Test:  Ho:  difference in coefficients not systematic

              chi2(4) = (b-B)'[(V_b-V_B)^(-1)](b-B)
                      =         0.48
         Prob>chi2 =        0.9754
         (V_b-V_B is not positive definite)
```

The IIA assumption is not violated for the multinomial logit model. Based on goodness-of-fit statistics and a test of IIA, we do not find a reason to prefer one model over the other. However, recently, we have alternative grounds to help make a decision.

Kropko (2008) conducted extensive simulation studies on both the multinomial logit and probit models, concluding that the multinomial logit nearly always provides more accurate results than the multinomial probit. Moreover, he found that these results even followed when the multinomial logit IIA assumption was severely violated. Recall that the assumption of IIA is not applicable to the multinomial probit and has been a foremost reason that it has been favored by many researchers. In any case, he concluded that the multinomial logit is a more reliable empirical model than is the multivariable probit, and suggested that it be used in its place whenever possible. When IIA assumptions have been satisfied by a model, there is no question as to the logit's power to better model the data than the multinomial probit model.

For now, Kropko's findings lead us to the conclusion that if our multinomial logit model is found to violate IIA assumptions, it is preferable to attempt to resolve the violation within the context of the logit model than to resort to modeling with a multinomial probit.

For additional discussion on this model, see Long and Freese (2006) or Hardin and Hilbe (2007).

Exercises

11.1. Use the data and model developed from Exercise 10.1. Model it as a multinomial model, and statistically compare it to the ordered response model.

a. Which model is statistically a better fit?

b. Is IIA violated?

    c. Interpreted as a multinomial model, how are the exponentiated coefficients interpreted?

11.2. Using the **gsssurvey** data, model the three-level *polview3* on predictors related to religious views and behavior.

    a. Using a multinomial model, determine if the amount of money spent on religious organizations has a bearing on one's political view.

    b. What is the risk ratio (probability) that a person who prays once a day or more is likely to be conservative rather than liberal? What is the probability of being a liberal if one never prays? Of being a conservative if one never prays? Of being a liberal if one prays once or more per day? Of being a conservative if one prays once or more per day?

    c. Test the models in item (a) and also in item (b) against proportional odds models on the same data. For each, which model better fits the data? Why?

11.3. Using the **gsssurvey** data, model the three-level *polview3* on predictors related to ethnic and socioeconomic status and behavior.

    a. Determine which predictors related to ethnic status and behavior provide statistical support for determining the probability of a person being politically liberal, moderate, or conservative.

    b. Determine which predictors related to socioeconomic status and behavior provide statistical support for determining the probability of a person being politically liberal, moderate, or conservative.

    c. What is the probability that a person who has a high socioeconomic index value is likely to be conservative rather than liberal? A moderate rather than a liberal?

    d. What is the relationship of years of education and political views?

    e. Test the model in (a) against a proportional odds model on the same data. Which model better fits the data? Why?

11.4. Use the **alligator** data set that comes from Agresti (2002). The following covariates are to be used in the model.

RESPONSE

    food   5-level categorical 1 = fish; 2 = invertebrate; 3 = reptile; 4 = bird; 5 = other

PREDICTORS

    gender   binary 1 = male; 2 = female

    lake   binary 1 = Trafford; 2 = George

    size   binary 1 = alligator ≤ 2.3meters; 2 = alligator ≥ 2.3 meters

a. Model *food* on *gender, lake,* and *size,* determining the risk ratios of each predictor.

b. Is there a difference in the preferred food of large compared to smaller alligators?

c. Does the lake in which the alligators live make a difference in which food is preferred?

d. Do male compared to female alligators prefer different foods?

e. What combination of predictor levels gives the highest probability of fish, invertebrates, reptiles, or birds being the favored food?

11.5. Create a synthetic multinomial logit model using the **syn_mlogit.do** procedure that is given in Chapter 11. Amend the program so that there are four levels and three parameter estimates in addition to the intercept. You may select your own parameter and intercept values. Users of packages other than Stata may create a similar macro appropriate for their own statistical application.

a. Run the model and confirm that a 20,000 observation multinomial logit model with the parameters and intercepts that have been assigned are in fact displayed. Level 1 is the reference.

b. Amend the model so that the highest level is the reference, that is, the reference is changed to Level 4. Adjust the defined parameters and intercepts accordingly.

c. Test the model to be assured it has not violated IIA assumptions.

d. Model the data as a multinomial probit; compare results.

11.6. Amend the synthetic multinomial logit model, **syn_mlogit.do**, so that the slopes of each level are proportional. Select your own parameters and intercepts.

a. Model the data using the amended synthetic multinomial *do* program, or macro.

b. Model the same data using a proportional odds model. Determine if the parameter estimates are the same as in the multinomial model, and that the slopes are in fact proportional. The intercepts you defined for each nonreference level should be reflected in the cut points.

R Code

Section 11.1 Unordered Logistic Regression

```
#11.1.2
library('foreign')
```

```
medpar<- read.dta('medpar_olrx.dta')  #read Stata format file
summary(medpar)
head(medpar)
str(medpar)

tab1<- table(medpar$white, medpar$type)
tab1
rowSums(tab1)
colSums(tab1)
sum(tab1)

(80*228)/(37*1054)
(80*86)/(1054*10)

library('nnet') #for multinomial models via neural networks
fit11_2a<- multinom(type ~ white, data=medpar)
summary(fit11_2a)
exp(coef(fit11_2a)) #ORs

(37*1054)/(80*228)
(37*86)/(10*228)

type2<- relevel(medpar$type, ref=2) #reference to 'Urgent'
fit11_2b<- multinom(type2 ~ white, data=medpar)
summary(fit11_2b)
exp(coef(fit11_2b)) #ORs

fit11_2c<- multinom(type ~ died + white + hmo + los,
  data=medpar)
summary(fit11_2c)
exp(coef(fit11_2c)) #ORs

df<- data.frame(died=rep(mean(medpar$died), 2), white=c(0,1),
  hmo=rep(mean(medpar$hmo),2), los=rep(mean(medpar$los),2))
df
cbind(white=c(0,1),predict(fit11_2c, newdata=df, type='probs'))

library('MASS')
fit11_2d<- polr(type ~ died + white + hmo + los, data=medpar,
  method='logistic')
summary(fit11_2d)

f<- df.residual(fit11_2d)-(nrow(medpar)-10)
f
X2<- deviance(fit11_2d)-deviance(fit11_2c)
X2   #LRT
pchisq(X2, f, lower.tail=FALSE) #P-value

syndata<- read.dta('mlogit1.dta') #read Stata data file
syndata$y<- as.factor(syndata$y)
```

```
head(syndata)
str(syndata)
table(syndata$y)
table(syndata$y)/length(syndata$y)

fit11_2e<- multinom(y ~ x1 + x2, data=syndata)
summary(fit11_2e)

y2<- relevel(syndata$y, ref=2)
fit11_2f<- multinom(y2 ~ x1 + x2, data=syndata)

y3<- relevel(syndata$y, ref=3)
fit11_2g<- multinom(y3 ~ x1 + x2, data=syndata)

p1<- predict(fit11_2e, newdata=syndata[1:10,], type='probs')
p2<- predict(fit11_2f, newdata=syndata[1:10,], type='probs')
p3<- predict(fit11_2g, newdata=syndata[1:10,], type='probs')
cbind(p1, p2, p3) #should be all same predictions
```

Section 11.2 Independence of Irrelevant Alternatives

Available after chapter written: library("mlogit"); hmftest()

Section 11.3 Comparison to Multinomial Probit

```
gss<- read.dta('gss_polviews2002.dta')
head(gss)
str(gss)

table(gss$politics3)
table(gss$politics3)/length(gss$politics3)
table(gss$agegrp)
table(gss$agegrp)/length(gss$agegrp)
table(gss$gradschool)
table(gss$gradschool)/length(gss$gradschool)

fit11_3a<- glm(party ~ agegrp2 + agegrp3 + gradschool,
   data=gss, family=binomial(link='logit'))
summary(fit11_3a)

fit11_3b<- multinom(party ~ agegrp2 + agegrp3 + gradschool,
   data=gss)
summary(fit11_3b)

fit11_3c<- glm(party ~ agegrp2 + agegrp3 + gradschool,
   data=gss, family=binomial(link='probit'))
summary(fit11_3c)
library('sampleSelection')
```

```
fit11_3c2<- probit(party ~ agegrp2 + agegrp3 + gradschool,
   data=gss)
summary(fit11_3c2)

#NOTE: mnp() requires a response with >2 categories.

0.3655915/0.2585123
-0.2352309/-0.166334

library('MNP')   #multinomial probit
fit11_3d<- mnp(politics3 ~ agegrp2 + agegrp3 + gradschool,
   data=gss)
summary(fit11_3d)

fit11_3e<- polr(politics3 ~ agegrp2 + agegrp3 + gradschool,
   data=gss, method='probit') #ordered multinomial probit
summary(fit11_3e)

#AIC and BIC not available for mnp()
#No Hausman test available for mnp()
```

12

Alternative Categorical Response Models

12.1 Introduction

In this chapter, we discuss those categorical logistic models that are not examined in the previous two chapters, but which are not panel models. The conditional logit model can be used as the foundation of several categorical response choice models, such as McFadden's alternative-specific choice model and rank-ordered logistic regression. All three models are constructed as panel models, where a case may consist of more than one line in the data. For example, a patient may be tested four times, with each testing incidence being represented as a single observation. The lines associated with a case may be identified by a variable, such as *id*, *group*, or *case*. For example:

id	age	sex	testvalue
1	37	1	5.42
1	37	1	3.56
1	37	1	2.45
2	28	0	6.78
2	28	0	5.78
2	28	0	4.98
3	33	1	5.53
...	...	...	...
100	44	1	4.55

The models we discuss in this chapter continue to be case specific, that is, one line per observation. They were developed to address specific problems that the more traditional models failed to accommodate. Few commercial software applications support the models we address in this chapter. Those that do as of 2008 are identified below, along with the section of the chapter in which they are discussed.

12.2 Continuation ratio (Stata, SAS, LogXact)

12.3 Stereotype logistic regression (Stata, SAS)

12.4 Heterogeneous choice logistic model (Stata, user authored)

12.2 Continuation Ratio Models

The continuation ratio model may be formulated using the following relationship (1980, S. F. Feinberg):

$$\ln[\Pr(Y = y_j|x)/\Pr(Y > y_j|x)] = \alpha_j - x\beta \tag{12.1}$$

where

$$j = 1, 2, \ 3,\ldots,J$$

or

$$\ln[\Pr(Y = y_j \mid x)] - \ln[\Pr(Y > y_j \mid x)] \tag{12.2}$$

In many ordered logistic regression modeling situations, the response levels are such that the lowest level must occur before the second, the second before the third, and so forth until the highest level. There is even more information lost in such a relationship, if modeled using multinomial methods—or even standard-ordered logistic levels. The Continuation Ratio model incorporates this type of priority into its algorithm. We can model the relationship of education level to religiosity, controlling for having children, age, and gender using the **edreligion** data set. We will use Stata's **ocratio** command to model the data. Moreover, we can assess if priority does make a difference by checking the respective model AIC goodness-of-fit statistics.

The data, a tabulation of the response, and of the response on religiosity, are as follows:

```
Contains data from http://www.stata-press.com/data/hh2/
edreligion.dta
  obs:           601
  vars:            5                              9 Nov 2006 11:33
  size:        7,212 (99.9% of memory free)
-------------------------------------------------------------------
              storage  display    value
variable name  type    format     label     variable label
-------------------------------------------------------------------
male           byte    %9.0g                =1 if male
age            float   %9.0g                in years
kids           byte    %9.0g                =1 if have kids
```

```
educlevel          byte    %8.0g             educlevel
religious          byte    %8.0g
---------------------------------------------------------------------
Sorted by:


. tab educlevel

  educlevel |      Freq.      Percent        Cum.
------------+-----------------------------------------
       AA  |        205        34.11        34.11
       BA  |        204        33.94        68.05
    MA/PhD |        192        31.95       100.00
------------+-----------------------------------------
     Total |        601       100.00

. tab educlevel religious

            |      religious
  educlevel |        0            1  |     Total
------------+-----------------------------+----------
       AA  |      104          101  |       205
       BA  |      127           77  |       204
    MA/PhD |      110           82  |       192
------------+-----------------------------+----------
     Total |      341          260  |       601
```

Both *kids* and *male* are binary, but *age* is categorical with nine levels. A few statisticians simply model *age* in this fashion as a continuous predictor, but we should definitely combine *age* 17.5-and-under into the under-22 age group.

```
. tab age

  in years |      Freq.      Percent        Cum.
------------+-----------------------------------------
     17.5  |          6         1.00         1.00
       22  |        117        19.47        20.47
       27  |        153        25.46        45.92
       32  |        115        19.13        65.06
       37  |         88        14.64        79.70
       42  |         56         9.32        89.02
       47  |         23         3.83        92.85
       52  |         21         3.49        96.34
       57  |         22         3.66       100.00
------------+-----------------------------------------
     Total |        601       100.00
```

We shall combine groups so that we have age groups of under 22, 27, 32, 37, and 57, with the lower values excluded from the higher. This can be done in

Stata using the **recode** command. Create a new age variable called *agegrp*, categorizing as delineated above.

```
. recode age (1 17.5=22) (2 27=27) (3 32=32)   (4 37=37)
(5 42/52=57), gen(agegrp)

. tab agegrp

  RECODE of |
    age (in |
     years) |      Freq.      Percent         Cum.
------------+-----------------------------------
         22 |        123        20.47        20.47
         27 |        153        25.46        45.92
         32 |        115        19.13        65.06
         37 |         88        14.64        79.70
         57 |        122        20.30       100.00
------------+-----------------------------------
      Total |        601       100.00
```

Since we are going to use the five-level *agegrp* predictor in our models, we should first make the levels into indicator variables.

```
. quietly tab agegrp, gen(agegrp)
```

The data now appear as:

```
Contains data from edreligion.dta
  obs:            601
  vars:            11                        30 Apr 2007 12:38
  size:        12,621 (99.9% of memory free)
-------------------------------------------------------------------
              storage display value
variable name  type   format  label      variable label
-------------------------------------------------------------------
male           byte   %9.0g              =1 if male
age            float  %9.0g              in years
kids           byte   %9.0g              =1 if have kids
educlevel      byte   %8.0g   educlevel

religious      byte   %8.0g
agegrp         float  %10.0g             RECODE of age (in years)
agegrp1        byte   %8.0g              agegrp==      22.0000
agegrp2        byte   %8.0g              agegrp==      27.0000
agegrp3        byte   %8.0g              agegrp==      32.0000
agegrp4        byte   %8.0g              agegrp==      37.0000
agegrp5        byte   %8.0g              agegrp==      57.0000
-------------------------------------------------------------------
Sorted by:
    Note:   dataset has changed since last saved
```

Model the data using an ordered logistic regression command:

```
. ologit educlevel religious kids male agegrp2-agegrp5, nolog

Ordered logistic regression              Number of obs   =         601
                                         LR chi2(7)      =      129.83
                                         Prob > chi2     =      0.0000
Log likelihood = -595.08761              Pseudo R2       =      0.0984

------------------------------------------------------------------------
educlevel |    Coef. Std. Err.     z  P>|z|  [95% Conf. Interval]
----------+-------------------------------------------------------------
religious |-.3626379 .1622982  -2.23 0.025 -.6807365    -.0445393
     kids |-.7836839 .2126397  -3.69 0.000  -1.20045    -.3669178
     male |  1.41435 .1664234   8.50 0.000  1.088166     1.740534
  agegrp2 | 1.120131 .2471355   4.53 0.000  .6357544     1.604508
  agegrp3 | 1.211077 .2884475   4.20 0.000  .6457302     1.776424
  agegrp4 |  1.77401 .3197456   5.55 0.000   1.14732       2.4007
  agegrp5 | 1.191035 .2944595   4.04 0.000  .6139051     1.768165
----------+-------------------------------------------------------------
    /cut1 | .1596998 .1936208               -.2197899     .5391896
    /cut2 | 1.848094 .2097492                1.436993     2.259195
------------------------------------------------------------------------
. aic
AIC Statistic =    2.006947              AIC*n =    1206.1752
BIC Statistic =   -2597.793
```

Note: The **abic** postestimation command that has been used previously to calculate AIC and BIC statistics does not work with the **ologit** command. **ologit** does not save the required global macros, or saved values, for **abic** to use in its calculations. **aic**, an older command, is used in its place. For consistency and comparative purposes, **aic** is used throughout this chapter.

Check the proportional-odds assumption using the Brant test:

```
. brant, detail

Estimated coefficients from j-1 binary regressions

                       y>1          y>2
religious     -.50454345   -.24489372
     kids     -1.1408243   -.34646367
     male       1.184672    1.6494569
  agegrp2       1.431636    1.1415567
  agegrp3      1.1788485    1.3950112
  agegrp4      1.5867963    2.0111911
  agegrp5      1.2236598    1.4261125
    _cons      .17755928   -2.5340327

Brant Test of Parallel Regression Assumption
```

```
    Variable |     chi2    p>chi2    df
-------------+-------------------------------
         All |    29.32     0.000     7  <=
-------------+-------------------------------
   religious |     1.56     0.211     1
        kids |     6.39     0.011     1  <=
        male |     4.79     0.029     1  <=
     agegrp2 |     0.55     0.457     1
     agegrp3 |     0.28     0.595     1
     agegrp4 |     1.01     0.314     1
     agegrp5 |     0.25     0.619     1
---------------------------------------------
```

A significant test statistic provides evidence that the parallel regression assumption has been violated.

I have placed arrows to the right of those predictors that violate the proportional-odds assumption. *kids, male,* and the model as a whole violate the assumption. We next must run a partial proportional-odds model, holding those predictors constant across categories that were classified as nonsignificant using brant.

```
. gologit2 educlevel religious kids male agegrp2-agegrp5,
nolog autofit lrforce
----------------------------------------------------------------
Testing parallel lines assumption using the .05 level of
significance...

Step  1:  Constraints for parallel lines imposed for agegrp5
          (P Value = 0.5606)
Step  2:  Constraints for parallel lines imposed for agegrp3
          (P Value = 0.6582)
Step  3:  Constraints for parallel lines imposed for agegrp4
          (P Value = 0.4813)
Step  4:  Constraints for parallel lines imposed for religious
          (P Value = 0.3738)
Step  5:  Constraints for parallel lines are not imposed for
          kids (P Value = 0.00104)
          male (P Value = 0.02703)
          agegrp2 (P Value = 0.03111)

Wald test of parallel lines assumption for the final model:
 ( 1)  [AA]agegrp5 - [BA]agegrp5 = 0
 ( 2)  [AA]agegrp3 - [BA]agegrp3 = 0
 ( 3)  [AA]agegrp4 - [BA]agegrp4 = 0
 ( 4)  [AA]religious - [BA]religious = 0

         chi2(  4) =      1.80
     Prob > chi2 =      0.7717
```

An insignificant test statistic indicates that the final model
does not violate the proportional odds/ parallel lines
assumption

If you re-estimate this exact same model with gologit2,
instead of autofit you can save time by using the parameter

```
pl(agegrp5 agegrp3 agegrp4 religious)
```

Generalized Ordered Logit Estimates   Number of obs   =   601
                                        LR chi2(10)   =   156.58
                                        Prob > chi2   =   0.0000
Log likelihood = -581.71431   Pseudo R2   =   0.1186

 (1)   [AA]agegrp5 - [BA]agegrp5 = 0
 (2)   [AA]agegrp3 - [BA]agegrp3 = 0
 (3)   [AA]agegrp4 - [BA]agegrp4 = 0
 (4)   [AA]religious - [BA]religious = 0
----

educlevel	Coef.	Std. Err.	z	P>|z|	[95% Conf. Interval]	
AA						
religious	-.3940791	.1636127	-2.41	0.016	-.714754	-.0734041
kids	-1.287969	.267434	-4.82	0.000	-1.81213	-.7638076
male	1.189787	.1929103	6.17	0.000	.8116901	1.567885
agegrp2	1.536946	.2914182	5.27	0.000	.9657769	2.108115
agegrp3	1.360434	.2999842	4.54	0.000	.7724762	1.948393
agegrp4	1.841107	.3287475	5.60	0.000	1.196774	2.485441
agegrp5	1.317171	.3043349	4.33	0.000	.7206854	1.913656
_cons	.1265897	.2140777	0.59	0.554	-.2929948	.5461743
BA						
religious	-.3940791	.1636127	-2.41	0.016	-.714754	-.0734041
kids	-.4248231	.2498083	-1.70	0.089	-.9144383	.0647921
male	1.654449	.202629	8.16	0.000	1.257304	2.051595
agegrp2	.9979147	.2956057	3.38	0.001	.4185381	1.577291
agegrp3	1.360434	.2999842	4.54	0.000	.7724762	1.948393
agegrp4	1.841107	.3287475	5.60	0.000	1.196774	2.485441
agegrp5	1.317171	.3043349	4.33	0.000	.7206854	1.913656
_cons	-2.307202	.272596	-8.46	0.000	-2.841481	-1.772924

```
. aic
AIC Statistic =    1.972427     AIC*n =   1185.4286
BIC Statistic =   -2579.749
```

The model accounts for predictor odds that are both proportional and
nonproportional across categories. The partial-proportional model results
in a slightly lower AIC statistic; it is not significantly lower. However, we

prefer the partial proportional-odds model because it accommodates the proportionality violations found in the standard and generalized models.

On the other hand, the model did not take into consideration the continuation-ratio nature of the response. We now model it using **ocratio**:

```
. aic
AIC Statistic =    1.972427               AIC*n =   1185.4286
BIC Statistic =   -2579.749

. ocratio educlevel religious kids male agegrp2-agegrp5

Continuation-ratio logit Estimates        Number of obs =       997
                                          chi2(7)       =    134.79
                                          Prob > chi2   =    0.0000
Log Likelihood = -592.6098                Pseudo R2     =    0.1021
---------------------------------------------------------------------
 educlevel |    Coef. Std. Err.    z  P>|z|  [95% Conf. Interval]
-----------+---------------------------------------------------------
 religious |-.3269944 .1458074 -2.24 0.025 -.6127716 -.0412172
      kids |-.6120951 .1911805 -3.20 0.001  -.986802 -.2373882
      male | 1.277602 .1471529  8.68 0.000  .9891879  1.566017
   agegrp2 | .9844616 .2241227  4.39 0.000  .5451893  1.423734
   agegrp3 | 1.120732 .2585095  4.34 0.000  .6140632  1.627402
   agegrp4 | 1.597828 .2866582  5.57 0.000  1.035989  2.159668
   agegrp5 | 1.122695 .2628734  4.27 0.000   .607473  1.637918
---------------------------------------------------------------------
 _cut1  | .1892129   .1800996         (Ancillary parameters)
 _cut2  | 1.238825   .2053886
---------------------------------------------------------------------

. aic
AIC Statistic =    1.204834      AIC*n =   1201.2196
BIC Statistic =   -5643.579
```

The AIC statistic shows that the continuation ratio model fits the data much better than alternative models. We can parameterize the model in terms of odds ratios by using the *eform* option.

```
. ocratio educlevel religious kids male agegrp2-agegrp5, eform

Continuation-ratio logit Estimates        Number of obs =       997
                                          chi2(7)       =    134.79
                                          Prob > chi2   =    0.0000
Log Likelihood = -592.6098                Pseudo R2     =    0.1021
---------------------------------------------------------------------
educlevel|Odds ratio Std. Err.    z  P>|z|  [95% Conf. Interval]
---------+-----------------------------------------------------------
religious|  .7210878 .1051399 -2.24 0.025   .541847   .9596207
     kids|  .5422137 .1036607 -3.20 0.001  .3727669   .7886851
     male| 3.588026 .5279884  8.68 0.000   2.68905    4.78754
```

```
agegrp2|  2.676371 .5998353   4.39 0.000 1.724935    4.152597
agegrp3|    3.0671 .7928745   4.34 0.000 1.847925    5.090631
agegrp4|  4.942288 1.416748   5.57 0.000  2.81789    8.668261
agegrp5|  3.073126  .807843   4.27 0.000 1.835787    5.144446
-----------------------------------------------------------------
_cut1 | .1892129 .1800996        (Ancillary parameters)
_cut2 | 1.238825 .2053886
-----------------------------------------------------------------
```

Interpretation of the odds ratios follows the same logic as the ordered logistic model. Predicted levels may be accessed using the **ocrpred** command, as done for **ologit**. Note that the number of observations in the **ocratio** model above has been inflated to 997 from 601. The reason is based on how levels are compared: level 1 versus level 2, 3, and level 2 versus level 3. This results in $[205 + (204 + 192)] + [204 + 192] = 997$.

12.3 Stereotype Logistic Model

Stereotype logistic regression may be considered as a medium between multinomial and ordered logistic regression. One generally uses the model when unsure of the ordering of levels, or when it is suspected that one or more levels can be collapsed. However, if the levels of the categorical response appear to have an order to them, but do not meet the criterion of proportionality of slopes that is a defining characteristic of the ordered logistic model—or the proportional odds model—then a choice may be made between the stereotype and generalized ordinal logistic models. The essential difference between these two, though, is that the stereotype logistic model places ordering constraints on an otherwise multinomial model.

The model was first designed by J.A. Anderson (1984). He proposed imposing a scale or score to each level of the response, y_k, appearing as:

$$\beta_j = -\phi_j \beta \quad \text{for } j = 1,\ldots,k. \tag{12.3}$$

When the above is included into the formula for the multinomial logistic model, it becomes the stereotype regression model:

$$\Pr(y \mid x) = \frac{\exp(\alpha_j - x' - \phi_j \beta)}{\sum_{j=1}^{k} \exp(\alpha_j - x' - \phi_j \beta)} \tag{12.4}$$

In addition to the above score statistic associated with the levels of y, Anderson also imposed an order on the scores such that:

$$1 = -\phi_1 > -\phi_2 > -\phi_3 > -\phi_k = 0 \tag{12.5}$$

This representation or parameterization of the model is one dimensional, where the score parameters, φ_j, are measures of the distance between levels of y_k. The one-dimensional model is a multinomial logistic model where the parameter estimates are constrained to be parallel, that is, $\beta_j = -\varphi_j\beta$. After modeling the data, we test the score values as to their equality; if we do not reject this hypothesis, then the categories may be collapsed.

For an example, we shall use the **heart01** data, with four-level *killip* as the response, and *anterior* and age levels as the explanatory predictors.

```
. slogit killip anterior age2-age4, nolog

Stereotype logistic regression      Number of obs   =        4503
                                     Wald chi2(4)    =       17.88
Log likelihood =   -3264.808         Prob > chi2     =      0.0013

 ( 1)   [phi1_1]_cons = 1
-------------------------------------------------------------------
   killip |      Coef. Std. Err.    z  P>|z| [95% Conf. Interval]
----------+--------------------------------------------------------
 anterior |    .729973 .1892416  3.86 0.000    .3590663    1.10088
     age2 |   .6840726 .2076392  3.29 0.001    .2771073   1.091038
     age3 |   1.000138 .2553464  3.92 0.000    .4996685   1.500608
     age4 |    1.45963 .3690928  3.95 0.000    .7362216   2.183039
----------+--------------------------------------------------------
  /phi1_1 |          1         .     .     .           .          .
  /phi1_2 |   .5382339 .1216708  4.42 0.000    .2997635   .7767043
  /phi1_3 |  -.1527598 .2942203 -0.52 0.604   -.7294209   .4239013
  /phi1_4 |          0 (base outcome)
----------+--------------------------------------------------------
 /thetal  |   5.316561   .33578 15.83 0.000    4.658444   5.974677
 /theta2  |   3.547394 .3390581 10.46 0.000    2.882852   4.211936
 /theta3  |   1.372428 .3712444  3.70 0.000    .6448026   2.100054
 /theta4  |          0 (base outcome)
-------------------------------------------------------------------
(killip=4 is the base outcome)
```

The /phi1_* levels represent the scales as defined above. Note that $-\varphi_1 = 1$ and $-\varphi_4 = 0$.

The scores ϕ_1 and ϕ_2 may be tested to determine if they are statistically different by:

```
. test [phi1_2]_cons  = [phi1_1]_cons

 ( 1)  - [phi1_1]_cons + [phi1_2]_cons = 0

            chi2( 1) =     14.40
          Prob > chi2 =     0.0001
```

The result of 0.0001 indicates that levels 1 and 2 are distinguishable. The same is the case when testing the other levels (not shown). Therefore, it is not statistically justified to collapse any of the response levels.

We may use the likelihood ratio test to come to the same results. This is done by estimating the model, saving the results, defining an equality constraint for any two contiguous levels, and again modeling the data, but with the defined constraints. A likelihood ratio test may then be calculated based on the constrained versus nonconstrained models.

```
. qui slogit killip anterior age2-age4
. est store ncmodel
. constraint define 1 [phi1_1]_cons=[phi1_2]_cons
. qui slogit killip anterior age2-age4, constraint(1)
. lrtest ncmodel
```

```
Likelihood-ratio test                   LR chi2(1)   =      59.03
(Assumption: . nested in ncmodel)       Prob > chi2 =      0.0000
```

The same test may be given to other levels following the same procedure, but exchanging levels, for example, [phi1_2]_cons = [phi1_3]_cons. The results (not shown) are the same. Both Wald and likelihood ratio tests confirm that it is not justified to collapse any of the contiguous levels of the response.

Other features of the stereotype logistic model go beyond detailed discussion in this text. For example, Anderson (1984) extended the basic one-dimensional stereotype model by constructing a two-dimensional model. One of the uses for this parameterization, where the parameters are defined as:

$$\beta_j = -\phi_j \beta - \varphi_j \gamma \quad \text{for } j = 1, \dots, k \tag{12.6}$$

is also the determination if response levels should be collapsed. A good discussion of this method can be found in Long and Freese (2006).

Another use of the stereotype model is to evaluate if the levels are in their proper order. This is an interesting use of the model, which can be determined by observing the values of the scores. If they do not descend in a sequential order, a re-ordering may be necessary.

The stereotype logistic model can be reparameterized so that estimates are odds ratios, although commercial software does not currently provide for its estimation. To calculate odds ratios for the stereotype model, we express (12.4) for two contiguous response levels, as:

$$\exp\{(\alpha_a - \alpha_b) - (\phi_a - \phi_b)x'\beta\} \tag{12.7}$$

which can be used to estimate the change in odds for a one-unit change in x_k. This is the standard interpretation of a logistic model. It is also understood that other predictors are held constant. The ratio of the odds of levels a to b for a one-unit increase in x_k is given as:

$$(\exp(\phi_a)/\exp(\phi_b))^{-\beta k} \tag{12.8}$$

or

$$\exp(\phi_a - \phi_b)^{-\beta k} \tag{12.9}$$

The score, or scale parameters, must be considered in the calculation of odds ratios, with the exception of the odds of the first level and base level (the default being the last or upper level). In this case, the odds are interpreted as a normal multinomial logit model. This is due to the score values of the two extreme levels as 1 and 0. Therefore,

$$\exp(\phi_a - \phi_b)^{-\beta k} => \tag{12.10}$$

$$\exp(\phi_b - \phi_a)^{\beta k} => \tag{12.11}$$

$$\exp(1-0)^{\beta k} => \tag{12.12}$$

$$e^{\beta k} \tag{12.13}$$

Interpret the odds for other levels as follows: for a one-unit change in x_k, the odds of level a to level b change by $\exp(\phi_b - \phi_a)^{\beta k}$, with the other predictors held constant.

12.4 Heterogeneous Choice Logistic Model

The heterogeneous choice logistic model was developed from the work of Keele and Park (2006), and Williams (2007). As you may recall from the initial chapters of this text, the Bernoulli and the more general binomial probability functions have a single parameter—which is normally referred to as the location parameter. The scale parameter, which is typically found in continuous families of distributions, is assigned a value of one (1). In the exponential family form of probability distribution functions, the scale is designated as $\alpha(\phi)$.

The heterogeneous choice logistic model adds a scale value to the distribution that can take values other than 1. Since both location and scale parameters are estimated, the model is also known as the **location-scale logistic model**. Where we have a standard logistic probability of $y = 1$ equaling:

$$\Pr(y_i = 1|x_i) = \mu_i = g^{-1}(x_i\beta) = 1/(1 + \exp(-x_i\beta)) \tag{12.14}$$

the heterogeneous logistic function appears as:

$$\Pr(y_i = 1|x_i) = \mu_i = g^{-1}(x_i\beta/\sigma_i) = 1/\{\sigma_i(1 + \exp(-x_i\beta))\} \tag{12.15}$$

with σ_i being the scale parameter. We use this model most effectively to adjust the variances in an otherwise proportional-odds model that fails tests of proportionality. That is, the variances existing in the predictors that contribute to nonproportional slopes in a proportional-odds model are adjusted so that proportionality is obtained.

The complete heterogeneous choice probability function may be expressed as (Williams, 2008):

$$\Pr(Y_i = j) = \prod_{j=1}^{M} \prod_{i=1}^{nj} \frac{(1/(1 + \exp(x_i\beta - \kappa_j))) - (1/(1 + \exp(x_i\beta - \kappa_{j-1})))}{\exp(z_i\gamma)}, j = 1, 2, ..., M \tag{12.16}$$

where $\exp(z_i\gamma) = \exp(\ln(\sigma_i)) = \sigma_i$ and $\kappa_0 = -\infty$ and $\kappa_M = \infty$.

The numerator in the above equation is the choice function whereas the denominator is the variance. In traditional probabilistic terms, these are the location and scale functions, respectively. Note should also be given that the choice equation has an intercept; the variance does not. κ's indicate the cut points, which may be recast to intercepts by adding them to the linear predictor, $x_i\beta$, rather than by subtracting. As it is, the cut points and intercepts have the same value, but opposite signs.

For an example, use data from the 1977/1989 General Social Survey, which can be found in Long and Freese (2006) and Williams (2007). The data, called **ordwarm2**, include a number of variables. Those of interest are

RESPONSE

warm   A 4-level choice variable answering the question, "Can a working mother have the same warm and secure relationship with her children as a mother who does not work?"

1: strongly disagree	(297)	3: agree	(856)
2: disagree	(723)	4: strongly agree	(417)

PREDICTORS

```
yr89    year of survey    0 = 1977    (1379)  1 = 1989   (914)
male    gender            0 = female  (1227)  1 = male  (1066)
white   ethnic identity   0 = nonwhite (283)  1 = white (2010)
age     age               18-89
ed      highest yr educ   0-20
```

We shall categorize the two continuous predictors: *age* into 4 levels, and *education* into five levels.

```
. use ordwarm2

. gen byte agegrp=1 if age<=30
. replace agegrp=2 if age>30 & age<=45
. replace agegrp=3 if age>45 & age<=60
. replace agegrp=4 if age>60
. lab define agegrp 1 "<=30" 2 ">30&<=45" 3 ">45&<=60" 4 ">60"
. lab values agegrp agegrp

. tab agegrp, gen(agegrp)
```

agegrp	Freq.	Percent	Cum.
<=30	565	24.64	24.64
>30&<=45	706	30.79	55.43
>45&<=60	544	23.72	79.15
>60	478	20.85	100.00
Total	2,293	100.00	

```
. gen byte educ=1 if ed<12
. replace educ=2 if ed==12
. replace educ=3 if ed>12 &ed<16
. replace educ=4 if ed==16
. replace educ=5 if ed>16
. lab define educ 1 "not HS grad" 2 "HS Grad" 3 "some college"
4 "Bachelor" 5 "Grad School"
. lab values educ educ
. tab educ, gen(educ)
```

educ	Freq.	Percent	Cum.
not HS grad	670	29.22	29.22
HS Grad	782	34.10	63.32
some college	449	19.58	82.90
Bachelor	210	9.16	92.06
Grad School	182	7.94	100.00
Total	2,293	100.00	

We keep the new categorical variables and resultant indicator variables, together with the binary predictors specified for the model. The data are saved as **ordwarmjh**.

```
. keep warm yr89 male white age agegrp1-agegrp4 ed educ1-educ5
. save ordwarmjh
```

The data are modeled as a proportional odds model, then evaluated to determine if its slopes are proportional. The model results are stored in memory for a future likelihood ratio test.

```
. ologit warm yr89 male white agegrp3-agegrp4 educ3-educ5, nolog
```

Ordered logistic regression				Number of obs	=	2293
				LR chi2(8)	=	303.06
				Prob > chi2	=	0.0000
Log likelihood = -2844.2397				Pseudo R2	=	0.0506

warm	Coef.	Std. Err.	z	P>|z|	[95% Conf.	Interval]
yr89	.5229807	.0798988	6.55	0.000	.3663819	.6795796
male	-.755782	.0788281	-9.59	0.000	-.9102823	-.6012817
white	-.3161896	.1172796	-2.70	0.007	-.5460534	-.0863259
agegrp3	-.4857694	.0963016	-5.04	0.000	-.674517	-.2970217
agegrp4	-.9425593	.1019109	-9.25	0.000	-1.142301	-.7428177
educ3	.4711666	.1022163	4.61	0.000	.2708263	.6715068
educ4	.4340494	.1338825	3.24	0.001	.1716445	.6964544
educ5	.778255	.147749	5.27	0.000	.4886722	1.067838
/cut1	-2.616874	.1415261			-2.894261	-2.339488
/cut2	-.787485	.1305152			-1.04329	-.5316798
/cut3	1.108769	.1316708			8506986	1.366839

```
. estimates store propodd
. brant
```

```
Brant Test of Parallel Regression Assumption
```

Variable	chi2	p>chi2	df	
All	48.78	0.000	16	
yr89	13.72	0.001	2	*
male	21.19	0.000	2	*
white	1.23	0.542	2	
agegrp3	1.05	0.592	2	
agegrp4	2.82	0.244	2	
educ3	0.23	0.889	2	
educ4	3.86	0.145	2	
educ5	1.35	0.509	2	

yr89 and *male* have significant *p*-values at the <0.05 level. The Brant test appears to demonstrate that the assumption of proportionality has been violated. These two predictors are then entered into a heterogeneous choice logistic model as variance adjusters. The model is given as:

```
. oglm warm yr89 male white agegrp3-agegrp4 educ3-educ5,
het(yr89 male) store(hclm) hc

Heteroskedastic Ordered Logistic Regression
                                    Number of obs   =       2293
                                    LR chi2(10)     =     331.59
                                    Prob > chi2     =     0.0000
Log likelihood = -2829.9731         Pseudo R2       =     0.0553
-----------------------------------------------------------------
      warm |      Coef. Std. Err.    z  P>|z| [95% Conf. Interval]
-----------+-----------------------------------------------------
choice     |
      yr89 |   .4516254 .0689427   6.55 0.000   .3165002  .5867507
      male |  -.6543821 .0704044  -9.29 0.000  -.7923721 -.5163921
     white |  -.2506121  .101772  -2.46 0.014  -.4500816 -.0511426
   agegrp3 |  -.4254286 .0838425  -5.07 0.000  -.5897569 -.2611004
   agegrp4 |  -.8033565 .0908578  -8.84 0.000  -.9814345 -.6252785
     educ3 |   .3823991 .0886468   4.31 0.000   .2086546  .5561435
     educ4 |   .3307407 .1158752   2.85 0.004   .1036294   .557852
     educ5 |   .6303187 .1259016   5.01 0.000   .3835561  .8770814
-----------+-----------------------------------------------------
variance   |
      yr89 |  -.1459944 .0459932  -3.17 0.002  -.2361393 -.0558494
      male |   -.188151 .0448299  -4.20 0.000  -.2760159 -.1002861
-----------+-----------------------------------------------------
     /cut1 |  -2.255561 .1399025 -16.12 0.000 -2.529765 -1.981357
     /cut2 |  -.6750258 .1165385  -5.79 0.000  -.903437 -.4466147
     /cut3 |   .9679339 .1185224   8.17 0.000  .7356343  1.200233
-----------------------------------------------------------------
```

The proportional-odds model results were stored by the name **propodd**; the heterogeneous choice logit model results were stored as **hclm**. Submitting these stored names to a likelihood ratio test results in:

```
. lrtest propodd hclm, stats force

Likelihood-ratio test                    LR chi2(2)   =      28.53
(Assumption: propodd nested in hclm)     Prob > chi2 =     0.0000
-----------------------------------------------------------------
    Model |  Obs  ll(null)  ll(model)  df       AIC        BIC
----------+------------------------------------------------------
  propodd | 2293  -2995.77   -2844.24  11   5710.479   5773.593
     hclm | 2293  -2995.77  -2829.973  13   5685.946   5760.535
-----------------------------------------------------------------
```

The **heterogeneous choice logistic model** (**hclm**) has lower values for both the AIC and BIC statistics than does the proportional-odds model, which was estimated using **ologit**.

This result indicates that the heterogeneous choice model is preferred. Moreover, since the difference in BIC statistics exceeds a value of 10, we may conclude that the heterogeneous model is *very strongly preferred* over the proportional-odds model.

Variations of the heterogeneous choice model have appeared in statistical literature. For example, Hauser and Andrew's (2006) *Logistic Response Model with Proportionality Constraints* reconstructs the model so that multiplicative scalars are defined in estimation to express a proportional change in the effect of certain predictors across changes in response levels. The scalars express the effect of choosing one level from a contiguous level. The relationship of link and linear predictor is given by the two authors as:

$$\ln(\mu_{ij}/(1 - (\mu_{ij})) = \beta_{j0} + \lambda_j \sum_{k=1}(\beta_k X_{ijk}) \tag{12.17}$$

The same authors also proposed another model, called *Logistic Response Model with Partial Proportionality Constraints*. This model has greater generality than their first model, analogous to the relationship of partial proportional odds and proportional odds. The link-linear predictor relationship is expressed as:

$$\ln(\mu_{ij}/(1 - (\mu_{ij})) = \beta_{j0} + \lambda_j \sum_{k=1}^{k'}(\beta_k X_{ijk}) + \sum_{k'+1}^{k}(\beta_{jk} X_{ijk}) \tag{12.18}$$

Alternatives exist to these models as well. A more detailed account of the heterogeneous choice model can be found in Williams (2007) and at http://www.nd.edu/~rwilliam/oglm/oglm_Stata.pdf.

12.5 Adjacent Category Logistic Model

The adjacent category logistic model, developed by Goodman (1978), has the following representation:

$$\ln[\Pr(Y = y_j \mid x) / \Pr(Y = y_{j+1} \mid x)] = \alpha_j - x'\beta_j, \quad j = 1,\ldots k \tag{12.19}$$

where the highest values of α and β, i.e., α_k and β_k, are 0. The parameters β_1, β_2, and so forth, are the respective parameter estimates for the log-odds of $(Y = y_1)$ to $(Y = y_2)$, the log-odds of $(Y = y_2)$ to $(Y = y_3)$, and so forth until the final category.

Exponentiating the parameter estimates result in odds ratios of $(Y = y_j)$ relative to $(Y = y_{j+1})$ for each 1-unit increase in the value of x.

The model has also been expressed with the greater level in the numerator (Manor, Matthews, and Power, 2000).

$$\ln\left[\Pr(Y = y_j \mid x) / \Pr(Y = y_{j-1} \mid x)\right] = \alpha_j - x'\beta_j, \quad j = 1,\dots k \quad (12.20)$$

Use General Social Survey data with a four-level measure of satisfaction as the response, designating level 4 as the reference level.

Satisfaction:	1: Very Dissatisfied	3: Moderately Satisfied
	2: A Little Satisfied	4: Very Satisfied

Predictors:

Income	1: <5,000	2: 5,000–15,000	3: 15,000–25,000	4: >25,000
Female	1: Male	0: Female		

LogXact 8, manufactured by Cytel, was used to model the data. The data and model come from Agresti (2002) and are used in the LogXact reference manual.

```
Model       satisfact=%Const+income+gender
Response    Adjacent Category Model: Categories(1, 2, 3, 4)
            Reference: 4
Link type  Logit
```

Number of observations in analysis	104
Number of records rejected	6
Number of groups	8

Summary Statistics

Statistics	Value	DF	*P*-Value
-2 loglikelihood	12.55	19	0.8608
Likelihood Ratio	81.69	5	7.325e-010

						Point Estimate		Confidence Interval and *P*-Value for Beta		
Model	Term	Category	Type	Beta	SE(Beta)			95	%CI	2*1-sided
								Lower	Upper	P-Value
%Const	satisfact=1		MLE	−0.5507	0.6795			−1.882	0.7811	0.4177
%Const	satisfact=2		MLE	−0.655	0.5253			−1.685	0.3745	0.2124
%Const	satisfact=3		MLE	2.026	0.5758			0.8974	3.155	0.0004344
income			MLE	−0.3888	0.1547			−0.6919	−0.08563	0.01195
gender			MLE	0.04469	0.3144			−0.5716	0.661	0.887

Income appears to significantly contribute to the model ($p = .012$), whereas gender does not ($p = .887$). For a unit increase in income, such as from

$5,000–$15,000 to $15,000–$25,000, there is exp(–.3888) = .67787 increase in the odds of choosing *satisfact==3* (moderately satisfied) compared to *satisfact==2* (a little satisfied). We may invert this to conclude that for each unit increase in income group, there is a 32 percent decrease in the odds of being a little satisfied compared to moderately satisfied.

The primary feature of the model is in comparing adjacent levels of categories of the response, rather than comparing a given level to a referent. It is most appropriate when the statistical interest is in interpreting individual level effects rather than on the cumulative level effects. Like the continuation ratio logistic model, the adjacent categories model can be applied to highly unbalanced data, as well as to data with a small sample size.

12.6 Proportional Slopes Models

In this section, we shall present a brief overview of proportional slopes models, which includes the proportional slopes logit model, otherwise known as the proportional-odds model or ordered logistic regression. Here we consider the ordered logistic or ordered logit model to be but one link among other possible ordered binomial links, analogous to the standard binomial links of generalized linear models.

Recall that the ordered logistic model assumes that the slopes of the various response levels are proportional. For the logistic model, an exponentiated slope, or model coefficient, is an odds ratio—hence the designation proportional odds model. Odds ratios, however, are not appropriate for nonlogistic binomial models. For example, exponentiating a probit model coefficient does not produce an odds ratio. These models are therefore simply referred to as proportional slopes models, or ordered binomial models.

The standard binomial links, unordered as well as ordered, are enumerated in Table 12.1. A nonstandard link, the Cauchit, is also shown. It is a link

TABLE 12.1

Proportional Slopes Model Link Functions

BINOMIAL LINK	FUNCTION	APPLICATION
Logit	$\ln(\mu/(1-\mu))$	Proportional odds
Probit	$\Phi^{-1}(\mu)$	y^* normally distributed
Cloglog	$\ln(-\ln(1-\mu))$	Higher levels more likely
Loglog	$-\ln(-\ln(\mu))$	Lower levels more likely
Cauchit	$\tan(\pi(\mu-0.5))$	y_j with many extreme values

available for SPSS-ordered binomial models and is also found in Stata's user-authored **oglm** command.

It should be noted that, at the time of this writing, SPSS has interchanged the traditionally defined link and inverse link functions between the complementary loglog (cloglog) and loglog (which SPSS terms negative loglog) links. The output displayed for one is in fact appropriate for the other. SPSS developers are aware of the discrepancy and have indicated that they will correct this error in a future release. SPSS versions 15 and 16, however, are subject to the error.

Also we should note that proportional slopes models are best used when the data are balanced and the number of observations is not small. This is in contrast to continuation ratio and adjacent categories models that do well with unbalanced data and smaller data sets.

We shall first discuss the proportional slope algorithms, followed by modeling with synthetic data. Finally, we discuss testing methods of the assumption of proportionality. The Brant test is only applicable for the ordered logit, or proportional odds model.

12.6.1 Proportional Slopes Comparative Algorithms

Proportional slopes models assume that the betas (β) are identical for all categories or levels of y. For all of the links, the linear predictor, $x\beta$, across levels is defined as:

$$x\beta_j = x\beta - k_j \tag{12.21}$$

with $i = j - 1$.

It must be emphasized that some software applications reverse the relationship of $x\beta$ and k_j that is expressed in Equation 12.19, that is, at times we find:

$$x\beta_j = \kappa_j - x\beta \tag{12.22'}$$

We shall base the remainder of our discussion on Equation 12.20. In fact, our previous discussion of the proportional odds model was based on 12.20 as well. I should also mention that all of the ordered response models we have previously discussed have at times been parameterized in such a reverse manner. Care must be taken when comparing results between applications.

The estimating algorithms for the logit and probit linked proportional slopes models are symmetric, which makes the calculations a bit easier. The complementary loglog, loglog, and Cauchit are nonsymmetrical. The inverse link functions, or cumulative distribution functions, of each link are used to estimate the log-likelihood function, fitted values, and cut points for the various models. We use the left side of Equation 12.20 to designate the linear predictor in the algorithms that follow.

For the logit and probit links, the inverse links are, respectively:

LOGIT: $\text{invlogit}(x\beta) = 1/(1 + \exp(-x\beta))$

or

$\text{invlogit}(x\beta) = \exp(x\beta)/(1 + \exp(x\beta))$

PROBIT: $\text{norm}(z) = $ cumulative standard normal distribution

where $\text{invlogit}()x)$ and $\text{norm}(z)$ are standard Stata functions. Similar functions exist in nearly all commercial statistical software applications.

Inverse link functions, or CDFs, for the other three links are, respectively:

COMPLEMENTARY LOGLOG: $1 - \exp(-\exp(x\beta))$
LOGLOG   : $\exp(-\exp(-x\beta))$
CAUCHIT   : $0.5 + (1/\pi) * \text{atan}(-x\beta)$

The ordered logit linear predictor and log-likelihood functions may be schematized for each proportional level as follows.

ORDERED LOGIT

Initialize
{
  y (Level 1)
    $x\beta_1 = (x\beta - \kappa_1)/\sigma$
    $\text{LL} = \ln(1/(1 + \exp(x\beta_1)))$

  y(Levels 2–(M–1))
    for $j = 2/M$ {
    $i = j - 1$
    $x\beta_j = (x\beta - \kappa_j)/\sigma$
    $x\beta_i = (x\beta - \kappa_i)/\sigma$
    $\text{LL} = \ln(1/(1 + \exp(x\beta_j))) - ln(1/(1 + \exp(x\beta_i)))$
    }

  y(Level M)
    $x\beta_j = (x\beta - \kappa_M)/\sigma$
    $\text{LL} = \ln(1/(1 + \exp(-x\beta_1)))$
}

The Cauchit linear predictor and log-likelihood are schematized as:

ORDERED CAUCHIT

```
{
   y (Level 1)
      xβ₁ = (xβ − κ₁)/σ
      LL = ln(.5 + (1/π) * atan(−xβ₁))

   y(Levels 2−(M−1))
      for j = 2/M {
      i = j− 1
      xβⱼ = (xβ − κⱼ)/σ
      xβᵢ = (xβ − κᵢ)/σ
      LL = ln(.5 + (1/π) * atan(−xβⱼ)) − (.5 + (1/π) * atan(−xβᵢ))
      }

   y(Level M)
      xβⱼ = (xβ − κ_M)/σ
      LL = ln(.5 + (1/π) * atan(−xβⱼ))
}
```

The Cauchit linear predictor and log-likelihood are schematized as:

$xβ_1 = (xβ − κ_1)/σ$

$LL = \ln(.5 + (1/π) * \text{atan}(−xβ_1))$

$xβ_j = (xβ − κ_j)/σ$

$xβ_i = (xβ − κ_i)/σ$

$LL = \ln(.5 + (1/π) * \text{atan}(−xβ_j)) − (.5 + (1/π) * \text{atan}(−xβ_i))$

$xβ_j = (xβ − κ_M)/σ$

$LL = \ln(.5 + (1/π) * \text{atan}(−xβ_j))$

The logic of the formulae is obvious, with the major difference between links being the various inverse link functions, or CDFs.

12.6.2 Modeling Synthetic Data

Code for creating a synthetic ordered logit model was provided in Section 10.1. A single line amendment to the algorithm was also given so that a synthetic ordered probit model could be created. The algorithm allows the user to define parameter estimates and cut points. For completeness, I repeat the ordered logit and probit key functions, that is, the lines that alter the output for ordered logit, ordered probit, ordered cloglog, and ordered loglog.

C is defined as code generating the user-defined coefficient values for variables $x1$ and $x2$, .75 and 1.25, respectively. Because the code represented by C is the same for all four models, it is used as the first part of the defining function for each model. An *err* function is used for all but the probit algorithm. It is defined below in Stata code. Again, the complete algorithm is located in Section 10.1. It should be relatively simple to convert this code into the higher level language of all the major statistical packages having programming capabilities.

```
C : gen y = .75*x1 + 1.25*x2
gen err = uniform()
```

LOGIT	:	$C + \ln(err/(1-err))$
PROBIT	:	$C + invnorm(uniform())$
COMP. LOGLOG:		$C - \ln(-\ln(1-err))$
LOGLOG	:	$C + \ln(-\ln(err))$

The code displayed in Section 10.1 may be amended so that a 50,000 observation simulated complementary loglog data set is created, with three parameter estimates and four cut points—one more each than was created earlier for the ordered logit model. Parameter estimates are defined as $x1 = .75$, $x2 = 1.25$, $x3 = -1.0$; cut points at each higher level are given as 2, 3, 4, and 5. The **do**-file is called **syn_oclog3.do**,

```
. do syn_oclog3

. * STATA DO-FILE CREATING AN ORDERED CLOGLOG MODEL WITH 3
DEFINED COEFFICIENTS
. * Hilbe, Joseph 29Apr 2008
. qui {

Ordered Cloglog Regression         Number of obs    =      50000
                                   LR chi2(3)       =   17793.59
                                   Prob > chi2      =     0.0000
Log likelihood = -52976.227        Pseudo R2        =     0.1438
------------------------------------------------------------------
     ys |      Coef. Std. Err.     z   P>|z|   [95% Conf. Interval]
--------+---------------------------------------------------------
     x1 |   .7476834 .0084458  88.53 0.000    .7311299    .7642368
     x2 |   1.249428 .0129584  96.42 0.000    1.22403    1.274826
     x3 |  -1.026288 .0237506 -43.21 0.000  -1.072838   -.9797379
--------+---------------------------------------------------------
  /cut1 |   1.985241 .0263626  75.31 0.000   1.933571    2.036911
  /cut2 |   2.986135 .0282326 105.77 0.000   2.930801     3.04147
  /cut3 |   3.971924 .0305436 130.04 0.000   3.912059    4.031788
  /cut4 |   4.969817 .0351323 141.46 0.000   4.900959    5.038675
------------------------------------------------------------------
```

Modeling the synthetic data results in parameter estimates and cut points that closely approximate the assigned values. Given that the data are randomly generated, deviations from the user-defined values are expected. $x1$ differs by .0023, $x2$ by .00057, and $x3$ by .026. Differences in cut values are: $cut1 = .0148$, $cut2 = .0139$, $cut3 = .0281$, and $cut4 = .0302$. Other runs will produce different values and deviations, centering around the user-defined values. We have run the algorithm 1,000 times, saving and taking the mean value of each parameter estimate and cutpoint. The deviations closely approximate 0.000.

In Chapter 9, a number of synthetic data sets were created with the purpose of showing the impact of leaving out an important predictor from the model, the impact of not creating an interaction when the data require it, and so forth. Here we compare modeling the data as an ordered logit, when in fact the data are ordered complementary loglog.

```
. ologit ys x1 x2 x3, nolog

Ordered logistic regression        Number of obs   =      50000
                                   LR chi2(3)      =   17031.21
                                   Prob > chi2     =     0.0000
Log likelihood = -53357.422        Pseudo R2       =     0.1376
------------------------------------------------------------------
     ys |     Coef. Std. Err.     z   P>|z|   [95% Conf. Interval]
--------+---------------------------------------------------------
     x1 |  .9561261 .0114708   83.35 0.000   .9336439    .9786084
     x2 |  1.594284 .0175299   90.95 0.000   1.559926    1.628642
     x3 | -1.316743 .0323959  -40.65 0.000  -1.380237   -1.253248
--------+---------------------------------------------------------
  /cut1 |  1.994816 .0341723                 1.92784    2.061793
  /cut2 |  3.317288 .0365041                 3.245742    3.388835
  /cut3 |  4.487771 .0391853                 4.410969    4.564573
  /cut4 |  5.574835 .0433442                 5.489882    5.659788
------------------------------------------------------------------
```

Note that the coefficients of the above-ordered logit model are substantially different from the "true" parameter. x1 and x2 are both some 27.5 percent greater than the true values (.956/.75; 1.594/1.25), and x3 is 31.7 percent greater. Cut points after the first are 11 percent, 12 percent, and 12 percent greater than the true values.

The caveat here is making certain to check the appropriateness of the link when modeling ordered binomial models, or any generalized linear models based model. A problem may exist, however, when one's software package does not have the appropriate model by which to estimate the data at hand. Even if a researcher does have a statistical package with the capability of appropriately modeling a set of data, it is important to actually search for the best model.

A more extreme example can be given when modeling ordered probit data with an ordered logit estimating algorithm. We assign a two-coefficient ordered probit synthetic model generator the values of *Beta1* = .50 and *Beta2* = –1.75. Cut points are given at 2, 3, and 4. The header script and statistics have been deleted.

ORDERED PROBIT

```
------------------------------------------------------------------
     ys |     Coef. Std. Err.     z   P>|z|   [95% Conf. Interval]
--------+---------------------------------------------------------
     x1 |  .4948366 .0076362   64.80 0.000   .4798701    .5098032
     x2 | -1.745119 .0140419 -124.28 0.000  -1.772641   -1.717598
--------+---------------------------------------------------------
```

```
/cut1 |   1.996893 .0221207                   1.953537   2.040249
/cut2 |   2.994247 .0242958                   2.946628   3.041866
/cut3 |   4.010731 .0280926                   3.95567    4.065791
------------------------------------------------------------------
b1 = .50; b2 = -1.75
Cut1=2; Cut2=3; Cut3=4

. estat ic

------------------------------------------------------------------
Model |   Obs  ll(null)  ll(model)   df        AIC        BIC
------+-----------------------------------------------------------
    . | 50000  -44729.8  -33047.35    5   66104.69   66148.79
------------------------------------------------------------------
```

Modeling the true ordered probit data by an ordered logit model results in:

ORDERED LOGIT

```
------------------------------------------------------------------
   ys |    Coef. Std. Err.     z  P>|z| [95% Conf. Interval]
------+-----------------------------------------------------------
   x1 |  .8778159 .0137748  63.73 0.000  .8508178    .904814
   x2 | -3.089639  .026903 -114.84 0.000 -3.142368   -3.03691
------+-----------------------------------------------------------
/cut1 |  3.522202 .0407767                   3.442281   3.602123
/cut2 |  5.258908 .0459671                   5.168814   5.349002
/cut3 |   7.12476 .0543069                    7.01832     7.2312
------------------------------------------------------------------

. estat ic

------------------------------------------------------------------
Model |   Obs  ll(null) ll(model)   df        AIC        BIC
------+-----------------------------------------------------------
    . | 50000  -44729.8 -33149.81    5   66309.62   66353.72
------------------------------------------------------------------
```

The differences in parameter estimates and cut points are significant. As expected, the respective AIC and BIC statistics indicate that the ordered probit is the preferred model.

We next address tests of proportionality for proportional slopes models.

12.6.3 Tests of Proportionality

The standard method of evaluating the proportionality of slopes for the proportional odds model is the Brant test. We discussed the test at some length in Sections 10.1 and 10.2. However, the Brant test is not appropriate for ordered binomial models other than the proportional odds model. Fortunately, an "approximate" likelihood ratio test has been used with evident success for evaluating non-logit proportionality. The test statistic is the *Chi2* distribution. A likelihood ratio test *p*-value, based on the *Chi2* statistic, tests the proportionality of slopes across all response levels, or categories. Unlike the Brant test, which provides proportionality information for each predictor in the model, this likelihood ratio test provides a single *p*-value for all predictors taken together. It does not, therefore, have the discriminating power of Brant, but it is nonetheless a valid method for evaluating if the proportionality assumption has been violated.

We shall evaluate the power of the Brant test to check the proportionality of slopes of an ordered-logit model, where there is a clear indication that there has been a violation of the IIA assumption. We create a synthetic-ordered logit model as before, with two coefficients assigned true parameter values of .75 and 1.25, and cut points of 2, 3, and 4.

```
. do syn_ologit

. * DO FILE FOR CREATING AN ORDERED LOGIT MODEL WITH DEFINED
COEFFICIENTS
. * Hilbe, Joseph 15Feb 2008
. qui {
Ordered logistic regression          Number of obs   =      50000
                                      LR chi2(2)      =   10929.44
                                      Prob > chi2     =     0.0000
Log likelihood = -54852.219           Pseudo R2       =     0.0906
--------------------------------------------------------------------
     ys |     Coef.  Std. Err.     z   P>|z|    [95% Conf. Interval]
--------+-----------------------------------------------------------
     x1 | .7529516  .0107529   70.02  0.000    .7318764    .7740269
     x2 |  1.26078  .0163425   77.15  0.000    1.228749    1.292811
--------+-----------------------------------------------------------
  /cut1 | 2.004859  .0292476                    1.947535    2.062184
  /cut2 | 2.994769  .0309955                    2.934019    3.055519
  /cut3 | 4.018492  .0333522                    3.953123    4.083861
--------------------------------------------------------------------
```

A Brant test is applied to the model, with the expectation that the slopes are all proportional. In fact, they have been defined to be proportional.

```
. brant

Brant Test of Parallel Regression Assumption

    Variable |    chi2   p>chi2     df
-------------+-----------------------------
         All |    1.19    0.879      4
-------------+-----------------------------
          x1 |    0.99    0.610      2
          x2 |    0.35    0.841      2
-------------------------------------------
```

A significant test statistic provides evidence that the
parallel regression assumption has been violated.

Nonsignificant *p>chi2* values at the 0.05 level indicate that the predictor
supports proportionality of the slopes—as expected.

We may apply the likelihood ratio test, but only after modeling the data
by a more general ordered binomial model than **oglm**. The command we
use, **omodel**, was authored by Rory Wolfe of the Royal Children's Hospital,
Melbourne, Australia, and William Gould, CEO of Stata Corporation (Wolfe
and Gould, 1998). Stata users may have to reset their memory; the algorithm
creates a large internal matrix, which may exceed the default setting of the
host application. Typing *set memory 40M* on the command line will solve the
problem.

```
. omodel logit ys x1 x2, nolog

Ordered logit estimates              Number of obs    =        50000
                                     LR chi2(2)       =     10929.44
                                     Prob > chi2      =       0.0000
Log likelihood = -54852.219          Pseudo R2        =       0.0906
------------------------------------------------------------------------
    ys |      Coef.  Std. Err.     z    P>|z|    [95% Conf. Interval]
-------+----------------------------------------------------------------
    x1 | .7529516  .0107529   70.02   0.000    .7318764     .7740269
    x2 |  1.26078  .0163425   77.15   0.000    1.228749     1.292811
-------+----------------------------------------------------------------
 _cut1 | 2.004859  .0292476          (Ancillary parameters)
 _cut2 | 2.994769  .0309955
 _cut3 | 4.018492  .0333522
------------------------------------------------------------------------

Approximate likelihood-ratio test of proportionality of odds
across response categories:
          chi2(4) =        1.32
      Prob > chi2 =      0.8572
```

The likelihood ratio test results in the same conclusion as did the Brant test.

What happens if we model the synthetic ordered logit data by an ordered
probit command—and test for the proportionality of slopes? We can do both
by using the **omodel** command.

```
. omodel probit ys x1 x2, nolog

Ordered probit estimates             Number of obs    =        50000
                                     LR chi2(2)       =     10813.57
                                     Prob > chi2      =       0.0000
Log likelihood = -54910.151          Pseudo R2        =       0.0896
```

```
---------------------------------------------------------------
    ys |    Coef. Std. Err.     z    P>|z|   [95% Conf. Interval]
-------+-------------------------------------------------------
    x1 |  .4425041 .0062524  70.77  0.000   .4302496     .4547585
    x2 |  .7424014 .0094596  78.48  0.000    .723861     .7609418
-------+-------------------------------------------------------
 _cut1 | 1.182034 .0169464          (Ancillary parameters)
 _cut2 | 1.773435 .0176178
 _cut3 |  2.35838 .0185571
---------------------------------------------------------------
```

Approximate likelihood-ratio test of equality of coefficients
across response categories:
           chi2(4) =   80.56
        Prob > chi2 =   0.0000

Not only are the parameter estimates substantially different from the true parameter values, but the likelihood ratio test on slope proportionality fails; in other words, the slopes are no longer proportional.

I have used synthetic data here because true parameters may be assigned and modeled to determine the effectiveness of the algorithm in estimating them. Likewise for the slopes; the model is defined such that the slopes are proportional. If a difference in estimates and slope values is discovered when modeling a proportional slopes model with a link other than that which was defined, it reinforces the point that care must be taken to ensure that we are using the appropriate model with the data at hand.

Exercises

12.1. Using the **affairs** data that is on this book's Web site, estimate the same data using the response and predictors employed in Exercise 10.2, with an adjacent categories model.

   a. Estimate the best-fitted model.

   b. Compare the ordered, multinomial, and adjacent categories of *ratemarr* on the same predictors used in Exercise 10.2.

12.2. Using the **edreligion** data set, model the three-level *educlevel* on *religious*, *kids*, *age*, and *male* using a continuation ratio model.

RESPONSE

   educlevel : 1 = AA degree, 2 = BA degree, 3 = MA/PhD degree

PREDICTORS

religious	binary – 1 = religious; 0 = not religious
kids	binary – 1 = have kids; 0 = not have kids
male	binary – 1 = male; 0 = female
age	continuous – 9 levels from 17.5 to 57

a. Determine the significant predictors of *educlevel*. Do people who call themselves religious tend to have a higher education? How can you tell?

b. Is there a significant difference in males or females regarding level of education?

c. Categorize *age* into indicator variables, and model instead of employing it as a continuous predictor. Are all levels significant? Further categorize age levels. Model on optimal categorization.

d. Compare the model with a proportional and partial proportional slopes logistic model. Which is the statistically better fitting model? Why?

12.3. Use the data below from Johnson and Albert (1999) relating class grades to SAT math scores. The first grade (G0) is for an Introduction to Statistics I course. The grade (G1) on the second semester Introduction to Statistics course is also given. Grades and scores for 30 students were recorded, with A = 1, B = 2, C = 3, D = 4, and F = 5.

St = Student G1 = Grade G0 = Previous grade
SAT = SAT Math Score

St	G1	SAT	G0	St	G1	SAT	G0	St	G1	SAT	G0
1	4	525	2	11	2	576	2	21	1	599	2
2	4	533	3	12	2	525	1	22	4	517	3
3	2	545	2	13	3	574	1	23	1	649	1
4	4	582	1	14	3	582	4	24	2	584	2
5	3	581	3	15	2	574	3	25	5	463	4
6	2	576	4	16	4	471	2	26	3	591	2
7	3	572	2	17	2	595	2	27	4	488	3
8	1	609	1	18	4	557	3	28	2	563	2
9	3	543	4	19	5	557	1	29	2	553	2
10	3	543	4	20	2	584	1	30	1	549	1

a. Model the data using an adjacent categories model.

b. Interpret the model with the aim of determining if SAT score and the grade in the first-level course are good predictors of grades in the second-level course.

13

Panel Models

13.1 Introduction

The term *panel model* is a generic term for a wide variety of clustered, nested, and longitudinal models. In fact, this chapter could well have been titled *Modeling Clustered Data*. The problem with clustered data is that they are not independent and therefore violate an essential criterion of maximum likelihood theory—that observations in a model are independent of one another. Clustered data, depending on how they are generated, tend to bias standard errors as well as parameter estimates.

Statistical methods designed to deal with panel models serve to adjust clustered data in such a manner that bias is minimized, and perhaps eliminated altogether. Generally, though, the attempt is to minimize the bias resulting from correlation in the data. Each panel or clustered model we address in this chapter is a means to deal with data that have been collected in a certain manner. A model is also employed to reflect what we wish to determine from the data. Keeping these two points in mind will help in selecting the most appropriate model to use for a given data situation.

This chapter presents an overview of three varieties of panel models. First we detail three types of population averaging models: generalized estimating equations (GEE), quasi-least squares (QLS) regression, and alternating logistic regression (ALR). All three are quasi-likelihood models that employ the generalized linear models algorithm to estimate the model mean responses, but adjust the variance-covariance matrix by a so-called working correlation structure reflecting the nature of the data. Alternating logistic regression was designed to accommodate estimating problems with a binary response logit GEE model. It estimates odds ratios using a combination of both GEE and standard logistic regression.

The second variety of panel models discussed is the fixed effects models. Fixed effects can take a number of forms, the basic two of which are unconditional and conditional fixed effects. Unconditional fixed effects models may be estimated using a standard logistic regression algorithm. Conditional logistic fixed effects models are also called conditional logistic regression. This type of model requires that the data be structured in a particular manner. Whereas the unconditional fixed effects method models the various fixed effects in the data, the conditional logit algorithm conditions them out, so that the effects are not themselves directly modeled. Generalizations of the conditional logit model also exist. Of these, we shall overview matched case-control and rank-ordered logistic regression models.

The third variety of panel models discussed is the random and mixed effects logistic models. Specifically, we discuss the logistic random effects model with both normal and beta-binomial random effects, and both random-intercept and random-coefficient mixed-effects models. These latter types of models are many times referred to as multilevel or hierarchical models. It should be noted that fixed, random, and mixed effects models are all subject-specific models, which differentiate them from population averaged models. We discuss the distinction in more detail in the following section.

The models discussed in this chapter are complex, and care must be taken when both constructing and interpreting them. Given that this is a survey text on logistic models, we shall only touch on the basics of estimation. The log-likelihood functions of each type of model will be given, together with details on how the data are structured for analysis. Most importantly, we shall discuss how the respective panel models are interpreted and evaluated as to their comparative worth.

13.2 Generalized Estimating Equations

Generalized estimating equations, more commonly known by the acronym GEE, was first described by Liang and Zeger (1986). The method was designed to enhance models of the generalized linear models (GLM) family. Specifically, logistic GEE models are aimed at modeling correlated data that would otherwise be modeled using standard GLM methods—which includes both binary response and proportional response logistic regression.

The key to understanding GEE models is in identifying the source and structure of the correlation. The source must be identifiable in terms of a cluster or group of items or subjects. For example, data on teaching methods used in schools throughout a large metropolitan area are likely clustered on schools. Teaching methods employed within each school are more likely similar than are teaching methods between schools. There is a correlation effect for within-group or within-subject clusters that violates the basic assumptions of likelihood theory. Recall that likelihood estimation methods require that each observation, or record, in the model is independent of the others. When there is a specific commonality to the items or subjects within each of the various groups in a model, then the correlation resulting from such commonality biases the standard errors, and even the parameter estimates, of the model.

We dealt with methods of adjusting the standard errors of logistic models with correlated data in Sections 5.6 to 5.8 when we discussed scaling, robust variance estimators, and bootstrapping. These were *post hoc* methods of adjusting standard errors to accommodate the extra dispersion in the data. The only method of adjustment we found that allows us to identify the probable source of the extra correlation, in other words, to identify the cluster covariate, is the cluster robust variance estimator.

The other source of correlation, which is algorithmically the same as when identifying a cluster or group variable, comes when dealing with longitudinal data. Suppose that we have data recording patient responses to a given treatment protocol. If each subject, or patient, is tested five times over the course of a year, it is likely that each subject in the study exhibits a more similar range of responses than do other patients in the study. That is, there is more correlation in the responses of a given patient over time than there is in the responses of other patients. Again, the within-subject variance, which likely exceeds the between-subject variance in longitudinal data, needs to be accounted for if we are to minimize the bias in the standard errors (and perhaps estimates) resulting from the within-subject correlation of responses. On the other hand, if the within-subject variance greatly exceeds the between-subject variance, the intra-class correlation coefficient (ICC) will be low, therefore imposing less of a requirement to adjust for biased standard errors.

Generalized estimating equations are a population averaging method of estimation. This differs from both fixed effects and random effects models, which are generally referred to collectively as subject-specific models. In the context of dealing with clusters and within-subject variance for logistic models, population averaging means that odds ratios refer to the average of measures within a panel or cluster. For subject-specific models, odds ratios relate to the individual subject, not to the average of measures. Standard logistic regression is a population averaging method; the odds of death for patients having an anterior rather than an inferior infarct relates to all patients in the study. The distinction between the two methods is important when interpreting estimate results and should become clear as we progress through the chapter.

GEE attempts to accommodate extra correlation in the data resulting from cluster or longitudinal effects. In the following subsection, a brief overview of the derivation and logic of GEE is presented. Subsequent subsections provide a description of GEE correlation structures, and an example GEE logistic model examining how it is interpreted and evaluated. In the final two subsections we look at two methods that have been used as alternatives to the standard GEE method. Both methods were developed to ameliorate certain problem areas with GEE. ALR relates specifically to the GEE logistic model; QLS is a method of handling GEE models that are otherwise not feasible.

13.2.1 GEE: Overview of GEE Theory

GEE models are an extension of generalized linear models (GLM), in which a specific type of correlation structure is incorporated into the variance function. The structure adjusts the model variance in such a manner as to minimize bias resulting from extra correlation in the data. Again, this correlation comes from the clustering effect of having the data represent groups of subjects or items, or because the data were collected on subjects over time.

Equation 13.1 below is the standard formula for the log-likelihood function of the single parameter exponential family of distributions, but with an extra summation to account for clusters. The inner summation calculates the log-likelihood function for each individual in the model. The outer summation sums over clusters, which may be subjects, each of which has a series of measures over time. In any case, Equation 13.1 is the panel model parameterization of the exponential distribution, which includes the binomial.

$$L(\mu_{ij}; y_{ij}, \phi) = \sum_{i=1}^{n} \sum_{i=j}^{n_i} \left\{ \frac{y_{ij}\theta_{ij} - b(\theta_{ij})}{\alpha(\phi)} + C(y_{ij}; \phi) \right\} \tag{13.1}$$

You may recall from Chapter 4 that the following relationships obtain:

θ_i is the canonical parameter or link function.

$b(\theta_i)$ is the cumulant, $b'(\theta_i) = $ mean; $b''(\theta_i) = $ variance.

$\alpha(\phi)$ is the scale parameter, set to 1 for discrete and count models.

$C(y_i; \phi)$ is the normalization term, guaranteeing that the probability function sums to unity.

The estimation of the log-likelihood function is based on its derivative, taken to 0.

$$\left\{ \frac{\partial L}{\partial \beta_j} = \sum_{i=1}^{n} \sum_{j=1}^{n_i} \frac{y_{ij} - \mu_{ij}}{\alpha(\phi)V(\mu_{ij})} \left(\frac{\partial \mu}{\partial \eta} \right)_{ij} x_{ij} \right\} = 0 \tag{13.1a}$$

Written in matrix form, in terms of panels, j, Equation 13.1a appears as:

$$\left\{ \sum_{i=1}^{n} x_{ji} D\left(\frac{\partial \mu}{\partial \eta} \right)_i (V(\mu_i))^{-1} \left(\frac{y_i - \mu_i}{\alpha(\phi)} \right) \right\}_{j=1...n} = 0 \tag{13.2}$$

where $D()$ denotes a diagonal matrix. $V(\mu_i)$ is also a diagonal matrix that is normally decomposed as:

$$V(\mu_i) = [D(V(\mu_{ij}))^{1/2} I_{(n_i \times n_i)} D(V(\mu_{ij}))^{1/2}] n_i \times n_i \tag{13.3}$$

Note that the variance function that is employed in standard logistic regression, $V(\mu_i)$, has been bifurcated so that an $n \times n$ correlation matrix can be inserted between two instances of the function. The square matrix I is the identity matrix, with 1s along the diagonal and 0s elsewhere. This gives us the same value as $V(\mu_i)$, but in square matrix format. The relationship is termed the independence correlation structure, and is the pooled form of panel model.

$$
\begin{array}{ccc}
\sqrt{V(\mu_i)} & \begin{matrix} 1 & 0 & 0 & 0 \end{matrix} & \sqrt{V(\mu_i)} \\
\sqrt{V(\mu_i)} & \begin{matrix} 0 & 1 & 0 & 0 \end{matrix} & \sqrt{V(\mu_i)} \\
\sqrt{V(\mu_i)} & \begin{matrix} 0 & 0 & 1 & 0 \end{matrix} & \sqrt{V(\mu_i)} \\
\sqrt{V(\mu_i)} & \begin{matrix} 0 & 0 & 0 & 1 \end{matrix} & \sqrt{V(\mu_i)}
\end{array}
$$

A key notion to GEE is the replacement of the I matrix with a more meaningful correlation structure. Traditionally symbolized as $R(a)$, there are a number of structures used by standard GEE software applications, including the exchangeable, m-dependent, autoregressive, unstructured, and others. The nature of the various traditional GEE correlation structures, which the developers of GEE call working correlation structures, will be detailed in the following subsection.

Equation 13.3 may be expressed as:

$$V(\mu_i) = [D(V(\mu_{ij}))^{1/2} R(\alpha)(n_i \times n_i) D(V(\mu_{ij}))^{1/2}]n_i \times n_i \qquad (13.4)$$

when a non-identity correlation structure is inserted into the variance equation. The above equation may be amended so that weights and a scale statistic can be made part of the variance. This formulation appears as:

$$V(\mu_i) = [\phi D(V(\mu_{ij}))^{1/2} W^{-1/2} R(\alpha)(n_i \times n_i) W^{-1/2} D(V(\mu_{ij}))^{1/2}]n_i \times n_i \qquad (13.5)$$

The values of the various correlation structures are based on Pearson residuals, which we defined in Chapter 7. They are defined as the raw residual divided by the square root of the variance. Residuals are summed across subjects within panels, and then summed across panels:

$$r_{ij} = \sum_{i=1}^{n} \sum_{j=1}^{ni} \frac{y_{ij} - \mu_{ij}}{\sqrt{V(\mu_i)}} \qquad (13.6)$$

The logistic regression Pearson residual, with a Bernoulli model variance function of $\mu(1 - \mu)$, is expressed as:

$$r_{ij} = \sum_{i=1}^{n} \sum_{j=1}^{ni} \frac{y_{ij} - \mu_{ij}}{\sqrt{[\mu_{ij}(1 - \mu_{ij})]}} \qquad (13.7)$$

FITTING ALGORITHM

GEE is a quasi-likelihood model. The fact that the likelihood has been amended by the expanded variance means that there is no true probability function from which the GEE likelihood function can be directly derived. Quasi-likelihood theory has demonstrated that models based on such likelihoods can produce unbiased estimates and standard errors (Wedderburn, 1974).

In any event, the GEE algorithm progresses by first estimating a standard GLM model, such as a logistic regression. Actually, the first iteration uses a variance function with an independence correlation structure, I. At the end of the first iteration values of μ, η, $V(\mu)$, and r are calculated. Depending on the selected type of model, the values of the correlation structure are calculated from the residuals. Each correlation structure employs the residuals in a different manner.

On the second iteration, the calculated correlation structure is sandwiched into the variance function, replacing the original I. Each subsequent iteration uses the updated values of μ, η, $V(\mu)$, r, and $R(\alpha)$, until convergence is achieved.

Note that some software applications use $x\beta$ rather than μ for maximizing the quasi-likelihood function. Recall that for the Bernoulli logistic model, $\mu = 1/(1 + \exp(-xb))$).

The estimation of β as a quasi-maximum likelihood model is based on the following:

$$\beta_{j+1} = \beta_j + \left[\sum_{i=1}^{J}\partial\mu_i/\partial\beta \ V^{-1}(\mu_i) \ \partial\mu_i/\partial\beta\right]^{-1} \left[\sum_{i=1}^{J}\partial\mu_i/\partial\beta \ V^{-1}(\mu_i)(y_i - \mu_i)\right] \quad (13.8)$$

A thorough examination of the derivation of GEE, of the GEE estimating algorithm, and of how each correlation structure is defined and interpreted can be found in Hardin and Hilbe (2003).

13.2.2 GEE Correlation Structures

A question arises regarding when to use a particular correlation structure. In their founding article on GEE, Liang, and Zeger (1986) claimed that estimation of a GEE model is robust to misspecifications of the working correlation structure. Subsequent research has demonstrated that this statement was rather strong. In fact, the specification of a correlation structure that is not feasible for the data results in a failure to converge, or if convergence is achieved, the standard errors may show extreme values that are far greater than should be expected. QLS regression is a method that has been developed recently to amend certain incorrectly specified structures to achieve estimation. We address QLS later in this chapter. In any case, it appears that employing the correct correlation structure to the algorithm can make a significant difference in the results.

First, several guidelines are provided in Table 13.1 that may assist in selecting the optimal correlation structure for a GEE model.

TABLE 13.1

Guidelines on Selecting the Best Correlation Structure

Use structure if criteria exist—
If the number of panels is small; no evidence of correlation—independence
If data relate to 1st level clustered data : exchangeable
If panel data related to measurements over time periods: autoregressive, Markov, unstructured, stationary, nonstationary
If time periods are assumed to be equal: autoregressive
If time periods are equal or unequal: Markov
If panels are small and data are balanced and complete: unstructured

We now give a brief overview of each major correlation structure, displaying:

1. A schematic of the correlation structure
2. Example GEE logistic model output using the structure
3. Brief discussion

GEE models with independence and exchangeable correlation structures are designed to model clustered data. However, unlike other structures, neither of these models incorporates time clustering into the algorithm. Most GEE applications assume that when measures are taken at various time intervals, the intervals are of equal length. The correlation structures require that the observations are equally spaced so that calculations based on lags correspond to a constant or equal time period. When this assumption is violated, using the *force* option in Stata, or similar options in other software applications, allows estimation to continue after adjusting for the lack of temporal equality. If a subject has missing values for a particular year, this fact is also a violation of the equality of time intervals requirement. Because GEE models are quasi-likelihood, and not directly derived from an underlying PDF, use of empirical or robust standard errors is appropriate.

Missing values is a problem when dealing with GEE and longitudinal models in general. Refer to the previous discussion of missing values in Section 5.10. For problems dealing with missing values in GEE, see Hardin and Hilbe (2003).

Examples are based on the German health survey taken for the 5-year period 1984–1988 **(rwn_1980)**. Variables include:

RESPONSE

outwork   1 = out of work; 0 = working

PREDICTORS

female   1 = female; 0 = male
married   1 = subject married; 0 = not married
*edlevel*1-4   1 = not HS grad (reference); 2 = HS grad; 3 = univ/college;
                   4 = grad school

13.2.2.1 Independence Correlation Structure Schematic

```
1
0    1
0    0    1
0    0    0    1
0    0    0    0    1
```

The independence correlation structure has an I in place of $R(\alpha)$ in the variance function. This indicates that the data are independent of one another, and that no clustering effect is specified. Use of the cluster robust estimator over subjects (*id*) does account for clustering, but only in a *post hoc* fashion. For many studies, this is all that is necessary. The independence correlation structure is used as the base for evaluating other structures.

Two example output models are displayed. The first is a standard logistic regression model, but with a robust, or empirical, variance based on the cluster variable, *id*. The second output is an independence GEE model. Both odds ratios and standard errors are identical. Note the values of the correlation structure, which are obtained by using the **xtcorr** command following the model, is an identity matrix.

Although the GEE model with an independence correlation structure assumes that the repeated measures or items within groups are independent of one another, the model nevertheless provides consistent estimators when the data are in fact correlated. Unfortunately, this consistency is offset by inefficiency. However, Glonek and McCullagh (1995) have demonstrated that the loss of efficiency is not always great. Due to this feature of the independence model, and due to its straightforward interpretation, the GEE model with an independence correlation structure is commonly used as (1) the reference model for more complex GEE correlation, and (2) as the basis for the derivation of model diagnostics such as the QIC goodness-of-fit statistic.

STANDARD LOGISTIC REGRESSION, WITH ROBUST VARIANCE ESTIMATOR

```
. logit outwork female married edlevel2-edlevel4, nolog robust
cluster(id) or

Logistic regression                    Number of obs    =       19609
                                       Wald chi2(5)     =     1174.52
                                       Prob > chi2      =      0.0000
Log pseudolikelihood = -10653.009      Pseudo R2        =      0.1559

                   (Std. Err. adjusted for 6127 clusters in id)
------------------------------------------------------------------
             |              Robust
     outwork | Odds Ratio Std. Err.     z  P>|z| [95% Conf. Interval]
---------+--------------------------------------------------------
      female |   7.147992 .4229753  33.24  0.000  6.365244  8.026996
     married |   1.235824 .0858986   3.05  0.002   1.07843  1.416189
    edlevel2 |   .7208191 .0828943  -2.85  0.004  .5753577  .9030559
    edlevel3 |   1.267492 .1295823   2.32  0.020  1.037344  1.548703
    edlevel4 |   .3669405 .0497218  -7.40  0.000   .281355  .4785602
------------------------------------------------------------------
```

GEE WITH INDEPENDENT CORRELATION STRUCTURE

```
. xtlogit outwork female married edlevel2-edlevel4, nolog pa
robust i(id) or corr(indep)

GEE population-averaged model    Number of obs      =       19609
Group variable:            id    Number of groups   =        6127
Link:                   logit    Obs per group: min =           1
Family:              binomial                   avg =         3.2
Correlation:      independent                   max =           5
                                 Wald chi2(5)       =     1174.52
Scale parameter:            1    Prob > chi2        =      0.0000

Pearson chi2(19609):  19823.05   Deviance           =    21306.02
Dispersion (Pearson): 1.010916   Dispersion         =    1.086543

                    (Std. Err. adjusted for clustering on id)
------------------------------------------------------------------
             |          Semi-robust
     outwork | Odds Ratio Std. Err.     z  P>|z| [95% Conf. Interval]
---------+--------------------------------------------------------
      female |   7.147992 .4229753  33.24  0.000  6.365244  8.026996
     married |   1.235824 .0858986   3.05  0.002   1.07843  1.416189
    edlevel2 |   .7208191 .0828943  -2.85  0.004  .5753577  .9030559
```

```
edlevel3 |   1.267492  .1295823   2.32  0.020  1.037344  1.548703
edlevel4 |    .3669405  .0497218  -7.40  0.000   .281355  .4785602
---------------------------------------------------------------------


. xtcorr

Estimated within-id correlation matrix R:

          c1         c2        c3        c4        c5
r1   1.0000
r2   0.0000    1.0000
r3   0.0000    0.0000    1.0000
r4   0.0000    0.0000    0.0000    1.0000
r5   0.0000    0.0000    0.0000    0.0000    1.0000
```

Notice the high odds ratio for *female*. For the subjects in the study, the odds of being out of work are seven times greater if *female* rather than *male*. It is possible that this reflects the domestic situation in Germany in the late 1980s.

13.2.2.2 Exchangeable Correlation Structure Schematic

$$
\begin{matrix}
1 & & & & \\
\alpha & 1 & & & \\
\alpha & \alpha & 1 & & \\
\alpha & \alpha & \alpha & 1 & \\
\alpha & \alpha & \alpha & \alpha & 1
\end{matrix}
$$

The exchangeable correlation structure is the default structure for nearly all commercial GEE applications, and is the most commonly used structure in research. It is generally the appropriate model to use with clusters of level-1 nested data. It is rarely used with longitudinal data, and for good reason—it ignores a time variable. The time panel, indicated as $t()$ in Stata, is not applicable to the exchangeable specification since all of the pairs of residuals contribute to the single parameter, α. In other words, it doesn't matter what the t index is since all contributions to it go into the same sum for α. A single parameter α is thereupon assigned to all diagonal elements of the correlation matrix.

Again, the exchangeable correlation structure assumes that correlations between subsequent measurements within a panel are the same, without regard to any time interval. This fact may be expressed as noting that observations or subjects within any cluster can be temporally exchanged or re-arranged without affecting the correlation between subjects—hence the correlation structure name of *exchangeable*. The Greek letter α, or alpha, in the above matrix is a single value scalar that does not vary between panels. We see this effect when modeling the example data using an exchangeable structure. Note that the independence structure is identical to the exchangeable with $\alpha = 0$. Also note that ρ (Greek rho) is also used to represent what we have referred to as the correlation parameter.

```
. xtlogit outwork female married edlevel2-edlevel4, nolog pa
robust i(id) or corr(exch)

GEE population-averaged model      Number of obs       =     19609
Group variable:              id    Number of groups    =      6127
Link:                     logit    Obs per group: min  =         1
Family:                 binomial                  avg  =       3.2
Correlation:        exchangeable                  max  =         5
                                   Wald chi2(5)        =   1135.44
Scale parameter:              1    Prob > chi2         =    0.0000

                              (Std. Err. adjusted for clustering on id)
-------------------------------------------------------------------
             |              Semi-robust
   outwork | Odds Ratio Std. Err.    z    P>|z| [95% Conf. Interval]
---------+---------------------------------------------------------
    female |   6.020553 .3295659  32.79  0.000   5.40806   6.702415
   married |   1.106834 .0628156   1.79  0.074  .9903184   1.237059
  edlevel2 |   .8031449  .089825  -1.96  0.050  .6450512   .9999852
  edlevel3 |   1.278551 .1311921   2.39  0.017  1.045625   1.563363
  edlevel4 |   .3508881 .0500778  -7.34  0.000  .2652695    .464141
-------------------------------------------------------------------

. xtcorr

Estimated within-id correlation matrix R:

          c1      c2      c3      c4      c5
r1    1.0000
r2    0.6424  1.0000
r3    0.6424  0.6424  1.0000
r4    0.6424  0.6424  0.6424  1.0000
r5    0.6424  0.6424  0.6424  0.6424  1.0000
```

Notice that the odds ratio of female has diminished by a unit from 7.5 to 6.

13.2.2.3 *Autoregressive Correlation Structure Schematic*

```
        1
      C^1    1
      C^2   C^1    1
      C^3   C^2   C^1    1
      C^4   C^3   C^2   C^1    1
```

The autoregressive correlation structure is best used when the panels are collections of data over time for the same person. The structure assumes that there is a specific decrease in correlation coefficient values with a corresponding increase in time or distance measurements between subjects within panel time intervals. Simply, correlation decreases with the increase in the distance or time between observations. Each off-diagonal value, from the main diagonal, is calculated by incrementally increasing the power, beginning with the square of the first off-diagonal. The incremental decrease in diagonal values reflects an increase of the powers from the first diagonal from the main. An example will make this relationship clear.

```
. xtlogit outwork female married edlevel2-edlevel4, nolog pa
robust or i(id) t(year) corr(ar1) force

GEE population-averaged model     Number of obs      =     18459
Group and time vars:     id year  Number of groups   =      4977
Link:                      logit  Obs per group: min =         2
Family:                 binomial                 avg =       3.7
Correlation:               AR(1)                 max =         5
                                   Wald chi2(5)       =   1081.17
Scale parameter:               1   Prob > chi2        =    0.0000

                         (Std. Err. adjusted for clustering on id)
-----------------------------------------------------------------
             |               Semi-robust
   outwork | Odds Ratio  Std. Err.      z   P>|z| [95% Conf. Interval]
---------+-------------------------------------------------------
    female |   7.014761  .4267328   32.02  0.000  6.226317  7.903045
   married |   1.133619  .0714074    1.99  0.046  1.001958  1.282582
  edlevel2 |   .8241348  .1047629   -1.52  0.128  .6423839  1.057309
  edlevel3 |   1.130885  .1368001    1.02  0.309  .8921767  1.433461
  edlevel4 |   .3461595  .0539872   -6.80  0.000  .2549895  .4699268
-----------------------------------------------------------------


. xtcorr

Estimated within-id correlation matrix R:

          c1       c2       c3       c4       c5
r1   1.0000
r2   0.7240   1.0000
r3   0.5242   0.7240   1.0000
r4   0.3796   0.5242   0.7240   1.0000
r5   0.2748   0.3796   0.5242   0.7240   1.0000
```

```
. di (.7240)^2
.524176

. di (.7240)^3
.37950342

. di (.7240)^4
.27476048
```

Large autoregressive AR1 matrices result in small coefficient values. However, this relationship does not hold for higher AR values, such as AR2. We shall discuss later the difference in AR1 and AR2, as well as other models that employ this type of designation.

The essential notion underlying the autoregressive structure is that the value of the response at each time interval is primarily influenced by its value at one measurement earlier. The AR1 structure constrains the correlation values between succeeding measures on an item or subject to decrease with increasing separation. The structure makes sense for many longitudinal situations, where it is generally expected that measures will be more dissimilar when taken farther apart in time and as a consequence will be less correlated than will be measures more closely associated in time. The AR1 correlation structure assumes that measures are balanced or that the time between measures is equal. For some types of longitudinal occasions, this is not a realistic supposition.

When we discuss quasi-least squares models in Section 13.2.6, we shall discover that the Markov correlation structure relaxes the AR1 assumption of the temporal equality of measures. In other respects it is the same as AR1. The Markov structure is currently implemented only in quasi-least squares applications.

Use of the autoregressive correlation structure requires that the user provide both a cluster ID and a time variable to the estimating algorithm. In Stata, and the equivalent option in SAS, the use of the *force* option is required with unequal time or space intervals. For the example model the intervals are generally equal, except when missing values result in the exclusion of a particular time interval. The model results are affected if *force* is used with data having unequal space time intervals. If the option is used and the model results are unchanged, this is evidence that the observations are equally spaced.

13.2.2.4 *Unstructured Correlation Structure Schematic*

```
        1
        C1    1
        C2    C5    1
        C3    C6    C8    1
        C4    C7    C9    C10   1
```

All correlations are assumed to be different in the unstructured correlation structure; in other words, the correlations are freely estimated from the data—there is no structure between any two contiguous values in the correlation matrix. For large matrices, this can result in the calculation of a large number of correlation coefficients.

As a result of having a different coefficient for each matrix cell, the unstructured structure fits the data well. The downside, however, is a loss of efficiency, and as a result, a loss of interpretability. This is particularly the case when the panels have more than three observations.

The number of correlation coefficients estimated by the unstructured model is based on the size of the panel having the greatest number of observations. The formula used to calculate the number of correlation coefficients may be expressed as:

$$K = n(n-1)/2 \tag{13.9}$$

where n is the number of observations in the largest panel. For example, the model below shows that the maximum observations per group is five. Therefore $n = 5$, and $K = 10$.

Due to the problem with maximum panel size, the unstructured correlation structure is appropriate when the size of panels is small, when there are relatively few predictors, and when there are no missing values.

```
. xtlogit outwork female married edlevel2-edlevel4, nolog pa
robust or i(id) corr(unstr)

GEE population-averaged model      Number of obs        =    19609
Group and time vars:     id year   Number of groups     =     6127
Link:                      logit   Obs per group: min =        1
Family:                 binomial                    avg =      3.2
Correlation:        unstructured                    max =        5
                                   Wald chi2(5)         =  1163.20
Scale parameter:              1    Prob > chi2          =   0.0000
```

(Std. Err. adjusted for clustering on id)

outwork	Odds Ratio	Semi-robust Std. Err.	z	P>|z|	[95% Conf. Interval]	
female	6.092716	.3316659	33.20	0.000	5.47614	6.778715
married	1.116064	.0614031	2.00	0.046	1.001977	1.24314
edlevel2	.804111	.088713	-1.98	0.048	.6477506	.9982153
edlevel3	1.290989	.1311404	2.51	0.012	1.057929	1.575392
edlevel4	.3532482	.0498111	-7.38	0.000	.2679496	.4657007

```
. xtcorr
```

Estimated within-id correlation matrix R:

```
         c1        c2        c3        c4        c5
r1   1.0000
r2   0.7025   1.0000
r3   0.6090   0.7295   1.0000
r4   0.5190   0.6365   0.7451   1.0000
r5   0.4682   0.5620   0.6445   0.7525   1.0000
```

Notice that the values of the correlations decrease the further they depart from the initial off-diagonal array of values. In this matrix, all off-1 diagonals are .7*, the off-2 diagonals are .6*, the off-3 diagonals are .5*, and the single off-4 diagonal value is .4*. This relationship rarely appears in this manner; but cells always differ in value from one another. [Note: An * indicates additional digits.]

13.2.2.5 Stationary or m-Dependent Correlation Structure Schematic

```
      1
      C1     1
      C2     C1     1
      0      C2     C1     1
      0      0      C2     C1     1
```

The stationary correlation structure specifies a constant correlation for each off-diagonal. The diagonals can be interpreted as lags. Correlations that are c lags apart are equal in value to one another, those that are $c + 1$ lags apart are also equal to one another, and so forth until a defined stop point, m, is reached. Correlations further away from m are defined as zero. This is the basis of the alternative name for this structure.

In large matrices, the correlation structure appears as a band of diagonal values. The schematic above displays an $m = 2$ structure. Values greater than the second off-diagonal have values of zero.

```
. xtlogit outwork female married edlevel2-edlevel4, nolog pa
robust or i(id)   force c(stat)
note:   some groups have fewer than 2 observations
        not possible to estimate correlations for those groups
        1150 groups omitted from estimation
```

```
GEE population-averaged model        Number of obs      =    18459
Group and time vars:       id year   Number of groups   =     4977
Link:                      logit     Obs per group: min =        2
Family:                    binomial                 avg =      3.7
```

```
Correlation:          stationary(1)                      max =        5
                                        Wald chi2(5)          = 1078.32
Scale parameter:                   1    Prob > chi2           =  0.0000

                         (Std. Err. adjusted for clustering on id)
----------------------------------------------------------------------
             |              Semi-robust
  outwork | Odds Ratio    Std. Err.    z   P>|z| [95% Conf. Interval]
---------+------------------------------------------------------------
   female |   7.601208   .4831115  31.91  0.000  6.710929  8.609593
  married |   1.236037   .0918056   2.85  0.004  1.068585  1.429729
 edlevel2 |   .7617445   .0953663  -2.17  0.030  .5959962  .9735878
 edlevel3 |   1.174442   .1350419   1.40  0.162  .9374695  1.471316
 edlevel4 |   .3525592   .0522451  -7.04  0.000  .2636903  .4713786
----------------------------------------------------------------------

working correlation matrix not positive definite
convergence not achieved

. xtcorr

Estimated within-id correlation matrix R:

         c1        c2        c3        c4        c5
r1   1.0000
r2   0.7213   1.0000
r3   0.0000   0.7213   1.0000
r4   0.0000   0.0000   0.7213   1.0000
r5   0.0000   0.0000   0.0000   0.7213   1.0000
```

We have a problem. The algorithm failed to converge, but still produced parameter estimates, standard errors, and an associated correlation matrix. These values cannot be trusted since they are not based on convergence. The correlation matrix, however, displays the characteristic of a stationary 1 model, that is, where $m = 1$. We defined the model with that stopping point, which is reflected in the correlation matrix, no matter the worth of the actual values. The reason why the model failed to converge will be discussed in Section 13.2.6, Quasi-Least Squares Regression.

13.2.2.6 Nonstationary Correlation Structure Schematic

```
          1
         C1     1
         C5    C2     1
          0    C6    C3     1
          0     0    C7    C4     1
```

The nonstationary correlation structure is similar to the stationary, except that the values of each lag or off-diagonal are not constant. The stationary structure, as discussed above, have equal off-diagonal values until the stopping criterion is reached. In a sense, the nonstationary is like the unstructured, but with the addition of a defined stop. Correlations values beyond the *m* are 0.

The nonstationary correlation structure is the same as the stationary except that the values of each lag or off-diagonal are not constant. Of course, correlation values beyond *m* are all 0.

Some statisticians use the nonstationary correlation structure when they have ruled out the others but still have a limit to the range of measurement error or lags in the data.

It should be noted that the model below sets *m* equal to 2. This means that panels with fewer than 3 observations will be discarded from estimation. When *m* = 1, panels with 1 observation only are discarded. Care must be taken to ensure that most panels have more values than *m*.

```
. xtlogit outwork female married edlevel2-edlevel4, nolog pa
robust or i(id) t(year) corr(non2) force

GEE population-averaged model      Number of obs      =    16495
Group and time vars:      id year  Number of groups   =     3995
Link:                       logit  Obs per group: min =        3
Family:                  binomial                 avg =      4.1
Correlation:                nonst                 max =        5
                                   Wald chi2(5)       =   118.37
Scale parameter:                1  Prob > chi2        =   0.0000
```

```
                   (Std. Err. adjusted for clustering on id)
------------------------------------------------------------------
            |            Semi-robust
    outwork | Odds Ratio Std. Err.     z    P>|z| [95% Conf. Interval]
------------+-----------------------------------------------------
     female |  8.792153  1.772673  10.78  0.000  5.922104  13.05312
    married |  1.195466  .2532439   0.84  0.399   .789262  1.810728
   edlevel2 |  1.150309  .4312629   0.37  0.709  .5516839  2.398496
   edlevel3 |  2.231941  1.115839   1.61  0.108  .8377839  5.946115
   edlevel4 |  .2091531  .1735366  -1.89  0.059  .0411352  1.063446
------------------------------------------------------------------

working correlation matrix not positive definite
convergence not achieved

. xtcorr

Estimated within-id correlation matrix R:

          c1        c2        c3        c4        c5
r1    1.0000
r2    0.6694    1.0000
```

```
r3   0.6229   0.7513   1.0000
r4   0.0000   0.6607   0.7737   1.0000
r5   0.0000   0.0000   0.6677   0.7794   1.0000
```

As was the case when modeling the stationary model, the nonstationary GEE model above failed to converge. The estimates, standard errors, and correlation values cannot be trusted.

The nonstationary correlation structure, as the stationary, is used with time data. However, the nonstationary is in fact rarely used with real data. At times researchers have employed it when the other structures have been ruled out. Because of its rather poor convergence and interpretation properties, several commercial applications have dropped it as an option.

Additional insight into how the GEE model is structured, and the nature and formal definition of its various correlation structures, can be found in Diggle, Liang, and Zeger (1994), Hardin and Hilbe (2003), and Twist (2003). For additional discussion on how GEE relates to count models, see Hilbe (2007a).

13.2.3 GEE Binomial Logistic Models

The relationship between observation-based logistic GEE models and binomial or grouped GEE logistic models is similar to what we found for standard binary and binomial logistic models. Recall that we showed how to convert a binary logistic to a grouped or binomial logistic model in Chapter 8. The parameter estimates and standard errors were identical in both parameterizations. We do not find the same equality of estimates or of standard errors between the two parameterizations of the GEE logistic model. The estimates and errors are close, but not exactly the same. The values of the respective correlation structures, however, do differ substantially. The **heart02grp** data were constructed from the **heart01** data following the methods outlined in Chapter 8. Estimated as GEE with exchangeable correlation structures, the models appear as:

OBSERVATION BASED

```
. use heart01

. xtgee death anterior hcabg kk2-kk4 age2-age4, nolog fam(bin)
i(center) c(exch) robust

GEE population-averaged model          Number of obs       =      4503
Group variable:              center    Number of groups    =        47
Link:                         logit    Obs per group: min  =         3
Family:                    binomial                    avg  =      95.8
Correlation:           exchangeable                    max  =       215
                                       Wald chi2(8)        =    183.80
Scale parameter:                  1    Prob > chi2         =    0.0000

                  (Std. Err. adjusted for clustering on center)
```

```
            |             Semi-robust
      death |     Coef.   Std. Err.      z    P>|z|    [95% Conf. Interval]
------------+----------------------------------------------------------------
   anterior |  .6434073   .2037555    3.16   0.002     .2440538    1.042761
      hcabg |  .7407087   .3049303    2.43   0.015     .1430562    1.338361
        kk2 |  .8115746    .177902    4.56   0.000     .4628931    1.160256
        kk3 |  .7743569    .320007    2.42   0.016     .1471548    1.401559
        kk4 |  2.657819   .2966113    8.96   0.000     2.076471    3.239166
       age2 |  .5040532   .3019612    1.67   0.095      -.08778    1.095886
       age3 |  1.522822   .2216656    6.87   0.000     1.088366    1.957279
       age4 |   2.19583   .2762431    7.95   0.000     1.654403    2.737256
      _cons | -5.063689   .2510388  -20.17   0.000    -5.555716   -4.571662
----------------------------------------------------------------------------
```

alpha= -.0005

GROUPED OR BINOMIAL BASED

. use heart02grp

. xtgee dead anterior hcabg kk2-kk4 age2-age4, nolog fam(bin
cases) i(center) c(exch) robust

```
GEE population-averaged model       Number of obs      =      895
Group variable:           center    Number of groups   =       47
Link:                      logit    Obs per group: min =        2
Family:                 binomial                   avg =     19.0
Correlation:        exchangeable                   max =       33
                                    Wald chi2(8)       =   183.20
Scale parameter:               1    Prob > chi2        =   0.0000
```

                    (Std. Err. adjusted for clustering on center)

```
            |             Semi-robust
       dead |     Coef.   Std. Err.      z    P>|z|    [95% Conf. Interval]
------------+----------------------------------------------------------------
   anterior |  .6438512   .2056067    3.13   0.002     .2408694    1.046833
      hcabg |  .7385437   .3058415    2.41   0.016     .1391054    1.337982
        kk2 |  .8131047   .1787598    4.55   0.000      .462742    1.163467
        kk3 |  .7793416   .3212124    2.43   0.015      .149777    1.408906
        kk4 |  2.668025   .2955455    9.03   0.000     2.088766    3.247283
       age2 |  .5031309   .3033768    1.66   0.097    -.0914766    1.097738
       age3 |  1.524378    .223337    6.83   0.000     1.086645     1.96211
       age4 |  2.200113    .278056    7.91   0.000     1.655133    2.745093
      _cons | -5.071416   .2533412  -20.02   0.000    -5.567956   -4.574877
----------------------------------------------------------------------------
```

alpha=.0032

Note that the values of α for both models approximate zero, indicating little correlation in the data. The model is likely statistically similar to the independence model.

GEE models incorporating a time element into the correlation structure may find even greater dissimilarities between the observation and grouped or aggregated parameterizations. Little research has been done in this area, but it would be an excellent subject for a dissertation.

We shall elaborate on this point at greater length shortly, but it should be understood that in contrast to other GLM/GEE family members, certain difficulties exist when modeling binary response GEE models. We address this concern in Section 13.2.5 and again in Section 13.2.6 when discussing quasi-least squares regression.

13.2.4 GEE Fit Analysis—QIC

The *quasi-likelihood information criterion* (QIC), or technically, the *quasi-likelihood under the independence model information matrix criterion*, was initially developed by Pan (2001) and later modified by Hardin and Hilbe (2003). The goodness-of-fit statistic is modeled after the *Akaike information criterion* (AIC), which is used to compare the fit of two maximum likelihood estimated models. However, since GEE is a quasi-likelihood model, the AIC statistic is not an appropriate measure of fit. Interestingly, though, the AIC still enjoys use in the literature and has proved to select the true correlation structure for certain models at a better rate than the QIC.

Pan developed two closely associated fit statistics: QIC and QICu. The QIC statistic is designed to evaluate correlation structures, for example, which correlation structure best fits the data, whereas the QICu statistic is designed to evaluate which predictors best explain the response within the model framework. The QICu statistic is simply the AIC statistic, but with the quasi-log-likelihood employed instead of the log-likelihood: $\text{QICu} = -2Q(g^{-1}(x\beta);I) + 2p$, where p are the number of predictors including intercept.

The QIC statistic may be expressed as:

$$\text{QIC}(R) = -2Q(g^{-1}(x\beta);I) + 2\text{trace}(A_I^{-1}V_R) \qquad (13.10)$$

with

$-2Q(g^{-1}(x\beta);I) = $ value of the estimated quasi-likelihood with an independence correlation structure, I. $\mu = g^{-1}(x\beta)$ is the fitted value of the model.

A_I is the independence structure variance matrix for the model. Some formulations of 13.10 use the symbol, Ω_I^{-1}, to refer to A_I^{-1}.

V_R is the modified robust or sandwich variance estimator of the model with hypothesized structure R(a). Some formulations use Σ_R as the symbol for V_R.

The AIC penalty term, $2p$, is expressed as $2\text{trace}(A_I^{-1}V_R)$ for the QIC statistic.

The QIC statistic may be calculated by hand using the following Stata code. We use a paradigm-format code, with y as the response, *xvars* as the predictors, and *id* as the variable identifying subjects over time or cluster. A binary logistic GEE model will be shown, with the quasi-likelihood function being derived as discussed earlier in this text.

We have placed comments to the right of each line to assist in interpretation. The code is adapted from Hardin and Hilbe (2003).

* INDEPENDENCE MODEL W NAÏVE SE

qui xtgee y xvars, i(id) corr(indep) fam(bin)	/// GEE independence model; naïve SE
matrix A = e(V)	/// variance-covariance matrix; SE = sqrt(diagonals)
matrix Ai = invsym(A)	/// invert the symmetric matrix
	/// note: *qui* = quietly; not display results on screen

* EXCHANGEABLE W ROBUST SE

qui xtgee y xvars, i(id) corr(exch) fam(bin) robust	/// GEE exchangeable; robust SE
matrix V = e(V)	/// variance-covariance matrix of exchangeable/robust model
matrix T = Ai*V	/// product of this variance matrix and Ai from first model
matrix t = trace(T)	/// sum of diagonal elements of T
scalar trac = t[1,1]	/// store t as a scalar; trace
qui predict double _mu, mu	/// _mu is fit, or probability of y = 1
qui gen double _QLL = (y * ln(_mu/(1−_mu)) + ln(1−_mu))	/// quasi-loglikelihood
qui summ _QLL, meanonly	/// single value for QLL
noi di "Trace = " trac	/// display trace
noi di "QIC = " −2*r(sum) + 2*trac	/// display QIC statistic; QLL = r(sum)
drop _mu _QLL	/// drop temporary variables.

The difference between Pan's original parameterization of the QIC statistic, QIC_P, and Hardin–Hilbe's version rests with the second term in Equation 13.10. Essentially, the term is the ratio of the robust covariance and the model–base covariance matrices. In the Hardin–Hilbe implementation, the model-based coefficients derive from estimation given an independence correlation structure, $\beta(I)$. Pan's model-based coefficients derive from the non-independence structure employed in the estimation, $\beta(R)$. Simulation studies by Hin and Wang (2008) have shown that QIC_{HH} selects the "true" correlation structure with approximately the same frequency for exchangeable and autoregressive structures as does QIC_P. Other structures were not evaluated. However, they argue that the second term of the QIC statistic, $2\text{trace}(A_I^{-1}V_R)$, is a significantly superior selector of the true correlation structure than either the QIC statistic or of the same term from Pan's parameterization. That is, the statistic they referred to as $T_{2HH}(R)$ is a more superior test statistic for GEE correlation structures than $QIC_P(R)$, $QIC_{HH}(R)$ or $T_{2P}(R)$.

Hin and Wang subsequently argued that a statistic they termed CIC(R), an acronym for *Correlation Information Criteria*, is equal to or perhaps a bit superior to $T_{2HH}(R)$, or $2\text{trace}(A_I^{-1}V_R)$. CIC(R) is defined as $T_P(R)$, that is, the trace statistic from Pan's parameterization. Again, only the exchangeable and AR1 structures were used for testing these statistics. The point appears to be that the quasi-likelihood function contributes relatively little to determining which correlation structure to use with a given GEE model. Rather, of interest is the ratio of the $R(\alpha)/R(I)$, regardless if a robust variance estimator is applied to the independence model (*P*) or not (HH), or whether these statistics are multiplied by 2 or not.

The quasi-likelihood function of the canonical binomial distribution, or logistic, is used to estimate Q() for GEE logistic models. It can be expressed as:

$$Q_{ij} = \sum_{i=1}^{n}\sum_{j=1}^{n_i} y_{ij}\ln(\mu_{ij}) + (m_{ij} - y_{ij})\ln(1 - \mu_{ij}) \qquad (13.11)$$

For the binary response logistic model, $m = 1$ and $y = (1/0)$. For the binomial or grouped logistic model, m is the binomial denominator while y is the numerator.

Additional analysis on the relationship of the QICu and QIC statistics has demonstrated that QICu approximates QIC when the GEE model is correctly specified. In either case, the QIC(u) value that is the lowest of all competing models is the preferred model. This is similar to how the AIC is used with maximum likelihood models. See Hardin and Hilbe (2003) for more details.

Vincent Carey of Harvard University developed the first QIC software in 2004 as part of his YAGS (*Yet Another GEE Solver*) package, which is an *R* library of GEE-related programs. James Cui (2007, 2008) of Deakin University, Australia, published a software implementation of the QIC/QICu procedures in Stata; the QIC() implementation is based on QIC_{HH}. The QIC statistic was recently made part of the SAS/STAT GENMOD procedure output as well.

For completeness sake, it should likely be mentioned that the *Rotnitzky–Jewell criterion* (1990) is yet another GEE fit statistic that is sometimes used to select the optimal correlation structure for a given model. Hin, Carey, and Wang (2007) engaged in a number of simulation studies to assess the comparative detection rate of the "true" correlation structure using the QIC and the R-J criterion. Neither statistic clearly comes out as the preferred fit statistic across all of the standard correlation structures. R-J appears to be slightly favored when the true structure is the exchangeable; QIC is preferred for other structures. However, given the complexity of the R-J criterion, the fact that it is not available in any commercial software (although it is displayed in YAGS output), and the limitations of its marginal preference over the QIC statistic for a single structure, the QIC is preferred to R-J. Note, however, that no differentiation of the QIC statistic, such as QIC_P and QIC_{HH}, was given in the Hin, Carey, and Wang article. When the difference was identified in Hin and Wang (2008), the authors made no reference to R-J, and used the second term of QIC_P to construct their proposed CIC(R) measure. For additional insight into the R-J statistic, see Shults et al. (2008). They provide an analysis of three versions of R-J, comparing their efficacy to the Shults–Chaganty criterion, which is based on the minimum value of the weighted error sum of squares. Like the H-C-W article, no one criterion was identified for selecting all of the GEE structures.

It should be clear that the development of fit criteria for the comparative assessment of GEE models is a subject of current active debate. The one item on which all researchers in this area of study agree is that there currently exists no reliable universal comparative fit statistic for GEE models. There does seem to be good evidence, though, that the CIC or some variation of $2\text{trace}(A_I^{-1}V_R)$ is superior to the QIC statistic. Since QIC software currently displays the trace statistic as part of its output, it is easy to use it, or twice its value, as a statistic for selection of the most appropriate GEE correlation structure. An example follows.

The *nodisplay* option in the Stata implementation, **qic**, is used to suppress a display of the model header and coefficients, showing only the relevant summary statistics. First shown are the independence and exchangeable model QIC results.

INDEPENDENCE CORRELATION STRUCTURE

```
. qic outwork female married edlevel2-edlevel4, nolog nodisp
fam(bin) i(id) corr(indep)

              QIC and QIC_u
```

Corr =	indep
Family =	bin
Link =	logit
p =	6
Trace =	16.596
QIC =	21339.210
QIC_u =	21318.019

EXCHANGEABLE CORRELATION STRUCTURE

```
. qic outwork female married edlevel2-edlevel4, i(id)
corr(exch) fam(bin)

              QIC and QIC_u

Corr =                     exch
Family =                    bin
Link =                    logit
p =                           6
Trace =                  15.023
QIC =                 21382.587
QIC_u =               21364.542
```

Two times the trace statistics above give us the values of $2\text{trace}(A_I^{-1}V_R)$ for each model:

INDEPENDENCE: 33.192   EXCHANGEABLE: 30.046

The QIC statistic selects the independence structure as the most appropriate for the data. The ratio statistic tells us that, of the two structures, the exchangeable is to be preferred. In Section 13.2.2.2 the same exchangeable model was shown to have a correlation parameter of 0.6424, which clearly indicates extra correlation in the data. The true correlation structure is therefore not the independence, or I. In this case, the 2trace statistic outperforms QIC.

13.2.4.1 QIC/QICu Summary–Binary Logistic Regression

Note: Smallest value is the preferred model or structure.

	QIC(R)	$2\text{trace}(A_I^{-1}V_R)$
INDEPENDENT	QIC = 21339.210	33.192
EXCHANGEABLE	QIC = 21382.587	30.046
UNSTRUCTURED	QIC = 21378.193	29.322
AUTOREGRESSIVE 1	QIC = 21356.251	59.822

The QIC result tells us that the independent correlation structure is preferred; the 2trace statistic indicates that the unstructured is best. Recall, however, that the autoregressive and unstructured models have *year* as the time panel variable. Moreover, since the stationary and nonstationary structures do not converge, we cannot use either test for them. Given our earlier discussion of the inadequacy of the QIC statistic, it may be wise to select the unstructured correlation with the model. Note that the unstructured QIC statistic is less than that of the exchangeable, therefore confirming our choice.

Regarding the unstructured correlation structure, ongoing work by Barnett et al. (2008) has demonstrated that the QIC is biased to the unstructured matrix, that is, it tends to conclude that the unstructured matrix is preferred when in fact the true structure is the independence, exchangeable, or autoregressive. This bias was not evident in our example model, but we have definitely noticed it in general. Moreover, in simulation studies, the QIC appears to be a more accurate predictor of the true structure when $\alpha > 0.5$. For weaker correlations, the QIC performs poorly—even compared to the AIC statistic. Barnett et al. conclude that when comparing the AIC and QIC, it is probably wise to use the AIC for models with weak correlations and QIC for models with stronger correlations. They also favor a Deviance Information Criterion (DIC) for use when models have a weak correlation. The DIC is based on Bayesian methods, however, and takes us beyond the scope of our discussion. In any case, we have discovered that some variety of 2trace statistic is likely superior to QIC for all values of α or ρ, but more research needs to be done.

It might be helpful to determine the value of the AIC for comparative purposes. Since it is not typically displayed in GEE results, we shall calculate it for the above GEE logistic models.

```
. qui xtgee outwork female married edlevel2-edlevel4, i(id)
  corr(exch) fam(bin)
. predict mue, mu          /// fit; or probability of outcome=1
. egen qlexch = sum(outwork *
  ln(mue/(1-mue)) + ln(1-mue))                    /// QL function
. su qlexch                                        /// display QL

Variable |    Obs      Mean    Std. Dev.        Min         Max
---------+-----------------------------------------------------------
  qlexch |  19609  -10676.27          0   -10676.27   -10676.27

. di 2*(-qlexch + (2*6))                   /// calculate AIC
21376.541                    ← AIC exchangeable
```

Using the same procedure as above for the exchangeable structure, we have:

$$\begin{aligned}
\text{Independence AIC} &= 21330.02 \\
\text{Unstructured AIC} &= 21372.871 \\
\text{AR1 AIC} &= 21353.842
\end{aligned}$$

The AIC selects the independence structure as did the QIC; **2trace**$(A_I^{-1}V_R)$ selects the unstructured matrix.

Always remember when using GEE models that tests such as the AIC, BIC, QIC, CIC, and 2trace$(A_I^{-1}V_R)$ do not always correctly select the true model, that is, the model that optimally reflects the structure of the data. If two or three tests, such as AIC, QIC, and 2trace, all select the same model, then we can be more assured that we have likely identified the correct structure. Actually, given the above simulation studies, we may conclude that for $\alpha < .5$ if the AIC and 2trace both select a specific structure, it is more likely correct; if $\alpha > .5$ and both QIC and 2trace select a given structure, it is more likely correct. Keep in mind, however, that this criterion is a tentative guideline. Future research may well lead to different conclusions.

Care must be taken when comparing QIC statistics between SAS GENMOD and Stata. QIC assumes that panels for GEE models having correlation structures capturing time intervals (AR, stationary, nonstationary, and unstructured) are balanced, that is, that the intervals between consecutive time periods are equal. When panels are unbalanced, both SAS and Stata provide a facility to force estimation, allowing viable parameter estimates to be calculated. However, there must be at least two observations in a panel in order to estimate a correlation structure at the base time interval, which for autoregressive, stationary, and nonstationary are AR1, stationary 1, and nonstationary 1, respectively. AR2, for example, must have at least three observations per panel. Both SAS and Stata exclude single observation panels when calculating level 1 correlation structures. However, SAS uses all observations when calculating β, the parameter estimates. Stata excludes all single level panels from calculation of both the correlation structure and parameter estimates. For AR2 models, Stata excludes from estimation all observations that are part of one and two observation panels. Obviously, the QIC statistics of the two applications can differ substantially when modeling correlation structures incorporating time intervals when the data consist of many single panel observations.

In addition, and independent of the above differences, it should be noted that SAS uses QIC_P rather than QIC_{HH} for its GEE test statistic. Variations in the displayed QIC values between SAS, on the one hand, and Stata and R on the other, will likely be due to the different parameterizations of QIC, or to the manner in which they handle panels as discussed in the previous paragraph.

13.2.5 Alternating Logistic Regression

Carey, Zeger, and Diggle (1993) argued that GEE with binary response logistic models may contain bias that cannot be eradicated from within standard GEE. A reason for the bias rests in the fact that the Pearson residuals, which we have been using to determine the various GEE correlation matrices, are not appropriate when dealing with binary data. Carey et al. proposed a model termed *alternating logistic regression* which aims to ameliorate this bias, which clearly affects logistic GEE models.

The alternating logistic regressions (ALR) algorithm models the association between pairs of responses with log odds ratios, instead of with correlations, as do standard GEE algorithms. The model is fit to determine the effect the predictors have on the pair-wise odds ratios. A result is that ALR is less restrictive with respect to the bounds on alpha than is standard GEE methodology. We shall see the importance of this later when we examine quasi-least squares regression.

Recall the definition of an odds ratio. It is the ratio of the probability of success ($y = 1$) to the probability of failure ($y = 0$). Of a pair of responses, the odds that $y_{ij} = 1$, given that $y_{ik} = 1$, is expressed as:

$$O\left(y_{ij}; y_{ik} = 1\right) = \frac{\Pr(y_{ij} = 1, y_{ik} = 1)}{\Pr(y_{ij} = 0, y_{ik} = 1)} \tag{13.12}$$

The odds that $y_{ij} = 1$, given that $y_{ik} = 0$ can be given as:

$$O(y_{ij}; y_{ik} = 0) = \frac{\Pr(y_{ij} = 1, y_{ik} = 0)}{\Pr(y_{ij} = 0, y_{ik} = 0)} \tag{13.13}$$

The odds ratio is the ratio of the two odds, which is then given as:

$$OR(y_{ij}, y_{ik}) = \psi_{ijk} =$$
$$[\Pr(y_{ij} = 1, y_{ik} = 1)/\Pr(y_{ij} = 0, y_{ik} = 1)]/[\Pr(y_{ij} = 1, y_{ik} = 0)/\Pr(y_{ij} = 0, y_{ik} = 0)]$$

or

$$OR(Y_{ij}, Y_{ik}) = \frac{\Pr(Y_{ij} = 1, Y_{ik} = 1)\Pr(Y_{ij} = 0, Y_{ik} = 0)}{\Pr(Y_{ij} = 1, Y_{ik} = 0)\Pr(Y_{ij} = 0, Y_{ik} = 1)} \tag{13.14}$$

Again, i in Equation 13.14 indicates a cluster, j is the first item of the pairs, and k is the second item of a pair.

Alternating logistic regression seeks to determine the correlation of every pair-wise comparison of odds ratios in the model. The logic of alternating logistic regressions is to simultaneously regress the response on the predictors, as well as modeling the association among the responses in terms of pair-wise odds ratios. The ALR algorithm iterates between a standard GEE logistic model in order to obtain coefficients, and a logistic regression of each response on the others within the same panel or cluster using an offset to update the odds ratio parameters. In other words, the algorithm alternates (hence the name) between a GEE-logistic model to obtain coefficients, and a standard logistic regression with an offset aimed at calculating pair-wise odds between members of the same panel, or $OR(Y_{ij}, Y_{ik})$. The algorithm may be constructed to initially estimate the log-odds ratios, subsequently converting them to odds ratios, or it can directly estimate odds and odds ratios.

Using the **heart01** data set, a standard GEE-logistic model is developed, together with a QIC statistic, followed by an ALR parameterization and its QIC statistic. The clustering effect is *center*, which is the hospital from which the data were gathered. Robust, or empirical, standard errors have been constructed for both models.

GEE LOGIT WITH EXCHANGEABLE CORRELATION STRUCTURE

```
. xtlogit death anterior hcabg kk2-kk4 age2-age4 , pa nolog
i(center)

GEE population-averaged model      Number of obs      =      4503
Group variable:            center  Number of groups   =        47
Link:                       logit  Obs per group: min =         3
Family:                  binomial                 avg =      95.8
Correlation:         exchangeable                 max =       215
                                   Wald chi2(8)       =    195.05
Scale parameter:                1  Prob > chi2        =    0.0000
-----------------------------------------------------------------
             |              Semi-robust
      death  |      Coef. Std. Err.      z  P>|z|  [95% Conf. Interval]
---------+-------------------------------------------------------
  anterior  |   .6434073 .2037555    3.16  0.002   .2440538  1.042761
     hcabg  |   .7407087 .3049303    2.43  0.015   .1430562  1.338361
       kk2  |   .8115746  .177902    4.56  0.000   .4628931  1.160256
       kk3  |   .7743569  .320007    2.42  0.016   .1471548  1.401559
       kk4  |  2.657819 .2966113    8.96  0.000   2.076471  3.239166
      age2  |   .5040532 .3019612    1.67  0.095    -.08778  1.095886
      age3  |  1.522822 .2216656    6.87  0.000   1.088366  1.957279
      age4  |   2.19583 .2762431    7.95  0.000   1.654403  2.737256
     _cons  | -5.063689 .2510388  -20.17  0.000  -5.555716 -4.571662
-----------------------------------------------------------------
```

The QIC and QICu statistics for the above model are determined by using the **qic** command.

```
. qic death anterior hcabg kk2-kk4 age2-age4, corr(exch)
nodisp i(center) nolog fam(bin)
```

```
                    QIC and QIC_u
```

Corr =	exch
Family =	bin
Link =	logit
p =	9
Trace =	9.182
QIC =	1292.017
QIC_u =	1291.653

ALTERNATING LOGISTIC REGRESSION

```
The SAS System
The GENMOD Procedure
Model Information
Data Set              WORK.HEART
Distribution          Binomial
Link Function         Logit
Dependent Variable    death

GEE Model Information

Log Odds Ratio Structure          Exchangeable
Subject Effect             center (47 levels)
Number of Clusters                          47
Clusters With Missing Values                47
Correlation Matrix Dimension               295
Maximum Cluster Size                       215
Minimum Cluster Size                         3

Algorithm converged.

GEE Fit Criteria
QIC        1291.6717
QICu       1291.6583

Analysis Of GEE Parameter Estimates
Empirical Standard Error Estimates
```

Parameter	Estimate	Standard Error	95% Confidence Limits		Z	Pr > |Z|
Intercept	-5.0679	0.2505	-5.5589	-4.5769	-20.23	<.0001
anterior	0.6429	0.2032	0.2446	1.0411	3.16	0.0016
hcabg	0.7524	0.2998	0.1648	1.3399	2.51	0.0121
killip 4	2.6810	0.2906	2.1114	3.2505	9.23	<.0001
killip 3	0.7871	0.3167	0.1664	1.4077	2.49	0.0129
killip 2	0.8147	0.1763	0.4690	1.1603	4.62	<.0001
killip 1	0.0000	0.0000	0.0000	0.0000	.	.

```
agegrp 4    2.1998   0.2746    1.6616    2.7381    8.01   <.0001
agegrp 3    1.5208   0.2202    1.0892    1.9523    6.91   <.0001
agegrp 2    0.5011   0.2994   -0.0857    1.0879    1.67   0.0942
agegrp 1    0.0000   0.0000    0.0000    0.0000     .        .
Alpha1      0.0291   0.0461   -0.0612    0.1194    0.63   0.5278
```

The comparative QIC statistics are

$$
\begin{array}{lll}
\text{GEE-logistic} & : & 1292.017 \\
\text{ALR} & : & 1291.672
\end{array}
$$

Statistically the fit tests inform us that the models fit equally well. With respect to the differences in parameter estimates and robust standard errors, the models are nearly identical. One is not preferred over the other. It is of interest to note that Twist (2003, 2008) argues that logistic GEE models are prime examples of the claim made by Liang and Zeger (1986), developers of the GEE method, that GEE analysis is robust against the wrong selection of correlation structure. However, such an argument is contrary to Diggle et al.'s (1994) caveat that binary response models are not appropriate for GEE unless amended via ALR methods.

Readers should know that there is still considerable discussion in the literature concerning logistic GEE models, and GEE models in general. Chaganty and Joe (2004, 2006) have argued that the GEE correlation structures are not correlations at all, but rather weight matrices. Their claim is based on the supposition that the range of correlations for multivariable binary distributions—i.e., Bernoulli distributions—are based on the marginal means, which they believe preclude the working correlation from being the true correlation of the data. Kim, Shults, and Hilbe (2009) have taken issue with these assertions, demonstrating that GEE working correlation structures are in fact true correlations.

In addition, Lindsey and Lambert (1998) and Chaganty and Mav (2008) have presented lists of difficulties they find with GEE, and in particular logistic-GEE models. Other authors have considered these claims to be based more on semantics than on statistics. Although these discussions take us beyond the scope of this text, the reader should be aware that this area of statistics is not settled. Neither, I might add, is the following model, quasi-least squares regression. This is particularly the case for binary response models.

13.2.6 Quasi-Least Squares Regression

Many times when modeling a GEE we find that the model fails to converge. This is particularly the case when a time panel is incorporated into the model. Regardless, the usual reason for non-convergence is that the correlation matrix is not positive definite. When this occurs, most software applications provide the user with an error message, and stop the converging process.

Justine Shults and fellow collaborators have been building on the previous work of Dunlop (1994), Crowder (1995), Chaganty (1997), Shults and Chaganty (1998), and Chaganty and Shults (1999) to provide those who run into such convergence problems with a way to appropriately model data within the population averaging framework. The models being constructed are called *quasi-least squares*, and have been developed to be robust to modeling with non-positive definite matrices. At this writing, several of the standard GEE families with canonical links and correlation structures have been incorporated into the quasi-least squares methodology:

Families

Gaussian (link = identity)
Bernoulli/Binomial (link = logit)
Poisson (link = log)

Correlation Structures

Autoregressive 1 (AR 1)
Stationary 1   (sta 1) [tridiagonal]
Exchangeable (exc) [equicorrelated]
Markov   (Markov)

Current software implementations exist in Stata by Shults et al. (2007), in MATLAB® by Ratcliffe and Shults (2008), in R by Xie and Shults (2008), and in SAS by Kim and Shults (2008). Other matrices are being developed, as well as extended options.

GEE models, as discussed in the previous section, estimate the correlation parameter, α, based on the Pearson residual. QLS, on the other hand, estimates α in such a manner that the estimating equations for α are orthogonal to the GEE estimating equations for β. Again, QLS models are alternatives to GEE when the model matrix is not positive definite, or when a Markov correlation structure is desired for the GEE-like model. The Markov structure, first introduced for GEE/QLS modeling by Shults and Chaganty (1998), relaxes the autoregressive (AR1) assumption that the time intervals between measures within panels are equal. Because many, perhaps most, real studies violate this assumption, the Markov structure is a useful addition to GEE/QLS estimation. Unfortunately, at this time, the Markov correlation structure is not available in any current GEE software. The QLS algorithm calls an underlying GEE procedure, making adjustments to the covariance matrix at each iteration such that a positive definite correlation matrix results.

The QLS algorithm is in fact a two-stage procedure, with the first stage estimated in a manner somewhat similar to that of the ALR algorithm. The procedure alternates between a GEE estimation of β and a subsequent estimation of the correlation parameter, α. The stage-one estimating equation for α consists of taking the first derivative of the specified correlation structure, which itself is sandwiched between vectors of Pearson residuals, based on the stage-one values of β,

$$\partial/\partial\alpha = \left\{ \sum_{i=1}^{n} Z_i'(\beta)R_i^{-1}Z_i'(\beta) \right\} = 0, \qquad (13.15)$$

with $Z_i(\beta)$ being the vector of $z_{ij} = (y_{ij} - \mu_{ij})/\sqrt{V(\mu_{ij})}$.

When the stage-one estimation process has converged, a stage-two estimate of α is calculated, which is based on stage-one estimates, and designed such that α is orthogonal to a subsequent estimation of β. The stage-two value of α is used for a final estimation of β in a GEE algorithm. The algorithm is explained in detail in Kim and Shults (2008). A thorough examination of QLS theory, methodology, and applications is currently in preparation by Hilbe and Shults and should be available in early 2010.

A QLS model using both the autoregressive and Markov correlation structures is given below. The software does not allow exponentiated coefficients, but does provide for robust, jackknifed, and bootstrapped standard errors.

The German Health Survey data, **rwm_1980** that was used for many examples in the section on GEE is used here for comparison. We model the AR1 structure again, not requesting odds ratios, and not using the force option, which is not applicable for quasi-least squares modeling.

AUTOREGRESSIVE STRUCTURE

```
. xtqls outwork female married edlevel2-edlevel4, i(id)
t(year) c(AR 1)
   f(bin 1) vce(robust)
```

```
GEE population-averaged model      Number of obs       =    19609
Group and time vars: id   00000S   Number of groups    =     6127
Link:                      logit   Obs per group: min  =        1
Family:                 binomial                  avg  =      3.2
Correlation:    fixed (specified)                 max  =        5
                                    Wald chi2(5)        =  1163.38
Scale parameter:                1   Prob > chi2         =   0.0000
```

```
                          (Std. Err. adjusted for clustering on id)
----------------------------------------------------------------------
             |             Semi-robust
   outwork   |    Coef.    Std. Err.     z    P>|z|   [95% Conf. Interval]
-------------+--------------------------------------------------------
    female   |  1.812054   .0546016   33.19  0.000    1.705037   1.919071
   married   |  .1092522   .0564717    1.93  0.053   -.0014303   .2199347
  edlevel2   | -.2267445   .1124875   -2.02  0.044   -.4472159   -.006273
  edlevel3   |  .2394851   .1030909    2.32  0.020    .0374307   .4415396
```

```
edlevel4 | -1.020864  .1412912   -7.23 0.000  -1.29779 -.7439386
   _cons | -1.630541  .0661374  -24.65 0.000 -1.760168 -1.500914
------------------------------------------------------------------

. xtcorr

Estimated within-id correlation matrix R:

          c1      c2      c3      c4      c5
r1  1.0000
r2  0.7377  1.0000
r3  0.5442  0.7377  1.0000
r4  0.4015  0.5442  0.7377  1.0000
r5  0.2962  0.4015  0.5442  0.7377  1.0000
```

Reiterating the discussion of autoregressive models earlier in the chapter, we note that each diagonal away from the main is a power from the preceding diagonal.

```
. di .7377 ^2
.54420129

. di .7377 ^3
.40145729

. di .7377 ^4
.29615504
```

MARKOV STRUCTURE

```
. xtqls outwork female married edlevel2-edlevel4 , i(id)
t(year) c(Markov) f(bin 1) vce(robust)

GEE population-averaged model  Number of obs      =      19609
Group and time vars:   id year Number of groups   =       6127
Link:                    logit Obs per group: min =          1
Family:               binomial                avg =        3.2
Correlation: fixed (specified)               max =          5
                                    Wald chi2(5)  =    1174.44
Scale parameter:            1       Prob > chi2   =     0.0000

                       (Std. Err. adjusted for clustering on id)
------------------------------------------------------------------
             |            Semi-robust
   outwork   |    Coef. Std. Err.     z  P>|z| [95% Conf. Interval]
---------+--------------------------------------------------------
    female   | 1.817712 .0544941  33.36 0.000  1.710905  1.924518
   married   | .1107938 .0562188   1.97 0.049  .0006071  .2209806
```

```
edlevel2 | -.2241368 .1109069  -2.02 0.043 -.4415104 -.0067632
edlevel3 |  .2487125 .1020215   2.44 0.015  .048754    .448671
edlevel4 | -1.032976 .1410568  -7.32 0.000 -1.309442 -.7565094
   _cons | -1.636607 .0658968 -24.84 0.000 -1.765762 -1.507452
----------------------------------------------------------------

. xtcorr

Estimated within-id correlation matrix R:

        c1      c2      c3      c4      c5
r1  1.0000
r2  0.7592  1.0000
r3  0.5764  0.7592  1.0000
r4  0.4376  0.5764  0.7592  1.0000
r5  0.3322  0.4376  0.5764  0.7592  1.0000
```

The Markov correlation structure is similar to the autoregressive, with differences existing because equality of time measures is not assumed. Markov is therefore a generalization of the autoregressive structure, and is generally the preferred structure of the two.

13.2.7 Feasibility

GEE models fail when the correlation matrix is not positive definite. This occurs when values of the matrix exceed structural assumptions, called Prentice bounds (1988). We use the term *feasible* to designate values that yield a positive definite correlation matrix.

The feasible region of a top-most non-diagonal value for specified correlation structures that can be adjusted by QLS is defined as:

EXCHANGEABLE: $-1/(n_m-1)$; 1, where n_m is the maximum value of n_i over $i = 1, 2, ..., m$, that is, it is the value of the panel with the greatest number of observations.

For the German health data (*rwm_1980*), we can calculate the highest value for a panel as follows:

```
. sort id
. by id: egen cnt=count(id)
. su cnt

    Variable |     Obs     Mean  Std. Dev.     Min     Max
-------------+----------------------------------------------
         cnt |   19609 3.865725  1.218388       1       5
```

$n_m = 5$. Therefore the left-side range is $-1/(5 - 1) = -0.25$. The range of feasibility is then $(-.25; 1)$. Since $\alpha = 0.6424$, α is feasible. See Section 13.1.2.2 for the exchangeable correlation structure.

AR-1: $-1; 1$
Note: A negative α, such as -0.90, produces within subject correlations that alternate in sign. For example, the correlation between the first and second measures of a subject will be:

$$(-0.90)^{|2-1|} = -0.90$$

Correlation between the 1st and 3rd measures will be $(-0.90)^{|3-1|} = 0.81$. Note the value of α for the QLS autoregressive structure is 0.7377, within the range of feasibility. The GEE model produced a value of $\alpha = 0.7240$, also a feasible value.

STATIONARY (TRIDIAGONAL): $(-1/c_m; 1/c_m)$ where $c_m = 2\sin[(\pi[n_m-1])]/ (2[n_m + 1])]$. For large n_m, the interval is approximately $(-0.5; 0.5)$. For the German data,

$$c_m = 2 * \sin((\_pi * 4)/(2 * 6) = 1.73205$$
$$\text{or } 1.73. \text{ Range} = -1/1.73; 1/1.73 \text{ or } -0.578; 0.578 \tag{13.16}$$

Recall that the GEE model with a stationary correlation structure given in Section 13.2.2.5 of this book failed to converge. For the model to converge, the correlation structure must be positive definite. And, to be positive definite, the values of the correlation matrix must be feasible for that structure. It is important to realize, however, that even if the Prentice bounds are satisfied, a compatible distribution that gives rise to the data for a particular correlation structure may not exist. That is, simply meeting Prentice bounds requirements does not guarantee convergence, but on the other hand, not resting within the bounds could result in nonconvergence.

For models with a time component, such as the stationary and nonstationary, only panels having more than the specified rank of the structure are retained for the estimation process. A stationary 1 structure drops all single observation panels. A note to this effect is given in the displayed output.

For ease of comparison, the GEE model is again displayed.

GEE

```
. xtgee outwork female married edlevel2-edlevel4 , i(id)
t(year) c(sta 1) f(bin 1) vce(robust) nolog force
note:   some groups have fewer than 2 observations
        not possible to estimate correlations for those groups
        1150 groups omitted from estimation
```

```
GEE population-averaged model      Number of obs      =    18459
Group and time vars:      id year  Number of groups   =     4977
Link:                       logit  Obs per group: min =        2
Family:                    binomial                avg =      3.7
Correlation:            stationary(1)                max =        5
                                    Wald chi2(5)       =  1078.32
Scale parameter:                 1  Prob > chi2        =   0.0000

                     (Std. Err. adjusted for clustering on id)
------------------------------------------------------------------
             |             Semi-robust
   outwork   |    Coef.   Std. Err.    z   P>|z|  [95% Conf. Interval]
-------------+----------------------------------------------------
    female   |  2.028307  .0635572  31.91 0.000   1.903737  2.152877
   married   |  .2119104  .0742742   2.85 0.004    .0663357   .357485
  edlevel2   | -.2721441  .1251946  -2.17 0.030   -.517521 -.0267672
  edlevel3   |  .1607932  .1149839   1.40 0.162   -.0645711  .3861576
  edlevel4   | -1.042537   .148188  -7.04 0.000   -1.33298 -.7520936
     _cons   | -1.902586  .0854885 -22.26 0.000   -2.07014 -1.735031
------------------------------------------------------------------
working correlation matrix not positive definite
convergence not achieved
r(430);

. xtcorr

Estimated within-id correlation matrix R:

        c1       c2       c3       c4       c5
r1  1.0000
r2  0.7213  1.0000
r3  0.0000  0.7213  1.0000
r4  0.0000  0.0000  0.7213  1.0000
r5  0.0000  0.0000  0.0000  0.7213  1.0000
```

The value of alpha is 0.7213. Recall that the range of feasibility calculated by the method described above is −0.578; 0.578. Since 0.7213 is outside the feasibility range, it fails to converge. However, this value of alpha is likely not the true value of alpha. It is displayed based on a model that failed to converge properly. The statistical results cannot be trusted. On the other hand, it is the only correlation value we have, and is in fact not too different from the true value. We can therefore use it as a guide to why a model does or does not converge.

Next, the same predictors are used with a quasi-least squares model. A caveat should be given that for models using the stationary correlation structure, convergence may take a considerable amount of time when the number of observations is large—as in this instance.

QLS

```
. xtqls outwork female married edlevel2-edlevel4 , i(id)
t(year) c(sta 1)
  f(bin 1) vce(robust)

Iteration 1: tolerance = .04119572
Iteration 2: tolerance = .00087612
Iteration 3: tolerance = .00003303
Iteration 4: tolerance = 1.746e-06
Iteration 5: tolerance = 8.004e-08
```

```
GEE population-averaged model     Number of obs      =     19609
Group and time vars: id __00000S  Number of groups   =      6127
Link:                      logit  Obs per group: min =         1
Family:                 binomial                 avg =       3.2
Correlation:    fixed (specified)                max =         5
                                  Wald chi2(5)       =   1178.90
Scale parameter:                1 Prob > chi2        =    0.0000
```

```
                       (Std. Err. adjusted for clustering on id)
---------------------------------------------------------------
            |             Semi-robust
    outwork |     Coef.  Std. Err.      z   P>|z|  [95% Conf. Interval]
---------+-----------------------------------------------------
     female |  1.930828  .0578594   33.37 0.000   1.817425   2.04423
    married |  .1626784  .0646077    2.52 0.012   .0360496   .2893073
   edlevel2 | -.3276806  .1150013   -2.85 0.004  -.553079  -.1022821
   edlevel3 |  .2326253  .1020695    2.28 0.023   .0325727   .4326779
   edlevel4 | -1.036127  .1377324   -7.52 0.000  -1.306078  -.7661769
      _cons | -1.762515  .0739718  -23.83 0.000  -1.907497  -1.617533
---------------------------------------------------------------
```

```
. xtcorr

Estimated within-id correlation matrix R:

          c1       c2       c3       c4       c5
r1    1.0000
r2    0.5296   1.0000
r3    0.0000   0.5296   1.0000
r4    0.0000   0.0000   0.5296   1.0000
r5    0.0000   0.0000   0.0000   0.5296   1.0000
```

Notice that the value of alpha is now 0.5296, just inside the outer feasibility or Prentice bound of 0.578. The model converges, but when alpha is close to the edge of the boundary, it typically consumes considerable processing time.

Since the QLS model converges, and the correlation matrix is feasible, the estimates and standard errors can generally be trusted. However, we have several caveats to give in general concerning binary response QLS models later in this section.

An alternative method of determining whether a correlation matrix is positive definite is to evaluate the matrix eigenvalues. If the eigenvalues are positive, the matrix is positive definite. For example, given the previously estimated model, we may calculate matrix eigenvalues by:

```
. matrix R = e(R)

. matrix symeigen x v = R

. matrix list v

v[1,5]
            e1          e2          e3          e4          e5
r1   1.9172729   1.5295877           1   .47041225   .08272711
```

All of the eigenvalues are positive. Therefore, the matrix is positive definite and the estimates are feasible.

We can also obtain the restrictions on alpha directly (in order for the matrix to be positive definite). Let \$max represent the dimension of the correlation matrix, which is 5.

```
. global max = 5

. scalar a = -1/(2*sin((scalar($max)-1)/(scalar($max) + 1)*_pi/2))

. scalar b = 1/(2*sin((scalar($max)-1)/(scalar($max) + 1)*_pi/2))

. scalar list a b
          a = -.57735027
          b =  .57735027
```

Since alpha is between the upper and lower bounds the matrix is positive definite.

If a stationary QLS model, or any QLS model for that matter, still does not converge even if its alpha is within the bounds of feasibility, it is due to causes other than the correlation structure.

MARKOV: (0; 1)

The Markov correlation structure is .7592, which is within the range of feasibility.

13.2.8 Final Comments on GEE

Prentice (1988) noted that there are additional constraints that alpha, the top-most correlation value, must satisfy for binary longitudinal data before assuring feasibility suitable for unbiased GEE model estimation.* The impact of the violation of these additional constraints is the topic of current debate and ongoing research. For example, Rochon (1998) argued that these potential violations will not typically have a serious impact on modeling GEE logistic models, but that they should be taken into account, especially at the design stage of a study. Chaganty and Joe (2004) disagreed with Rochon, arguing that due to the potential violation of bounds, the working structures should not "be confused with the true correlation structure" but instead should be considered a weight matrix. Shults et al. (2006) suggested an adjusted QLS approach that forces the Prentice bounds to be satisfied if the final estimate of alpha is found to violate the estimated bounds. Another charge against the binary GEE/QLS model is that upper and lower constraints for alpha depend on both the coefficients and covariates, and as such the upper bound could conceivably tend to zero as a predictor goes to infinity. Of course, if the GEE/QLS logistic model consists entirely of binary/categorical predictors, this criticism loses its force.

In any case, the impact of, and best way to address, the violation of bounds—particularly for logistic models—are issues that have not yet been resolved. In fact, our remodeling of the stationary GEE model that failed to converge used a QLS logistic model. The range of feasibility appeared to be satisfactorily adjusted when we used QLS, resulting in convergence. However, other data specific considerations are required to establish a range of feasible correlation values for binary response data. The QLS model converged properly, and the values are feasible; we know therefore that we were likely in the bounds. A thorough discussion of the criteria required to establish appropriate bounds for binary GEE/QLS models can be found in Shults, Sun, et al. (2008). It should be noted, though, the bounds discovered for a particular binary response model and correlation structure usually differs only a little from the constant preestablished bounds listed in this section, for instance, the stationary 1, or tridiagonal bounds of $(-1/c_m; 1/c_m)$.

We should mention that QLS appears to be superior to ALR for modeling otherwise logistic GEE models. First, although ALR is in general less restrictive on the bounds of alpha than is GEE, QLS is also less restrictive. In particular, the use of the AR1 structure with QLS logistic models is minimally restrictive. Second, QLS is more efficient an estimator than is ALR. Taken together, these two reasons tend to lead one to prefer QLS over ALR when modeling otherwise GEE logistic models.

---

* Unique to correlated Bernoulli data is the fact that the probability of $y_{ij} = 1$, $\Pr(y_{ij} = 1)$, and the correlation between measurements of y_{ij} and y_{ik}, or $\text{Corr}(y_{ij}, y_{ik})$, thoroughly determines the bivariate probability distribution of both y_{ij} and y_{ik}. With p being defined as $\Pr(y = 1)$, the bivariate distribution may be characterized as: (See Equation 13.17)

The advantage of GEE over random-effects models relates to the ability of GEE to allow specific correlation structures to be assumed within panels. Parameter estimates are calculated without having to specify the joint distribution of the repeated observations.

In random-effects models, a between-subject effect represents the difference between subjects conditional on them having the same random effect. Such models are thus termed conditional or subject-specific models. GEE parameter estimates represent the average difference between subjects, and thus are known as marginal, or population-averaged models. Which model to use depends on the context or the goals of the study.

Current GEE software typically comes with a set of specified correlation structures. Nearly all applications have the same structures that were employed in the first SAS GEE macro authored in the late 1980s. However, other correlation matrices can be used as well, and applications such as SAS and Stata allow for user-designed matrices to be implemented in place of the standard structures. The Markov structure, discussed above, is a useful extension of the traditional correlation offerings, but as yet is only implemented in QLS software. Because of its value to the modeling of longitudinal data, we suspect that it will be found in commercial statistical software in the near future. Other structures will likely be implemented as well. For example, a familial correlation structure has recently been developed to help understand complex family relationships. This structure can also be useful for genetic studies. Additionally, a mixed correlation structure where distinct R(a)s are constructed for each cluster or set of clusters in the data would be helpful when modeling many different types of complex clustered data. Although, to reiterate, most statistical software allows the user to implement these structures by having a "user-defined" structure option, it would be helpful to have more useful matrix options such as Markov built into the software.

The only text specifically devoted to generalized estimating equations is Hardin and Hilbe (2003).

The second edition of Diggle, Liang, and Zeger's 1994 foundational text, with Patricia Heagerty as an added author (2002), has an excellent chapter covering GEE methods. Twist (2004) also explains GEE in an articulate manner. These texts are recommended for those who want to learn more about this area of statistics, and longitudinal logistic models in particular. Those interested in quasi-least squares regression and its relationship to GEE and other population averaging methods are referred to Hilbe and Shults, *Quasi-Least Squares Regression*, forthcoming in 2010, or to any of the citations listed in this section.

$$\Pr(y_{ij}, y_{ik}) =$$

$$p_{ij}^{y_{ij}}(1-p_{ij})^{1-y_{ij}}\, p_{ik}^{y_{ij}}(1-p_{ik})^{1-y_{ij}} \left\{ 1 + \mathrm{Corr}(y_{ij}, y_{ik}) \frac{(y_{ij}-p_{ij})(y_{ik}-p_{ik})}{[(p_{ij}\, p_{ik}\,(1-p_{ij})(1-p_{ik})]^{.5}} \right\}$$

$$(13.17)$$

13.3 Unconditional Fixed Effects Logistic Model

There are a variety of fixed effects logistic models, each based on how the data are structured and on what the model is aimed to tell us. Fixed effects models come in two major types: unconditional and conditional. The conditional model is statistically the more interesting, and is what most statisticians refer to when using the term "fixed effects regression" model. When applied to the logistic distribution, the conditional fixed effects model is usually known as conditional logistic regression.

The unconditional fixed effects model is specified by including a separate fixed effect for each identified panel or group in the data. Standard logistic regression commands can be used to estimate the unconditional fixed effects model. The effects are modeled like any categorical predictor, with an excluded panel as the reference.

Each effect is a constant within each group or panel in the data. For example, in an analysis of treatment protocols or types of patients that characterize hospitals in a large metropolitan area, each hospital can be regarded as a fixed panel. The value of *hospital* is the same for all members of the panel.

The log-likelihood for the unconditional fixed effects logistic model can be expressed as follows:

$$L(x_{it}\beta; y_{it}) = \sum_{i=1}^{n} \sum_{t=1}^{ni} \left\{ y_{it}(v_i + x_{it}\beta) - \ln(1 + \exp(v_i + x_{it}\beta)) \right\} \qquad (13.18)$$

For the unconditional fixed effects model, v_i is estimated as a vector of indicator variables representing each panel in the data.

For an example of using an unconditional fixed effects logistic model we shall use the **wagepan** data, which are taken from a national economic survey related to wages as well as a host of other indicators. The survey was taken over an 8-year period.

Model *black* (black = 1; other = 0) on the following predictors.

south	1 = lives in South; 0 = lives elsewhere in the U.S.
agric	1 = works in agricultural industry; 0 = not work in industry
married	1 = married; 0 = not married
lwage	natural log of yearly wage in thousands of dollars.

We give the model an eight-level fixed effects covariate, representing the years 1980 through 1987. Each year is separated into indicator variables with 1980 as the reference. These effects are called *d81* through *d87*. Each year consists of 545 observations, and are considered to be panels within which the years do not change. They are thereby fixed for each panel. Additionally, it is appropriate to use a cluster robust variance estimator to adjust standard errors for the year panel effect.

The unconditional fixed effects logit model appears as:

```
. logit black south agric married educ lwage d81-d87, nolog or
cluster(year)

Logistic regression                              Number of obs =      4360
                                                 Wald chi2(4)   =         .
                                                 Prob > chi2    =         .
Log pseudolikelihood = -1439.9278                Pseudo R2      = 0.0776

                              (Std. Err. adjusted for 8 clusters in year)
-----------------------------------------------------------------------
             |              Robust
   black |  Odds Ratio  Std. Err.       z    P>|z|   [95% Conf. Interval]
--------+--------------------------------------------------------------
   south |   3.054709   .0890237    38.32   0.000    2.885115   3.234271
   agric |   .4596153   .0937318    -3.81   0.000      .30818   .6854639
 married |   .3620182   .0139206   -26.42   0.000    .3357371   .3903564
    educ |   .9733442   .0092098    -2.86   0.004    .9554598   .9915635
   lwage |   .7086386   .0635019    -3.84   0.000    .5944942    .844699
     d81 |   1.135082   .0132438    10.86   0.000    1.109419   1.161338
     d82 |   1.194605   .0210417    10.10   0.000    1.154068   1.236566
     d83 |   1.340502   .0331919    11.83   0.000       1.277   1.407161
     d84 |   1.446363   .0427927    12.47   0.000    1.364876   1.532714
     d85 |    1.50772   .0515015    12.02   0.000    1.410084   1.612117
     d86 |   1.568054   .0615655    11.46   0.000    1.451914   1.693485
     d87 |   1.666033   .0768363    11.07   0.000    1.522043   1.823646
-----------------------------------------------------------------------
```

Note that the odds ratio of each panel gets larger with each year. We used a robust variance estimator to adjust standard errors to account for the extra correlation resulting from the year effect.

Unconditional fixed effects models are best used when the data are gathered over a number of fixed units, such as years, hospitals, persons, schools. The number of groups or panels should not be large; however, the panels themselves should have more than a few observations, preferably greater than 16. Studies have demonstrated that more bias exists in the estimates and standard errors as the panel size becomes smaller. See Katz (2001) for an excellent discussion of this topic.

The modeling and interpretation of the estimates are the same as for standard logistic regression. It is not always necessary to use a cluster robust estimator when modeling unconditional fixed effects data, but it is important to evaluate models with and without such adjustments to determine if their application results in significantly different standard errors. If so, then employing robust variance estimates is warranted.

Other features of the unconditional fixed effects model will be discussed in the following section on conditional logistic regression. In particular, we shall expand on the differences between unconditional and conditional parameterizations.

13.4 Conditional Logistic Models

Conditional logistic regression is also known as conditional fixed effects logistic regression, matched case-control logistic regression, or McFadden's Choice Model. The estimation algorithm is identical in all three cases. The differences among them are based on how the data are structured and for what purpose the analysis is being used.

Conditional models, in all of their varieties, are types of panel model; there is typically more than one line of data or information about a subject or item. The panel representing a grouping variable, for example, a person being tested over time, is identified by a panel indicator. In Stata's conditional fixed effects logistic model, **xtlogit** (with *fe* option), the panel or group identifier is *i()*. A user provides the *i()* option with the name of the variable indicating group membership and *fe* to indicate that the data are to be modeled as a conditional fixed effects logit model. Stata's other conditional logit model is called **clogit**. Users tend to employ **clogit** when modeling matched case-control data, but **xtlogit** will likewise model data in case-control format. The panel or grouping indicator for **clogit** is *group()*.

Sections 13.4.1 through 13.4.3 present a brief overview of several of the major parameterizations or versions of the conditional fixed effects model. We begin with the traditional fixed effects logistic model.

13.4.1 Conditional Fixed Effects Logistic Models

First, it is important to describe the structure of fixed and random effects models in general. It will then be easier to understand the differences between them, and it will also give a better insight into the conditional fixed effects model.

Let us represent a panel model given the following schema:

$$y_{ij} = \beta_0 + \beta_i x_{ij} + \gamma_i + \varepsilon_{ij} \qquad (13.19)$$

y is the response, and in the case of a logistic panel model, a binary response.

i indicates separate individuals or items.

j represents the observation number within *i*. If the study is a longitudinal model, *j* represents the time a measurement on *i* has occurred.

ε is the error term, usually with a mean of 0 and variance of σ^2.

γ is a panel indicator. For fixed effects models, γ contains the fixed characteristics of an individual or item. For a random effects model, γ is a random covariate having a known probability distribution. In most applications γ is distributed as Gaussian, or normal.

If γ is a fixed effect, modeling may take either of two essential forms. The unconditional fixed effects model estimates the fixed parameters as a set of indicator predictors. The conditional fixed effects model conditions the fixed effects out of the estimation.

A central concern of panel models in general is how between-item or person variation differs from within-item variation. The term item is used here to indicate that panels may be of inanimate or non-human events. However, we usually think of them in terms of individuals. For our discussion then we shall use the person reference, but be aware that panels are commonly applied to non-human events or entities.

Rephrasing the above, a central concern in panel analysis is the relationship of between-person to within-person variation. Certainly, for fixed individual characteristics such as gender, height, date of birth, and so forth, there is no within-person variation. There is, however, between-person variation. Fixed effects models, though, emphasize the within-person variation for those predictors that are not fixed. Fixed covariates are conditioned out of the estimation process. The problem with conditional fixed effects models, which we simply refer to as fixed effects models from this point, is that it ignores between-person variation. Doing this allows fixed effects, γ, to be conditioned from estimation. If between-person variation was considered, then the algorithm could not condition out the fixed effects. The downside, however, of ignoring between-person variation is the inflation of model standard errors. Random effects models solve this problem by accounting for both within- and between-person variation. γ is not a container for fixed effects, but rather the effects are assumed to follow a specified probability distribution, the parameters of which are estimated by the model.

Again, understanding the relationship of within- to between-person variation is essential to comprehending the distinctions between fixed and random effects models. It bears on the different types of fixed effects models as well. As a consequence, it might be helpful to elaborate on the above discussion.

Let us suppose that we have a longitudinal study where individuals are measured over a 12-month period. Each individual, i, in the study, is measured, j, 12 times—once per month for a year. Each value for j represents the month that measurements were taken, with the final month being j_{12}, or J. Certain characteristics of each individual, sex, ethnic background, age (when study started), religion, country of birth, and so forth, if entered into the model, do not vary within i. The values for each j within a given i are the same. Of course, some studies may be involved with sex changes, in which case sex too would vary. Likewise, the same is the case for religion. People sometimes change their religious belief. However, this is rare for studies that are not directly concerned with these personal characteristics. If religion commonly changes in our study, or we are studying people who have, or have a tendency, to change religions, then religion is not considered to be a fixed effect.

Even for predictor values that are likely to change over the course of a year, some, such as income, may have minimal within-person variation compared with the between-person variation. The fixed effects model is best suited when there is substantial within-person variation.

We have touched only briefly on random effects models. However, it might be useful at this point to indicate a few of the benefits—and downsides—of each.

We have mentioned that fixed effects models are most useful for detecting within-person variation. With fixed effects being conditioned out by eliminating within-person fixed values, all that is left is within-person differences. Random effects models, on the other hand, balance both variances, but do so by assuming that the random effect is distributed in a given manner. Usually, for a given model, there are only one or two types of distributions that are appropriate for estimating random effects and parameter estimates.

Random effects models, as we shall discuss in Section 13.5, assume that the effects are randomly distributed according to the parameters of an underlying distribution. In the case of the logistic random effects model, the distribution is usually either normal or beta. Beta is the conjugate prior of the binomial distribution, and therefore can be better mixed with the latter, giving us a beta-binomial model. In any case, the assumption of having effects follow a particular distribution is a major one. When it appears that such a distribution is not appropriate to the data, a conditional fixed effects model may yield less biased estimates. Fixed effects models are in this respect less restrictive than random effects models, and as such usually model the data in a way that comes closest to reality.

A difference that must be emphasized between unconditional and conditional fixed effects models relates to the size of the panels or groups in the model. We touched on this in the last section, but it is important to reiterate. Since the unconditional fixed effects model provides parameters for each panel in the model, if the number of groups, items, or persons in each panel is less than 16, there is increased bias in the estimates and standard errors

(Katz, 2001). On the other hand, when there is a large number of panels in the data, using an unconditional fixed effects model leads to difficulties in handling and interpretation. When there is a large number of panels, it is preferred to use a conditional fixed effects model, which conditions out the panels. They are not displayed in the output as they are with the unconditional parameterization.

The **heart01** data are used to demonstrate how the conditional fixed effects model can be implemented. Here we condition on the hospital to which patients have come when they have experienced—or thought they have sustained—some type of cardiovascular distress. Hospitals are referred to as centers in the data, which as we may recall, comes from the national Canadian registry on cardiovascular disease, FASTRAK.

```
. xtlogit death anterior hcabg kk2-kk4 age3-age4, i(center)  fe
or nolog


Conditional fixed-effects logistic regression
Group variable: center
                                      Number of obs     =      4274
                                      Number of groups  =        35
                                      Obs per group: min =        19
                                                      avg =     122.1
                                                      max =       215
                                      LR chi2(7)        =    218.02
Log likelihood  = -552.30782          Prob > chi2       =    0.0000
-------------------------------------------------------------------
    death |      OR   Std. Err.     z   P>|z|  [95% Conf. Interval]
----------+--------------------------------------------------------
 anterior |    1.908   .3223057   3.82  0.000   1.370221   2.656845
    hcabg | 2.123426    .771463   2.07  0.038   1.041806   4.328004
      kk2 | 2.429385   .4477265   4.82  0.000   1.692868   3.486339
      kk3 | 2.611708   .7267183   3.45  0.001    1.51382   4.505831
      kk4 | 19.36256   7.280319   7.88  0.000   9.266406   40.45893
     age3 |  3.67474    .749447   6.38  0.000   2.463927   5.480566
     age4 | 7.364812   1.566507   9.39  0.000   4.854111   11.17413
-------------------------------------------------------------------


. abic
AIC Statistic =    .2617257            AIC*n       = 1118.6156
BIC Statistic =    .2648242            BIC(Stata)  = 1163.1378
```

There are 35 hospitals, represented as panels, in the model, with an average number of observations in each panel given as 122.1. Notice the extremely high odds ratios of the top levels of killip and age group. The software dropped 12 (229 obs) groups from the analysis because they had either all positive or all negative responses.

13.4.2 Matched Case-Control Logistic Model

We have previously discussed the fact that the data in a conditional fixed effects model is grouped in terms of panels. The generic conditional fixed effects model specifies a variable, which is constant for all members of the group, thereby identifying it as separate from other panels. A matched case-control model, which uses the same estimating algorithm as was used in the previous section, is structured so that the members of each panel or group are contiguous, that a panel indicator exists, and that there are two observations per panel, one with a response of 1, the other with a response value of 0. This is called a 1:1 matched case-control model. A 1:M matched case-control model may be developed by extending the number of 0 responses per panel. The same estimating algorithm is used in both cases, with a single response having the value of 1. Other observations or members of a panel have response values of 0. The data are set up in a manner similar to the form below:

1:1 MATCHED CASE-CONTROL

	y	x1	x2	id
1.	1	0	2	101
2.	0	1	3	101
3.	1	0	1	102
4.	0	1	2	102
5.	1	1	3	103
6.	0	0	0	103
7.	1	1	0	104
8.	0	0	2	104
9.	1	1	3	105
10.	0	1	2	105
11.	1	0	1	106
12.	0	0	1	106
13.	1	1	2	107
14.	0	0	2	107
15.	1	1	3	108
16.	0	1	1	108
17.	1	0	3	109
18.	0	0	1	109
19.	1	0	3	110
20.	0	1	3	110

1:M MATCHED CASE-CONTROL

	y	X1	X2	id
1.	1	0	2	101
2.	0	1	3	101

```
 3.    0        0        1     101
 4.    1        1        2     102
 5.    0        1        3     102
 6.    0        0        0     102
 7.    0        1        0     102
 8.    1        0        2     103
 9.    0        1        3     103
10.    0        1        2     103
11.    0        0        1     103
12.    0        0        1     103
```

The values of y indicate the matched case-control where 1 can indicate a *case* and 0 a *control*. The *id* variable indicates matched persons or items; however, *id* may be different persons that are matched, or the same person over time. *X1* and *X2* are predictors that may vary within the panel or group, or over time when the study is longitudinal. In other fixed effects implementations, *id* can represent an individual or item within a cluster or over time. It makes no difference as far as the estimating equations are concerned.

The log-likelihood of the 1:M matched case-control model may be expressed as:

$$\sum_{i=1}^{N} \frac{\exp(u_i \beta)}{\sum_{j=1.ni} \exp(v_{ij} \beta)} \tag{13.20}$$

with the 1:1 model being reduced to:

$$\sum_{i=1}^{N} \frac{\exp[(u_i - v_{i1}) \beta]}{1 + \exp[(u_i - v_{i1}) \beta]} \tag{13.21}$$

where u_i is the vector of case covariates and v_{ij} is the vector of control covariates. Note in the 1:1 function that the single control record for each panel is subtracted from the case record. This allows for a much simpler estimation.

Modeling the above 1:1 matched case-control data using a conditional fixed effects logistic model is displayed as:

```
. xtlogit y x1 x2, nolog i(id) or fe

Conditional fixed-effects logistic regression
Group variable: id
                                        Number of obs      =        20
                                        Number of groups   =        10
```

```
                              Obs per group: min =          2
                                             avg =        2.0
                                             max =          2

                              LR chi2(2)            =     0.73
Log likelihood  = -6.5660682  Prob > chi2           =  0.6939
---------------------------------------------------------------
    y |       OR  Std. Err.     z  P>|z|  [95% Conf. Interval]
------+--------------------------------------------------------
   x1 | .8286659  .7268057  -0.21  0.830   .148528     4.623286
   x2 | 1.456267  .6677226   0.82  0.412  .5928585       3.5771
---------------------------------------------------------------
```

Modeling the 1:M matched case-control data results in the following model output:

```
. xtlogit y x1 x2, nolog i(id) or fe

Conditional fixed-effects    logistic regression
Group variable: id
                              Number of obs      =         12
                              Number of groups   =          3

                              Obs per group: min =          3
                                             avg =        4.0
                                             max =          5

                              LR chi2(2)            =     2.68
Log likelihood   = -2.7       Prob > chi2           =   0.2613
---------------------------------------------------------------
    y |       OR  Std. Err.     z  P>|z|   [95% Conf. Interval]
------+--------------------------------------------------------
   x1 | .028343  .0800645  -1.26  0.207   .0001117     7.193728
   x2 | 4.81772  5.652797   1.34  0.180    .483167      48.0381
---------------------------------------------------------------
```

For a final example, we model 1:1 matched case-control data from a study on low birthweight babies. The data set comes from Hosmer and Lemeshow (2000) and is named **lowbirth.dta**. The panel indicator is given in the variable *pairid*.

The data include the following variables, with *low* as the response:

Pairid	Case-control pairid
Low	Low birthweight baby: < 2.5 kg (5lb 8oz)
Lwt	Mother: last menstrual weight
Smoke	Mother: smoked during pregnancy
Ptd	Mother: previous preterm baby
Ht	Mother: has hypertension
Ui	Uterine irritability

The data are modeled as:

```
. use lowbirth

. xtlogit low lwt smoke ptd ht ui, i(pairid) fe nolog or

Conditional fixed-effects logistic regression
Group variable (i): pairid
                                            Number of obs      =      112
                                            Number of groups   =       56

                                            Obs per group: min =        2
                                                           avg =      2.0
                                                           max =        2

                                            LR chi2(5)         =    25.16
Log likelihood  = -26.236872                Prob > chi2        =   0.0001
----------------------------------------------------------------------------
    low |        OR   Std. Err.      z   P>|z|  [95% Conf. Interval]
--------+-------------------------------------------------------------------
    lwt | .9850298   .0080246   -1.85   0.064  .9694268   1.000884
  smoke | 4.391033   2.467844    2.63   0.008  1.459403   13.21169
    ptd | 5.315326   3.969518    2.24   0.025  1.229836   22.97272
     ht | 10.27138   10.29757    2.32   0.02   1.439651   73.28252
     ui | 3.837782   2.662818    1.94   0.053  .9850994    4.95135
----------------------------------------------------------------------------
```

13.4.3 Rank-Ordered Logistic Regression

Rank-ordered logistic regression is known by several alternative names, including exploded logit and the choice-based method of conjoint analysis. It has also been called the Plackett–Luce model. The method was initially constructed by Beggs, Cardell, and Hausman (1981). An excellent presentation of this model can be found in Allison and Christakis (1994).

The model is set up by having each alternative choice listed as a separate observation. If there are five choices, there will be five separate observations; however, they will be linked by a common *id* number. If the model only addresses the most favored alternative, instead of listing all of the alternatives, it is identical to a standard conditional logistic regression model.

Suppose that five individuals are given a list of vacation destinations and are asked to rank them in the order in which they would prefer to visit. The destinations, or choices, are

A = Hawaii   B = Las Vegas   C = Europe   D = New York   E = Bahamas

Each of the five individuals may well have different orders of destination preferences, such as:

1: A → B → C → D → E
2: E → A → B → C → D
3: B → E → C → D → A
4: A → B → D → E → C
5: A = B → D → C → E

Note also that individual five has made a choice that the first and second selected destinations are equally preferred. This is a tie.

Rank-ordered logistic regression (ROLR, also referred to as ROLM) models the rankings given by individuals, where the choices may be different as well as tied. The model is quite unlike the previous models we have discussed, but it is one that plays an important role in marketing research. The model was first discussed by Punj and Staelin (1978) as a generalization of the conditional logit for ranked outcomes model.

The formal structure of the ROLR model can be expressed as:

$$\Pr\left(y_i = s \mid x\right) = \frac{\exp(x\beta_{c|b})}{\left\{\sum_{k=1} \exp(x\beta_{k|b})\right\} - \exp(x\beta_{c|b})} \tag{13.22}$$

with subscripts c = choice, b = base level, and k = order value of c, x = predictor values, and β are the coefficients.

An example should clarify the nature of the model and how the data are structured. We use an example from Long and Freese (2006), but have amended the dataset to accord with the requirements of the rank-ordered modeling algorithm.

The data, called **rankorderwls.dta**, are abstracted from the 1992 Wisconsin longitudinal study on job values. The data consist of information on what study subjects believed were the most important considerations in choosing one job over another. The categories or characteristics of a job are enumerated as:

- High esteem
- Variety
- Autonomy
- Security

The subjects were asked to rank these considerations from 1 to 4, with 1 representing the most important consideration. The data are structured so that each individual ranking is given a separate line in the data. For example, the person given ID number 13 made the following ranking choices:

```
. use rankorderwls

. l id rank esteem variety  autonomy security if id==13
```

```
     +------------------------------------------------------------+
     | id    rank    esteem    variety    autonomy    security |
     |------------------------------------------------------------|
29.  | 13     4        1          0           0           0 |
30.  | 13     2        0          1           0           0 |
31.  | 13     1        0          0           1           0 |
32.  | 13     3        0          0           0           1 |
     +------------------------------------------------------------+
```

The data tell us that subject 13 ranked autonomy as the most important consideration when seeking employment. Variety was ranked second, security as third, and esteem the least important consideration. In other words, this person wants control and variety over his job (subject 13 is a male).

As should be expected, most people ranked certain considerations together. They could not rank one higher than the other. Subject 1 ranked *variety* and *security* equally as the primary consideration. *Esteem* was the third choice, followed by *autonomy*.

. l id rank esteem variety autonomy security if id==1

```
     +------------------------------------------------------------+
     | id    rank    esteem    variety    autonomy    security |
     |------------------------------------------------------------|
1.   | 1      3        1          0           0           0 |
2.   | 1      1        0          1           0           0 |
3.   | 1      4        0          0           1           0 |
4.   | 1      1        0          0           0           1 |
     +------------------------------------------------------------+
```

Subject 12 ranked three considerations equally as first, with security being the least important consideration.

. l id rank esteem variety autonomy security if id==12

```
     +------------------------------------------------------------+
     | id    rank    esteem    variety    autonomy    security |
     |------------------------------------------------------------|
25.  | 12     1        1          0           0           0 |
26.  | 12     1        0          1           0           0 |
27.  | 12     1        0          0           1           0 |
28.  | 12     4        0          0           0           1 |
     +------------------------------------------------------------+
```

Two panel constant covariates are also included in the model, but as interactions. Since they are constant within each subject (panel), they fall out of the estimation process.

fem 1 = female; 0 = male

hn mental ability standardized test scores ranging from −2.84291 to 2.941332. Values are in standard deviations.

The entire data set of 3226 subjects, each consisting of four observations, can be modeled using Stata's **rologit** command as:

```
. rologit rank esteem variety autonomy hashi haslo esteemXfem
esteemXhn varietyXfem varietyXhn autonomyXfem autonomyXhn,
group(id) reverse nolog

Rank-ordered logistic regression
Group variable: id
                                   Number of obs      =     12904
                                   Number of groups   =      3226

Ties handled via the exactm method
                             Obs per group: min =          4
                                           avg =       4.00
                                           max =          4

Log likelihood = -6127.559         LR chi2(11)        =   1947.39
                                   Prob > chi2        =    0.0000
```

rank	Coef.	Std. Err.	z	P>|z|	[95% Conf. Interval]	
esteem	−1.01720	.05498	−18.50	0.000	−1.12497	−.90943
variety	.52822	.05258	10.04	0.000	.42515	.63129
autonomy	−.15167	.05103	−2.97	0.003	−.25170	−.05164
hashi	.17804	.03747	4.75	0.000	.10459	.25149
haslo	−.20641	.04258	−4.85	0.000	−.28987	−.12295
esteemXfem	−.14979	.07830	−1.91	0.056	−.30326	.00367
esteemXhn	.13753	.03942	3.49	0.000	.06025	.21481
varietyXfem	−.16402	.07283	−2.25	0.024	−.30676	−.02127
varietyXhn	.25904	.03709	6.98	0.000	.18633	.33174
autonomyXfem	−.14017	.07183	−1.95	0.051	−.28096	.00061
autonomyXhn	.21338	.03616	5.90	0.000	.14250	.28426

The *reverse* option was issued to tell the command that rank 1 is the highest choice, with 4 the least desired rank. The default is to have the highest numerical rank regarded as the most important. In addition, notice that *security* was made the reference level. Interpretation of the remaining characteristics is then in terms of the choice of security. The *has** variables refer to how the subject perceives his or her present job in relation to the four characteristics. If *hashi* on characteristic *variety* is 1, this indicates that the subject believes his/her current job is high on variety. If *haslo* = 1 when *variety* = 1, then the subject believes his current job has low variety. If both *hashi* and *haslo* are equal to 0, then the subject believes his job is moderate in *variety*.

```
. l rank fem esteem variety  autonomy security hashi haslo if
id==13
```

```
    +----------------------------------------------------------+
    | rank  fem  esteem  variety  autonomy  security  hashi  haslo |
    |----------------------------------------------------------|
29. |  4    0     1        0         0         0        0      1  |
30. |  2    0     0        1         0         0        1      0  |
31. |  1    0     0        0         1         0        0      0  |
32. |  3    0     0        0         0         1        0      1  |
    +----------------------------------------------------------+
```

Subject 13 believes that his (*fem* == 0) current job is moderate in autonomy, high in variety, and low in esteem and security.

If one is seriously modeling the above data it would be desirable to enter interaction effects for the various job characteristics. Other variables in the data, such as gender and mental test score, can also be brought into the model as interactions. *Security*, however, is the reference and is therefore not used in any interactions. We model the basic elements of the data for ease of interpretation.

The **rologit** command does not provide an option to display odds ratios. However, the **listcoef** command with the *percent* option provides the percent change in the odds of ranking a particular characteristic in relationship to the reference for a one-unit increment in a predictor value, with other predictor values maintained as constant. The **listcoef** command, written by Long and Freese (2006), can be used with a number of discrete response models. The *help* option provides information relevant to interpretation under the table of estimates, *p-values*, and values of percent change.

```
. listcoef, percent help
```

```
rologit (N=12904): Percentage Change in Odds
   Odds of: ranked ahead vs ranked behind
-------------------------------------------------------
       rank |       b         z      P>|z|       %
-------------+-----------------------------------------
      esteem |   -1.01720  -18.500    0.000    -63.8
     variety |    0.52822   10.044    0.000     69.6
    autonomy |   -0.15167   -2.972    0.003    -14.1
       hashi |    0.17804    4.751    0.000     19.5
       haslo |   -0.20641   -4.847    0.000    -18.7
  esteemXfem |   -0.14979   -1.913    0.056    -13.9
   esteemXhn |    0.13753    3.488    0.000     14.7
 varietyXfem |   -0.16402   -2.252    0.024    -15.1
  varietyXhn |    0.25904    6.983    0.000     29.6
autonomyXfem |   -0.14018   -1.951    0.051    -13.1
 autonomyXhn |    0.21339    5.900    0.000     23.8
-------------------------------------------------------
```

```
   b = raw coefficient
   z = z-score for test of b=0
P>|z| = p-value for z-test
   % = percent change in odds for unit increase in X
SDofX = standard deviation of X
```

The model can be interpreted by its interactions and characteristics. The most difficult interpretations are perhaps the characteristic by standardized test score interactions.

Each non-reference interaction has a non-0 value only if the characteristic in interaction has a value of 1. Multiplying any value by 0 always results in a 0 value.

Regardless, for the three *characteristic*hn* interactions, we can assert the following:

A one standard deviation increase in *hn* increases the odds ratio of selecting:

1. esteem over security (the reference) by 14.7 percent. —esteemXhn

2. variety over security by 29.6 percent. —varietyXhn

3. autonomy over security by 23.8 percent —autonomyXhn

holding all other predictors constant. The same logic maintains for changing from *male*(0) to *female* (1).

The *hashi* covariate can be interpreted from the table as meaning that if the subject believes that his or her job is high rather than moderate in any particular characteristic, the odds of selecting that characteristic over another increase by 19.57 percent. If the subject believes that his or her job is low rather than moderate, the odds of selecting that characteristic over another decrease by 18.7 percent.

Lastly, it is possible to calculate predicted probabilities for each alternative following the modeling of data by either a conditional or a ranked order logistic regression. However, one must specify the values of the predictors he or she wishes to use as a basis of calculating probabilities. The command does not provide probabilities for each observation in the data, as does *predict* following **logit, logistic, ologit**, and so forth.

To compare the probabilities of selecting or ranking each characteristic rather than the reference choice *security*, we can use another command, authored by Long and Freese (2006), called **asprvalue**. In fact, we shall compare females versus males for those who rated their current job as high in the associated characteristic. This is done in two parts. First, quietly calculate predicted probabilities for women, saving the results. Then calculate them for men, and request that the difference between the two probabilities be displayed.

CALCULATE PROBABILITIES COMPARING WOMEN AND MEN

```
. qui asprvalue, x(fem=1 hashi=1) base(security) save

. asprvalue, x(fem=0 hashi=1) base(security) diff
```

```
rologit: Predictions for rank

              Current         Saved          Diff
   esteem    .09200718     .08876531      .00324187
  variety    .43670157     .41536212      .02133945
 autonomy    .2202711      .21456315      .00570795
 security    .25102016     .28130943     -.03028926
```

For women who believe that their current job is high in relation to its associated job characteristic, the probability is .089 that they select *esteem* as their highest choice among alternatives. Men have a probability of .092 of selecting *esteem* as the top choice, i.e., they rank *esteem* as first only 9 percent of the time. There is only a slight difference in how men and women view esteem—selecting it as first fewer times than other alternatives.

Both men (.437) and women (.415) selected *variety* as their most preferred choice. There is little difference between men and women for any of the job characteristics.

Additional statistics were displayed for the command, but are not displayed here. See Long and Freese (2006) for an excellent presentation of the model.

The Stata reference manual for **rologit** provides a good description of the relationship of rank-ordered and conditional fixed effects logistic regression. The two models can yield identical results in the case when there are no tied choices.

13.5 Random Effects and Mixed Models Logistic Regression

13.5.1 Random Effects and Mixed Models: Binary Response

Recall the generic formula that was given at the beginning of Section 13.4.1 on fixed effects models. The formula, appropriate for both single-level fixed and random effects logistic models, was given as:

$$y_{ij} = \beta_0 + \beta_i x_{ij} + \gamma_i + \varepsilon_{ij} \tag{13.23}$$

where y is the binary response, i indicates separate individuals or items, and j represents the observation number within i. If the study is a longitudinal model, j represents the time when a measurement on i occurred. ε is the error term, which, for binary logistic models, is distributed as a logistic cumulative density function with a mean of 0 and variance of $\pi^2/3$—(~3.29).

γ is the panel or subject indicator. For random effects models, γ is a random covariate representing a larger population that is distributed as a known probability distribution. Random effects are usually specified as being distributed as Gaussian (normal). However, the most important criterion for determining the distribution of the random effect is whether the joint likelihood of the binomial distribution and distribution of the random effect can be solved analytically. For the binomial-logistic distribution, the mixture distribution is either Gaussian, as previously mentioned, or beta. The beta distribution is the conjugate prior to the binomial, therefore providing a well-defined joint distribution, which has parameters that may be estimated.

There are two types of random effects models that we shall discuss—the random-intercept model and the random coefficient or slopes model. The random-intercept model is usually what most statisticians choose when they are deciding to use either a conditional fixed effects or random effects model. Essentially, the difference between random intercepts and random coefficients is that the former allows the level of the response, y_{ij}, to vary over clusters or groups after controlling for predictors. Random coefficients are added to random intercepts, allowing the effects of the predictors to vary over clusters.

In other words, a random intercept is partitioned by clusters, each of which is distributed according to a specified probability function. The distribution provides a value to the manner in which the subjects or observations within each cluster are correlated. In this sense, each cluster has a separate random effect, or intercept. Associated with each random intercept is the model linear predictor, $x'\beta$ (or βx), which is the same across clusters. The random coefficient, or random slopes, model also allows the model coefficients, β, to vary. The manner in which the coefficients vary across clusters is again specified by a given probability distribution.

For the random effects model it is standard to use the symbol ξ_i rather than v to designate the random effects. Moreover, it is also common to use the letter z_{ij} to represent the covariates or predictors corresponding to the random effects, ξ_i. z_{ij} may be employed in the model to designate both random intercepts and random coefficients. ξ_i, as random effects, is usually distributed, as mentioned above, as a multivariate normal with a mean of 0 and a square variance matrix Σ. Σ is the designation of multivariate normal matrix, as σ^2 is the symbol commonly used for the vector of trace elements representing the normal distribution variance. For random-intercept models, z_{ij} is 1, which is effected in the model by multiplying $z_{ij} = 1$ by the identity matrix, I.

Given the definitions discussed thus far in this section, Equation 13.23 can be re-expressed for random effects models in general as:

$$y_{ij} = \beta_0 + \beta_i x_{ij} + z_{ij}\xi_i + \varepsilon_{ij} \tag{13.24}$$

Again, for random-intercept models, $z_{ij} = 1$. As in Equation 13.23, the error term has a mean of 0 and, since we are representing logistic random effects models, a variance of $\pi^2/3$.

It should be noted that the random effects are not directly estimated, but rather their parameters are estimated and summarized, These summary values are called variance components. The variance component values are typically displayed following the modeling of a random effects logistic model, but only if the random effect is multivariate normal. If it is beta, the model is known as a beta-binomial. Estimation of the beta-binomial is difficult, and many times results in a failure to converge. It is recommended that Gaussian random effects be used with these models unless there is evidence that a beta-binomial better models the data.

Using this symbolism, the random effects logistic probability function, expressed in exponential family form, can be given as:

$$f(y_i | X_i, \xi_i) = \exp\left\{ \sum\nolimits_{j=1} [y_{ij}(x_{ij}\beta + z_{ij}\xi_i) - \ln(1 + \exp(x_{ij}\beta + z_{ij}\xi_i))] \right\} \quad (13.25)$$

For a binomial logistic model, such as a grouped random effects logistic model with m as the binomial denominator, the probability function in exponential family form is

$$f(y_i | X_i, m_{ij}, \xi_i) = \exp\left\{ \sum\nolimits_{j=1} \left[y_{ij}(x_{ij}\beta + z_{ij}\xi_i) \right. \right.$$

$$\left. \left. - m_{ij}\ln(1 + \exp(x_{ij}\beta + z_{ij}\xi_i)) + \ln\binom{m_{ij}}{y_{ij}} \right] \right\} \quad (13.26)$$

The log-likelihood function of the Bernoulli model may therefore be expressed as:

$$L(x_{ij}\beta | y_i, \xi_i) = \sum\nolimits_{j=1} [y_{ij}(x_{ij}\beta + z_{ij}\xi_i) - \ln(1 + \exp(x_{ij}\beta + z_{ij}\xi_i))] \quad (13.27)$$

Tying together the familiar terminology we used when discussing standard binary response logistic regression, $\ln(\mu/(1 - \mu))$, the logistic link function, appears as $x_{ij}\beta + z_{ij}\xi_i$ when incorporating random effects into the formula. It is important to remember that ξ_is are independent across subjects as well as being independent of X.

Also following the same logic that we have used thus far in our discussion, the predicted probability of $y = 1$ is

$$\mu_{ij} = (1 + \exp(-(x_{ij}\beta + z_{ij}\xi_i)))^{-1} \quad (13.28)$$

To model the random coefficients logistic model, as well as its subset, the random-intercept model, a joint distribution of the above described random effects logistic distribution and the multivariate normal distribution must be derived. The best way is to integrate ξ_i out of the joint distribution. The calculations take us beyond the scope of this text, as do the actual estimating algorithms.

The logistic-Gaussian random effects model cannot be estimated by normal maximum likelihood methods. Alternatives that are commonly used are various types of quadrature, including adaptive and non-adaptive Gauss–Hermite quadrature, and simulation techniques. Monte Carlo methods are also popular in econometric software. The log-likelihood for the adaptive Gauss–Hermite quadrature estimation method is given in Appendix G.

Recall that the power, yet also the downside, of the fixed effects model is its inability to measure between-person or between-items variation. It concentrates on detecting the variation existing within effects or within panels. Constant variables within a panel have no variation, and are conditioned out of the estimation process. This is not the case, however, for random effects models, where the variability of both within and between subjects is estimated.

It should be noted that the *Hausman test* that we used to evaluate multinomial models can also be used for comparing random effects and fixed effects models. The test is based on the differences in parameter estimates, with the result informing the user which model is preferred with the data. Various assumptions hold for the test, so one may find that test results display an ancillary message indicating that a violation of test assumptions has occurred. The Hausman test on the **heart01** data example we have employed for discussing these models provides just such a message. Given that it is not appropriate for evaluating these models, it is not displayed here. The Stata code for comparing fixed and random effects in general can be given as:

```
.  quietly xtlogit y xvars, i(clustervar) fe
.  est sto fixed
.  quietly xtlogit y xvars, i(clustervar) re
.  est sto ranef
.  hausman fixed ranef
< hausman test results displayed>
```

The first example discussed is of a simple random-intercept model. Using the **heart01** data, we model *death* on the predictors we have used throughout the text, but indicate that *center*, or hospital, be regarded as the random effect or intercept. The random effect here is assumed to follow a Gaussian or normal distribution.

```
. xtlogit death anterior hcabg kk2-kk4 age3-age4, i(center) re
nolog or

Random-effects logistic regression   Number of obs    =    4503
Group variable: center               Number of groups =      47

Random effects u_i ~ Gaussian        Obs per group: min =       3
                                                    avg =    95.8
                                                    max =     215
```

```
                                     Wald chi2(7)        =  190.03
Log likelihood   =   -637.937        Prob > chi2         =  0.0000
-----------------------------------------------------------------------
   death |       OR  Std. Err.    z  P>|z|  [95% Conf. Interval]
---------+-------------------------------------------------------------
anterior | 1.897376   .3185573  3.81  0.000   1.365344   2.636725
   hcabg | 2.224794   .7885332  2.26  0.024   1.110705   4.456367
     kk2 | 2.290287   .4148257  4.58  0.000   1.605895   3.266351
     kk3 |  2.25287   .6113718  2.99  0.003   1.323554   3.834693
     kk4 | 15.20232   5.504659  7.52  0.000   7.476447    30.9118
    age3 | 3.559014   .7158089  6.31  0.000   2.399558   5.278714
    age4 | 7.047647   1.476139  9.32  0.000   4.674754   10.62501
---------+-------------------------------------------------------------
 /lnsig2u |-3.306946  1.707472                -6.653529   .0396375
---------+-------------------------------------------------------------
 sigma_u | .1913841   .1633915                .0359091   1.020016
     rho | .0110109   .0185939                .0003918   .2402681
-----------------------------------------------------------------------
Likelihood-ratio test of rho=0: chibar2(01) =      0.45 Prob >=
chibar2 = 0.252

. abic
AIC Statistic =    .287336             AIC*n      = 1293.874
BIC Statistic =    .2921217            BIC(Stata) = 1351.5865
```

The estimates have been parameterized in terms of odds ratios for ease of interpretation. The odds ratios are each interpreted as a standard logistic model. Therefore, if the model is well specified, we can conclude that a patient has some 90 percent greater odds of *death* within 48 hours of admission if the patient has sustained an *anterior*-site rather than an *inferior*-site heart attack, holding the other predictors constant.

The random effects logistic model, with a Gaussian random effect, is displayed showing two important diagnostic statistics: σ_v and ρ. The latter is a statistic having values between 0 and 1, indicating the proportion of the total variance given by the panel-level variance component. ρ is defined as:

$$\rho = \frac{\sigma_v^2}{\sigma_v^2 + \sigma_\varepsilon^2} \tag{13.29}$$

where σ_v is the panel level standard deviation and σ_ε is the standard deviation of the error term, ε. The squares of these respective terms are the panel level variances. We mentioned in this section that ε has a mean of 0, and for the logistic distribution, a variance of $\pi^2/3$. The standard deviation, therefore, is the square root of $\pi^2/3$, or $(\pi^2/3)^{.5}$. The numeric value of this statistic, which is also referred to as σ_ε, is 1.8138.

σ_v is calculated by the estimating algorithm, σ_ε is given as a constant, and ρ can easily be calculated. Recalling that the variance is the square of the standard deviation, ρ may be calculated as:

```
. di (0.1913841^2) / (0.1913841^2 + _pi^2/3)
.01101095
```

The use of ρ comes in the line underneath the table of estimates. The software performs a likelihood ratio test of $\rho = 0$. It is a test of random effects (panel) model versus a standard logistic regression model (pooled) with independent observations. It is therefore a correlation test of the independence of the panels. The closer ρ is to 0, the less the within panel variability differs from the between panel variability. If ρ is close to 0, then the data should not be modeled as a panel model, much less a random effects model. ρ is commonly referred to as the residual intraclass correlation coefficient, and is an estimate of the within-subject correlation. Here the p-value > .05 indicates that there is not a significant random effect in the data; in other words, a pooled model is preferable to the random effects model.

The same model can be estimated using mixed effects models, where parameter estimates are considered as the fixed effects and the variance component(s) are the random effect(s). The coefficients have been re-parameterized to odds ratios, but the underlying fixed effects are nevertheless still in the model. Notice that the results are identical to the above "random effects" algorithm; the only difference aside from how the variance component(s) are displayed is in the ability of the mixed effects algorithm to model more random intercepts as well as random coefficients.

The same likelihood ratio test is appropriate for this version of the model. It is a test to determine if the model is better fitting with or without ξ_i.

```
. xtmelogit death anterior hcabg kk2-kk4 age3-age4 ||
center:, or

Mixed-effects logistic regression    Number of obs     =      4503
Group variable: center               Number of groups  =        47

                                     Obs per group: min =         3
                                                    avg =      95.8
                                                    max =       215

Integration points =    7            Wald chi2(7)      =    190.03
Log likelihood =    -637.937         Prob > chi2       =    0.0000
-----------------------------------------------------------------------
   death | Odds Ratio  Std. Err.    z    P>|z|   [95% Conf. Interval]
---------+-------------------------------------------------------------
anterior |   1.897376  .3185572   3.81   0.000   1.365345    2.636724
   hcabg |   2.224794  .7885333   2.26   0.024   1.110705    4.456368
```

```
     kk2 |   2.290287   .4148256   4.58   0.000    1.605895    3.266351
     kk3 |    2.25287   .6113718   2.99   0.003    1.323554    3.834694
     kk4 |   15.20233   5.504664   7.52   0.000    7.476454    30.91182
    age3 |   3.559014   .7158087   6.31   0.000    2.399558    5.278713
    age4 |   7.047649    1.47614   9.32   0.000    4.674756    10.62502
---------------------------------------------------------------------------
```

```
---------------------------------------------------------------------------
Random-effects Parameters | Estimate  Std Err   [95% Conf. Int]
--------------------------+------------------------------------------------
center: Identity          |
             sd(_cons) |   .19138     .16338    .03591    1.01997
---------------------------------------------------------------------------
LR vs. logistic regression: chibar2(01) = 0.45 Prob>=chibar2 =
0.2523
```

We can model over *anterior,* which as a binary predictor means that there are two coefficients: anterior infarct and inferior infarct. Adding a random coefficient to the model results in a likelihood statistic indicating that the effects are inherent in the data.

```
. xtmelogit death anterior hcabg kk2-kk4 age3-age4 || center:
anterior , or
```

```
Mixed-effects logistic regression   Number of obs      =      4503
Group variable: center              Number of groups   =        47

                                    Obs per group: min =         3
                                                   avg =      95.8
                                                   max =       215

Integration points =    7           Wald chi2(7)       =    188.97
Log likelihood = -636.11043         Prob > chi2        =    0.0000
```

```
---------------------------------------------------------------------------
   death | Odds Ratio  Std. Err.    z   P>|z|  [95% Conf. Interval]
---------+-----------------------------------------------------------------
anterior |   1.753575    .336127   2.93   0.003    1.204384    2.553192
   hcabg |    2.22631   .7907471   2.25   0.024    1.109823    4.465991
     kk2 |    2.31914   .4224588   4.62   0.000    1.622821    3.314236
     kk3 |   2.353373   .6461029   3.12   0.002    1.374039    4.030718
     kk4 |   16.36526   5.949523   7.69   0.000      8.0255    33.37135
    age3 |   3.604954    .728607   6.34   0.000    2.425834    5.357207
    age4 |   7.183435   1.511038   9.37   0.000    4.756436    10.84882
---------------------------------------------------------------------------
```

```
------------------------------------------------------------------
Random-effects Parameters | Estimate  Std Err  [95% Conf. Int]
--------------------------+---------------------------------------
center: Independent       |
            sd(anterior)  | .439606    .157773   .217555 .888297
            sd(_cons)     | 2.28e-0    .239952      0        .
------------------------------------------------------------------
LR vs. logistic regression: chi2(2) = 4.10 Prob > chi2 = 0.1288
```

Note: LR test is conservative and provided only for reference.

Observing the standard deviations, it is clear that the randomness of the model is in *anterior*.

By using the *var* and cov(*unst*) options with the same model as above, the covariance between the constant and *anterior* is displayed. Only the lower table is different, and is shown next.

```
------------------------------------------------------------------
Random-effects Parameters   | Estimate Std Err   [95% Conf. Int]
----------------------------+-------------------------------------
center: Unstructured        |
            var(anterior)   | .284700 .241236   .054092 1.49844
            var(_cons)      | .010099 .035969   9.39e-0 10.8598
        cov(anterior,_cons) | -.053622 .113016 -.275131 .167886
------------------------------------------------------------------
LR vs. logistic regression: chi2(3) = 4.41 Prob > chi2 = 0.2209
```

Random effects models typically have less sampling variability than similar fixed effects models. Some authors recommend that one should use random effects models for all panel models; others prefer fixed effects. Mundlak (1978) has even argued that random effects models in general are special cases of fixed effects models. But, as we have seen in our discussion, selecting which panel model to use is largely based on the number of panels we have in our data, and how we wish to deal with between effects variation. As a general rule, though, if we can extend the interpretation of the effects beyond the data being modeled, a random effects model is preferred; if the effects are limited to the data being modeled, then a fixed effects model is likely more appropriate.

Of course, both of these methods are subject-specific models. This means that when we are evaluating the odds of killip level 2 with respect to level 1, we do so for each individual being evaluated at level 2. If we wish to model averages of the effects, then selecting a population averaging method such as GEE is appropriate. Then we are speaking of killip level 2 patients in general. In this respect, standard logistic regression is a population averaging method.

13.5.2 Alternative AIC-Type Statistics for Panel Data

A note should be given on alternative goodness-of-fit tests that have been devised for random and mixed effects models. We have seen that most tests relate to the viability of modeling the data with a panel versus pooled structure. In addition, we have discussed that standard AIC fit statistics are inapplicable for use with cluster or panel models. It was for this reason that the QIC statistic was developed for marginal or population averaging models.

Building on the prior work of Hodges and Sargent (2001), Vaida and Blanchard (2005) constructed two varieties of AIC fit statistic applicable for random and mixed effects models. Each statistic addresses a primary focus of research:

1. Inference related to the interpretation of parameter estimates
2. Inference related to ancillary cluster parameters

For the first research emphasis, the standard AIC formula of:

$$\text{AIC} = -2 * \text{log-likelihood} + 2 * k \tag{13.30}$$

where k is the degrees of freedom, calculated by the number of predictors in the model, is amended to the form:

$$m\text{AIC} = -2 * \text{marginal log-likelihood} + 2 * k \tag{13.31}$$

where k is the number of fixed parameters, including mean parameters (predictors) and variance components. This parameterization of AIC is called marginal AIC.

The second formulation of AIC, related to understanding the effect of clustering, is defined as:

$$c\text{AIC} = 2 * \text{conditional log-likelihood} + 2 * k \tag{13.32}$$

where $k = \omega + \sigma^2$. ω reflects the number of predictors in the model and σ^2 is the Gaussian variance.

Vaida and Blanchard added adjustments to k for both parameterizations based on whether the model was a fixed effects or random effects, or even on how a model is to be estimated.

The formulae can get rather complex. However, the end result is goodness-of-fit statistics that better relate specific types of clustered models to one another. Specifically, the goal of the tests is to find the optimal balance between the bias and variance of model estimates, and therefore predictions.

The downsides of using the mAIC and cAIC tests to assess the comparative fit of clustered models are

1. The formulae are specific to comparisons of similar models; that is, they are not to be used for comparing a fixed effects to a random effects model. The problem, therefore, is specificity of application.
2. The m/cAIC formula for each of the many varieties of random, fixed, and mixed effects models differ from one another. Implementing this type of statistic in a research situation can therefore be a tedious process.
3. The formulae, with appropriate adjustments included, are complex.

On the upside — the tests provide a means to evaluate the comparative worth of panel models. Prior to this work, only the QIC statistic of Pan and Hardin and Hilbe, which are used for GEE and related marginal models, have proven to be useful as goodness-of-fit tests with this type of model. It is hoped that future simulation studies provide evidence of more parsimonious AIC-type tests for subject-specific panel models. See Vaida and Blanchard (2005), Lee and Nelder (2001), and Lee, Nelder, and Pawitan for additional and related discussion.

13.5.3 Random-Intercept Proportional Odds

The proportional slopes models discussed in Section 12.6 assumed the independence of observations throughout the model. If there was evidence that data were collected across various groups or sites, or was longitudinal in nature, then robust variance estimators were applied to adjust the model standard errors. In this chapter we have thus far discussed enhanced methods to accommodate clustering and longitudinal effects, but have done so for binary and binomial response models. This section presents a brief overview of how random-intercept modeling can be applied to proportional slopes models. In particular, we examine how to model ordered categorical response data using a random-intercept proportional odds model.

The **medpar** data will be used for an example. The same example data are used when discussing the proportional odds model. Recall that the **medpar** file contains Medicare patient data from the state of Arizona. Gathered in 1991, all observations are related to a single diagnostic related group (DRG). The problem, though, is that data were collected from patients across 54 different hospitals. A proportional odds model of the data provides the following output.

```
. ologit type died white hmo los, nolog

Ordered logistic regression        Number of obs   =        1495
                                   LR chi2(4)      =      115.61
```

```
                                     Prob > chi2      =      0.0000
Log likelihood = -977.66647          Pseudo R2        =      0.0558

-------------------------------------------------------------------------
   type |      Coef. Std. Err.       z  P>|z|   [95% Conf. Interval]
--------+----------------------------------------------------------------
   died | .5486314 .1278436       4.29 0.000   .2980626        .7992002
  white | -.5051675 .1963398     -2.57 0.010  -.8899864       -.1203486
   hmo | -.7107835 .1985937      -3.58 0.000  -1.10002         -.321547
   los | .0577942 .0068349        8.46 0.000   .0443979        .0711904
--------+----------------------------------------------------------------
  /cut1 | 1.393847 .2104434                     .981385        1.806308
  /cut2 | 3.046913 .2321391                    2.591928        3.501897
-------------------------------------------------------------------------
```

Parameterized in terms of odds ratios, the table of estimates appears as:

```
. ologit, or  /// header and cut statistics deleted

-------------------------------------------------------------------------
   type | Odds Ratio Std. Err.      z   P>|z|  [95% Conf. Interval]
--------+----------------------------------------------------------------
   died |   1.730882 .2212822    4.29   0.000  1.347246        2.223762
  white |   .6034045 .1184723   -2.57   0.010  .4106613        .8866113
   hmo |   .4912592  .097561    -3.58   0.000  .3328644        .7250266
   los |   1.059497 .0072416     8.46   0.000  1.045398        1.073786
--------+----------------------------------------------------------------
```

To adjust for the possible clustering effect of having data collected from different sites, we employ a robust variance estimate grouped on hospitals. The hospital ID is given in the variable, *provnum*. We are looking to see if there is a violation of the independence of observation criterion in this data. If the standard errors appreciably differ from one another, this is an indication that the assumption is violated. If it has, then modeling the data using a robust variance estimate is preferred, but it may also be the case that the model requires additional adjustment, such as via a random effect.

The model adjusted by the clustering effect of hospitals is given as:

```
. ologit type died white hmo los, nolog vce(cluster provnum)

Ordered logistic regression          Number of obs    =        1495
                                      Wald chi2(4)     =       35.33
                                      Prob > chi2      =      0.0000
Log pseudolikelihood = -977.66647 Pseudo R2           =      0.0558

          (Std. Err. adjusted for 54 clusters in provnum)
```

```
--------------------------------------------------------------------
         |                 Robust
   type  |     Coef.  Std. Err.      z    P>|z|    [95% Conf. Interval]
---------+----------------------------------------------------------
   died  |  .5486314  .1662668    3.30   0.001    .2227544     .8745084
  white  | -.5051675  .1959724   -2.58   0.010   -.8892664    -.1210686
    hmo  | -.7107835  .2263512   -3.14   0.002   -1.154424    -.2671432
    los  |  .0577942  .0120382    4.80   0.000    .0341997     .0813887
---------+----------------------------------------------------------
  /cut1  |  1.393847  .2879427                     .8294892     1.958204
  /cut2  |  3.046913  .3509648                     2.359034     3.734791
--------------------------------------------------------------------
```

Note the inflated standard errors for predictors *died, hmo* and *los. Died* has been inflated by 30 percent, *hmo* by 14 percent, and *los* by a full 76 percent. *White* remained effectively unchanged. This tells us that there is a difference between hospitals with respect to patient death during the hospital admission, whether the patient belongs to an HMO (Health Maintenance Organization) and length of stay. Which hospital a patient is admitted to makes a difference in these predictors. On the other hand, there appears to be no racial preference in hospital admission—at least among 1991 Arizona Medicare patients. This may reflect practices of non-discrimination in admissions or may simply be a consequence of the fact that few minority Medicare patients entered the hospital for this particular DRG.

It is important to remember that the odds ratios in the proportional odds model as above parameterized provide marginal or population-averaged effects. This is the same interpretation as for standard logistic regression, although this fact is seldom recognized. All subjects sharing the same pattern of covariates have the same probability of success or same probability of being in a particular level compared to another level(s). In order to obtain subject-specific results we use a random effects or random-intercept model.

If we treat *hospital* as a random effect, the proportional odds model can be estimated as:

```
. gllamm type died white hmo los, i(hospital) link(ologit)
adapt nolog

number of level 1 units = 1495
number of level 2 units = 54

Condition Number = 64.07977

gllamm model

log likelihood = -795.53284
```

```
------------------------------------------------------------------
      type |     Coef.  Std. Err.    z   P>|z|  [95% Conf. Interval]
-----------+------------------------------------------------------
type       |
      died |   .545592  .1506324   3.62 0.000   .2503579    .840826
     white | -.7305823  .2496819  -2.93 0.003  -1.21995   -.2412147
       hmo | -.4250105  .2241625  -1.90 0.058  -.864361     .01434
       los |  .0408815   .009255   4.42 0.000   .0227421   .0590209
-----------+------------------------------------------------------
_cut11     |
      _cons |  1.157956 .3712715   3.12 0.002   .430277    1.885634
-----------+------------------------------------------------------
_cut12     |
      _cons |  3.579339 .3946536   9.07 0.000  2.805832    4.352846
------------------------------------------------------------------


Variances and covariances of random effects
------------------------------------------------------------------


***level 2 (hospital)

     var(1):  3.3516367  (.87750193)
------------------------------------------------------------------
```

A table of odds ratios for the proportional odds model is displayed by:

```
. gllamm, eform      /// table of exponentiated estimates only

------------------------------------------------------------------
      type |  exp(b)  Std. Err.    z   P>|z|   [95% Conf. Interval]
-----------+------------------------------------------------------
type       |
      died |  1.72563 .2599357   3.62 0.000   1.284485    2.318281
     white | .4816284 .1202539  -2.93 0.003    .295245    .7856729
       hmo | .6537629 .1465491  -1.90 0.058   .4213207   1.014443
       los | 1.041729 .0096412   4.42 0.000   1.023003   1.060797
-----------+------------------------------------------------------
_cut11     |
-----------+------------------------------------------------------
_cut12     |
------------------------------------------------------------------
```

First, it should be mentioned that the **gllamm** command is not part of official Stata; rather it is a user command authored by Sophia Rabe-Hesketh of the University of California, Berkeley. However, **gllamm** enjoys widespread use in the Stata community and is thought of as a Stata command by most Stata users. The command provides users with a wide variety of random-intercept and coefficient models, plus a host of other capabilities. The command, together with other major Stata commands for mixed models, is featured in the text by Rabe-Hesketh and Skrondal (2008).

We notice that the standard errors for the robust and random-intercept models are fairly close in value (see Table 13.2). On the other hand, the coefficients on *died* are nearly identical, on *los* are close, but the coefficients on *white* and *hmo* considerably differ.

TABLE 13.2

Proportional Odds Coefficients

	Robust variance	Random-intercept	Unadjusted
died	0.549 (.166)	0.546 (.151)	0.549 (.128)
white	−0.505 (.196)	−0.731 (.250)	−0.505 (.196)
hmo	−0.711 (.226)	−0.425 (.224)	−0.711 (.199)
los	0.058 (.012)	0.041 (.009)	0.058 (.007)

Racial differences in admission type appear evident in the random-intercept model compared to the original proportional odds model, but this is not apparent when a robust variance estimator is applied to the model, as earlier mentioned. Interestingly, the random-intercept model appears to interchange the effect of *white* and *hmo* on the type of admission. The differences in effect may simply reflect the difference in interpretation, in other words, that the unadjusted and robust proportional odds models are population-averaged models whereas the random-intercept model is subject specific. Probabilities, as well as odds ratios, relate to each subject identified as such on the command line. This difference in interpretation is particularly apparent when cut points are compared in Table 13.3.

TABLE 13.3

Proportional Odds Cut Points

	Unadj/Robust	R.Intercept
Cut1	1.394	1.158
Cut2	3.047	3.579

Finally, mention should be made of the random-intercept variance (Rabe-Hesketh and Skrondal, 2008). In our example model, the estimated variance is 3.3516. If the response, *type*, is considered as a partition of an underlying latent continuous variable, y_{ij}^*, then the estimated residual intraclass correlation of y_{ij}^* is

$$\rho = \frac{\Psi_{11}}{\Psi_{11} + \pi^2/3} \tag{13.33}$$

Note that the equation is similar to Equation 13.27, with a value of $\pi^2/3 = 3.2899$, and ψ_{11} provided to us in the output as 3.3516, $\rho = .5046$.

Note also that random coefficient proportional odds models may be constructed using the **gllamm** command. I do not discuss them here, but interested readers are referred to Rabe-Hesketh and Skrondal (2008). The logic is nearly the same as was used for random coefficient models, but applied to ordered response models.

For an excellent overview of the differences between fixed, random, and mixed effects, together with GEE, see Gardiner, Luo, and Roman (2008). The texts, Rabe-Hesketh and Skrondal (2008), *Multilevel and Longitudinal Modeling Using Stata* (Stata Press), Gelman and Hill (2007), *Data Analysis Using Regression and Multilevel/Hierarchical Models* (Cambridge University Press), Baltagi (2008), *Econometric Analysis of Panel Data*, 4th edition (Wiley), and Little et al. (2006), *SAS for Mixed Models* (SAS), are all extremely well written and informative.

Exercises

13.1. Using the **drugprob** data given in Exercise 8.4, add the additional four continuous predictors to panel models as indicated in the following questions.

ADDED CONTINUOUS PREDICTORS

depress	subjects rated on a standardized scale of depression
esteem	subjects rated on a standardized scale of self-esteem
stress	subjects rated on a standardized scale of feelings of stress
income	12 levels in \$10,000s, i.e., \$10,000—\$120,000 yearly income

PANEL COVARIATES

personid	subject indicator
year	year of test (7)

a. Model *drabuse* on the binary predictors given in Exercise 8.4 together with the above continuous predictors. Model as:

GEE using all described correlation structures. Use the time panel effect for the appropriate models.

If the GEE model using an autoregressive 1 or stationary 1 correlation structure fails to converge, assess its range of feasibility and remodel using a quasi-least squares model.

Unconditional and Conditional fixed effects, with *personid* as the panel indicator. Random effects, with *personid* as the random effect.

Random-intercept and random-coefficient models

b. Execute the comparative fit tests used to determine which of the above models best fits the data.

c. Construct the best-fitted model for the data. You may drop predictors, create interactions, transform values of the continuous predictors, or do whatever manipulation is required to best fit the data.

13.2. Use the **wheezehh** data set on the effect of *smoking* on *wheezing*. The data come from a subset of data in Ware, Dockery, Speizer, and Ferris (1984), and include the following covariates:

RESPONSE

  wheeze   1 = patient wheezes; 0 = not wheezes

PREDICTORS

  smoking   1 = smoker; 0 = not a smoker
  age   ages 9–12

PANEL

  id   subject
  t   years (4)

a. Model *wheeze* on *smoke* and levels of *age* using a GEE model. Find the best correlation structure. You may need to adjust the levels of age to construct a meaningful model.

b. Use a quasi-least squares regression for modeling *wheeze* on *smoke* and *age*, with a Markov correlation structure. Compare it with an AR1 model. Which is the most appropriate model to use? Why?

c. Model *wheeze* on *smoke* and *age* levels using a random coefficient model. *Id* is the intercept, and *t* is the panel used for estimating the random coefficient. Evaluate the model fit.

d. Model *wheeze* on *smoke* and *age* levels using a random-intercept model. Compare with the GEE model.

e. How are the random effects model estimates interpreted differently than the GEE estimates?

13.3. Use the **madrasl** data, which are based on the Madras, India Longitudinal Schizophrenia Study. The study was described in Thara et al. (1994) and in Rabe-Hesketh and Everitt (2007). The subset given here was first analyzed by Diggle et al. (2002). Thought disorder was evaluated as 1 = yes; 0 = no at baseline (0) and at 2, 6, 8, and 10 months after commitment. Covariates in the model are

RESPONSE

y	binary 1 = have disease; 0 = not have disease

PREDICTORS

month	categorical 0, 2, 6, 8, 10
early	binary 1 = age of onset of disease < 20 yrs; 0 = ≥ 20 yrs
month_early	categorical interaction 0, 2, 6, 8, 10

PANEL

id	44 panels of 5 observations each
const	constant of 1; used for random-intercept model

The study goal is to determine if the range of the disease (schizophrenia) differs from those patients with early (<20 yrs) compared to late (≥ 20 yrs) onset times.

a. Model the data using a random effects logistic regression model. Give an appropriate answer to the study goal.

b. Using *const* as the intercept and *month* as the slope, model as a random slopes logistic model. [Note: Stata users can employ either **xtmelogistic** or **gllamm**.]

c. Can this model be estimated using GEE? Why?

13.4. Use the **schiz2** data. Model *imps2* on *treatment* and *sex*, with *id* as a subject identifier and *week* as the week of evaluation.

a. Model the above as a random effects model, with *id* as the random effects. Evaluate the worth of the predictors.

b. Model as a random-intercept model with *id* as the intercept. Evaluate the worth of the predictors. Compare results with the random effects model above.

    c. Model as a GEE with *week* as the time panel.

    d. Model as a QLS with the Markov correlation structure.

13.5. Use the **cluslowbwt** data set, as described in Exercise 14.6.

    a. Model *low* on *smoke*, *lwtkg*, *age*, and levels of *race* with *race* = 1 as the referent. Model as a random effects model with *id* as the random effect.

    b. Model the same covariates as in (a) as a GEE model with an exchangeable correlation structure.

    c. How do the random effects and GEE models differ in terms of the interpretation of the parameter estimates?

    d. Find the best-fitting panel model for the above response and predictors. Categorize the two continuous predictors if it results in a better-fitted model.

13.6. Use the **unionocc** dataset to model union membership on nine levels of occupation and three levels of ethnic background.

Response : union 1 = yes; 0 = no

Predictors:

occ1 = 1 if professional, technical	nr = personal ID (8 levels)
occ2 = 1 if mgr, official, proprietor	year = 1980–1987
occ3 = 1 if sales	
occ4 = 1 if clerical	white = reference for hisp and black
occ5 = 1 if craftsman, foreman	(not listed, but implied, as predictor)
occ6 = 1 if operative	
occ7 = 1 if laborer, farmer	
occ8 = 1 if farm laborer, foreman	
occ9 = 1 if service	
hisp = 1 hispanic	
black = 1 black	

    a. Model as a logistic GEE; construct best-fitted model.

    b. For best-fitted model in (a), calculate which occupation most likely has the most union membership. Which ethnic category has the worst union membership?

    c. Determine if the observations in the model are equally spaced. How do you tell?

d. Model as a QLS with Markov structure. Determine if it fits better than AR1.

e. Model as an ALR. Determine if the model is superior to standard GEE methods.

f. Model as a random coefficient model with personal ID as the intercept and year as the random slope.

13.7. Do additional outside research on the CIC statistic. Check the Web to determine if the statistic is being used in preference to the QIC statistic to evaluate the appropriateness of a correlation structure for a given GEE model.

R Code

Section 13.1 Generalized Estimating Equations

Sections 13.1.1–13.1.3

```
library('foreign')
german<- read.dta('rwm_1980.dta') #read Stata format file
summary(german)
head(german)
str(german)

library(MASS)
fit13_1a<- glm(outwork ~ female + married + edlevel2 +
  edlevel3 + edlevel4, data=german,
  family=binomial(link='logit'))
summary(fit13_1a)2
exp(coef(fit13_1a)) #ORs

library('gee')
fit13_1b<- gee(outwork ~ female + married + edlevel2 +
  edlevel3 + edlevel4, data=german, id=id,
  corstr='independence', family=binomial)
summary(fit13_1b)
exp(coef(fit13_1b)) #ORs

fit13_1c<- gee(outwork ~ female + married + edlevel2 +
  edlevel3 + edlevel4, data=german, id=id,
  corstr='exchangeable', family=binomial)
summary(fit13_1c)
exp(coef(fit13_1c)) #ORs
```

```
library('geepack') #another package for geeglm() and geese()
  fit13_1d<- geeglm(outwork ~ female + married + edlevel2 +
  edlevel3 + edlevel4,data=german, id=id, corstr='ar1',
  family=binomial(link='logit'))
summary(fit13_1d)
exp(coef(fit13_1d)) #ORs

fit13_1e<- gee(outwork ~ female + married + edlevel2 +
  edlevel3 + edlevel4, data=german, id=id,
  corstr='unstructured', family=binomial)
summary(fit13_1e)
exp(coef(fit13_1e)) #ORs

fit13_1f<- geeglm(outwork ~ female + married + edlevel2 +
  edlevel3 + edlevel4, data=german, id=id,
  corstr='unstructured', family=binomial(link='logit'))
summary(fit13_1f)
exp(coef(fit13_1f)) #ORs

fit13_1g<- gee(outwork ~ female + married + edlevel2 +
  edlevel3 + edlevel4, data=german, id=id, corstr='stat_M_
  dep', Mv=1, family=binomial)
summary(fit13_1g)
exp(coef(fit13_1g)) #ORs

fit13_1h<- gee(outwork ~ female + married + edlevel2 +
  edlevel3 + edlevel4, data=german, id=id, corstr='non_stat_M_
  dep', Mv=1, family=binomial)
summary(fit13_1h)
exp(coef(fit13_1h)) #ORs

heart<- read.dta('heart01.dta') #read Stata format file
heart$anterior <- heart$anterior[,drop=TRUE] #drop empty
  levels that complicate tabulations
heart$center<- factor(heart$center) #convert to factor from
  numeric levels
heart$killip<- factor(heart$killip) #convert to factor from
  numeric levels

fit13_1i<- gee(death ~ anterior + hcabg + kk2 + kk3 + kk4 +
  age3 + age4, data=heart, id=center, corstr='exchangeable',
  family=binomial)
fit13_1i

heartgrp<- read.dta('heart02grpx.dta')
head(heartgrp)
ix<- order(heartgrp$center) #get indices for center order
heartgrp2<- heartgrp[ix,] #sort by center
```

```
fit13_1j<- gee(cbind(dead, cases-dead)~ anterior + hcabg +
   kk2 + kk3 + kk4 + age3 + age4, data=heartgrp2, id=center,
   corstr='exchangeable', family=binomial)
fit13_1j
```

Section 13.1.4 QIC

```
library(YAGS)
rwm<- read.dta('rwm_1980.dta') #read Stata format file
data(rwm)
fit13_qica<- yags(outwork ~ female + married + edlevel2 +
   edlevel3 + edlevel4, data=rwm, id=id, corstr='independence',
   family=binomial)
fit13_qica


fit13_qicb<- yags(outwork ~ female + married + edlevel2 +
   edlevel3 +  edlevel4, data=rwm, id=id,
   corstr='exchangeable', family=binomial)
fit13_qicb
```

Section 13.1.5 ALR

```
# Standard GEE
heart<- read.dta('heart01.dta') #read Stata format file
fit13_1_5<- gee(death ~ anterior + hcabg + kk2 + kk3 + kk4 +
   age2 + age3 + age4, data=heart, id=center,
   corstr='exchangeable', family=binomial)
fit13_1_5

# ALR of same model; SAS ALR model of data in book

library(YAGS)
heart<- read.dta('heart01.dta') #read Stata format file
fit13_alr<- yags(death ~ anterior + hcabg + kk2 + kk3 + kk4 +
   age2 + age3 + age4, data=heart, id=center,
   corstr='exchangeable', family=binomial)
fit13_alr
```

Section 13.1.6 QLS*

```
setwd("C:/My Documents/Joe")

library('foreign') #foreign import package library
library('qlspack')
```

---

* My thanks to Jichun Xie, Univ. of Pennsylvania, for her contribution to this section of
  R code.

```
rwm<- read.dta('rwm_1980.dta') #works for STATA v.5-9

attach(rwm)

id.list<- cluster.size(id)
n<- id.list$n
cnt<- rep(n,n)
length(cnt) #number of observations
summary(cnt) #range, mean and quantiles
sd(cnt) #standard error

qls.ar1<- qls(formula = outwork ~ female + married + edlevel2
   + edlevel3 + edlevel4, data = rwm,
      id = id, family = binomial, time = year, correlation =
         "ar1")
      #fit the model with the AR1 structure
summary(qls.ar1) #show more details

alpha<- qls.ar1$geese$alpha
corr.mat<- cormax.ar1(alpha,id,time="NA")
corr.mat #get the correlation matrix
eigen(corr.mat) #get the eigenvalues for the correlation matrix

n.max<- max(n)
a = -1/(2*sin((n.max-1)/(n.max + 1)*pi/2))
b = 1/(2*sin((n.max-1)/(n.max+ 1)*pi/2))
scope<- c(a,b)
names(scope)<- c("a","b")
print(scope) #get the boundary

qls.markov<- qls(formula = outwork ~ female + married +
   edlevel2 + edlevel3 + edlevel4,
      data = rwm, id = id, family = binomial, time = year,
correlation = "markov")
      #fit the model with the Markov structure
summary(qls.markov) #show more details

alpha<- qls.markov$geese$alpha
corr.mat<- cormax.markov(alpha,id,time="NA")
eigen(corr.mat)
qls.tri<- qls(formula = outwork ~ female + married + edlevel2
   + edlevel3 + edlevel4 data = rwm,
      id = id, family = binomial, time = year, correlation =
   "tridiagonal") #fit the model with the Markov structure
summary(qls.tri) #show more details

alpha<- qls.tri$geese$alpha
corr.mat<- cormax.tri(alpha,id,time="NA")
eigen(corr.mat)
```

Section 13.2 Unconditional Fixed Effects Logistic Regression

```
library(MASS)
wage<- read.dta('wagepan.dta') #read Stata format file
data(wagepan)
fit13_2<- gee(black ~ south + agric + married + educ + lwage +
  factor(year),
        data=wage, id=year, corstr='independence',
        family=binomial) # 1980, the
        earliest year, is the referent
fit13_2
exp(coef(fit13_2))
```

Section 13.3 Conditional Fixed Effects Logistic Regression

```
# 13.3.1 conditional fixed effects logistic regression
heart<- read.dta('heart01.dta') #read Stata format file
cfelr <- lme (death~ anterior + hcabg + kk2 + kk3 + kk4 +
  age3 + age4, data=heart, fixed=~1|center)
summary(cfelr)
exp(coef(cfelr)) #ORs
```

Section 13.4 Random Effects and Mixed Effects Logistic Regression

```
# Random Effects LR
library(nlme4)
heart<- read.dta('heart01.dta') #read Stata format file
rndint1 <- lme (death~ anterior + hcabg + kk2 + kk3 + kk4 +
  age3 + age4, data=heart, random=~1|center)
summary(rndint1)
exp(coef(rndint1)) #ORs

(0.1913841^2) / (0.1913841^2 + _pi^2/3)

# Random Intercept LR; same as random effects
library(nlme)
heart<- read.dta('heart01.dta') #read Stata format file
rndint2 <- lme (death~ anterior + hcabg + kk2 + kk3 + kk4 +
  age3 + age4, data=heart, random=~1|center)
summary(rndint2) \
exp(coef(rndint2)) #Ors

# Random Coefficient LR : Not included

# Proportional odds
library(MASS)
medpar<- read.dta('medpar.dta') #read Stata format file
propo13 <- polr(factor(type)~ died + white + hmo + los,
  data=medpar)

exp(coef(propo13)) #OR
```

14

Other Types of Logistic-Based Models

In this chapter, we address three types of models that cannot be easily catego-rized into the discussions of the previous chapters. We first provide an overview of the survey model, which may take the form of most of the models we have discussed. The difference is in how observations in the model data are sampled, weighted, and clustered. The second model we cover is **scobit**, or skewed logis-tic regression, for which the mean probability is not centered around 0.5. Logit (logistic) and probit models are symmetrically distributed about 0.5, while com-plementary loglog and log log binomial models are asymmetrically distributed about the same value. Scobit is unique in shifting the center of the distribution away from 0.5 to capture the greatest impact on the range of probability values.

The final set of models we discuss are collectively called discriminant anal-ysis. We concentrate our attention on three parameterizations: Dichotomous discriminant analysis based on ordinary least squares, and two maximum likelihood applications commonly referred to as canonical linear discrimi-nant analysis and linear logistic discriminant analysis. All three discriminant analysis models employ a logistic distribution to determine the probability of observations being in a particular response group or level.

14.1 Survey Logistic Models

The majority of the models we have discussed in this text can be amended in such a manner that they may be used for survey data. Generally, all that is required is to provide the model with the capability of using sampling or survey weights and to allow for grouping options to identify clustering in the data.

Survey data refer to the manner in which sample data are collected from a greater population. Survey models attempt to minimize the bias that can exist when sample data are collected. All survey models must adhere to two criteria when collecting sample data from a greater population:

1. All members of the study population must have an opportunity to be selected into the sample study.

2. The probability of being in the study is clearly defined for each group or stratum of the population.

The first criterion specifies that inclusion in the study is based on being randomly selected. The second criterion assumes that the population is divided into various levels, or areas, and that the probability of being selected from a level is based on the population size of the level compared to the study population as a whole. Probability or sampling weights are assigned to each group or stratum in the study population.

Suppose that the target or study population has 1,000,000 members. If we consider only the study population, each member has a 1/1,000,000th probability or chance of being selected at random. Now suppose that the population represents the number of people living in a U.S. city. Let us further divide the city into six districts. The population of each district is

DISTRICT (in thousands)

1. (A) 100   4. (D) 250
2. (B) 200   5. (E) 200
3. (C) 150   6. (F) 100

What is the probability that a person from the study population being selected at random is from district B? It is 200/1000 = 0.2, or 20 percent. Let us suppose also that the ethnic background of those living in each of the districts differs from one another. Suppose that the following ethnic percentages exist in the first three districts as:

District 1 (A) White = 50%, Hispanic = 30%, Black = 10%, Asian = 05%, Other = 05%
District 2 (B) White = 25%, Hispanic = 30%, Black = 20%, Asian = 15%, Other = 10%
District 3 (C) White = 30%, Hispanic = 20%, Black = 10%, Asian = 35%, Other = 05%

If we have a survey that attempts to obtain an unbiased sample of responses, we must weight by both the *District* and, within each district, the *Ethnic* population. We can weight by count or percentage, that is, if we choose to sample 1,000 people from the study population, the percentages of people of each ethnic background within each district must be apportionately weighted. Probability weights are defined, therefore, as the inverse of the probability of being a member of each level: district and ethnic class within district.

The weights are interpreted as the number of persons (or sampling elements like blocks) in the study population which each sampled case represents. We need to select 52 (actually 52.5) Asian persons from District C to answer survey questions to fairly represent the study population of 1,000,000. If we happen to select 10 Asians from District C to answer survey questions, the results may be biased. We would necessarily have to undersample another level. We use probability weights to represent the appropriate number of individuals sampled at each level in the model.

Instead of basing probability weights on individuals, we have indicated that sampling units may be considered as clusters within the greater population. For example, a PSU (**primary sampling unit**) can be identified as the ethnic background per district in the population of 1,000,000 we have described. Whites in District B represent a PSU, Whites in District C another PSU, and Other in District 6 yet another. We could define 30 PSU within the six strata of districts. Other sampling plans, of course, may be defined.

Sampling fraction is another term commonly used in survey modeling. This term means essentially the same as the probability of being included in the survey for a given strata. For the example data given above, there are 45,000 white citizens in District C (1000*150*.3). Out of a population of 1,000,000, and a survey sample of 1,000, this means that we randomly select 45 whites from District C. Each person selected represents 1000 people, with a sampling fraction of .001.

We can use another example to better understand the meaning of some of the most popular terms used in survey modeling. The example comes from Vittinghoff et al. (2005), which provides an excellent overview of survey methods in biostatistics.

In a similar manner, state census surveys generally sample on counties, county census tracts, blocks, and households. Sampling at levels greater than households is typically done on clusters (cluster sampling); only when it reaches households are individuals considered as a sampling unit.

Two examples will be given to further elucidate the concepts involved in survey modeling. The first is a simple binary response logit survey model; the second is a more complex proportional odds survey model. The data used in both examples come from genuine government survey data.

EXAMPLE 14.1—A LOGISTIC SURVEY MODEL

For our first example, we use the National Health and Nutrition Examination Survey (NHANES III) that was conducted from 1988 to 1994. The first survey was performed from 1971 to 1974, and the second from 1976 to 1980. Sponsored by the National Center for Health Statistics (NCHS), the study was aimed to better understand health and nutrition statistics in the United States. The data have been often used for examples of survey modeling and are found in several commercial software packages. We have amended several of the variables for ease of modeling.

The response we shall attempt to understand is *hbp*, or high blood pressure. It is a binary variable defined below.

RESPONSE

hbp	High blood pressure	1 = avg systolic BP > 140
		0 = avg systolic BP <= 140

SURVEY VARIABLES

wtpfhx6	Probability weight:	225.93 – 139744.9
sdpstra6	Strata	1–49
sdppsu6	PSU	1–2

PREDICTORS

hsageir	Age	continuous: 20–90 in years
hssex	Sex	1 = male; 0 = female
dmaracer	Race	1 = white; 2 = black; 3 = other
bmpwtlbs	Body weight	continuous: pounds
bmphtin	Height	continuous: inches
pepmnk1r	Avg Systolic BP	continuous: mm Hg
pepmnk5r	Avg Diastolic BP	continuous: mm Hg
har1	Smoke>100 cig. in life	1 = yes; 0 = no
har3	Smoke now	1 = yes; 0 = no
smoke	Smoking	1 = smoke < 100 cigarettes in life
		2 = smoke > 100 in life and not smoke now
		3 = smoke > 100 in life and smoke now
tcp	Serum cholesterol	continuous: mg per 100ml

The variable black was defined by the author from the *dmaracer* variable. It takes the following values:

Black	Black ethnic background	1 = black; 0 = white

The data are found in the **nhanes3** data set.

We first model the data using a basic logistic regression model with continuous predictors age (*hsageir*), height (*bmphtin*), and diastolic blood pressure (*pepmnk5r*), and binary predictors sex (*hssex*) and race = black (*black*).

```
. use nhanes3

. logit hbp hsageir hssex black bmphtin  pepmnk5r, nolog or

Logistic regression                        Number of obs   =      15961
                                           LR chi2(5)      =    6758.59
                                           Prob > chi2     =     0.0000
Log likelihood = -4857.2227                Pseudo R2       =     0.4103
---------------------------------------------------------------------
     hbp | Odds Ratio  Std. Err.     z   P>|z| [95% Conf. Interval]
---------+-----------------------------------------------------------
 hsageir |   1.105727  .0022556  49.27  0.000  1.101315   1.110157
   hssex |   .7699371  .0556134  -3.62  0.000  .6683009   .8870302
```

```
   black |   1.287747  .0768981    4.23  0.000   1.145515   1.447639
 bmphtin |    .9472949  .0090101   -5.69  0.000    .9297991   .9651199
pepmnk5r |   1.140672  .0035474   42.32  0.000   1.133741   1.147646
----------------------------------------------------------------------
```

The model appears to be well fitted and has a correctly predicted percentage of 86 percent, but does not pass the Hosmer–Lemeshow goodness-of-fit *Chi2* test (29.07) at the 0.05 *p* level (0.0003). We know that the data were collected by survey methods, so there should be rather substantial correlation in the data. Some of the correlation can be adjusted by using robust standard errors, which we attempt in the following model. It was also discovered that there is an interaction effect between sex and diastolic blood pressure.

```
. gen sexxdbp  = hssex*pepmnk5r

. logit hbp hsageir hssex black bmphtin pepmnk5r sexxdbp ,
nolog or cluster(sdpstra6)
```

```
Logistic regression                   Number of obs   =      15961
                                      Wald chi2(6)    =    2889.36
                                      Prob > chi2     =     0.0000
Log pseudolikelihood = -4846.1536     Pseudo R2       =     0.4116
```

                    (Std. Err. adjusted for 49 clusters in sdpstra6)
```
------------------------------------------------------------------
         |              Robust
     hbp | Odds Ratio  Std. Err.    z   P>|z|  [95% Conf. Interval]
---------+--------------------------------------------------------
 hsageir |   1.106141  .0027414  40.70  0.000   1.100781   1.111527
   hssex |   .1105448  .0570398  -4.27  0.000    .0402095   .3039122
   black |   1.295102  .0888878   3.77  0.000   1.132095   1.48158
 bmphtin |   .9476322  .0112972  -4.51  0.000    .9257468   .970035
pepmnk5r |    1.12748  .0057677  23.45  0.000   1.116232   1.138842
 sexxdbp |   1.025029  .0066246   3.83  0.000   1.012127   1.038096
---------+--------------------------------------------------------
```

The important predictors are *age, black, height,* and the interaction of *sex* and *diastolic blood pressure.*

We now model the data using methods to adjust for the manner in which the data were collected, which gives the data a specific type of correlation.

In Stata, survey models must be defined prior to modeling. With PSU, probability weights, and stratum covariates specified for the data, the following code attaches them to the model for estimation.

```
. svyset sdppsu6 [pw=wtpfhx6], strata(sdpstra6)

      pweight: wtpfhx6
          VCE: linearized
  Single unit: missing
     Strata 1: sdpstra6
```

```
        SU 1: sdppsu6
        FPC 1: <zero>
```

The survey model is designated as such by the prefatory *svy*, which calls the information saved in *svyset*.

```
. svy: logit hbp hsageir hssex black bmphtin pepmnk5r sexxdbp
, nolog or (running logit on estimation sample)
```

Survey: Logistic regression

```
Number of strata   =    49        Number of obs      =        15961
Number of PSUs     =    98        Population size    =    166248125
                                  Design df          =           49
                                  F(   6,      44)    =       158.80
                                  Prob > F           =       0.0000
```

| | | Linearized | | | |
hbp	Odds Ratio	Std. Err.	t	P>|t|	[95% Conf. Interval]
hsageir	1.11938	.0044785	28.19	0.000	1.110416 1.128416
hssex	.0573905	.04626	-3.55	0.001	.0113594 .2899504
black	1.505573	.1123603	5.48	0.000	1.295893 1.749181
bmphtin	.9506724	.0146707	-3.28	0.002	.9216429 .9806163
pepmnk5r	1.139253	.0084681	17.54	0.000	1.122362 1.156398
sexxdbp	1.034229	.0100782	3.45	0.001	1.014174 1.054682

A goodness-of-fit test authored by Kellie Archer and Stanley Lemeshow (2006) has been specifically designed to evaluate logistic survey models. Called **svylogitgof**, the test performs an F-adjusted mean residual test of the model. *p*-values are interpreted in the same manner as in the traditional Hosmer–Lemeshow test. Also see Graubard, Korn, and Midthune (1997).

```
. svylogitgof
F-adjusted test statistic = 6.0293887
p-value                   = .00002374
```

The goodness-of-fit test results indicate that the model may not be well fitted. Unfortunately, the test does not produce a table as does the Hosmer–Lemeshow test, which allows the statistician to observe differences and trends in differences between the observed and predicted probabilities at various levels in the data.

14.1.1 Interpretation

Respondents to the survey have an 11 percent increase in the odds of having high systolic blood pressure with each year increase in age. Blacks have half again greater odds of having high systolic blood pressure than whites. There is a 5 percent decrease in the odds of having high systolic blood pressure with each additional inch in height. Finally, males having increasingly higher diastolic pressure have 3 percent greater odds of having higher systolic blood pressure than females with low diastolic blood pressure.

Aside from the biological uncertainties and correlations ($R = 0.34$ between *hbp* and diastolic blood pressure) involved in the above model, it is evident that the survey adjustments made to the data to account for the manner in which the data were collected made little difference to the model itself. We expected substantial correlation in the data because a survey was conducted, but in fact, there was little. This does not always occur. In fact, other modeling examples using the same NHANES survey do exhibit the expected correlation—with survey techniques substantially amending the unadjusted model.

EXAMPLE 14.2—A PROPORTIONAL ODDS SURVEY MODEL

For an example of a more complex survey model, we use a U.S. government study on car crash injuries. The number of predictors used in the published model was far more than needed, and other aspects of the model were flawed as well. We shall employ a more parsimonious model, which is sufficient for our purposes.

The study is called "The Role of Vertical Roof Intrusion and Post-Crash Headroom in Predicting Roof Contact Injuries to the Head, Neck, or Face During FMVSS No. 216 Rollovers; An Updated Analysis" and is stored in the file, **crashinjury**. The covariates of interest include:

RESPONSE

- *Injury*: degree of injury 1 = none, 2 = minor, 3 = severe, 4 = death

PSU AND WEIGHTS

- *psu, psustrat, simplewgt*: Cluster, strata, and weight for each observation
  Cluster weights : *psu* (27 cluster weights, from 2–82)
  Stratum weights : *psustrat* (12 strata weights, from 1–12)
  Prob weight : *simplewgt* (from .16 to 4278.68)

PREDICTORS

- *rollover*: The number of quarter turns, 0–17
- *ltv*: 1 if the vehicle is a light truck or van, 0 if the vehicle is a passenger car
- *diintrude*: 1 if roof intrusion is positive, 0 if it is zero
- *diposthr*: 1 if post-crash headroom is positive, 0 if it is zero or negative

A tabulation of the response, *injury*, shows a four-level categorical variable with increasing levels of injury.

```
. use crashinjury

. tab injury
```

```
   injury |       Freq.       Percent         Cum.
------------+-----------------------------------------
Not injured |       1,907        67.17         67.17          /// 1
      Minor |         588        20.71         87.88          /// 2
     Severe |         191         6.73         94.61          /// 3
      Death |         153         5.39        100.00          /// 4
------------+-----------------------------------------
      Total |       2,839       100.00
```

We can first model *injury* using a standard proportional odds model.

```
. ologit injury diintrude ltv diposthr rollover , nolog

Ordered logistic regression            Number of obs    =        2743
                                       LR chi2(4)       =      201.26
                                       Prob > chi2      =      0.0000
Log likelihood = -2465.5969            Pseudo R2        =      0.0392


-----------------------------------------------------------------------
   injury |     Coef. Std. Err.     z   P>|z|   [95% Conf. Interval]
----------+------------------------------------------------------------
 diintrude |   .4430073 .1108086   4.00 0.000   .2258264    .6601881
       ltv |   -.033818 .0831923  -0.41 0.684   -.196872     .129236
  diposthr |  -.7647335 .0972648  -7.86 0.000  -.9553691    -.574098
  rollover |   .0910363 .0160218   5.68 0.000   .0596341    .1224384
----------+------------------------------------------------------------
     /cut1 |   .9163087 .1500308               .6222537    1.210364
     /cut2 |   2.25006 .1565087               1.943309    2.556812
     /cut3 |  3.150036 .1678616               2.821033    3.479038
-----------------------------------------------------------------------
```

Testing the proportionality of slopes using the Brant test tells us that the proportionality assumption has not been violated.

```
. brant

Brant Test of Parallel Regression Assumption

    Variable |      chi2    p>chi2     df
-------------+----------------------------
         All |     13.32     0.101      8
-------------+----------------------------
   diintrude |      0.14     0.931      2
         ltv |      3.18     0.204      2
    diposthr |      4.84     0.089      2
    rollover |      3.31     0.191      2
----------------------------------------

A significant test statistic provides evidence that the
parallel regression assumption has been violated.
```

Knowing that the data have been collected as a survey, and weights have been determined for the PSU sampling clusters and strata, we can model the data as a survey model. First, the PSU, strata, and probability weights are assigned to the model by issuing the following command.

```
. svyset psu [pw=simplewgt], strata(psustrat)

      pweight: simplewgt
          VCE: linearized
  Single unit: missing
     Strata 1: psustrat
        SU 1: psu
       FPC 1: <zero>
```

The survey model for ordered logit is given as:

```
. svy: ologit injury diintrude ltv diposthr rollover , nolog
(running ologit on estimation sample)

Survey: Ordered logistic regression

Number of strata    =      12    Number of obs      =      2743
Number of PSUs      =      27    Population size  = 193517.93
                                 Design df          =        15
                                 F(  4,      12)  =     12.29
                                 Prob > F           =    0.0003
-----------------------------------------------------------------
            |             Linearized
   injury   |    Coef. Std. Err.    t  P>|t| [95% Conf. Interval]
------------+----------------------------------------------------
 diintrude  |  .8875137 .4370708  2.03 0.060 -.0440806   1.819108
      ltv   | -.4966259 .1932894 -2.57 0.021 -.9086125  -.0846392
 diposthr   | -1.159013 .5143516 -2.25 0.040 -2.255328  -.0626988
 rollover   |  .1380345 .0596172  2.32 0.035  .0109635   .2651056
------------+----------------------------------------------------
    /cut1   |  1.594247 .8008267  1.99 0.065 -.1126747   3.301169
    /cut2   |  3.613993 .6744746  5.36 0.000  2.176384   5.051601
    /cut3   |  5.071164 .6227192  8.14 0.000  3.743869   6.398458
-----------------------------------------------------------------
```

The SAS code for modeling the same data can be expressed as:

```
proc surveylogistic data=nhtsadata.jh ;
     weight simplewgt;
     cluster psu;
     strata psustrat;
     class injury ;
     model injury=diintrude ltv diposthr rollover /link=logit ;
```

```
        title `Ordered Survey Proportional Odds Model for assessing
            Injury level on crash data';
run;
```

The odds ratios for the survey model can be obtained by the command:

```
. ologit, or  /// [header and cut information has been deleted]

-----------------------------------------------------------------
           |                 Linearized
    injury | Odds Ratio Std. Err.    t    P>|t| [95% Conf. Interval]
-----------+-----------------------------------------------------
diintrude |   2.429083 1.061681   2.03 0.060 .9568768   6.166356
      ltv |   .6085806 .1176322  -2.57 0.021 .4030831   .9188437
 diposthr |   .3137957 .1614013  -2.25 0.040 .1048392   .9392263
 rollover |   1.148015 .0684415   2.32 0.035 1.011024   1.303569
-----------+-----------------------------------------------------
    /cut1 |   1.594247 .8008267              -.1126747   3.301169
    /cut2 |   3.613993 .6744746               2.176384   5.051601
    /cut3 |   5.071164 .6227192               3.743869   6.398458
-----------------------------------------------------------------
```

Adjusting the model by the apportionate sampling weights results in parameter estimates that differ considerably from the unadjusted proportional odds model. In any case, the parameter estimates and odds ratios are interpreted like any other proportional odds model.

Natarajan et al. (2008) present a good overview of more complex survey designs and propose a simple method for their estimation. A well-written R function called **svyglm** can be used for estimating data from stratified cluster sampling designs using a generalized linear models approach. It can be found at: http://faculty.washington.edu/tlumley/survey/html/svyglm.html.

14.2 Scobit-Skewed Logistic Regression

Logistic regression assumes that the majority of fitted probability in a model rests symmetrically around a mean of 0.5. Non-canonical linked models such as complementary loglog and loglog are asymmetric, but still assume a mean fitted value of 0.5.

Skewed logistic regression was initially designed by J. Nagler (1994) to provide flexibility of where the logistic weight is distributed. The impact of the predictors can, therefore, be skewed away from 0.5, hence the name of the procedure. Nagler also gave the procedure the name of *scobit*, comparing it with *logit* and *probit*. The procedure is generally known by both names and is a popular method used in political science research.

The **scobit** function is defined as:

$$Pr(y|x) = 1 - 1/(1 + \exp(xb))^{\alpha}$$

where the fitted y, or μ, is the probability of 1.

Skewed logistic regression, or scobit regression, not only estimates μ, the probability of success, but it also estimates an ancillary or location parameter that reflects the skew of the distribution. The point of the maximum impact of the scobit fitted values is constrained by the above formula to rest in the probability range of 0 to 0.632+, where 0.632+ is $1 - 1/e$. e, as you may recall, is calculated as 2.718+. Note that + at the end of decimals indicates more digits are calculated but not displayed. e is often designated in formulae as $\exp()$.

The scobit model has itself been adjusted to allow for situations in which the maximum impact is outside the range specified above (Achen, 2002). A so-called *power logit* is constructed by reversing the response values of 1 and 0. This is similar to reversing response values to change from a complementary loglog to a loglog model. The model predicts failure rather than success. To model success, the coefficients must have their signs reversed as well, but care needs to be taken when calculating resultant confidence intervals.

We use the **heart01** data, modeling *death* on the list of predictors we have used in previous examples. First, however, we should again look at the distribution of the response variable, death. It appears that there is less than 5 percent probability of success, that is, dying, among the patients sustaining a heart attack and entering the hospital. We hope that the sample of patients in this study is representative of the nation as a whole, but for now, we can only conclude that the descriptive statistics in the model are valid for the group in this study:

```
. tab death

     Death |
  within 48 |
  hrs onset |      Freq.      Percent        Cum.
-----------+-----------------------------------
         0 |      5,146        95.51        95.51
         1 |        242         4.49       100.00
-----------+-----------------------------------
     Total |      5,388       100.00
```

It is certainly possible that *death* is better modeled as skewed than symmetric about a mean of 0.5. We employ a robust variance estimator:

```
. scobit death anterior hcabg kk2-kk4 age2-age4, nolog vce(robust)
```

```
Skewed logistic regression                    Number of obs    =    4503
                                              Zero outcomes    =    4327
Log pseudolikelihood = -631.4663             Nonzero outcomes  =     176
```

death	Coef.	Robust Std. Err.	z	P>|z|	[95% Conf. Interval]
anterior	1.038279	.2995395	3.47	0.001	.4511923 1.625366
hcabg	1.0814	.6059206	1.78	0.074	-.1061824 2.268983
kk2	1.20453	.3151514	3.82	0.000	.5868441 1.822215
kk3	1.694347	.8064317	2.10	0.036	.1137694 3.274924
kk4	7.12694	2.84551	2.50	0.012	1.549843 12.70404
age2	.4843345	.3925374	1.23	0.217	-.2850248 1.253694
age3	1.988824	.4002964	4.97	0.000	1.204257 2.77339
age4	3.37586	.6937686	4.87	0.000	2.016098 4.735621
_cons	-2.697439	.4945119	-5.45	0.000	-3.666664 1.728213
/lnalpha	-2.757289	.4821015	-5.72	0.000	-3.702191 1.812388
alpha	.0634636	.0305959			.0246694 .1632639

```
Likelihood-ratio test of alpha=1:  chi2(1) =   10.72   Prob >
chi2 = 0.0011
```

Note: likelihood-ratio tests are recommended for inference
with scobit models.

The likelihood ratio test evaluates if the scobit model is significantly different from a logistic model ($\alpha = 1$). With a *Chi2* p-value under 0.05, we find that, in this case, the model indeed differs from a similar logistic model. As was the case with the logistic model we estimated on these data, *age2* is not a significant contributor to the model, nor is *hcabg*. Actually, we previously found that the significance of *hcabg* depended on how we modeled the data, for example, robust standard errors versus model standard errors, clustering, and so forth. Although the results are not displayed here, the model standard errors differ little from the robust we used on the above model.

It may be appropriate to model the data as before, but parameterized by clustering the data on the basis of the hospitals admitting the patients. Because the model is no longer strictly speaking a likelihood model—patients are no longer independent—the likelihood ratio test cannot be used to evaluate its relationship to a logistic model.

```
. scobit death anterior hcabg kk2-kk4 age2-age4, nolog
cluster(center)

Skewed logistic regression                    Number of obs    =    4503
                                              Zero outcomes    =    4327
Log pseudolikelihood = -631.4663             Nonzero outcomes  =     176
```

```
                        (Std. Err. adjusted for 47 clusters in center)
-----------------------------------------------------------------------
             |              Robust
    death    |    Coef.   Std. Err.    z    P>|z|   [95% Conf. Interval]
---------+-------------------------------------------------------------
 anterior |  1.038279  .3258267   3.19  0.001    .3996704   1.676887
    hcabg |    1.0814  .5388816   2.01  0.045    .0252116   2.137589
      kk2 |   1.20453  .2980076   4.04  0.000    .6204454   1.788614
      kk3 |  1.694347  .8920634   1.90  0.058   -.0540655   3.442759
      kk4 |   7.12694  2.341554   3.04  0.002    2.537578    11.7163
     age2 |  .4843345  .4008649   1.21  0.227   -.3013462   1.270015
     age3 |  1.988824  .3778834   5.26  0.000    1.248186   2.729462
     age4 |   3.37586  .7202213   4.69  0.000    1.964252   4.787468
     _cons | -2.697439  .5644558  -4.78  0.000   -3.803752  -1.591126
---------+-------------------------------------------------------------
 /lnalpha | -2.757289  .5424061  -5.08  0.000   -3.820385  -1.694193
---------+-------------------------------------------------------------
    alpha |  .0634636   .034423                  .0219194   .1837475
-----------------------------------------------------------------------
```

Note that *hcabg* now significantly contributes to the model at the <0.05 level. However, *kk3* does not appear to contribute; and we cannot test its contribution using a likelihood ratio test.

An argument can be made that although using a clustering variable with robust variance estimates amends the model standard errors, the parameter estimates are left unadjusted. Since the estimates are the same as the non-clustered model with which we did employ a likelihood ratio test, the argument states that the test is valid for the clustered model as well. This line of reasoning appears to make sense and is used by many statisticians, but it is nevertheless wrong to claim that it is valid to use a likelihood ratio test on the clustered data. On the other hand, there is no other test we can use at present, and the results of any similar test would most likely be the same unless standard errors were part of the test formula. We can therefore use the test, but with the caveat that it is theoretically inappropriate for this type of data. The better approach is to ignore the clustering effect and use the test comparing the scobit and logit models. In doing so, however, we keep in mind how clustering affected the standard errors, and make no apodictic or final determination on the significance of *hcabg* and *kk3*.

14.3 Discriminant Analysis

For many years, discriminant analysis was the standard statistical procedure with which statisticians modeled binary as well as categorical response data. Discriminant analysis is a multivariate method used to both predict and classify data. It remained the predominant method for modeling categorical data until the 1990s.

Unlike the majority of multivariate models, discriminant analysis is structured as a multinomial model, with a response consisting of two or more levels, and a number of predictors, which could be continuous, categorical, or binary. We shall discuss three parameterizations of discriminant analysis in this chapter: a dichotomous or binary response discriminant analysis in Section 14.3.1, a canonical linear discriminant analysis in Section 14.3.2, and a categorical response linear logistic discriminant analysis in Section 14.3.3.

14.3.1 Dichotomous Discriminant Analysis

First, we address a binary response model that is based on an ordinary least squares algorithm. I wrote this implementation in the late 1980s and created a Stata command for the model in 1991. It was published in January 1992 (Hilbe, 1992a).

Dichotomous discriminant analysis, as standard categorical response discriminant analysis, is based on the discriminant function, which is a linear function of the covariates that maximizes the ratio of the between-group and within-group sum of squares. In effect, it is the attempt to separate the groups or levels of the response in such a manner that they are as distinct as possible. For the binary response model, which is the subject of this section, the procedure is based on the assumptions that both group covariance matrices are closely the same, and that the independent predictors are multivariately normal. Although discriminant analysis is fairly robust to violations of this assumption, simulation studies have demonstrated that binary logistic regression does a better job at classification when the violations exist. On the other hand, logistic regression fails to converge when there is perfect prediction. Discriminant analysis has no such restriction and is still used in logistic regression situations where perfect prediction occurs, and when it is not feasible to amend the model covariates.

It should also be noted that when there are significantly fewer covariate patterns in the model data than observations, m-asymptotic structured data for logistic regression generally yields superior correct classification rates than does either discriminant analysis or individual case logistic regression. A summary of when the discriminant analysis model is preferred to logistic regression is given in Table 14.1:

TABLE 14.1

Discriminant Analysis Preferred to Logistic Regression

1. When there is perfect prediction between response and predictors

2. When there is near equality of group covariance matrices

3. When the predictors are all fairly normally distributed

Models having binary or categorical predictors serve to undermine the assumptions upon which discriminant analysis is based. If both groups are multivariately normal, a discriminant analysis may be preferred to a logistic model. Also, any time perfect prediction occurs, discriminant analysis is preferred.

The logic of the OLS-based binary response discriminant model used here is detailed in Hilbe (1992a). We use the **auto** data set that comes with the Stata package, LogXact, and several other statistical applications. The variables used in our model consist of the following:

RESPONSE

foreign	1: foreign	0: domestic <response>

PREDICTORS

mpg	continuous	miles per gallon
length	continuous	length of car in meters
weight	continuous	weight of car in kilograms

The model, now called **discrim2**, is used to model *foreign*. Note that all three predictors are continuous:

```
. discrim2 foreign mpg length weight, anova detail
```

Dichotomous Discriminant Analysis

Observations	= 74	Obs Group 0 =	52
Indep variables	= 3	Obs Group 1 =	22

Centroid 0	=	-0.4971	R-square =	0.3751
Centroid 1	=	1.1750	Mahalanobis =	2.7962
Grand Cntd	=	0.6779		

Eigenvalue	=	0.6004	Wilk's Lambda =	0.6249
Canon. Corr.	=	0.6125	Chi-square =	33.1521
Eta Squared	=	0.3751	Sign Chi2 =	0.0000

Variable	Discrim Function Coefficients	Unstandardized Coefficients
mpg	0.1541	-0.0921
length	0.0260	-0.0156
weight	0.0028	-0.0017
constant	-16.1689	10.0083

Discriminant Scores v Group Variable

```
                             Analysis of Variance
           Source              SS        df     MS            F     Prob > F
---------------------------------------------------------------------------------
Between groups         43.2275933    1   43.2275933      43.23      0.0000
Within groups          72.0000066   72    1.00000009
---------------------------------------------------------------------------------
     Total             115.2276     73    1.57846027
```

Bartlett's test for equal variances: chi2(1) =   5.9171
Prob>chi2 = 0.015

```
PRED     = Predicted Group        DIFF     = Misclassification
LnProb1  = Probability Gr 1       DscScore = Discriminant Score
                                  DscIndex = Discriminant Index
```

	foreign	PRED	DIFF		LnProb1	DscIndex	DscScore
1.	0	0			0.4083	0.3710	0.1171
2.	0	0			0.3884	0.4542	0.0673
3.	0	1	*		0.7150	-0.9200	0.8891
4.	0	0			0.2260	1.2308	-0.3971
5.	0	0			0.0296	3.4916	-1.7491
6.	0	0			0.0638	2.6865	-1.2676
7.	0	1	*		0.8045	-1.4149	1.1851
8.	0	0			0.1947	1.4200	-0.5102
9.	0	0			0.0637	2.6881	-1.2686
10.	0	0			0.1671	1.6064	-0.6217
11.	0	0			0.0176	4.0208	-2.0656
70.	1	1			0.8833	-2.0237	1.5492
71.	1	1			0.5083	-0.0332	0.3588
72.	1	1			0.9432	-2.8103	2.0196
73.	1	1			0.9318	-2.6140	1.9022
74.	1	0	*		0.3861	0.4636	0.0617

A few of the important test statistics are the eigenvalues, which is a means of evaluating the discriminating power of the model. Values over .40 are preferred. The eigenvalue of this model is .60. Wilk's λ measures the degree of difference between group means. It can range from 0.0 to 1.0. Lower values indicate a model with superior discriminating power. For example, a λ of .20 indicates that the difference between the two groups account for some 80 percent of the variance in the predictors. The value of .62 in this model tells us that only 38 percent of the predictor variance is accounted for. η^2 (not eta, the linear predictor) indicates the ratio of the total variance (SS_t) in the discriminant scores that can be explained by the difference between the two groups. η^2 is defined as $1 - \Lambda$, where $\Lambda = SS_w/SS_t$.

Two other statistics that play a valuable role in model interpretation are the χ^2 and Bartlett's test. χ^2 is defined as:

$$\chi^2 = -(n - (p + 2)/2 - 1) * \ln(\Lambda)$$

where p is the number of predictors in the model, and is also the degrees of freedom, and n is the number of observations in the model. For this model the value is

```
. di -(74 - ((3+2)/2)-1) * ln(.6249)
33.146537
```

which is the same as the displayed output. The significance of χ^2 with p degrees of freedom is 0.000:

```
. di chiprob(3, 33.146537)
2.999e-07
```

indicating that we cannot reject the assumed hypothesis that the variances are homogeneous, which is what we desire of a discriminant analysis.

Bartlett's test for the equality of variances is also shown to have a significance of under 0.05 (0.015). The result is the same interpretation as we have for the χ^2 test.

Lastly, we can count the number of *s in the prediction table, only partially shown in the output above, to determine the correct prediction rate. There are 15 out of 74 incorrectly predicted cases, based on the logistic distribution. This gives a $1 - 15/74 = 79.73$ percent correct prediction rate.

We now compare the results of the discriminant analysis with a logistic regression:

```
------------------------------------------------------------------
foreign |     Coef. Std. Err.    z    P>|z|   [95% Conf. Interval]
--------+---------------------------------------------------------
    mpg | -.1685852 .0919688  -1.83 0.067  -.3488407    .0116703
 length |  .0000339 .0589596   0.00 1.000  -.1155247    .1155925
 weight | -.0039076 .0019293  -2.03 0.043  -.0076889   -.0001264
  _cons |  13.70478 7.70354    1.78 0.075  -1.393882   28.80344
------------------------------------------------------------------
```

Compare the above coefficients with the unstandardized discriminant coefficients. The values are rather close.

The classification rate for the logistic model is given as:

```
. lstat

Logistic model for foreign

                    -------- True --------
Classified |            D                ~D  |        Total
-----------+----------------------------------+-----------
     +     |           13                6  |           19
     -     |            9               46  |           55
-----------+----------------------------------+-----------
   Total   |           22               52  |           74

Classified + if predicted Pr(D) >= .5
True D defined as foreign != 0
----------------------------------------------------------
Sensitivity                        Pr( +| D)     59.09%
Specificity                        Pr( -|~D)     88.46%
Positive predictive value          Pr( D| +)     68.42%
Negative predictive value          Pr(~D| -)     83.64%
----------------------------------------------------------
False + rate for true ~D           Pr( +|~D)     11.54%
False - rate for true D            Pr( -| D)     40.91%
False + rate for classified +      Pr(~D| +)     31.58%
False - rate for classified -      Pr( D| -)     16.36%
----------------------------------------------------------
Correctly classified                             79.73%
----------------------------------------------------------
```

Recall that the discriminant classification rate was 79.73 percent. It is calculated as the sum of the D+ (13) and ~D− (46) cells, which is 59/74 = 79.73. This value is identical to the logistic classification—also 79.73 percent. In this instance, the models are both good models of the data, giving near identical coefficients and identical prediction rates of some 80 percent.

14.3.2 Canonical Linear Discriminant Analysis

Canonical discriminant analysis was originated by R.A. Fisher (1936) and P.C. Mahalanobis (1936), although they differed in how they approached classification. The table presented in the previous subsection is, in a sense, a representation of the two approaches. Fisher developed discriminant function coefficients as the basis of classification, which is also referred to as Fisher coefficients; Mahalanobis based his manner of classification on unstandardized coefficients, which are often referred to as canonical discriminant function coefficients. The method used in this subsection is essentially based on Mahalanobis' work, but there is no doubt that both methods have greatly overlapped in the sophisticated discriminant analysis procedures found in software such as Stata, SAS, SPSS, and other leading applications. Fisher coefficients and canonical loadings are both presented in the output of these packages, and both methods of classification usually tend to emphasize one approach or another.

Variable	Discrim Function Coefficients	Unstandardized Coefficients
mpg	0.1541	-0.0921
length	0.0260	-0.0156
weight	0.0028	-0.0017
constant	-16.1689	10.0083

Stata has two commands that provide linear coefficients that aim to separate groups in a distinct a manner as possible for the purposes of both prediction and classification. **discrim** *lda*, for linear discriminant analysis, emphasizes the predictive feature of discriminant analysis. **candisc** is based on the same algorithm as discrim lda, but is structured to provide classifications. We shall use the **candisc** command to develop a classification table as we did in the previous section, comparing it with the OLS regression based dichotomous discriminant analysis classification table and the table developed on the basis of a logistic regression model.

```
. candisc mpg weight length, group(foreign)

Canonical linear discriminant analysis

    |                              | Like-
    | Canon. Eigen-   Variance     | lihood
Fcn | Corr.  value   Prop. Cumul.  | Ratio    F      df1 df2 Prob>F
----+-----------------------------+----------------------------------
  1 | 0.6125 .600383 1.0000 1.0000 | 0.6249  14.009   3   70  0.0000 e
----------------------------------------------------------------------
Ho: this and smaller canon. corr. are zero;        e = exact F


Standardized canonical discriminant function coefficients

                 | function1
    -------------+----------
             mpg | -.4934496
          weight | -1.069307
          length | -.2864726
```

Canonical structure

```
                  | function1
-----------------+-----------
            mpg  |  .5522397
         weight  | -.9500428
         length  | -.8957762
```

Group means on canonical variables

```
        foreign  | function1
-----------------+-----------
       Domestic  |  -.497135
        Foreign  |  1.175046
```

Resubstitution classification summary

```
+---------+
| Key     |
|---------|
| Number  |
| Percent |
+---------+
                  | Classified
True foreign      | Domestic     Foreign  |     Total
------------------+-------------------------+---------
        Domestic  |       39          13  |        52
                  |    75.00       25.00  |    100.00
                  |                        |
         Foreign  |        2          20  |        22
                  |     9.09       90.91  |    100.00
------------------+-------------------------+---------
           Total  |       41          33  |        74
                  |    55.41       44.59  |    100.00
                  |                        |
          Priors  |   0.5000      0.5000  |
```

Compare the "Group means on canonical variables" listing above with the *centroids* displayed in the output of the previous subsection, and compare the classification rate of the two methods. **candisc** is based on matrix statistics independent of any type of regression application. **discrim2** uses OLS regression and its corresponding ANOVA table to determine the various classification statistics and discriminant coefficients.

The centriods and group means are identical, and the classification rates of both are identical at 79.43 percent. It is interesting, however, to note that the values in the classification cells differ. The correctly predicted values of 15, produced in both tables, are sums of 6 and 9, and 2 and 13, respectively.

14.3.3 Linear Logistic Discriminant Analysis

Linear logistic discriminant analysis attempts to ameliorate the problem of group multivariate normality that plagues many discriminant analysis models. Instead, the logistic parameterization is based on the assumption that the likelihood ratios of each group employ an exponential format, in a manner similar to multinomial logistic regression. The method was originated by Albert and Lesaffre (1986) with the intent of using a multinomial logistic model for discriminant classification purposes.

Modeling the same data as performed with the dichotomous discriminant analysis above, we display a classification matrix to determine if this method has a superior classification rate.

```
. discrim logistic mpg weight length, group(foreign) nolog

Logistic discriminant analysis
Resubstitution classification summary

          +---------+
          | Key     |
          |---------|
          | Number  |
          | Percent |
          +---------+
                    | Classified
True foreign        | Domestic    Foreign  |    Total
-------------------+----------------------+---------
        Domestic   |       40         12  |       52
                   |    76.92      23.08  |   100.00
                   |                      |
         Foreign   |        3         19  |       22
                   |    13.64      86.36  |   100.00
-------------------+----------------------+---------
           Total   |       43         31  |       74
                   |    58.11      41.89  |   100.00
                   |                      |
          Priors   |   0.5000     0.5000  |
```

The correct prediction rate of the above classification table can be calculated as the sum of the Domestic-Domestic (40) and Foreign-Foreign (19) cells, which total to 59. This 59 correctly predicted of the 74 cases gives us 59/74 = 79.73 percent—the same value as produced by the dichotomous discriminant analysis and logistic regression models in the previous subsection.

```
mlogit foreign mpg length weight, nolog

Multinomial logistic regression            Number of obs   =       74
                                           LR chi2(3)      =    35.72
                                           Prob > chi2     =   0.0000
Log likelihood = -27.175156               Pseudo R2       =   0.3966

------------------------------------------------------------------
     foreign |    Coef. Std. Err.    z   P>|z| [95% Conf. Interval]
-------------+----------------------------------------------------
Foreign      |
        mpg  | -.1685852 .0919688 -1.83 0.067 -.3488407    .0116703
     length  |  .0000339 .0589596  0.00 1.000 -.1155247    .1155925
     weight  | -.0039076 .0019293 -2.03 0.043 -.0076889   -.0001264
      _cons  |  13.70478  7.70354  1.78 0.075 -1.393882    28.80344
------------------------------------------------------------------
(foreign==Domestic is the base outcome)
```

A comparison with the logistic model in the previous subsection shows that the coefficients are nearly identical, except that the sign is reversed for the constant. We also observed that the logistic coefficients were similar to the unstandardized discriminant coefficients.

If we develop a classification table based on the multinomial coefficient table, we will discover that the table is identical to the classification table displayed above. This comes as no surprise because the logistic discriminant analysis table is derived from the multinomial coefficients. The interesting point is in the similarity of results between the respective methods of logistic/multinomial regression and linear canonical discriminant analysis. Their methods of estimation differ substantially, but the results proved the same. Statisticians love to see this type of consistency, which confirms the validity of the methods.

Exercises

14.1. Using the data and model from Exercise 5.8, remodel the data using each of the three types of discriminant analyses discussed in the text.

   a. Compare the predicted probabilities of the binary logistic model and the discriminant analysis.

   b. Interpret the meaning of the canonical and unstandardized coefficients. In what way do they differ from logistic parameter estimates?

14.2. Use the same **edreligion** data as given in Exercise 12.2. Model as a logistic discriminant analysis.

    a. Compare the results of the multinomial and discriminant analysis. Which is the best-fitted model?

    b. Interpret the Wilk's λ statistic for the discriminant model.

    c. Compare the proportion of correctly predicted values for each of the levels of response between the multinomial and discriminant models.

14.3. Use the **schiz2** data as described in Exercise 13.4. Here we use the *imps4* covariate as the response, with *treatment* and *sex* as the predictors. Model as a logistic discriminant analysis.

    a. Test the assumptions of the logistic discriminant analysis to evaluate if the model is appropriate to model these data.

    b. Model using a logistic discriminant analysis. Calculate the probabilities for each of the four levels.

    c. Compare the fit with proportional odds and multinomial models of the data.

14.4. Using the **crashinjury** data analyzed in this chapter, model as a survey proportional slopes model with a probit link.

    a. Evaluate if the probit link is preferable to the logit.

    b. Dichotomize injury so that 0 = not injured, and 1 = injured. Model as a survey logistic regression, determining the odds ratios of the estimates. Compare with a model on the same data, but not adjusted for survey sampling. Which is the better fitted model? How can you tell?

14.5. Using the **medpar** data, estimate *died* on *los, age80, type2* and *type3*, where *type1* is the reference of the categorical covariate, *type*.

    a. Use both a logit and a skewed (**scobit**) logistic model, comparing the differences in their odds ratios, and AIC-BIC fit statistics.

    b. Compare the fit statistics for both the logit and scobit models. Explain why their values might appear as they do.

    c. What is the essential difference in the logic of logit and scobit models?

14.6. Use the **cluslowbwt** data to model a survey analysis on data abstracted from Hosmer and Lemeshow (2000). The relevant covariates in the model are

RESPONSE

low   binary   1: birthweight under 2500 grams;
            0: weight > 2500 grs

PREDICTORS

smoke	binary	1: smokes; 0: not smokes
lwtkg	continuous	36.4 to 123.6 kg
age	continuous	14–43 years
race	categorical	1: white; 2: black; 3: other

PANEL

id   2 to 4

a. Model *low* on smoke as the risk factor, with *smoke, lwtkg, age,* and *race* as confounders. Race = 1 is the referent level.

b. Model the same as in (a) but as a survey model with *id* as the subject identifier. Determine if there is a significant correlation effect resulting from the survey nature of the data.

c. As a survey model, determine if *lwtkg* or age should be categorized. If so, categorize and remodel with the lowest level as the referent.

15

Exact Logistic Regression

15.1 Exact Methods

Exact statistical analysis refers to the manner in which the parameter estimates and standard errors are calculated. Exact statistical methods construct a statistical distribution that can be determined completely. This is unlike the methods used by maximum likelihood and ordinary least squares regression. The exact statistical algorithm is highly iterative, involving at times many thousands of permutations.

The method is used primarily when the data to be analyzed are very small and/or unbalanced. In either case, correct parameter estimates and standard errors via IRLS are mistaken. In fact, in many unbalanced situations, the maximum likelihood estimates cannot even be obtained. Regardless, the *p*-values that derive from exact methods are superior to those based on asymptotic methods.

We'll use the **hiv** data set to show an example of an exact statistical logistic model. [Note: **hiv** may also be obtained using the Web facility in Stata by typing: *webuse hiv_n.*]

```
. describe

Contains data from hiv.dta
  obs:              8
  vars:             4                          14 Apr 2005
10:52
  size:            64 (99.9% of memory free)
-------------------------------------------------------------------
variable  storage  display  value
name      type     format   label        variable label
-------------------------------------------------------------------
infec     byte     %8.0g    number of patients infected with hiv
cases     byte     %8.0g    cases with same covariate pattern
cd4       byte     %8.0g    hiv marker
cd8       byte     %8.0g    hiv marker
-------------------------------------------------------------------
```

The complete data set consists of the following:

```
        +--------------------------------+
        | infec    cases    cd4    cd8  |
        |--------------------------------|
    1.  |    1        1       0      2  |
    2.  |    2        2       1      2  |
    3.  |    4        7       0      0  |
    4.  |    4       12       1      1  |
    5.  |    1        3       2      2  |
        |--------------------------------|
    6.  |    2        7       1      0  |
    7.  |    0        2       2      0  |
    8.  |    0       13       2      1  |
        +--------------------------------+
```

We model *infec* on *cd4* and *cd8*, with *cases* as the binomial denominator. *cases* represents the number of cases having identical covariate patterns. *infec* represents the number of infected cases with the same pattern of covariates.

I use Stata's **glm** command since neither the **logistic** nor **logit** commands allow grouped models. Either way, the result is a logistic regression model.

```
. glm infec cd4 cd8 , fam(bin cases)
Generalized linear models                  No. of obs      =        8
Optimization       : ML                    Residual df     =        5
                                           Scale parameter =        1
Deviance        =   4.471460696            (1/df) Deviance =    .8943
Pearson         =   3.840665018            (1/df) Pearson  =    .7683

Variance function: V(u) = u*(1-u/c)
Link function    : g(u) = ln(u/(c-u))

                                           AIC             =   2.4622
Log likelihood    = -6.848646175           BIC             =  -5.9257
-------------------------------------------------------------------------
             |                OIM
   infec |       Coef. Std. Err.     z  P>|z|  [95% Conf. Interval]
---------+---------------------------------------------------------------
    cd4  | -2.541669  .8392231   -3.03 0.002 -4.186517   -.8968223
    cd8  |  1.658586  .821113     2.02 0.043   .0492344   3.267938
   _cons |   .5132389  .6809007   0.75 0.451  -.8213019    1.84778
-------------------------------------------------------------------------
```

Next, model the same data using **exlogistic**, Stata's exact logistic regression command. Note that one must specifically request calculation of the coefficients and constant.

```
. exlogistic infec cd4 cd8, binomial(cases) estc coef

Exact logistic regression              Number of obs  =        47
Binomial variable: cases               Model score    =  13.34655
                                       Pr >= score    =   0.0006
-----------------------------------------------------------------
  infec |     Coef.   Suff.   2*Pr(Suff.)  [95% Conf. Interval]
--------+--------------------------------------------------------
    cd4 | -2.387632      10       0.0004  -4.699633   -.8221807
    cd8 |  1.592366      12       0.0528  -.0137905    3.907876
   _cons |  .4749287     14       0.7034  -1.027262    2.115926
-----------------------------------------------------------------
```

To determine the consistency between the only two software applications that provide both exact logistic and Poisson modeling capability, I show output for modeling the same data using Cytel's **LogXact**. SAS has a similar capability, based on Cytel code.

```
Binary Regression

regression (type=logit, model(hiv = cd4 cd8), frequency=freq,
estimate(cd4 cd8), method=exact);


Summary Statistics
Statistics      Value DF      P-Value
Devianc         4.471 5       0.4837
Likelihood Ratio     23.65  3        2.952e-005

Parameter Estimates
            Point Estimate     Confidence Interval and P-Value for
Beta                                   95       %CI      2*1-sided
ModelTerm Type Beta  SE(Beta) Type   Lower    Upper     P-Value
%Const MLE   0.5132 0.6809  Asymp  -0.8213  1.848     0.451
CD4    MLE  -2.542 0.8392  Asymp  -4.187   -0.8968   0.002457
       CMLE -2.388 0.8006  Exact  -4.7     -0.8222   0.0004138
CD8    MLE   1.659 0.8211  Asymp  0.04923  3.268     0.04339
       CMLE  1.592 0.8047  Exact  -0.0137  3.908     0.05278
```

Note that LogXact does not exactly estimate the constant. Notice also that the exact p-values of *cd4* and *cd8* are 0.0004138 and 0.05278, respectively, and the constant 0.7034. Both applications produced the same results. The asymptotic p-values, seen in the Stata output as well as in the LogXact output, indicated by "asymptotic," are 0.002457 and 0.04339. *cd4* significantly contributes to the model regardless of method, but this is not the case with *cd8*. A value lower than 0.05 is usually regarded as the cutpoint defining significance. The exact p-value of *cd8* is not significant at the 0.05 level, whereas it did indicate significance using asymptotic estimating methods.

We see in the table of parameter estimates below how the *p*-value of *cd8* changes when a robust or empirical adjustment is made. We discussed this earlier. The robust variance estimate is nearly identical to the exact value: exact = 0.05278; robust = 0.053. With rounding, the two values are identical. The *cd4 p*-values are close as well.

Only the table of estimates is displayed below. The headings are identical.

```
. glm infec cd4 cd8 , fam(bin cases) robust nohead

------------------------------------------------------------------------
             |               Robust
   infec |      Coef. Std. Err.      z   P>|z|   [95% Conf. Interval]
--------+---------------------------------------------------------------
    cd4 |  -2.541669 .7694029   -3.30 0.001   -4.049671   -1.033667
    cd8 |   1.658586 .8562788    1.94 0.053   -.0196893    3.336862
  _cons |    .5132389 .3531111    1.45 0.146   -.1788461    1.205324
------------------------------------------------------------------------
```

Comparing the parameter estimate robust *p*-values between the asymptotic and exact methods gives us:

p-Values	Asymptotic	Asymptotic-robust	Exact
cd4	.002	.001	.0004
cd8	.043	.053	.0528
constant	.451	.146	.7034

The parameter estimate *p*-values for the asymptotic robust *p*-value and exact *p*-value are similar. We have performed other tests showing that this similarity exists for the majority of logistic and Poisson models.

Recall that both *cd4* and *cd8* have three levels:

cd4	0	8	cd8	0	16
	1	21		1	25
	2	18		2	6

Rather than treat these two covariates as continuous, we categorize them so that they each have three indicator variables. The first categorized level, which was level 0 when uncategorized, is now level 1 and will be regarded as the reference. The following code creates indicator variables *cd4_1*, *cd4_2*, and *cd4_3*, and *cd8_1*, *cd8_2*, and *cd8_3*:

```
. tab cd4, gen(cd4_)
```

```
blood serum |
   level of |
       CD4; |
ordinal 0, |
      1, 2 |        Freq.       Percent          Cum.
------------+------------------------------------------
        0 |           8         17.02         17.02
        1 |          21         44.68         61.70
        2 |          18         38.30        100.00
------------+------------------------------------------
    Total |          47        100.00
```

```
. tab cd8, gen(cd8_)
```

```
blood serum |
   level of |
       CD8; |
ordinal 0, |
      1, 2 |        Freq.       Percent          Cum.
------------+------------------------------------------
        0 |          16         34.04         34.04
        1 |          25         53.19         87.23
        2 |           6         12.77        100.00
------------+------------------------------------------
    Total |          47        100.00
```

```
. save hivex            /// save new data format
```

We now model *hiv* on the levels of *cd4* and *cd8*.

```
. logit hiv cd4_2 cd4_3 cd8_2 cd8_3, nolog
```

```
Logistic regression                   Number of obs   =          47
                                      LR chi2(4)      =       20.22
                                      Prob > chi2     =      0.0005
Log likelihood = -18.515956           Pseudo R2       =      0.3532
--------------------------------------------------------------------
     hiv |     Coef. Std. Err.      z  P>|z|  [95% Conf. Interval]
---------+----------------------------------------------------------
   cd4_2 | -1.203973 1.132843   -1.06 0.288 -3.424304    1.016359
   cd4_3 | -20.21283 1.443376  -14.00 0.000  -23.0418   -17.38387
   cd8_2 |  .2231436 1.036822    0.22 0.830  -1.80899    2.255277
   cd8_3 |  19.23201         .       .     .         .           .
   _cons |  .2876821 .7637626    0.38 0.706 -1.209265    1.784629
--------------------------------------------------------------------
Note: 15 failures and 3 successes completely determined.
```

The model cannot converge properly since there are either too few *cd4_3* and *cd8_3* observations with respect to their respective references, or there is perfect prediction at certain levels. Observe the greatly inflated coefficients on *cd4_3* and *cd8_3*. This is a clear indication that convergence problems have occurred. A message is displayed under the output that perfect prediction has occurred. In fact, these results give us reason to doubt the reliability of the results we obtained when leaving the two *cd* predictors uncategorized.

```
. tab cd4_1 cd4_3

   cd4== |       cd4==      2.0000
  0.0000 |         0              1 |     Total
---------+---------------------------+----------
       0 |        21             18 |        39
       1 |         8      =>       0 |         8
---------+---------------------------+----------
   Total |        29             18 |        47


. tab cd8_1 cd8_3

   cd8== |       cd8==      2.0000
  0.0000 |         0              1 |     Total
---------+---------------------------+----------
       0 |        25              6 |        31
       1 |        16      =>       0 |        16
---------+---------------------------+----------
   Total |        41              6 |        47
```

We try estimating the data as an exact statistics model. Recalling that the constant, or intercept, is estimated separately and is not conditioned on the number of successes of *hiv*, it is not estimated.

```
. exlogistic hiv cd4_2 cd4_3 cd8_2 cd8_3, coef


Exact logistic regression                Number of obs  =          47
                                         Model score    =    15.08336
                                         Pr >= score    =      0.0024
----------------------------------------------------------------------
     hiv |      Coef.    Suff.  2*Pr(Suff.) [95% Conf. Interval]
---------+------------------------------------------------------------
   cd4_2 | -1.113069        8      0.5921    -4.012823    1.419422
   cd4_3 | -2.935362*       1      0.0145         -Inf    -.5388759
   cd8_2 |   .2115246       4      1.0000    -2.147421    2.921878
   cd8_3 |  2.247257*       4      0.0580    -.0700209       +Inf
----------------------------------------------------------------------
(*) median unbiased estimates (MUE)
```

Exact statistical values are calculated for both *cd4_2* and *cd8_2*, whereas median unbiased estimates (MUE) are used to derive estimates for the two higher codes. MUEs are discussed in a following section. Note that the confidence intervals of *cd4_3* and *cd8_3* both extend to infinity, although in opposite directions.

The same model is estimated using LogXact.

```
===============================================================
Parameter Estimates

          Point Estimate Confidence Interval and P-Value for Beta

                                              95%CI    2*1-sided
Model Type    Beta    SE(Beta)  Type   Lower    Upper    P-Value
%Const MLE    ?        ?        Asymp    ?        ?         ?
cd4_2  MLE    ?        ?        Asymp    ?        ?         ?
       CMLE  -1.113   1.084     Exact  -4.013   1.419    0.5921
cd4_3  MLE    ?        ?        Asymp    ?        ?         ?
       MUE   -2.935   NA        Exact  -INF    -0.5389   0.01448
cd8_2  MLE    ?        ?        Asymp    ?        ?         ?
       CMLE   0.2115  1.01      Exact  -2.147   2.922    1.000
cd8_3  MLE    ?        ?        Asymp    ?        ?         ?
       MUE    2.247   NA        Exact  -0.0700  +IN      0.0579
===============================================================
```

The results are the same as we have for the Stata estimation. Note, moreover, that the MUE values for *cd4_3* and *cd8_3* given in Stata are nearly identical to the MUE statistics produced by LogXact. The *p*-values are also the same, but the constant was not estimated. Typically, LogXact will revert to MUE if CMLE (conditional maximum likelihood estimation) exact statistics cannot be calculated. Both applications produce the same results.

This modeling situation clearly points out the value of estimation using exact methods. The asymptotic model failed to produce meaningful estimates for *cd4_3* and *cd8_3*, and may have thereby also adversely affected the estimates of related parameters (*cd4_2* and *cd8_2*). In addition, employing MUE estimates when the exact methods failed allowed acceptable estimates to be obtained when we otherwise would have had nothing.

We have observed in this section that exact statistics are valuable when dealing with small data situations—producing better estimates when, because of sample size considerations or unbalanced levels, we cannot be confident of asymptotic results. We have also seen that exact methods can produce meaningful and accurate results when asymptotic methods fail altogether. It is an important advance in logistic regression modeling, and in statistical modeling in general. Unfortunately, exact methods are presently limited to Bernoulli/binomial and Poisson models. Exact methods have a longer history of providing exact statistics for a wide range of tabulation and nonparametric analyses. Interested readers are advised to review StatXact software, produced by Cytel.

We now discuss the alternatives to exact statistical modeling that are employed by Stata, Cytel, and SAS. We have already observed how MUE has been implemented, but we have not discussed what it is. Monte Carlo sampling is another method commonly used to approximate exact estimates, but it is usually performed when memory problems inhibit estimation of exact models, not when there is a distributional problem. For pedagogical purposes, we shall temporarily return to modeling the **hiv** data with uncategorized predictors.

15.2 Alternative Modeling Methods

It is not uncommon for the exact statistics estimating algorithm to freeze, or for a message to appear on the screen or output page that there is insufficient RAM to continue with the estimation process. Exact modeling requires enormous memory resources, and unless the model is relatively small, or at most moderately sized, it will not be possible to obtain exact estimates. LogXact and Stata both have an alternative estimation process to use when calculating exact statistical values are not possible. A brief presentation of each method will be given in this section.

15.2.1 Monte Carlo Sampling Methods

When exact methods fail while using LogXact or SAS as a result of memory limitations, it is recommended that estimation be attempted using Monte Carlo sampling methods. There are two such methods employed by the software: the Direct Monte Carlo sampling method of Mehta et al. (2000) and the Markov Chain Monte Carlo (MCMC) sampling method, which is based on the Gibbs sampling algorithm proposed by Forster et al. (1996). However, Mehta et al. (2000) demonstrated that the MCMC method can lead to incorrect results and recommend that it only be used when all other methods fail. We suggest that it not be used even in those circumstances, and that there are methods not based on Monte Carlo sampling that can be used to approximate exact statistical results. We discuss these methods in the two next sections.

The numerical algorithms underlying both MCMC and Direct Monte Carlo estimation take us beyond the scope of this text. The latter is the default Monte Carlo method, which approximates exact statistics values as the number of observations grow. Of course, this is helpful since it is the larger exact model, which results in estimation failure. Refer to the LogXact Reference Manual for details of the calculations.

The **hiv** example data will be used to compare the results calculated using exact statistics with the results calculated using Direct Monte Carlo methods. The exact results, together with standard asymptotic results, are given as:

EXACT LOGISTIC REGRESSION

		Point Estimate		Confidence Interval and P-Value for Beta			
Model					95%	%CI	2*1-sided
Term	Type	Beta	SE(Beta)	Type	Lower	Upper	P-Value
%Const	MLE	0.5132	0.6809	Asymp	-0.8213	1.848	0.451
CD4	MLE	-2.542	0.8392	Asymp	-4.187	-0.896	0.00245
	CMLE	-2.388	0.8006	Exact	-4.7	-0.822	0.00041
CD8	MLE	1.659	0.8211	Asymp	0.04923	3.268	0.04339
	CMLE	1.592	0.8047	Exact	-0.0137	3.908	0.05278

DIRECT MONTE CARLO LOGISTIC SAMPLING METHOD

LOGXACT
Summary Statistics

Statistics	Value	DF	P-Value
Deviance	4.471	5	0.4837
Likelihood Ratio	23.65	3	2.952e-005

Parameter Estimates
Point Estimate   Confidence Interval and P-Value for Beta

Model				95	%CI	P-Value	
Term	Type	Beta	SE(Beta)	Type	Lower	Upper	2*1-sided
							SE
%Const	MLE	0.5132	0.680	Asymp	-0.8213	1.848	0.45
CD4	MLE	-2.542	0.839	Asymp	-4.187	-0.8968	0.00245
	CMLE	-2.388	0.798	Monte	-4.744	-0.8238	0.000
CD8	MLE	1.659	0.821	Asymp	0.04923	3.268	0.0433
	CMLE	1.589	0.795	Monte	-0.01461	3.915	0.052

Comparing the exact parameter estimates with the Monte Carlo and asymptotic gives us:

	ASYMPTOTIC	(SE)	EXACT	(SE)	MONTE CARLO	(SE)
cd4	−2.542	(.8392)	−2.388	(.8006)	−2.388	(.7986)
cd8	1.659	(.8211)	1.592	(.8047)	1.589	(.7953)

Notice how close the estimates and standard errors are between exact and Monte Carlo results. Again, the larger the model data set, the closer the values between exact and Monte Carlo.

15.2.2 Median Unbiased Estimation

Stata has Monte Carlo estimation capabilities, and it is possible to use Stata's higher level programming language to emulate the Monte Carlo results provided in LogXact and SAS. Stata's sole alternative approximation to exact statistics is median unbiased estimation (MUE), which is based on the methods first described by Hirji, Tsiatis, and Mehta (1989). This method was used in Section 15.1 when we categorized the *cd* predictors. LogXact automatically employs MUE when exact statistical estimation fails. Technically, LogXact converts to MUE when the sufficient statistic of the parameter, β, being estimated rests at one of the two extremes of its range of values. When that occurs, exact values cannot be estimated. Strangely, however, LogXact labels MUE estimation as exact, but it is not what is usually meant by exact estimation.

The calculations for MUE are, like Monte Carlo sampling estimation, beyond the scope of this text. Refer to the version 10 Stata Reference Manual under the command **exlogistic** for details.

The same example data are modeled using MUE methods below:

```
. exlogistic infec cd4 cd8, binomial(cases) estc coef mue(cd4
cd8) nolog

Exact logistic regression                    Number of obs =          47
Binomial variable: cases                     Model score   = 13.34655
                                             Pr >= score   =    0.0006
---------------------------------------------------------------------------
   infec |     Coef.   Suff. 2*Pr(Suff.) [95% Conf. Interval]
---------+-----------------------------------------------------------------
     cd4 | -2.346197*      10   0.0004 -4.699633   -.8221807
     cd8 |  1.551565*      12   0.0528 -.0137905    3.907876
   _cons |   .4749287      14   0.7034 -1.027262    2.115926
---------------------------------------------------------------------------
(*) median unbiased estimates (MUE)
```

Notice that no standard errors are displayed. However, they are internally calculated to calculate *p*-values and confidence intervals.

MUEs may be alternatively calculated by using the mid-*p* value rule. This method halves the probability value of a discrete variate prior to adding it to the incremented *p*-values in the calculation. When all predictors are binary, this may be the preferred method.

```
. exlogistic infec cd4 cd8, binomial(cases) estc coef mue(cd4
cd8) midp nolog

Exact logistic regression            Number of obs =          47
Binomial variable: cases             Model score   =   13.34655
                                     Pr >= score   =    0.0005
--------------------------------------------------------------------
   infec |    Coef.    Suff.  2*Pr(Suff.) [95% Conf. Interval]
---------+----------------------------------------------------------
    cd4 | -2.327051*     10      0.0002   -4.355021   -.9533896
    cd8 |   1.53306*     12      0.0307    .1270188    3.563178
  _cons |  .4749287      14      0.5002   -.875449    1.924891
--------------------------------------------------------------------
(*) median unbiased estimates (MUE)
mid-p-value computed for the MUEs, probabilities, and CIs
```

The estimates and *p*-values of the various methods for the example **hiv** data are compared in Table 15.1.

TABLE 15.1

Comparison of Exact and Approximated Exact Statistics

	cd4		cd8	
	Estimate	*p*-Value	Estimate	*p*-Value
Asymptotic	−2.542	.0025	1.659	.0434
Exact	−2.388	.0004	1.592	.0528
Direct MC	−2.388	.0004	1.589	.0528
MUE	−2.346	.0004	1.552	.0528
MUE_mp	−2.327	.0002	1.533	.0307

Table 15.1 indicates that Direct Monte Carlo Sampling methods more closely approximate exact parameter estimates and *p*-values than do median unbiased estimate methods. However, the exact, Monte Carlo, and MUE *p*-values are all the same. This degree of approximation will not always be the case, but this example does demonstrate that both Monte Carlo and MUE can be excellent estimators of exact statistical values. They are certainly both superior to those produced by asymptotic means.

One may conclude that perhaps Monte Carlo or MUE methods should be used with all logistic models. However, this would be a mistake. Monte Carlo does not always work this well—and this is likewise the case for MUE. There are no objective tests to determine when bias is creeping into Monte Carlo and MUE results other than observing results that we know by statistical common sense to be mistaken. We may try modeling samples of the data using exact methods and compare them with Monte Carlo and MUE results. If they are consistent, then we may take this as evidence that Monte Carlo and MUE estimates based on the larger data set, from which the samples were taken and modeled, will likely produce reliable estimates. In the previous section of this chapter, we found confirmation of this reasoning when we categorized both *cd* predictors. The MUE values of *cd4_3* and *cd8_3* were the same as the exact values of these levels produced by LogXact. I have also tested this method with rather successful results on simulated data with known parameters (unpublished). But, as with most general statements regarding tactics used in statistical modeling, care must always be taken, and confirmation employed, when used in specific data situations.

15.2.3 Penalized Logistic Regression

Heinze and Schemper (2002) employed a method derived by David Firth (1993) to overcome what is called the problem of "separation" in logistic regression. This is a condition in the data that results in at least one parameter in a maximum likelihood estimation becoming infinite. The model may converge, but with one or more parameter estimates showing huge positive or negative values. The standard errors and confidence intervals either are absent or show similar huge (infinite) values.

The problem of separation occurs when the responses and nonresponses are perfectly separated by a predictor (risk factor) or by a particular linear combination of predictors. Rather than concluding that the model is estimating infinite parameters in the real world, such values indicate that no maximum likelihood estimates exist for the parameters in question.

Problems of this sort occur more often in small and medium-sized data sets. In particular, we should not be surprised about seeing this type of effect when the number of observations in the model is small and there are few binary or categorical predictors. We saw this happening with the HIV data (**hiv**) above when we attempted to model it. In fact, the infinite parameters appear in exact logistic models when the predictors *cd4* and *cd8* are factored into separate levels.

To deal with infinite parameters, the method proposed by Firth (1993), and modified for logistic models by Heinze and Schemper (2002), penalizes the logistic log-likelihood by half of the logarithm of the determinant of the information matrix. This penalty term is known as Jeffrey's invariant prior. The resulting value of the fit ($y=1$) is given as:

$$\Pr(y_i = 1 \mid x_i, \beta) = \mu_i = \{1 + \exp(-\textstyle\sum_{r=1}^{k} x_{ir}\beta_r)\}^{-1} \tag{15.1}$$

where r is the predictor, or risk factor, and k is the total number of model predictors.

The usual score equation, $U(\beta_r) = \sum_{i=1}^{n}(y_i - \mu_i)x_{ir} = 0$, is modified to appear as:

$$U(\beta_r)^* = \textstyle\sum_{i=1}^{n} (y_i - \mu_i + h_i(1/2 - \mu_i)x_{ir} = 0, \left(r = 1, 2, 3, \ldots, k\right) \tag{15.2}$$

The Firth-adjusted estimates, β, are obtained as in standard maximum likelihood estimation, with iteration continuing until convergence is achieved. The log-likelihood, however, is an adjusted logistic log-likelihood to account for the bias resulting from separation.

Joseph Coveney (2008) constructed a Stata command to estimate Firth-based penalized logistic regression models. Called **firthlogit**, the command is designed to estimate separation. We use it below to re-model the **hiv** data, with categorical predictors *cd4* and *cd8* factored into three indicator variables each, with level 1 as the referent. Moreover, the data have been converted to observation-level format from its original grouped form. The new binary response variable is called *hiv*, with the data saved as **hiv1**.

```
. firthlogit hiv cd4_2 cd4_3 cd8_2 cd8_3, or nolog

                                        Number of obs  =        47
                                        Wald chi2(4)   =      6.22
Penalized log likelihood = -18.501484   Prob > chi2   =    0.1834
--------------------------------------------------------------------
    hiv | Odds Ratio Std. Err.    z   P>|z|  [95% Conf. Interval]
--------+-----------------------------------------------------------
  cd4_2 |   .3603535 .3768914  -0.98  0.329  .0463941     2.798949
  cd4_3 |   .0180626  .032079  -2.26  0.024  .000556      .5868249
  cd8_2 |   1.141116 1.093832   0.14  0.890  .1743408     7.468963
  cd8_3 |   25.19302 44.31278   1.83  0.067  .8018177     791.5619
--------------------------------------------------------------------
```

All levels of the categorical predictors are estimated. Only *cd8_3* displays an odds ratio that is high, but not infinitely so, as well as a high upper 95 percent confidence boundary. Compare this output to an exact logistic regression model on the same data. Note that exact estimates are not calculated for two levels; median unbiased estimates are used for both level 3 predictors. A –inf value is given for the *cd4_3* lower bound confidence interval and +*inf* for the *cd8_3* upper bound confidence interval. The same values are low and high, respectively, for the penalized model, but not infinite. The associated *p*-value is 0.067.

```
. exlogistic hiv cd4_2 cd4_3 cd8_2 cd8_3,  nolog
note: CMLE estimate for cd4_3 is -inf; computing MUE
note: CMLE estimate for cd8_3 is +inf; computing MUE

Exact logistic regression        Number of obs  =        47
                                  Model score    =  15.08336
                                  Pr >= score    =    0.0024
--------------------------------------------------------------------
    hiv | Odds Ratio   Suff.  2*Pr(Suff.) [95% Conf. Interval]
--------+-----------------------------------------------------------
  cd4_2 |   .3285492       8     0.5921   .0180823     4.134728
  cd4_3 |  .0531115*       1     0.0145          0     .5834037
```

```
cd8_2 |     1.23556          4        1.0000    .116785      18.57614
cd8_3 |    9.461749*         4        0.0580    .9323743        +Inf
-----------------------------------------------------------------
```
(*) median unbiased estimates (MUE)

Next is displayed the command and coefficient tables of the exact and penalized models, respectively; the header statistics have been dropped. Note again, for **exlogistic**, that *cd4_3* and *cd8_3* are median unbiased estimates, not exact values. The lower bound *cd4_3* confidence interval for the MUE odds ratio is displayed as 0. The natural log of 0 is infinite. Again, the penalized model appears to be a much better fit than the other methods we have discussed in this chapter.

```
. exlogistic hiv cd4_2 cd4_3 cd8_2 cd8_3, mem(80m) nolog estc
coef
```

```
---------------------------------------------------------------
    hiv |     Coef.     Suff.   2*Pr(Suff.) [95% Conf. Interval]
--------+------------------------------------------------------
  cd4_2 |  -1.113069        8      0.5921   -4.012823    1.419422
  cd4_3 |  -2.935362*       1      0.0145        -Inf    -.5388759
  cd8_2 |   .2115246        4      1.0000   -2.147421    2.921878
  cd8_3 |   2.247257*       4      0.0580   -.0700209       +Inf
  _cons |   .2876821       14      1.0000   -1.489135    2.208517
---------------------------------------------------------------
```
(*) median unbiased estimates (MUE)

```
. firthlogit hiv cd4_2 cd4_3 cd8_2 cd8_3,  nolog
```

```
---------------------------------------------------------------
    hiv |   Coef. Std. Err.      z   P>|z|   [95% Conf. Interval]
--------+------------------------------------------------------
  cd4_2 |  -1.02067 1.045894  -0.98  0.329   -3.070583    1.029244
  cd4_3 | -4.013914 1.775994  -2.26  0.024   -7.494798   -.5330289
  cd8_2 |  .1320064 .9585634   0.14  0.890   -1.746743    2.010756
  cd8_3 |  3.226567 1.758931   1.83  0.067    -.220874    6.674008
  _cons |  .2431532 .7142204   0.34  0.734   -1.156693    1.642999
---------------------------------------------------------------
```

Heinze and Schemper (2002) suggested the use of an alternative manner of calculating the likelihood ratio test. The standard method has the model being estimated with the log-likelihood statistic saved, followed by a re-estimation less a predictor coefficient of interest. A reduced or nested model log-likelihood is calculated and compared with the log-likelihood for the full model. Heinze and Schemper, on the other hand, constrain the coefficient of interest to 0 and leave it in the model. The Heinze–Schemper statistic is computed as:

$$LR_{H\text{-}S} = -2\{LL^P_{UC} - LL^P_C\}$$

where the LL^P_{UC} is the unconstrainted penalized model log-likelihood and LL^P_C the constrained penalized model log-likelihood. Note the analogous form of the statistic to the standard form of the likelihood ratio test.

```
. quietly firthlogit hiv cd4_2 cd4_3 cd8_2 cd8_3, nolog

. estimates store constr

. constraint define 1 cd8_3

. firthlogit hiv cd4_2 cd4_3 cd8_2 cd8_3, nolog constraint(1)

                                        Number of obs =        47
                                        Wald chi2(3)  =      7.76
Penalized log likelihood = -21.291445   Prob > chi2   =    0.0513

( 1)   [xb]cd8_3 = 0
-----------------------------------------------------------------------
    hiv |     Coef. Std. Err.    z  P>|z|  [95% Conf. Interval]
--------+--------------------------------------------------------------
  cd4_2 |   -.415469 .8947016 -0.46 0.642 -2.169052     1.338114
  cd4_3 |  -2.228629 1.151974 -1.93 0.053 -4.486458      .0291991
  cd8_2 |  -.8630783 .7805336 -1.11 0.269 -2.392896      .6667394
  cd8_3 |  (dropped)
  _cons |   .4385633   .67357  0.65 0.515 -.8816097     1.758736
-----------------------------------------------------------------------

. lrtest constr .

Likelihood-ratio test                   LR chi2(1)    =      5.58
(Assumption: . nested in const          Prob > chi2   =    0.0182
```

It appears that the penalized logistic model is preferred to both Monte Carlo and Median Unbiased estimates as alternatives to exact logistic regression. In fact, it appears somewhat conclusive that penalized logistic models may be preferred to exact methods when separation occurs in the data. More research needs to be done on the comparative worth of these models given various data situations. At this time, it is likely best to model small, unbalanced, and possible separated data using all four methods, looking for consistency of results.

Exercises

15.1 What are the benefits and pitfalls of using exact logistic regression compared with standard asymptotic estimation? Why would one use Monte Carlo or median unbiased regression instead of exact methods?

15.2 For readers having access to LogXact, model the data from Exercise 5.3 using exact statistics. Compare the results. Why is there a difference in both the parameter estimates and standard errors?

15.3 Use the **absent** data set described in Exercise 5.4. Model slow on *aborig* and *grade4* using both a logistic regression and exact logistic regression. You may either create a new variable with *grade4* as equal to 1 and other grade levels as 0, or leave in the model as only *grade4*.

   a. Why does the logistic model not allow a parameter estimate for *grade4*?

   b. What problems are experienced when modeling the data using an exact logistic regression? Why?

   c. Interpret the results of the exact logistic model.

15.4. Model the following data from Heinze (2006) using a penalized logistic model (**firthlogit**). The data appear in four columns, with four variables each:

| Variable 1: obs | | Variable 2: fibrinogen | |
Variable 3: globulin		Variable 4: response	
1 2.52 38 0	2 2.56 31 0	3 2.19 33 0	4 2.18 31 0
5 3.41 37 0	6 2.46 36 0	7 3.22 38 0	8 2.21 37 0
9 3.15 39 0	10 2.60 41 0	11 2.29 36 0	12 2.35 29 0
13 5.06 37 1	14 3.34 32 1	15 3.15 36 0	16 3.53 46 1
17 2.68 34 0	18 2.60 38 0	19 2.23 37 0	20 2.88 30 0
21 2.65 46 0	22 2.28 36 0	23 2.67 39 0	24 2.29 31 0
25 2.15 31 0	26 2.54 28 0	27 3.93 32 1	28 3.34 30 0
29 2.99 36 0	30 3.32 35 0		

   a. With *response* as the response variable, and creating new variables called *iqr* as the difference between the 75th and 25th percentile of the variable *fibrinogen*, and *fibrinogen_iqr* as *fibrinogen* divided by *iqr*, model as:

   firthlogit *response fibrinogen_iqr globulin*, or

   b. Test the variable *fibrinogen_iqr* by constraining it to a value of zero and performing a penalized likelihood ratio test.

Conclusion

We have covered quite a bit of modeling material in this text. Aside from the points learned in the above discussions, keep in mind that a statistical model is just that—a model. It is not meant to be a perfect replication of the data. Building a parsimonious model, including all relevant predictors, will help ensure that the model results will be compatible to other similar data situations, that is, that the model is saying something about the population as a whole. Otherwise, our model is only as good as our current data.

Also recall that the basic logistic regression can be approached as an independent maximum likelihood (ML) binary response model, or as a member of the family of generalized linear models (GLM). Each approach has its respective advantages. Current ML implementations usually (Stata, SAS, SPSS) incorporate m-asymptotics, allowing use of Hosmer and Lemeshow (H&L) fit statistics, including the H&L goodness-of-fit test. The decile table that is part of the latter test is a valuable tool in assessing model fit. m-asymptotics refers to the structuring of data in the model—observations are based on covariate patterns. The GLM approach (Stata-glm; SAS-GENMOD; SPSS-GENLIN; R-glm) allows use of a variety of other residual statistics, which are usually based on traditional n-asymptotics. n-asymptotic data structure is the normal manner in which data are typically collected—on an observation-by-observation basis. When continuous predictors exist in a model, there will be so many covariate patterns that the use of m-asymptotics will make no difference. m- and n-asymptotics will be identical, or closely so, and the former will give no advantage in assessing fit. Moreover, the GLM approach allows easy use of grouped or binomial logistic models, provides easy use of alternative binary response models (such as **probit**, **cloglog**), and allows the use of deviance statistics. These three capabilities are generally unavailable in ML implementations of logistic regression. Remember, though, that the models estimated using either approach are identical; the differences exist when dealing with the analysis of fit. I have attempted to demonstrate the use of both ML and GLM approaches in this text.

Appendix A: Brief Guide to Using Stata Commands

Stata has been used for the majority of examples in the text. I selected Stata for this purpose because it is one of the most popular and inexpensive statistical software environments on the commercial market. Moreover, it is comparatively simple to use. Most importantly, though, nearly all of the range of logistic-based models discussed in this text can be estimated using Stata.

As users of Stata know, the official capabilities of the software—i.e., what comes on the installation disk—are complemented by a host of user-authored commands and functions that, for the most part, give every appearance of being created by the company. The software is unique in this capability, and users have exploited Stata's higher level programming language by creating a wide variety of commands, many of which cannot be found in other commercial software. I have written a number of these user commands, the first in 1990, being what is now called **logistic**, one of Stata's primary logistic regression commands. In fact, many user commands end up being adopted into official Stata.

We shall use both commercial and user-authored commands and functions in this text. Many user commands were written to accompany articles in the *Stata Technical Bulletin* (STB), which was published from 1991 to 2000, and its offspring, the *Stata Journal* (SJ), which began in 2000. The commands and other related programs and data sets accompanying articles published in the bimonthly STB journals and quarterly SJ journals can be downloaded from the Stata Web site: *http://www.stata.com*.

Several hundred user-created Stata commands, functions, and data management tools have been published on the Statistical Software Components Web site of the Boston College Department of Economics (*http://ideas.repec.org/s/boc/bocode.html*). These programs can also be accessed and downloaded directly from within Stata by typing:

```
. ssc install <command name>
```

on Stata's command line. A list of all new commands posted to the site can be obtained by typing:

```
. ssc new
```

The most popular programs can be identified by:

```
. ssc hot
```

All of these programs can be downloaded directly from within Stata.

Other files and commands are also available from within Stata. Information about these alternative sources can be obtained by clicking on the Support button on Stata's main page.

Stata users have access to both a command line method of executing requests as well as a menu system. In the text, we use the command line method so that these commands can easily be duplicated by the reader if desired. Moreover, the commands are generally such that they can be understood for other statistical packages as well.

It is important to remember that this text is not a book on programming with Stata or any statistical package. A statistical package must be employed to display examples of the statistical methodologies we discuss. Stata is used due to its ease and scope of use. The important thing is to understand the data management and statistics involved.

A.1 Basic Statistical Commands

Stata commands are formatted according to the logic of the procedure. Descriptive commands are generally formatted as:

command [*varlist*] [*if*] [*in*] [*weight*] [, *options*]

For example, using the **heart01** data set, the variables in the data file may be described using the *describe* command. We can type:

```
. use heart
. describe
```

or more simply, . d, to get a listing of all variables in the data.

```
. d

Contains data from heart.dta
  obs:        5,388
  vars:           7                                    15 Feb 2008 22:12
  size:      64,656 (99.4% of memory free)
```

```
-----------------------------------------------------------------
variable   storage  display   value
  name      type    format    label    variable label
-----------------------------------------------------------------
death       byte    %4.0g              Death within 48 hrs onset
anterior    byte    %9.0g     anterior 1=anterior; 0=inferior
hcabg       byte    %4.0g              Hist of 1=CABG;0=PTCA
center      int     %12.0g             Provider or Center
killip      byte    %8.0g              Killip level
agegrp      byte    %8.0g     agegrp   4 groups of age
age         byte    %9.0g              Patient age
-----------------------------------------------------------------
Sorted by:  center
```

The output provides us with the number of observations, number of variables, memory size, and a listing of each variable in the data. Each variable is accompanied with a storage type, display format code, the name of the labeling code, and the variable descriptor. Also seen underneath the table of variables is a note indicating that the data are sorted by the variable *center*.

The **describe** command has several options, which can be found using the **help** command.

Type:

```
. help describe
```

to obtain a list of the various features and options of the **describe** command. As you will find, though, **describe** is one of the few commands that does not employ the [*if*] [*in*] qualifiers. All Stata commands come with an accompanying **help** file. I strongly recommend using the **help** command frequently. It is easy to forget the many attributes of a particular command—but they can be found using the **help** facility. Commands that are associated with the one being helped are often found in the **Also See** section at the bottom of the help file.

Another well-used Stata command is **list**. As the name implies, the command lists the values of each variable entered on the command line following **list**. If **list** is used by itself, the values of all variables are listed.

[*if*] and [*in*] are commonly used with the **list** command to limit the range of values listed. We could therefore request:

```
. list age if age>50
```

which lists the values of *age*, but only if they are greater than 50. Care must be taken to check for missing values, which are stored in Stata as high (unlike

SAS, which stores them low). If there were missing values of age, they would be listed as well. To eliminate missing values from the listing above, we could type:

```
. list age if (age>=50 & age~=.)    // ! is also a not symbol
```

[Note that comments can be made to the side of the code following double forward slashes.]

We may also combine expressions. For example:

```
. list age if age>=50 & death==1
```

gives a listing of the values of *age* for observations limited by being greater than or equal to 50 and where *death* is equal to 1.

We can also give a listing of the values of *death, anterior, hcabg, killip,* and *age* for *center* number 5507. *center* is the same as hospital or provider number.

```
. list death anterior hcabg killip age if center == 5507
```

	death	anterior	hcabg	killip	age
3066.	0	.	0	3	98
3067.	0	.	0	2	76
3068.	0	Inferior	0	1	73
3069.	0	Anterior	0	1	74
3070.	0	.	0	1	66
3071.	0	Anterior	1	1	62
3072.	0	Anterior	0	1	49
3073.	0	Anterior	0	1	70
3074.	0	.	0	1	97

All values of *death* in the above output are 0, meaning that they are alive. Missing values are indicated in Stata with a dot. It should be noted that *anterior* is recorded as a byte variable—see the results of the **describe** command. We have labeled the numeric values as 1 = Anterior and 0 = Inferior. If the list command is used with the *nolabel* option, only numeric values are displayed, unless the variable is internally stored as a string.

The [in] qualifier is used to limit the range of observations used by the initial command. For instance:

```
. list age if age>=50 & death==1 in 1/100

        +-----+
        | age |
        |-----|
   2.   |  98 |
   4.   |  79 |
  12.   |  69 |
  69.   |  96 |
  99.   |  76 |
        +-----+
```

limits the observation numbers affected by the command to be from 1 through 100.

The **summarize** command is another important command. Type **help summarize** to learn all of its options.

Summarize may be shortened to **su**, and can be used to summarize one or more variables. To summarize the variable *age*, we can simply type:

```
. summarize age
```

or

```
. su age
      Variable |       Obs        Mean    Std. Dev.       Min        Max
-------------+--------------------------------------------------------
           age |      5388    65.80976    14.42962         40        100
```

The *detail* option is a particularly useful **summarize** option.

```
. su age, detail

                                  age
-----------------------------------------------------------------
         Percentiles       Smallest
  1%          40               40
  5%          43               40
 10%          46               40         Obs                5388
 25%          54               40         Sum of Wgt.        5388

 50%          66                           Mean           65.80976
                             Largest       Std. Dev.      14.42962
 75%          76              100
 90%          84              100         Variance       208.2138
 95%          93              100         Skewness        .1598778
 99%          99              100         Kurtosis        2.428674
```

The 50 percent percentile is the median value of *age*—66. We can see that it is nearly equal to the mean (65.8). Other standard descriptive statistics are included in the displayed output.

The user also has access to the above values. Each of the major statistics are stored in returned results. To see which values are stored after each command, type:

```
. return list
```

following the command. The results of the above **return list** command following **su age, detail** are

```
scalars:
                 r(N)  =   5388
             r(sum_w)  =   5388
              r(mean)  =   65.80976243504084
               r(Var)  =   208.2138493588113
                r(sd)  =   14.42961708981951
          r(skewness)  =   .159877836604101
          r(kurtosis)  =   2.428674024638072
               r(sum)  =   354583
               r(min)  =   40
               r(max)  =   100
                r(p1)  =   40
                r(p5)  =   43
               r(p10)  =   46
               r(p25)  =   54
               r(p50)  =   66
               r(p75)  =   76
               r(p90)  =   84
               r(p95)  =   93
               r(p99)  =   99
```

The *r()* results on the left side of the equal sign can be used in subsequent calculations. For example, we may calculate the value of the second standard deviation above the mean by:

```
. di r(mean) + 2*r(sd)
94.668997
```

di is the shortened version of **display**, which is used when we want to use Stata as a calculator. There are other uses of the command as well, which can be determined by typing **help display**. This capability is used often throughout the text.

Another commonly used Stata command is **tabulate**, or **tab**:

```
. tab death

    Death |
within 48 |
hrs onset |        Freq.       Percent         Cum.
----------+-------------------------------------------
        0 |        5,146         95.51        95.51
        1 |          242          4.49       100.00
----------+-------------------------------------------
    Total |        5,388        100.00
```

provides a count of the number of observations having the various values of the variable.

Using the *summarize* option allows us to get mean and standard deviation values of the summarized variable for each level of the tabulated variable.

```
. tab death, sum(age)

    Death |
within 48 |            Summary of age
hrs onset |        Mean    Std. Dev.         Freq.
----------+-------------------------------------------
        0 |   65.290517    14.267272          5146
        1 |    76.85124     13.4298           242
----------+-------------------------------------------
    Total |   65.809762    14.429617          5388
```

If we type two variables after the **tab** command, the result is a cross tabulation of the two variables.

```
. tab death killip

    Death |
within 48 |                Killip level
hrs onset |      1          2          3          4 |     Total
----------+-------------------------------------------+---------
        0 |   3,671        958        262         37 |     4,928
        1 |      97         71         30         27 |       225
----------+-------------------------------------------+---------
    Total |   3,768      1,029        292         64 |     5,153
```

Column and/or row percentages can be displayed by using the *col* and *row* options.

We can also use the *summarize()* option to obtain mean and standard deviations of the summarized variable for each cell of the two-way tabulation. This would give us eight mean/SD values for the above tabulation. Other options are available. For example, we can obtain the p-value associated with the *Chi2* statistic (χ^2) statistic for the above two-way tabulation, as well as the traditional asymptotic *Chi2*, by:

```
. tab death killip, chi exact

Enumerating sample-space combinations:
stage 4:   enumerations = 1
stage 3:   enumerations = 37
stage 2:   enumerations = 1987
stage 1:   enumerations = 0
```

```
    Death |
within 48 |                  Killip level
hrs onset |       1          2          3          4 |     Total
----------+-----------------------------------------------+----------
        0 |   3,671        958        262         37 |     4,928
        1 |      97         71         30         27 |       225
----------+-----------------------------------------------+----------
    Total |   3,768      1,029        292         64 |     5,153
```

```
          Pearson chi2(3) = 288.4391    Pr = 0.000
          Fisher's exact  =                  0.000
```

A number of alternative test statistics may be requested. In addition, Stata has a variety of different tabulation commands. However, we shall use only a few, which will be explained in the context of the example. Remember that each command has returned results, where the values of *r()* are stored for subsequent use. However, when a new command is executed, the *r()* results are reset with new values. The values stored in *r()* can be saved into a user created variable using the generate command. For example, suppose that we wish to save the value of the above Pearson *Chi2* statistic into a new variable. Let's call it *mychi2*. We can:

```
. gen mychi2 = r(chi2)

. di mychi2
288.43915
```

or

```
. scalar mychi2 = r(chi2)
```

can also be used, and is more efficient.

Of course, *mychi2* is a new variable with 5153 observations, all having the same value of 288.43915. By issuing the **di** command, only the value of the first observation is duplicated. In any case, we can use *mychi2* just as any other variable—it is indistinguishable from the other variables. I may want to label it, though. We can:

```
. label var mychi2 "Pearson Chi2 from last xtab"
                                /// lab can also be used
```

We can verify that it worked by describing the variable. Type **de**(scribe) to see all variables in memory; here We only want to see *mychi2*.

```
. de mychi2

              storage display value
variable name  type   format  label variable label
----------------------------------------------------------------
mychi2               float  %9.0g            Pearson Chi2 from last xtab
```

Suppose that we wish to divide the data set into two halves, indicating each half by a new variable having values of 0 for the first half, and 1 for the second. There are several ways to accomplish this task, but an easy way is to give each observation in the data set an individual identifier; call it *id*.

```
. gen id = _n

. su id

Variable |       Obs        Mean    Std. Dev.       Min        Max
---------+----------------------------------------------------------
      id |      5388      2694.5    1555.526         1       5388
```

Stata internally codes the observations in the data by a system variable called _n. The values run from 1 to the total number of observations. In this data set, we have 5388 observations. This value could also have been obtained by typing:

```
. di _N
5388
```

where _N is the system stored total number of observations. Regardless, each observation has a sequentially ascending numeric value associated with it. Both *id* and *_n*[100] have the same values, as can be seen by typing:

```
. di _n[100]
100

. di id[100]
100
```

or

```
. l id in 100

        +-----+
        |  id |
        |-----|
100. |  100 |
        +-----+
```

Back to the task of bifurcating the data set: We saw the displayed mean value
of *id* as 2694.5. We shall create a variable called *group*, with a value of 0 for
values of *id* ranging from 1 to 2694, and 1 for values 2695 to 5388.

```
. gen byte group= 0 if id>0 & id<2694.5
. replace group = 1 if group ==.
. tab group

    group |      Freq.     Percent        Cum.
----------+-----------------------------------
        0 |      2,694       50.00       50.00
        1 |      2,694       50.00      100.00
----------+-----------------------------------
    Total |      5,388      100.00
```

When we created the *group* variable, assigning the value of 0 to the lower
range of *id*, the non-0 values were automatically coded as missing by Stata.
The second command simply revalues the missing values to 1. Note that
I specifically assigned the storage code for *group* to byte. This is recommended for all variables having 1/0 values. Less memory is consumed, and
the application runs a bit more quickly.

A faster way, but trickier, is to generate the values of *group* using only a
single command. First, we delete *group* from memory:

```
. drop group

. gen byte grp = id>2694.5

. tab grp

      grp |      Freq.     Percent        Cum.
----------+-----------------------------------
        0 |      2,694       50.00       50.00
        1 |      2,694       50.00      100.00
----------+-----------------------------------
    Total |      5,388      100.00
```

When a specific numeric value is not assigned after the equal sign, Stata
makes it a 1. Instead of having values outside the range defined on the
right-hand side assigned as missing, Stata gives them a 0. For example, given
a variable called *grade* with values of 1, 2, 3, and 4, and labeled as D, C, B, and

A, respectively, the following code generates a new variable called *a_grade* with a value of 1 if grade is A, and 0 otherwise.

```
. gen a_grade = grade==4
```

Other basic commands, grouped roughly into three areas of similar use, are:

clear:	clears the memory of all previous variables and results
discard:	drops program command definitions from memory
drop:	eliminates a variable(s) or observation(s) from memory
quietly:	suppresses output when placed in front of a command
dir:	displays files in the active directory
expand:	duplicates the data set by the values being expanded
gsort:	allows both ascending and descending sorting
preserve:	saves data in temporary memory, allowing other data to be used
restore:	restores data saved by **preserve**
sample:	draws a random sample of observations from the data
save:	saves a data set to memory; the data can be used later by typing **use <filename>**
sort:	sorts the data by the ascending values of the variable(s) given after the command
use:	brings a data file into memory
webuse:	load a data set stored on Stata's Web site
count:	enumerates the observations, given specified conditions
egen:	extended generate; see **help egen** to learn of the many functions supported by **egen**
tabstat:	produces a variety of statistics for variables listed after the command
order:	reorders the variables in memory to the order specified on the command line
encode:	converts strings into numeric values, and vice versa
recode:	recodes the values of categorical variables

A.2 Basic Regression Commands

We shall be primarily dealing with various types of regression models in this text, in particular, models that are based on the logistic function. Stata regards all regression commands in a similar manner:

```
command <varlist> [if] [in] [weight], level(#) vce(vcetype)
<other_options>
```

where *level()* refers to 1-α for the Confidence Interval, with the default set as 95 percent. *vce()* instructs the estimating algorithm on how to calculate standard errors. Options include *oim, robust, bootstrap, jackknife,* and others that are related to the type of procedure involved. *oim* specifies estimation via the observed information matrix, which is the default for the majority of maximum likelihood procedures. We shall discuss these methods when addressing standard errors.

The first variable following the command—for instance, **logistic**—is the response or dependent variable. The others following the response are explanatory predictors. In some contexts, the second variable after the command is the exposure or explanatory predictor of interest, with the remaining predictors being regarded as adjusters or confounders. There is some variability involved with how various regression commands are structured, but the basic logic is as indicated above.

A.2.1 Logit

The regression commands that will be of special interest to us in this text are **logit, logistic,** and **glm**. These are described in considerable detail when first used, so I need only give general guidelines here as to their structure and use.

Stata's basic logistic regression command is **logit**, which is hardcoded into the executable in binary format. Its structure is

```
logit <depvar> [predictors] [if] [in] [weight] [, options]
```

with the foremost options being:

level(#)	set confidence level; default is level(95)
or	display odds ratios rather than coefficients
noconstant	suppresses the constant
offset(varname)	a nonestimated variable in model; coefficient constrained to 1
vce(vcetype)	oim, robust, cluster(clustvar), bootstrap, jackknife

A.2.2 Logistic

The most used command for logistic regression is **logistic**, which was originally written in 1990 to offer Stata users the ability to incorporate Hosmer and Lemeshow fit statistics into their analysis. It was not made part of the commercial package until 1993. The default output displays odds ratios rather than coefficients, unlike the **logit** command.

The real value of both the **logit** and **logistic** commands rests with their range of post estimation statistics. After modeling one's data, the statistician may type:

estat clas	displays a classification table
estat gof	displays the Pearson and Hosmer and Lemeshow fit statistics; with the table and group options, provides a table comparing expected versus observed probabilities
lroc	displays a ROC graph and C-statistic (SAS terminology)
lsens	graphs sensitivity and specificity versus probability cutoff.

A host of residuals are also available using the **predict** command. The options include the generation of standardized Pearson and deviance statistics, score statistics, and several types of influence statistics. These residuals are based on covariate patterns, called *m*-asymptotics, rather than on individual observations, *n*-asymptotics.

A.2.3 GLM

Lastly, the final primary command we shall use for logistic regression estimation is the **glm** command. **glm** is an acronym for generalized linear models, of which logistic regression is a foremost member.

Unlike the **logit** and **logistic** command, **glm** can estimate aggregated or grouped logistic models. We shall generally refer to these models as binomial logistic models. Stata does have a command called **glogit**, but the command is in reality a frequency weighted **logit** command. It is not an estimation that is based on the full binomial distribution. These distinctions are discussed in considerable depth in the text.

GLM is a family of models, each of which belongs to the one-parameter exponential family of distributions. Standard GLM distributions, called families, include the Gaussian (normal), gamma, inverse Gaussian, Bernoulli/binomial, Poisson, geometric, and negative binomial (with a defined ancillary parameter). Each of these families has associated link, inverse link, and variance functions, which together define a specific GLM model.

Binary responses are assumed to follow a Bernoulli distribution, which is a special case of the binomial distribution. The Bernoulli has a binomial denominator equal to one. Logistic regression, when estimated using GLM methodology, is identified as belonging to the binomial family and logit link. If the model has a binary response, it is more exact to say that it belongs to the Bernoulli family, but nearly all software applications use the more generic binomial family. When the model is aggregated, or grouped, the binomial family is appropriate, but both a numerator and denominator must be identified. For binary response models, the denominator is assumed to have a value of 1.

The GLM binomial family is associated with the logit, probit, complementary loglog, and loglog links. We are primarily interested in the first of these links. The GLM command is structured as:

```
glm  [depvar] [predictors] [if] [in] [weight] [, fam( )
link() <eform> offset() lnoffset() vce()  <other_options>
```

For binary response logistic models we complete the fam() and link() functions as:

```
fam(bin)   link(logit)
```

However, we do not actually need to specify a link function. With the *bin*(*omial*) family, the logit link is the default.

For an aggregated or grouped logistic model, we indicate the binomial denominator as a second term in the family function:

```
fam(bin <denom>)
```

For example, we could convert the **heart01** data to a grouped format by combining identical covariate patterns to a single observation. A method of implementing this conversion is shown in the text. The caveat is that all model predictors must be binary or categorical. Continuous predictors must be transformed to categorical format. For example, suppose that we use the **heart01** data with *death* as the binary response and *anterior* and *killip* levels as predictors. *anterior* is binary; *killip* is a four-level categorical variable. We create a new response, *died*, which will be the binomial numerator, and a binomial denominator called *cases*. *cases* is the number of observations in the individual-level data having the same pattern of covariates, or predictor values. *died* is the number for which *death*==1 for the same covariate pattern. Note that this meaning of *cases* as the binomial denominator is to be understood in context, and should not be confused with its typical use in epidemiology or in health science as the number of diseased people in the data under study. An example of how we use it as a binomial denominator throughout the text may help.

	death	anterior	killip	group
obs 1	1	0	3	1
obs 2	1	0	2	2
obs 3	0	0	2	2
obs 4	0	1	1	3
obs 5	1	1	4	4
obs 6	0	0	3	1
obs 7	1	0	2	2
obs 8	1	1	4	4

The *group* variable is one that I created to identify similar covariate patterns. The variables used to construct a covariate pattern are only those employed as model predictors. The response is not included. Here we only have two predictors with which to construct covariate patterns. There are two observations or cases having anterior==0 and killip==3. This pattern is indicated as a 1 in the group variable. Only one of the two elements of this covariate pattern (*group==1*) has a value of 1 for *death*. For group==1, the binomial numerator is 1, the denominator is 2. One of two observations with this covariate pattern has death==1.

To convert the individual-level data to grouped or aggregated data, we can use the following Stata commands:

```
. keep death anterior killip          /// restrict vars to only what is to
                                           be used in the model

. egen grp = group(anterior killip)
. egen cases = count(grp), by(grp)
. egen died = total(death), by(grp)
. sort grp
. by grp: keep if _n==1
```

We first restrict the variables in the data to what is in the model. There may be many other variables in the data set. In fact, this is not a necessary command, but it does help clarify the model terms.

I briefly mentioned the **egen** command above. It is similar to **generate** in that a new variable is created—with the name being the term directly to the right of the word **egen**.

The new variable, *grp*, is defined by the *group()* function as a number assigned to each covariate pattern. This is shown in the above table. *grp* has identical values to what is shown in the table under *group*.

egen *cases* uses the *count()* function to create the variable *cases*, which is a count of the number of observations belonging in each *grp*. **egen** *died* uses the *total()* function to obtain the number of *death* == 1 for each covariate pattern, or *grp*.

Stata requires that prior to performing a function on the data for each value of a variable, the data must be sorted by that variable. We wish to combine all observations within each level of *grp*, so that there is one observation per covariate pattern. To do this, we must first sort the data on *grp*. A listing of the data just after sorting is

```
. list                  /// l can also be used
```

	death	anterior	killip	grp	cases	died
1.	0	0	2	1	3	2
2.	1	0	2	1	3	2
3.	1	0	2	1	3	2
4.	0	0	3	2	2	1
5.	1	0	3	2	2	1

```
|----------------------------------------------------------------|
6. |      0            1           1        3          1        0  |
7. |      1            1           4        4          2        2  |
8. |      1            1           4        4          2        2  |
   +------------------------------------------------------------+
```

Finally, we collapse the individual-level data to group data. The final command instructs the computer to keep the first observation for each value of *grp*. We now have the data structured as:

. list

```
+--------------------------------------------------------------+
   | death    anterior    killip    grp    cases    died |
   |--------------------------------------------------------------|
1. |    0          0           2       1       3        2  |
2. |    0          0           3       2       2        1  |
3. |    0          1           1       3       1        0  |
4. |    1          1           4       4       2        2  |
   +--------------------------------------------------------------+
```

The important variables in the new grouped data are *anterior* and *killip*, since they are the model predictors, in addition to *died* and *cases*, the values of the binomial numerator and denominator, respectively.

Finally, prior to modeling the data, we want to level or factor the categorical predictor *killip*. Stata has several methods of doing this. The easiest method, and the one I like, is to create dummy variables.

. tab killip, gen(killip)

Here we tabulate *killip*, and generate a new variable that takes the names of *killip1*, *killip2*, *killip3*, and *killp4*. This is accomplished automatically by the software.

The data now appear as:

. list

```
+----------------------------------------------------------------------+
   | death anterior killip grp cases died    k1    k2    k3    k4 |
   |----------------------------------------------------------------------|
1.|    0       0        2    1     3     2     0     1     0     0  |
2.|    0       0        3    2     2     1     0     0     1     0  |
3.|    0       1        1    3     1     0     1     0     0     0  |
4.|    1       1        4    4     2     2     0     0     0     1  |
   +----------------------------------------------------------------------+
```

The individual-level, or binary response model, is estimated using the command:

```
. glm death anterior killip2-killip4, fam(bin)
                                    /// link(logit) is assumed
```

The grouped or binomial logistic model is estimated using:

```
. glm died anterior killip2-killip4, fam(bin cases)
```

killip1 has been assigned the role as referent. The hypen between levels of *killip* tells the command to incorporate all variables in memory that are between *killip2* and *killip4* into the model. Here, there are three such variables—*killip2*, *killip3*, and *killip4*.

The model results are identical, with the exception of the number of observations in the data.

Care must be taken when there is a missing category in the levels of the predictor being converted into indicator variables. For example, suppose that the categorical variable, *agegrp*, has level values of 1, 2, and 4. When the following command is issued,

```
. tab agegrp,gen(age)
```

indicators *age1*, *age2*, and *age3* will be created, but will be mapped to *agegrp*==1, *agegrp*==2, and *agegrp*==4 respectively. This gives the impression that *age3* correseponds to *agegrp*=3 when it is in fact *agegrp*=4.

Stata also has the capability of automatically creating dummies by use of the xi: prefix command. We can do this by:

```
. xi:    glm died anterior i.killip, fam(bin cases)
```

New variables are created by the software for any variable name directly following the "*i.*" indicator. Each newly created variable is prefixed by "*_I*" and suffixed by "*_**", where * is a sequentially ascending number. The above command creates variables named *_Ikillip_1* through *_Ikillip_4*. Many authors prefer the *xi:* method of generating dummy predictors. I like having control myself.

It should be mentioned that in the case above where *agegrp* was missing a level with a value of 3, *xi: i.agegrp* labels them as *_Iagegrp_2* and *_Iagegrp_4*, thereby letting the user know that *agegrp* = 3 is missing.

It should also be mentioned that a reference level may be changed by using the *char* command. For example, killip level 2 can be assigned as the reference level when using the xi: method by the command:

```
. char killip[omit]2
```

The *char* command is issued prior to employing the *xi: logit* command.

We have gone through the exercise of showing how the various **egen,** etc. commands can be used to create and manipulate variables in Stata. Together with an overview of other commands, we have covered most of the functions and basic commands that we shall use in this text. There are a few others, however, that should be mentioned.

Prior to providing an overview of logistic-based model commands, it should be mentioned that the code used above to convert observation-based binary response models to aggregated binomial form can be shortened by using the **collapse** command.

```
. drop if anterior==. | kk2==.    /// deletes observations with
                                      missing values

. gen count=1

. collapse (sum) count (sum)  death, by(anterior killip)

. tab killip, gen(kk)

. glm death anterior kk2-kk4, nolog fam(bin count) eform
```

The results are identical to the six-step algorithm earlier described. The **collapse** command is, however, not as intuitive as the longer method, but nevertheless, produces an accurate conversion in fewer steps.

A.3 Overview Logistic-Based Model Commands

The **logit, logistic,** and **glm** commands play important roles in the examples given throughout the first part of the text. When discussing models that extend the basic binary and binomial or proportional logistic models, we use a wide variety of commands. However, the functions we have discussed will largely be the same.

A few of the more important Stata commands that are used in the estimation of extended logistic models are summarized below. Commands that are not official Stata commands, but instead ones created and published by users, are indicated by <U>.

Generalized Linear Models

glm	binary/binomial logistic one of many family/links; many fit statistics

Binary Response

logit	binary logistic regression; many fit/residual stats (as of version 10)
logistic	binary logistic regression; odds ratio default; many fit/ residual stats
glogit	grouped logistic regression—a weighted logit command
binreg	extensions to the binomial family, e.g., log binomial
scobit	skewed logistic regression

Categorical Response

mlogit	multinomial logistic regression
ologit	ordered logistic regression; proportional odds
gologit	generalized ordered logistic regression <U>
gologit2	partial proportional slopes regression <U>
oglm	proportional; partial proportional; heterogeneous choice model <U>
ocratio	continuation ratio model <U>
slogit	stereotype logistic regression
asclogit	alternative-specific conditional logit model (McFadden's choice)
rologit	rank-ordered logistic regression

Panel Models for Longitudinal and Cluster Data

clogit	conditional logistic regression (fixed effects)
xtgee	generalized estimating equations (GEE)
xtlogit	panel logistic: conditional fixed effects; random effects, GEE
xtmelogit	multilevel mixed effects logistic regression
xtmixed	multilevel mixed effects linear regression

Other Logistic Models

exlogistic	exact logistic regression—binary and binomial
nlogit	nested logistic regression

A.4 Other Utility and (Post)Estimation Commands

Important ancillary commands are commonly used by statisticians using Stata when engaged in modeling. Although they are generally described in the text as they are encountered, it might be helpful to list them here for easy reference. We have already discussed the following:

clear	count	describe	discard	display	dir
drop	egen	encode	expand	generate	gsort
help	label	list	order	preserve	quietly
recode	replace	restore	return	sample	save
sort	summarize	tabstat	tabulate	use	webuse

Others that are commonly used are listed below.

A.4.1 Model Construction

fracplot:	plots data and fit from most recently fitted fractional polynomial model
fracpoly, mfp:	univariable and multivariable fractional polynomial model
ladder:	provides comparative statistics to determine the optimal transformation of a continuous variable to normality; **gladder** provides a graphical representation of ladder, showing how the distribution would appear if transformed
lincheck <U>:	provides a quick-and-dirty check of whether a continuous covariate in a general linearized model (GLM) is linear in the link function
lintrend <U>:	examines the linearity assumption for an ordinal or interval variable against long odds of a binary outcome
locpr <U>:	semi-parametrically estimates probability/proportion as a function of one regressor
lpartr <U>:	smoothed logistic regression partial residual plots
mean:	provides the mean, standard deviation and confidence intervals for requested variables; may be standardized
mulogit <U>:	generates multivariable and univariable odds ratios (ORs) and 95 percent confidence intervals (CIs) for variables in varlist
nestreg:	fits nested models by sequentially adding blocks of variables and then reports comparison tests between the nested models
oddsrisk <U>:	converts logistic odds ratios to risk ratios
sktest:	performs skewness and kurtosis test for normality
swilk:	performs Shapiro–Wilk and Shapiro–Francia tests for normality

ulogit <U>:	compares the log-likelihood of a logistic regression model containing only the intercept with that of a model having a single predictor
uvrs and **mvrs** <U>:	regression spline models used with logistic-based commands

A.4.2 Model Testing

abic <U>:	author written: displays two different types of AIC and BIC tests after estimation
estat:	post-estimation statistics
estat ic:	displays number of observations, log-likelihood null and model statistics, degrees of freedom, AIC, BIC
estat sum:	displays observations, mean, std dev, maximum and minimum for each variable in model
lincom:	computes point estimates, standard errors, t or z statistics, p-values, and confidence intervals for linear combinations of coefficients after any estimation command; results can be optionally displayed as odds ratios
linktest:	a specification test of the appropriateness of the link function for 1-equation models
lroc:	displays a ROC graph and C-statistic (SAS terminology)
lrtest:	likelihood ratio test to determine if a variable(s) significantly contributes to the model; can also be used to decide on which of two nested models is a better fit
lsens:	graphs sensitivity and specificity versus probability cutoff
test:	a Wald test to evaluate the statistical relationship between variables and parameters; has many functions; **testparm** also used for similar tests

A.4.3 Model Interpretation and Postestimation

adjust:	calculates various statistics for given values of the predictors following estimation
ci:	calculates the confidence intervals for means, proportions, or counts
correlate:	displays correlation values between all variables listed on the command line; if no variables are listed, the correlations between all variables in memory are displayed; the option _coef displays the correlations between the coefficients of the last estimation; if *co*, short for *covariance*, is also added as an option, a variance-covariance matrix of coefficients is displayed; model standard errors are obtained as the square root of the diagonal terms, or trace, of the V-C matrix

The following model displays logistic regression parameter estimates and associated statistics, but without the usual Stata header (*nohead*) or iteration log (*nolog*).

```
. glm death anterior kk2-kk4, nohead fam(bin) nolog

------------------------------------------------------------------
            |               OIM
      death |    Coef.  Std. Err.     z   P>|z|   [95% Conf. Interval]
------------+-----------------------------------------------------
   anterior |  .7087039   .164241   4.32  0.000    .3867975    1.03061
        kk2 |  .9558424  .1768621   5.40  0.000    .6091991   1.302486
        kk3 | 1.190278   .2616342   4.55  0.000    .6774846   1.703072
        kk4 | 2.909672   .3298733   8.82  0.000    2.263132   3.556212
      _cons | -4.03746   .1477489 -27.33  0.000   -4.327042  -3.747878
------------------------------------------------------------------

. corr, _coef co

              | death:
              | anterior      kk2       kk3       kk4      _cons
------------+------------------------------------------------------
death         |
   anterior |  .026975
        kk2 | -.002108   .03128
        kk3 | -.004515   .012893  .068452
        kk4 | -.001558   .012662  .012801   .108816
      _cons |  -.01583  -.011304 -.009891  -.011626   .02183
```

The standard error of anterior is .164241, the square root of .026975, the value displayed in the above Variance-Covariance (V-C) matrix.

```
. di sqrt(.026975)
.16424068
```

estimates: displays stored statistics that were saved by the immediately previous estimation command

For example, typing **ereturn** *list* following the **glm** logistic regression model displayed above gives the following partial results. We show only half of the 27 scalars stored by the **glm** command and displayed by **ereturn**. Stored macros, matrices, and functions are also stored and displayed, but not shown below.

```
. ereturn list
```

```
scalars:
                   e(ll)  =  -688.1118353235119
            e(converged)  =  1
                 e(rank)  =  5
                    e(k)  =  5
                 e(k_eq)  =  1
                 e(k_dv)  =  1
                   e(ic)  =  4
                    e(N)  =  4503
                 e(chi2)  =  122.1575352594994
                    e(p)  =  1.84829523462e-25
                   e(rc)  =  0
                  e(aic)  =  .3078444749382687
                e(power)  =  0
```

dotplot: dotplot of varname, with one column per value of groupvar; can be
used to plot predicted probabilities for each of the outcome groups

. dotplot predprob, over(outcome) mean

predict: is used following a regression command to calculate a variety of fit
statistics and residuals; the command syntax is

predict <newvar> [if] [in] [, statistic *]

<newvar> is a placeholder indicating the name of a variable you select to hold
the values specified by *statistic*; following logit or logistic, the fitted value, or
probability of a success (1), can be obtained by, for example:

. predict fit, pr

We do not have to type *pr* in this case since it is the default option. Also, we
could have used any name not already in use, or reserved by Stata, for the
fitted value. Some other commonly used options include:

xb   : linear predictor
stdp   : standard error of the prediction
hat   : hat matrix diagonal, also called Pregibon leverage statistic
residuals   : Pearson residuals; using m-asymptotics
rstandard   : standardized Pearson residuals; m-asymptotics
ddeviance   : Hosmer–Lemeshow Delta-D influence statistic; m-asymptotics
score   : first derivative of the binomial log-likelihood, wrt the linear
              predictor, $x_i\beta$

The **glm** command has, in part, different **predict** options. Following the estimation of data using **glm**, we can use **predict,** but care must be taken regarding the option names. A few that differ from the other logistic commands are:

mu	: fitted value, the default
eta	: the linear predictor
response	: raw residual—difference between the response and fitted value
deviance	: deviance residuals
pearson	: Pearson residuals
standa	: a second option after *deviance* or *pearson,* standardizing the residuals
anscombe	: Anscombe residuals

SEARCH

findit	: used to locate information on a wide variety of statistical command and procedures. This is a very useful command, Example: findit logistic

A.5 Random Numbers

Stata has a wide variety of probability distributions, density functions, and mathematical functions. However, there are few random number generators. Thirteen random number generators were published by Hilbe and Linde-Zwirble (1995), and updated by Hilbe (1998), providing generation of *t*, *chi2*, *F*, lognormal, Poisson, negative binomial (NB2), binomial, gamma, inverse Gaussian, exponential, Weibull, beta binomial, and a 3-parameter generalized logistic random numbers. Synthetic data sets can also be developed for all members of the family of generalized linear models (GLM). Specialized versions of the GLM distributions were developed and published with the primary random number generators. We use some of these functions in this text, as well as other generators which are based on them. Stata users can gain access to these generators by typing **search rnd** at the command line.

Random number generators for the uniform and normal distributions come with the Stata package. We will show how these can be used to limit or extend the range of random numbers generated from these functions.

An example is a random number generator that produces a user-specified number of binomial variates with a mean of 0.5 and a binomial denominator of 1. We then summarize the resultant data.

```
. ssc install rnd                // installs the rnd commands

. rndbin 10000 .5 1
( Generating . )
Variable xb created.

. su

Variable |        Obs        Mean   Std. Dev.        Min        Max
---------+--------------------------------------------------------
      xb |      10000       .5013    .5000233          0          1
```

A logistic regression data set can be created with specified coefficients by first generating, for example, two predictors, defining a linear predictor, converting the linear predictor to a fitted probability, and submitting the results to the random number generator. We shall set the number of observations at 20,000. We could also set a seed number so that the same random numbers would be generated each time it was run [example: set seed 12345]. We choose not to do this.

The random variates will be generated so that the synthetic model coefficients are

Predictor 1: $\times 1 = $ .5
Predictor 2: $\times 2 = -1.25$
Constant   = 1

```
. set obs 20000
. gen x1 = invnorm(uniform())    /// predictor 1
. gen x2 = invnorm(uniform())    /// predictor 2
. gen xb = 1 + .5*x1 - 1.25*x2   /// linear predictor
. gen mu=1/(1+exp(-xb))          /// fitted value; i.e. prob==1
. gen byte den = 1               /// binomial denominator
. rndbinx mu den                 /// logistic
```

Note: **invlogit()** can be used instead of $1/(1 + \exp(-xb))$. Also note that the combined functions, *invnorm(uniform())*, that are used to create random numbers for synthetic data sets, can be interpreted as follows. *uniform()* generates uniformly distributed pseudorandom numbers on the interval, or range, 0 to near 1.0, with a mean of zero and standard deviation (SD) of 1.0. *invnorm()* randomly generates numbers from the inverse cumulative standard normal distribution. When used together as above, the *uniform()* function generates numbers from the uniform distribution which are used as the values to draw from a standard normal distribution using *invnorm()*. If a value is added to the joint functions, e.g., 20, the result is a distribution with a mean of 20. If the joint functions are multiplied by a value, e.g., 10, the result is a distribution with a standard deviation of 10.

```
. set obs 10000
. gen myrand = invnorm(uniform())
                              /// std. norm. random numb.
. gen myrand_m50 = 50 + invnorm(uniform())
                              /// with mean = 50
. gen myrand_m50sd20 = 50 + invnorm(uniform())*20
                              /// with std. dev. = 20
. sum
```

```
     Variable |    Obs      Mean   Std. Dev.      Min       Max
--------------+--------------------------------------------------
       myrand | 10000  -.0005992  1.000659  -3.918931  3.641588
   myrand_m50 | 10000    49.9985   .9946398   46.62004   54.1677
myrand_m5~20  | 10000   49.85687  20.10344  -21.59205  128.3178
```

The model appears as:

```
. logit bnlx  x1 x2, nolog
```

```
Logistic regression                    Number of obs   =      20000
                                       LR chi2(2)      =    5122.46
                                       Prob > chi2     =     0.0000
Log likelihood = -10011.331            Pseudo R2       =     0.2037
------------------------------------------------------------------
   bnlx |     Coef. Std. Err.      z  P>|z|    [95% Conf. Interval]
--------+---------------------------------------------------------
     x1 |   .5061092 .0184201   27.48 0.000    .4700066    .5422119
     x2 |  -1.237988 .0218744  -56.60 0.000   -1.280861   -1.195115
  _cons |   1.002126 .0187857   53.35 0.000    .9653069    1.038945
------------------------------------------------------------------
```

Note how close the parameter estimates are to the definitions we specified. Without a seed, these values will change slightly with each run—as is expected. A probit model can be specified by defining mu as a probit inverse link rather than a logistic.

The **genbinomial** random number generator (Roberto Gutierrez) was used to create synthetic models in earlier printings of this book. It is based on the logic of the **rnd** commands—in this case on **rndbinx** (Hilbe). The commands are on this book's Web site. For example, given *x1*, *x2*, and *xb*,

```
. genbinomial y, xbeta(xb) n(1)
. logit y  x1 x2, nolog
```

```
Logistic regression                    Number of obs   =      20000
                                       LR chi2(2)      =    5216.24
                                       Prob > chi2     =     0.0000
Log likelihood = -9969.6396            Pseudo R2       =     0.2074
```

```
------------------------------------------------------------
      y |      Coef.  Std. Err.      z    P>|z|   [95% Conf. Interval]
--------+---------------------------------------------------
     x1 |   .5264213  .0185289   28.41   0.000    .4901053    .5627372
     x2 |  -1.248699  .0219913  -56.78   0.000   -1.291801   -1.205597
  _cons |   1.005618  .0188561   53.33   0.000    .9686605    1.042575
------------------------------------------------------------
```

Binomial logistic models can also be defined using both of these commands. The binomial denominator is specified by the user as part of the command. Typing

```
. help rnd
. help genbinomial
```

will provide the necessary information on how to use these models.

Note: After this section was written, Stata enhanced its random number capabilities by including a new suite of random number generators with the official software. These have been used in this printing, whereas the user-authored **genbinomial** (Gutierrez) was employed in the first two printings. All lead to similar results.

```
. gen varname = invnorm(runiform())
```

A.6 Conclusion

All of the above official Stata commands can be accessed using the **menu** system. You may want to use it instead of the command line for house-keeping, and even for modeling. There are many aspects of the Stata commands and language that we did not cover, such as functions, dates, editing, and so forth. The functions we use throughout the book are fully explained, or are intuitively clear. Type *help functions* if you need clarification.

If you have a copy of Stata, additional support can be obtained using the manuals, or by accessing the Web. If you are not using Stata, but have reviewed this appendix to better understand the examples or to reproduce examples using other software applications, you can obtain additional information by going to the Stata Web site, *http://www.stata.com*, and clicking to the Frequently Asked Questions section. Other information on using Stata can be found using the support section of the same site. Resources for learning Stata, and on relating Stata commands to those of other packages, can be found on *http://www.stata.com/links/resources1.html*.

Appendix B: Stata and R Logistic Models

The following is a list of Stata commands that are related to the logistic models discussed in this text.

B.1 Basic Logistic Models

Binary logistic regression	logit
Binary logistic regression, odds ratios, and fit statistics	logistic
GLM-based binary logistic models	glm, f(bin)
GLM-based binomial logistic models	glm, f(bin #)
Binomial models—generalized link	binreg

B.2 Categorical Response, Ordered Response Logistic Models

Proportional odds logistic regression	ologit
Proportional odds binomial; logit an option	oglm <U>
Generalized ordered logistic regression	gologit <U>
Partial proportional ordered logistic models	gologit2 <U>
Stereotype ordinal logistic regression	soreg <U>
Rank ordered logistic regression	rologit
Alternative-specific conditional logit (McFaddens Choice) model	asclogit

B.3 Categorical Response, Unordered Response Logistic Models

Multinomial logistic regression	mlogit
Stereotype logistic regression	slogit
Multinomial conditional logit regression	mclgen, mclest <U>

B.4 Panel Logistic Models

Fixed effects, random effects, and GEE panel models	xtlogit
GEE—logistic models family	xtgee
Conditional logistic regression	clogit
Conditional ordered logit regression	gllamm, l(ologit)
Multilevel mixed effects logistic regression	xtmelogit
Random intercept logistic; random parameter logistic	gllamm <U>

B.5 Other Types of Logistic-Based Models

Nested logistic regression	nlogit
Skewed logistic regression	scobit
Logistic discriminant analysis	discrim logistic
Exact logistic regression	exlogistic
Search web for type of statistical procedure	findit

R logistic model commands can be found in both the packages downloaded with the software from a mirror site or as part of a user-created package that can be installed into your R application. R functions that duplicate many of the examples in the text are found at the ends of most chapters.

Appendix C: Greek Letters and Major Functions

Greek letters are commonly used in statistics, as well as in mathematics in general, to designate various statistical qualities and functions. Below is a list of Greek letters that play traditionally assigned roles in statistical notation. Our interest is, of course, in notation that is related to generalized linear models and logistic models in general.

For each Greek letter, pronunciation, symbol, and a common role for which it is used in statistical literature are displayed.

alpha	α	linear constant; negative binomial ancillary parameter; correlation
beta	β	parameter estimate
gamma	γ	random effect parameters
delta	δ	NB-1 ancillary parameter
epsilon	ε	linear model error term
zeta	ζ	random effect
eta	η	linear predictor
theta	θ	canonical link
iota	ι	subscript indicating individuals; interaction
kappa	κ	cluster number; lag
lambda	λ	mean; for count models
mu	μ	mean; all models
nu	ν	shape parameter; random effect
xi	ξ	random effect; log of dispersion
omicron	o	odds
pi	π	constant: 3.14159...; probability of success in logit models
rho	ρ	correlation; fraction of variance
sigma	$\sigma; \varsigma$	population standard deviation; (ς used at end of word only)
tau	τ	rate parameter; fixed effect
upsilon	υ	matrix element; random intercept
phi	$\varphi; \phi$	scale parameter; dispersion
chi	χ	main term of *Chi2* statistic; χ^2
psi	ψ	variance, in random effects
omega	ω	number of parameters for cAIC and mAIC fit statistics

SIGMA	Σ	summation term; log-likelihood functions; variance matrix
PI	Π	multiplicative function for probability function
GAMMA	Γ	when logged, the log-gamma function, common term in log-likelihoods
PHI	Φ	cumulative normal probability; probit link
OMEGA	Ω	full covariance matrix

Appendix D: Stata Binary Logistic Command

The following code is a skeleton form showing the traditional GLM-based IRLS method of estimation. The program is written with version 9 code, which allows users of version 9 and above to use the command with data. The command allows a wide variety of options, including survey models, weights, offsets, clustering, and robust variance estimators. The *or* option produces odds ratios.

Output includes the deviance and Pearson *Chi2* statistics and their respective dispersion statistics, and the AIC and BIC goodness-of-fit statistics. The likelihood statistic is also displayed; if a model employs robust variance estimators and clustering, a pseudo-likelihood is displayed instead.

D.1 Main Binary Logistic Regression Model: (logisticjh)

```
*! Version 1.0.0
* LOGISTIC REGRESSION:  Joseph Hilbe :  16Apr2008
program logisticjh, properties(svyb svyj svyr)
    version 9
    syntax [varlist] [if] [in] [fweight pweight aweight
      iweight] [, ///
        Level(cilevel) OR Robust noLOG ///
        OFFset(passthru) EXposure(passthru) ///
        CLuster(passthru) FROM(string asis) * ]
    gettoken lhs rhs : varlist
    marksample touse
    mlopts mlopts, 'options'
    if ("'weight'" != "") local weight "['weight' 'exp']"
    if ('"'from'"' != '""') local initopt '"init('from')"'

    ml model lf jhlogit_ll (xb: 'lhs' = 'rhs', 'offset'
      'exposure') ///
        if 'touse' 'weight', ///
        'mlopts' 'robust' 'cluster' ///
        title("Logistic Regression") ///
      maximize 'log' 'initopt'
    ml display, level('level') 'or'
qui {
* AIC
    tempvar aic
```

```
        local nobs e(N)
        local npred e(df_m)
        local df = e(N) - e(df_m) -1
        local llike e(ll)
        gen 'aic' = ((-2*'llike') + 2*('npred'+1))/'nobs'

* PRELIMINARY STATS
tempvar y lp mu musq  ymean
predict `lp', xb
gen double `mu' = 1/(1+exp(-`lp'))
gen `y' = `lhs'

* DEVIANCE
tempvar dev
gen double `dev' =ln(1/`mu') if `y'==1
replace `dev' = ln(1/(1-`mu')) if `y'==0
replace `dev'= sum(`dev')
local deviance = 2*`dev'[_N]
local nobs e(N)
local npred e(df_m)
local df = `nobs' - (`npred'+1)
local ddisp = `deviance'/ `df'

* PEARSON
tempvar pchi
gen `pchi' = (`y'-`mu')^2/(`mu'*(1-`mu'))
replace `pchi' = sum(`pchi')
local pchi2 = `pchi'[_N]
local dpchi2 = `pchi2'/`df'

* BIC
local bic = -2*`llike' + ln(`nobs')*(`npred'+1)
}
* DISPLAY
di in gr _col(1) "AIC Statistic = " in ye %11.3f 'aic' _col(53)
   in gr "BIC Statistic = " in ye %10.3f 'bic'
di in gr _col(1) "Deviance = " in ye %11.3f 'deviance' _col(53)
   in gr "Dispersion = " in ye %10.3f 'ddisp'
di in gr _col(1) "Pearson Chi2 = " in ye %11.3f 'pchi2' _col(53)
   in gr "Dispersion = " in ye %10.3f 'dpchi2'
end
```

D.2 Log-Likelihood Function (called by logisticjh)

```
*   LOGISTIC REGRESSION : LOGLIKELIHOOD FUNCTION
* Joseph M. Hilbe: Logistic Regression Models : 16Apr2008
```

```
program define jhlogit_ll
   version 9.1
   args lnf xb
   qui replace `lnf' = ln(invlogit(`xb')) if $ML_y1 == 1
   qui replace `lnf' = ln(invlogit(-`xb')) if $ML_y1 ==0
end
```

D.3 Example Output (hearto1 data)

```
. logisticjh death anterior hcabg kk2-kk4,  nolog or

Logistic Regression                   Number of obs   =       4503
                                       Wald chi2(5)    =     125.89
Log likelihood = -686.28751           Prob > chi2     =     0.0000

--------+----------------------------------------------------------
  death | Odds Ratio  Std. Err.    z    P>|z|  [95% Conf. Interval]
--------+----------------------------------------------------------
anterior |   2.05989   .3391911  4.39   0.000   1.491699    2.844506
   hcabg |  2.051369   .7098967  2.08   0.038   1.041063    4.042133
     kk2 |  2.573692   .4558432  5.34   0.000   1.818841    3.641819
     kk3 |  3.244251   .8504558  4.49   0.000    1.94079    5.423136
     kk4 |  18.55299   6.123173  8.85   0.000   9.715961    35.42761
--------+----------------------------------------------------------
AIC Statistic =          0.307     BIC Statistic =      1423.050
Deviance      =       1372.575     Dispersion    =         0.305
Pearson Chi2  =       4426.951     Dispersion    =         0.984
```

Appendix E: Derivation of the Beta Binomial

$$f(y;\mu) = f(y|p)f(p)$$

$$= [(m\ y)p^y(1-p)^{m-y}] * \left| \frac{\Gamma(A+B)}{\Gamma(A)\Gamma(B)}\ p^{A-1}(1-p)^{B-1} \right|$$

$$= \frac{\Gamma(A+B)\ \Gamma(m+1)}{\Gamma(A)\Gamma(B)\ \Gamma(y+1)\ \Gamma(m-y+1)} p^{y+A-1}\ (1-p)^{m-y+B-1}$$

By integrating with respect to p, we can obtain the margin distribution of Y, and therefore the probability density function of the beta-binomial.

$$f(y) = \int_0^1 \frac{\Gamma(A+B)\ \Gamma(m+1)}{\Gamma(A)\Gamma(B)\ \Gamma(y+1)\ \Gamma(m-y+1)} p^{y+A-1}\ (1-p)^{m-y+B-1} dp$$

$$f(y) = \int_0^1 \frac{\Gamma(A+B)\ \Gamma(m+1)}{\Gamma(A)\Gamma(B)\ \Gamma(y+1)\ \Gamma(m-y+1)}\ \frac{\Gamma(y+A)\ \Gamma(m-y+B)}{\Gamma(m+A+B)}$$

$$\frac{\Gamma(m+A+B)}{\Gamma(y+1)\ \Gamma(m-y+1)\ \Gamma(A)\Gamma(B)}\ p^{y+A-1}\ (1-p)^{m-y+B-1} dp$$

$$= \frac{\Gamma(A+B)\ \Gamma(m+1)}{\Gamma(A)\Gamma(B)\ \Gamma(y+1)\ \Gamma(m-y+1)}\ \frac{\Gamma(y+A)\ \Gamma(m-y+B)}{\Gamma(m+A+B)}$$

$$* \int_0^1 \frac{\Gamma(m+A+B)}{\Gamma(A)\Gamma(B)\ \Gamma(y+1)\ \Gamma(m-y+1)}\ p^{y+A-1}\ (1-p)^{m-y+B-1} dp$$

where the second group of terms (after the * sign) is equal to 1.

$$= \frac{\Gamma(A+B)\ \Gamma(m+1)}{\Gamma(y+1)\ \Gamma(m-y+1)\ \Gamma(A)\Gamma(B)}\ \frac{\Gamma(y+A)\ \Gamma(m-y+B)}{\Gamma(m+A+B)}$$

Appendix F: Likelihood Function of the Adaptive Gauss–Hermite Quadrature Method of Estimation

F.1 Random Effects Logistic Model with Multivariate Normal Effects

$$L(x_{it}\beta; y_{it}, v_i\sigma) \approx \Sigma \ln\left[\sqrt{2\sigma_i} \Sigma_{m=1} w_m^* \exp\{(\alpha_m)^2\} \frac{\exp\left\{-\left(\sqrt{2\sigma_i}\alpha_m^*\right)+\mu_i\right)^2/2\sigma_v^2\right\}}{\sqrt{2\pi}\sigma_v} \right.$$

$$\left. \Pi\ F(y_{it}, x_{it}\beta + \sqrt{2\sigma_i}\alpha_m^*)+\mu_i) \right]$$

where w_i is the quadrature weight supplied by either the user or algorithm. $w_i = 1$ if w is not specified.

This method of estimation is generally appropriate when the number of random effects is less than 20.

Appendix G: Data Sets

DATA SETS USED IN TEXT		USER-CREATED COMMANDS		
NAME	Data Defined or First Found	COMMAND	FIRST USED	AUTHOR
heart01	Ch 2	ulogit	5.1	Hilbe
heartr	Ch 5.2.5	lpartr	5.2.3	Hilbe
anterior	Ch 5.5.2			
lbw	Ch 5.9	lstand	5.3	Hilbe
lr_model	Ch 5.9	oddsrisk	5.5.1	Hilbe
felon	Ch 5.10	genbinomial	5.6	Gutierrez
imputed	Ch 5.10	rndbinx	5.6	Hilbe
medpar	Ch 5.11	abic	5.8	Hilbe
titanic	Ch 6.1	ice	5.10	Royston
overex	Ch 9.2.1.1	misschk	5.10	Long/Freese
cancer	Ch 9.2.1.6	logitem	5.11	Cleves/ Tosetto
gss2002_10	Ch 10.1	postgr3	6.1	Long/Freese
medpar_olr	Ch 10.1	xi3	6.1	Long/Freese
gss_polviews2002	Ch 11.3	lrdrop1	7.3.7	Wang
edreligion	Ch 12.1	williams	9.4.2	Hilbe
ordward2	Ch 12.3	syn_ologit.do	10.1	Hilbe
ordwarmjh	Ch 12.3	distinct	10.1	Longton/Cox
rwm_1980	Ch 13.1.2	gologit2	10.2	Williams
heart02grp	Ch 13.1.3	**prtab**	11.1.2	Long
wagepan	Ch 13.2	mlogtest	11.1.2	Long/Freese
lowbirth	Ch 13.3.2	syn_mlogit1. do	11.1.2	Hilbe
rankorderwls	Ch 13.3.3	aic	12.2	Hilbe
nhanes3	Ch 14.1	oglm	12.3	Williams
auto	Ch 14.3.1	omodel	12.5.3	Wolfe
crashinjury	Ch 14.1	syn_oclog3.do	12.6.2	Hilbe
hiv	Ch 15.1	qic	13.1.4	Cui
hiv1	Ch 15.2.3			

DATA SETS USED IN TEXT		USER-CREATED COMMANDS		
NAME	Data Defined or First Found	**COMMAND**	**FIRST USED**	**AUTHOR**
TABLES MODELED				
lrm3_4_2a	Ch 3.4.2	xtqls	13.1.6	Shults
lrm5_5_tbl1	Ch 5.5	listcoef	13.3.3	Long/Freese
lrm5_5_tbl2	Ch 5.5	asprvalue	13.3.3	Long/Freese
lrm_5_2_2	Ch 5.2.2	gllamm	Ex 13.3	Rabe-Hesketh
lrm9_1	Ch 9.1	svylogitgof	14.1	Archer/Lemeshow
lrm9_2	Ch 9.2	discrim2	14.3.2	Hilbe
lrm_mcc	Ch 13.3.2			
lrm13_4_2	Ch 13.4.2			

DATA SETS USED IN EXERCISES		
Chapter	Name	Exercise#
2	drugprob	2.8
5	affairs	5.2
5	kyp	5.3
5	absent	5.4
5	chd	5.8
6	environment	6.4
6	genestudy	6.5
6	medpar	6.6
7	shuttle	7.4
7	heartprocedure	7.5
8	doll	8.3
8	drugproblem	8.4
8	hiv	8.5
8	oakridge	8.6
10	lrm10_1	10.1
10	autoorder	10.3
11	gsssurvey	11.2
11	alligator	11.4
12	edreligion	12.2
13	wheezehh	13.2

DATA SETS USED IN EXERCISES		
Chapter	Name	Exercise#
13	madras1	13.3
13	schiz2	13.4
13	cluslowbwt	13.5
13	unionocc	13.6
14	crashinjury	14.4

Appendix H: Marginal Effects and Discrete Change

At times, one may desire to understand the relationship of an estimated predictor to its predicted probability, with other predictors held at a specified value. More specifically, we may wish to quantify the relationship of how the probability changes given a unit (or more) change in the predictor value. This relationship is commonly referred to as a marginal effect. One frequently finds logistic models in econometric, social science, and educational research literature in which marginal effects are reported in preference to odds ratios. Since marginal effects were not discussed in the main part of the text, we give a brief overview in this appendix of how to calculate marginal effects and when they should be used in binary logistic models. Note that marginal effects may be produced for any regression model; the logistic model is but one variety.

A marginal effect for a binary logistic model is the slope of the binomial-logistic probability distribution relating a given continuous predictor, x_j, to $\Pr(y = 1 \mid x)$, where x is the matrix of predictors. Recall that the predicted probability of $y = 1$ is calculated from the logistic probability function, $1/(1 + \exp(-xb))$, where xb is the logistic linear predictor. The predictor, x_j, is linear in the logit, but is nonlinear with respect to the probability of $y = 1$. Because of the nonlinearity in the relationship of x_j and $\Pr(y = 1 \mid x)$, the marginal effects differ across values of the predictor. This is not the case for standard linear regression.

Marginal effects are also conceived of as instantaneous rates of change; one predictor is calculated for the effect, while the other predictors in the model are held at a given value, usually at their mean. This definition follows from the nature of a slope—a rate of change. Here we will use the mean, but understand that other values may be used as well.

Recall that the logistic probability function can be formulated as:

$$F = \Lambda(\beta'x) = 1/(1 + \exp(-xb)) \text{ or } \exp(xb)/(1 + \exp(xb)) \quad \text{or} \quad \Pr(y = 1 \mid x)$$

where xb is the logistic linear predictor. The logistic density function is

$$f = \Lambda(\beta'x) * [1 - \Lambda(\beta'x)]$$

The marginal effect is calculated as:

MARGINAL EFFECTS

$$f(\beta'x)\beta$$

or

$$\frac{\partial \Pr(y=1\mid x)}{\partial x_j}$$

The basic strategy in calculating a marginal effect is to obtain three values: the coefficient of the predictor of interest, the probability of $y=1$ with predictors held at a specified value, and the probability of $y=0$, which is given as $1 - \Pr(y=1\mid x)$. I will demonstrate the calculations using the **heart01** data. For simplification, we will then model two predictors: one binary (*anterior*) and the other continuous (*age*). We shall also discuss later why using marginal effects with binary predictors in a logistic model is not advised. For this, we use a discrete change model, which I will demonstrate later. Also note that it is important to be certain to delete all missing values from predictors in the model.

For this example, we calculate the marginal effect on *age*, with *anterior* held at its mean.

```
. use heart01

. drop if anterior==.        /// anterior has 692 missing values

. logit death anterior age, nolog
```

Logistic regression	Number of obs	=	4696
	LR chi2(2)	=	144.62
	Prob > chi2	=	0.0000
Log likelihood = -713.68569	Pseudo R2	=	0.0920

death	Coef.	Std. Err.	z	P>|z|	[95% Conf. Interval]
anterior	.7266114	.1578098	4.60	0.000	.4173099 1.035913
age	.0567263	.00545	10.41	0.000	.0460445 .0674081
_cons	-7.603426	.4293466	-17.71	0.000	-8.44493 -6.761922

```
. su anterior age
```

Variable	Obs	Mean	Std. Dev.	Min	Max
anterior	4696	.4525128	.4977929	0	1
age	4696	65.48871	14.35942	40	100

LINEAR PREDICTOR, xb, AT PREDICTOR MEAN VALUES

```
. di .4525128*_b[anterior] + 65.80976*_b[age] - 7.603426
-3.5414801
```

```
Pr(Y=1|X)                          /// Probability of y=1
. di 1/(1+exp(3.5414801))
.02815476
```

```
Pr(Y=0|X)                          /// Probability of y=0
. di 1 - 1/(1+exp(3.5414801))
.97184524
```

```
Coef(age) * Pr(Y=1|X) * Pr(Y=0|X)
. di .0567263 * .02815476 * .97184524
.00155215    <= marginal effect of age
```

The marginal effect of *age* is 0.0015. To check, we employ Stata's marginal effects command, **mfx**, using its default values—which entails using mean values for predictor values. We use it to check the results of the hand-calculated value of *age*.

```
. mfx

Marginal effects after logit
      y   = Pr(death) (predict)
          =  .02766071
```

variable	dy/dx	Std Err	z	P>|z|	[95% C.I.]		X
anterior*	.020560	.00466	4.41	0.000	.0114	.0297	.45251
age	.001525	.00013	11.43	0.000	.0012	.0017	65.488

```
(*) dy/dx is for discrete change of dummy variable from 0 to 1
```

Note the rounding error, which differs at only at the 1/100,000ths place.

H.1 Alternative Method of Calculating Marginal Effects

An alternative method is to exploit the derivative aspect of the marginal effect. The logistic probability function is $1/(1 + \exp(-xb))$, or $\exp(xb)/(1 + \exp(xb))$. The derivative of the logistic function with respect to xb is, therefore, $\exp(xb)/(1 + \exp(xb))^2$.

```
. use heart01
```

```
. drop if anterior==.

. logit death anterior age, nolog

Logistic regression                      Number of obs    =         4696
                                         LR chi2(2)       =       144.62
                                         Prob > chi2      =       0.0000
Log likelihood = -713.68569              Pseudo R2        =       0.0920

------------------------------------------------------------------
    death |     Coef.   Std. Err.      z   P>|z|  [95% Conf. Interval]
----------+-------------------------------------------------------
 anterior |  .7266114   .1578098    4.60   0.000  .4173099   1.035913
      age |  .0567263     .00545   10.41   0.000  .0460445   .0674081
    _cons | -7.603426   .4293466  -17.71   0.000  -8.44493  -6.761922
------------------------------------------------------------------

. qui sum anterior                   /// quietly summarize
                                         anterior

. qui replace anterior=r(mean)       /// constrain anterior to
                                         its mean value

. qui sum age                        /// quietly summarize age

. qui replace age=r(mean)            /// constrain age to its
                                         mean value

. l anterior age in 1                /// verify that the values
                                         are at their means

        +--------------------+
        | anterior       age |
        |--------------------|
   1.   | .4525128   65.48872 |
        +--------------------+

. predict xb, xb                            /// predict xb based on
                                                mean values

. gen dfdxb= exp(xb)*(1+exp(xb))^(-2)  /// use derivative of
                                                probability

. l dfdxb in 1

        +----------+
        |    dfdxb |
        |----------|
   1.   | .0268956 |
        +----------+
```

MARGINAL EFFECT: ANTERIOR

```
. di _b[anterior]* dfdxb
.01954265                      /// compare with mfx - what
                                   should not be used - binary
```

MARGINAL EFFECT: AGE

```
. di _b[age]* dfdxb
.00152569                      /// same value as displayed in
                                   mfx output - continuous
```

It is important to note that Greene (2007) and others have found that when assessing the statistical significance of a predictor, one should base *p*-values on parameter estimates or odds ratios, not on marginal effects. In other words, the significance values displayed in a table of marginal effects as generated by programs such as **mfx** are not to be used as an accurate indication of the contribution of the predictor to the model. The *z*-values and corresponding *p*-values do not usually vary too much from parameter estimate-based statistics, but the confidence intervals are typically quite different.

Second, Stata has excellent marginal effects applications. Moreover, the text by Long and Freese (2006) listed below has detailed sections about using marginal effects with different types of discrete response models. Greene (2007) gives a nice presentation of marginal effects for the full range of models.

Finally, for logistic models, it is preferred to employ marginal effects for continuous predictors only. For binary and categorical predictors, it is recommended to use the method of discrete change, as described next.

H.2 Discrete Change

Marginal effects is not appropriate for use with binary predictors. This is particularly the case when the response is also binary. We use a discrete change instead to show the change in the predicted probability of binary predictor *x* as *x*==0 changes to *x*==1. This relationship has been formalized as:

$$\frac{\Delta \Pr(y = 1 \mid x)}{\Delta x_j}$$

Δ is a common symbol for change and is frequently used to indicate a derivative in differential calculus.

```
. use heart01

. logit death anterior age, nolog
```

```
Logistic regression                     Number of obs    =        4696
                                        LR chi2(2)       =      144.62
                                        Prob > chi2      =      0.0000
Log likelihood = -713.68569             Pseudo R2        =      0.0920
```

```
---------------------------------------------------------------------
    death |    Coef.   Std. Err      z    P>|z|  [95% Conf. Interval]
----------+----------------------------------------------------------
 anterior |  .7266114  .1578098    4.60   0.000   .4173099  1.035913
      age |  .0567263    .00545   10.41   0.000   .0460445  .0674081
    _cons | -7.603426  .4293466  -17.71   0.000   -8.44493 -6.761922
---------------------------------------------------------------------
```

```
. sum age
```

```
    Variable |    Obs      Mean    Std. Dev.        Min        Max
-------------+-------------------------------------------------------
         age |   4696  65.48871    14.35942         40        100
```

```
. replace age=r(mean)    /// replaces age with mean value

. replace anterior=0     /// set anterior as equaling 0

. predict p0, p          /// obtain predicted values given
                             anterior=0 and age at its mean
                             value

. replace anterior=1     /// set anterior as equaling 1

. predict p1, p          /// obtain predicted values given
                             anterior=1 and age at its mean
                             value
```

DISCRETE CHANGE FOR ANTERIOR

```
. di p1 - p0
.02056034                        /// value of discrete change.
                                     Compare to mfx output
```

Compare with the Stata output calculated for *anterior* using **mfx** with the default settings. The two values are identical.

```
. mfx

Marginal effects after logit
      y  = Pr(death) (predict)
         =  .02766071
------------------------------------------------------------------
variable |   dy/dx  Std Err    z    P>|z|  [ 95% C.I. ]        X
---------+--------------------------------------------------------
anterior*| .020560    .0046  4.41  0.000  .0114   .0297  .45251
     age | .001525    .0001 11.43  0.000  .0012   .0017  65.488
------------------------------------------------------------------
(*) dy/dx is for discrete change of dummy variable from 0 to 1
```

Long and Freese (2006) have developed a number of tools for assessing both the marginal effects and discrete change for logistic models. For those interested in using them in their research reports, I recommend the sources given in the References section.

References

Achen, C.H. (2002). Toward a new political methodology: Microfoundations and ART. *Annual Review of Political Science* 5: 423–450.

Agresti, A. (1984). *Analysis of Ordinal Categorical Data*. New York: John Wiley & Sons.

Agresti, A. (2002). *Categorical Data Analysis*, 2nd edition. New York: John Wiley & Sons.

Agresti, A. and J.B. Lang (1993). A proportional odds model with subject-specific effects for repeated ordered categorical responses. *Biometrika* 80: 527–534.

Ai, C. and E.C. Norton (2003). Interaction terms in logit and probit models. *Economic Letters* 80: 123–129.

Aiken, L.S. and S.G. West (1991). *Multiple Regression: Testing and Interpreting Interactions*. Newbury Park, CA: Sage.

Aitchison, J. and J. Bennet (1970). Polychotomous quantal response by maximum ndicant. *Biometrica* 57: 253–262.

Aitchison, J. and S. Silvey (1957). The generalization of probit analysis to the case of multiple responses. *Biometrica* 44: 131–140.

Akaike, H. (1973). Information theory and an extension of the maximum likelihood principle. In *Second International Symposium on Information Theory*. Petrov, B. and F. Csaki (eds.), Budapest, Akademiai Kiado, pp. 267–281.

Akaike, H. (1974). A new look at the statistical model identification. *I.E.E.E. Transactions on Automatic Control* AC 19: 716–723.

Albert, A. and E. Lesaffre (1986). Multiple group logistic discrimination. *Computers and Mathematics with Applications* 12A: 1209–224.

Albright, R.L, S.R. Lerman, and C.F Manski (1977). Report on the development of an estimation program for the multinomial probit model.

Allison, P.D. (1987). Introducing a disturbance into logit and probit regression models. *Sociological Methods and Research* 15: 355–374.

Allison, P.D. (1999). Comparing logit and probit coefficients across groups. *Sociological Methods and Research* 28: 185–208.

Allison, P.D. (2001). *Missing Data*. Thousand Oaks, CA: Sage.

Allison, P.D. (2005). *Fixed Effects Regression Methods for Longitudinal Data Using SAS*. Cary, NC: SAS Press.

Allison, P.D. and N. Christakis (1994). Logit models for sets of ranked items. In P. Marsden (ed.), *Sociological Methodology*. 24: 19–228. Oxford, U.K.: Blackwell.

Ananth, C.V. and D.G. Kleinbaum (1997). Regression models for ordinal responses: A review of methods and applications. *International Journal of Epidemiology* 26: 1323–33.

Anderson, J.A. (1984). Regression and ordered categorical variables. *Journal of the Royal Statistical Society B* 46: 1–30.

Anderson, J.A. and P.R. Philips (1981). Regression, discrimination, and measurement models for ordered categorical variables. *Applied Statistics* 30: 22–31.

Archer, K.J. and S. Lemeshow (2006). Goodness-of-fit for logistic regression model fitted using survey sample data. *Stata Journal* 6 (1): 97–105.

Armstrong, B.G. and M. Sloan (1989). Ordinal regression models for epidemiologic data. *American Journal of Epidemiology* 129: 191–204.

Baltagi, B.H. (2008). *Econometric Analysis of Panel Data*. Chichester, U.K.: Wiley.

Barnard, G.A. (1949). Statistical inference. *Journal of the Royal Statistical Society Series B* 11: 115–149.

Barnard, G.A., G.M. Jenkins, and C.B. Winsten (1962). Likelihood inference and time series. *Journal of the Royal Statistical Society A* 125 (3): 321–372.

Barnett, A.G., N. Koper, A.J. Dobson, F.K.A. Schmiegelow, and M. Manseau (2008). Selecting the correct variance-covariance structure for longitudinal data in ecology: a comparison of the Akaike, quasi-information and deviance information criteria. Unpublished manuscript.

Baum, C.F. (2006). *An Introduction to Modern Econometrics Using Stata*. College Station, TX: Stata Press.

Beggs, S., S. Cardell, and J. Hausman (1981). Assessing the potential demand for electric cars. *Journal of Econometrics* 16: 1–19.

Bell, D., J. Walker, G. O'Connor, J. Orrel, and R. Tibshirani (1989). Spinal deformation following multi-level thoracic and lumbar laminectomy in children. Unpublished research manuscript. University of Toronto Medical School, Toronto, Canada.

Ben-Akiva, M. and S. Lerman (1985). *Discrete Choice Analysis*. Cambridge, MA: MIT Press.

Berkson, J (1944). Application of the logistic function to bio-assay. *Journal of the American Statistical Association* 39: 357–365.

Breslow, N.E. and N.E. Day (1980). *Statistical Methods in Cancer Research. Vol. 1: The Analysis of Case-Control Studies*. Lyon, France: International Agency on Cancer.

Cameron, A.C. and P.K. Trivedi (1998). *Regression Analysis of Count Data*. Cambridge, U.K.: Cambridge University Press.

Carey, V., S.L. Zeger, and P. Diggle (1993). Modelling multivariate binary data with alternating logistic regressions. *Biometrika* 80 (3): 517–526.

Carlin J.B., N. Li, P. Greenwood, and C. Coffey. (2003). Tools for analyzing multiple imputed datasets. *Stata Journal* 3 (3): 226–244.

Carroll, R.J., D. Ruppert, and L.A. Stefanski (1995). *Measurement Error in Nonlinear Models*. London: Chapman & Hall/CRC Press.

Chaganty, N.R. (1997). An alternative approach to the analysis of longitudinal data via generalized estimating equations. *Journal of Statistical Planning and Inference* 63: 39–54.

Chaganty, N.R. and H. Joe (2004). Efficiency of generalized estimating equations for binary responses. *Journal of the Royal Statistical Society B* 66 (4): 851–860.

Chaganty, N.R. and H. Joe (2006). Range of correlation matrices for dependent Bernoulli random variables. *Biometrika* 93 (1): 197–206.

Chaganty, N.R. and D. Mav (2008). Estimation methods for analyzing longitudinal data occurring in biomedical research. In Khattree, R. and D. Naik (eds.), *Computational Methods in Biomedical Research*. Boca Raton, FL: Chapman & Hall/CRC Press, pp. 371–400.

Chaganty, N.R. and J. Shults (1999). On eliminating the asymptotic bias in the quasi-least squares estimate of the correlation parameter. *Journal of Statistical Planning and Inference* 76: 127–144.

Cheng, S. and S.J. Long (2007). Testing for IIA in the multinomial logit model. *Sociological Methods & Research* 35 (4): 583–600.

Clayton, D. and M. Hills (1993). *Statistical Models in Epidemiology.* Oxford, U.K.: Oxford University Press.

Cleves, M. and A. Tosetto (2000). Logistic regression when binary outcome is measured with uncertainty. *Stata Technical Bulletin* 55: 20–23.

Collett, D. (1991). *Modelling Binary Data.* London: Chapman & Hall.

Collett, D. (2003). *Modelling Binary Data,* 2nd edition. London: Chapman & Hall.

Cornfield, J. (1951). A method of estimating comparative rates from clinical data. Applications to cancer of the lung, breast and cervix. *Journal of the National Cancer Institute* 11: 1269–1275.

Coupé, T. (2005). Bias in conditional and unconditional fixed effects logit estimation: a correction. *Political Analysis* 13 (3): 292–295.

Coveney, J. (2008). Logistic regression by means of penalized maximum likelihood estimation in cases of separation. *2008 Summer North American Stata Users Group Meeting.* Presentation and software; personal correspondence prior to meeting. Chicago, IL: July 25, 2008.

Cox, D.R. and E.J. Snell (1989). *Analysis of Binary Data,* 2nd edition. London: Chapman & Hall/CRC Press.

Cramer, J.S. (2003). *Logit Models from Economics and Other Fields.* Cambridge, U.K.: Cambridge University Press.

Crowder, M. (1995). On the use of a working correlation matrix in using generalized linear models for repeated measures. *Biometrika* 82: 407–410.

Cui, J. (2007a). QIC program and model selection in GEE analysis. *Stata Journal* 7: 209–220.

Cui, J. (2007b). Selection of working correlation structure and best model in GEE analyses of longitudinal data. *Communications in Statistics—Simulation and Computation* 36: 987–996.

Cui, J. and L. Feng (2008). Correlation structure and model selection for negative binomial distribution in GEE. *Statistics in Medicine.*

Daganzo, C. (1979). *Multinomial Probit: The Theory and Its Application to Demand Forecasting.* New York: Academic Press.

Diggle, P.J., K.-Y. Liang, and S. Zeger (1994). *Analysis of Longitudinal Data.* New York: Oxford University Press.

Diggle, P.J., P. Heagerty, K.-Y. Liang, and S. Zeger (2002). *Analysis of Longitudinal Data,* 2nd edition. New York: Oxford University Press.

Doll, R. and A.B. Hill (1966). Mortality of British doctors in relation to smoking; observations on coronary thrombosis. In W. Haenszel (ed.), *Epidemiological Approaches to the Study of Cancer and Other Chronic Diseases* 19: 204–268. National Cancer Institute monograph, Bethesda, MD.

Dunlop, D. (1994). Regression for longitudinal data: A bridge from least squares regression. *The American Statistician* 48: 299–303.

Dupont, W.D. (2002). *Statistical Modeling for Biomedical Researchers.* Cambridge, U.K.: Cambridge University Press.

Echambadi, R. and J.D. Hess (2007). Mean-centering does not alleviate collinearity problems in moderated multiple regression models. *Marketing Science* 26 (3): 438–445.

Fagerland, M.W., D.W. Hosmer, and A.M. Bofin (2008). Multinomial goodness-of-fit tests for logistic regression models, *Statistics in Medicine,* Vol. 27: 4238–4253 Wiley InterScience DOI: 10.1002/sim.3202.

Faraway, J.J. (2006). *Extending the Linear Models with R: Generalized Linear, Mixed Effects and Nonparametric Regression Models.* Boca Raton, FL: Chapman & Hall/CRC Press.

Feinberg, S. (1980). *The Analysis of Cross-Classified Categorical Data,* 2nd edition. Cambridge, MA: MIT Press.

Finney, D.J. (1952). *Probit Analysis.* Cambridge, U.K.: Cambridge University Press.

Finney, D.J. (1971). *Probit Analysis,* 3rd edition. Cambridge: Cambridge University Press.

Firth, D. (1993). Bias Reduction of Maximum Likelihood Estimates. *Biometrika* 80: 27–38.

Fisher, R.A. (1922). On the Mathematical Foundations of Theoretical Statistics. *Phil. Trans. Royal Soc. A* 222: 326.

Fisher, R.A. (1935). The case of zero survivors. *Annals of Applied Biology* 22: 164–165.

Fisher, R.A. (1936). The use of multiple measurement in taxonomic problems. *Annals of Eugenics* 7: 179–188.

Fisher, R.A. (1937). The logic of inductive inference (with discussion). *Journal of the Royal Statistical Society* 98: 39–54.

Fleiss, J.L. (1981). *Statistical Methods for Rates and Proportions,* 2nd edition. New York: John Wiley & Sons.

Fleiss, J.L., B. Levin, and M.C. Park (2003). *Statistical Methods for Rates and Proportions,* 3rd edition. New York: John Wiley & Sons.

Forster, J.J., J.W. McDonald, and P.W.F. Smith (1996). Monte Carlo exact conditional tests for first-time urinary tract infection. *Journal of the Royal Statistical Society B* 58: 445–453.

Forthofer, D. and R.G. Lehnen (1981). *Public Program Analysis. A New Categorical Data Approach.* Belmont, CA: Lifetime Learning Publications.

Francis, B., M. Green, and C. Payne (ed.) (1993). *The GLIM System, Release 4 Manual.* Oxford, U.K.: Oxford University Press.

Gardiner, J.C., Z. Luo and L.A. Roman (2008). Fixed effects, random effects, mixed effects and GEE: What are the differences? *Statistics in Medicine.*

Gelman, A. and J. Hill (2007). *Data Analysis Using Regression and Multilevel/Hierarchical Models.* Cambridge, U.K.: Cambridge University Press.

Gertig, D., D. Hunter, and D. Spiegelman (2000). Personal communication to Cytel Software.

Glantz, S.A. (2002). *Primer of Biostatistics,* 5th edition. New York: McGraw Hill.

Glonek, G.J.N. and P. McCullagh (1995). Multivariate logistic models. *Journal of Royals Statistical Society B* 57: 533–546.

Goodman, L.A. (1978). *Analyzing Qualitative/Categorical Data.* Reading, PA: Addison-Wesley.

Gould, W. (2000). Interpreting logistic regression in all its forms. *Stata Technical Bulletin* 53: 19–29.

Graubard, B.I., E.L. Korn, and D. Midthune (1997). Testing goodness-of-fit for logistic regression with survey data. In *Proceedings of the Section on Survey Research Methods, Joint Statistical Meetings.* Alexandria, VA: American Statistical Association, pp. 170–174.

Greene, W.H. (2007) *LIMDEP version 9.0 Econometric Modeling Guide, Vol. 1.* Plainview, NY: Econometric Software, Inc.

Hannan, E.J. and B.G. Quinn (1979). The determination of the order of an autoregression. *Journal of the Royal Statistical Society B* 41: 190–195.

Hardin, J.W. and J.M. Hilbe (2003). *Generalized Estimating Equations*. London: Chapman & Hall/CRC Press.

Hardin, J.W. and J.M. Hilbe (2007). *Generalized Linear Models and Extensions*, 2nd edition. College Station, TX: Stata Press.

Hardin, J.W. and J.M. Hilbe (2008). Generalized estimating equations. In D'Agostino, R., L. Sullivan, and J. Massaro (eds.), *Encyclopedia of Clinical Trials*. New York: John Wiley & Sons.

Hastie, T.J. and R.J. Tibshirani (1990). *Generalized Additive Models*. London: Chapman & Hall.

Hauser, R.M. and M. Andrews (2006). Another look at the stratification of educational transitions: The logistic response model with partial proportionality constraints. *Sociological Methodology* 36 (1): 1–26.

Hausman, J.A. (1978). Specification tests in econometrics. *Econometrica* 46: 1251–1271.

Hausman, J.A. and D. McFadden (1984). Specification tests for the multinomial logit model. *Econometrica* 52: 1219–1240.

Heinze, G. (2006). A comparative investigation of methods for logistic regression with separated or nearly separated data. *Statistics in Medicine* 25: 4216–4226.

Heinze, G. and M. Schemper (2002). A solution to the problem of separation in logistic regression. *Statistics in Medicine* 21: 2409–2419.

Hilbe, J.M. (1991). Additional logistic regression extensions. *Stata Technical Bulletin* 1: 21–23. (May 1991).

Hilbe, J.M. (1991). An enhanced Stata logistic regression program. *Stata Technical Bulletin* 4: 16–18. (Nov 1991).

Hilbe, J.M. (1992). Regression based dichotomous discriminant analysis. *Stata Technical Bulletin* 5: 13–17 (Jan 1992).

Hilbe, J.M. (1992). Smoothed partial residual plots for logistic regression. *Stata Technical Bulletin* 10: 27 (Nov 1992).

Hilbe, J.M. (1993). Generalized linear models. *Stata Technical Bulletin* 11: 20–28 (Jan 1993).

Hilbe, J.M. (1993). Generalized additive models. *The American Statistician* 47: 59–64.

Hilbe, J.M. (1994). Generalized linear models. *The American Statistician* 48: 255–265.

Hilbe, J.M. (1997). Logistic regression: Standardized coefficients and partial correlations. *Stata Technical Bulletin* 35: 21–22.

Hilbe, J.M. (1998). Correction to random number generators. *Stata Technical Bulletin* 41: 23.

Hilbe, J.M. (2007a). *Negative Binomial Regression*. Cambridge: Cambridge University Press. (2008, 2nd printing with corrections and ammendments).

Hilbe, J.M. (2007b). The co-evolution of statistics and Hz. In Sawilowsky, S. (ed.), *Real Data Analysis*. Charlotte, NC: Information Age Publishing.

Hilbe, J.M. and J.W. Hardin (2008). Generalized estimating equations for longitudinal panel analysis. In Menard, S. (ed.). *Handbook of Longitudinal Research: Design, Measurement, and Analysis*. Burlington, MA: Academic Press/Elsevier.

Hilbe, J.M. and W. Linde-Zwirble (1995). Random number generators. *Stata Technical Bulletin* 28: 20–22.

Hilbe, J.M. and W.H. Greene (2008). Count response regression models. In Rao, C.R., J.P. Miller, and D.C. Rao (eds.). *Handbook of Statistics 27: Epidemiology and Medical Statistics*. Amsterdam: Elsevier, pp. 210–252.

Hilbe, J.M. and J. Shults (in publication). *Quasi-Least Squares Regression*. London: Chapman & Hall/CRC.

Hin, L.-Y., V. Carey, and Y.-G. Wang (2007). Criteria for working-correlation-structure selection in GEE: Assessment via simulation. *The American Statistician* 61: 360–364.

Hin, L.-Y. and Y.-G. Wang (2008). Working-correlation-structure identification in generalized estimating equations. *Statistics in Medicine*.

Hirji, K.F., A.A.Tsiatis, and C.R. Mehta (1989). Median unbiased estimation for binary data. *The American Statistician* 43: 7–11.

Hodges, J.S. and D.J. Sargent (2001). Counting degrees of freedom in hierarchical and other richly parameterized models. *Biometrika* 88: 367–379.

Hoffmann, J.P. (2004). *Generalized Linear Models: An Applied Approach*. Boston: Allyn & Bacon.

Holtbrugge, W. and M.A. Schumacher (1991). Comparison of regression models for the analysis of ordered categorical data. *Applied Statistics* 40: 249–59.

Hosmer, D.W. and S. Lemeshow (2000). *Applied Logistic Regression*, 2nd edition. New York: John Wiley & Sons.

Johnson,V.E. and J.H. Albert (1999). *Ordinal Data Modeling*. New York: Springer-Verlag.

Kahn, H.A. and C.T. Sempos (1989). *Statistical Methods in Epidemiology*. Oxford, U.K.: Oxford University Press.

Katz, E. (2001). Bias in conditional and unconditional fixed effects logit estimation. *Political Analysis* 9: 379–384.

Keele, L.J. and D.K. Park (2005). "Ambivalent About Ambivalence: A Re-Examination of Heteroskedastic Probit Models." Ohio State University, http://www.polisci.ohio-state.edu/faculty/lkeele/hetprob.pdf.

Kim, H., J.M. Hilbe, and J. Shults (2008). On the designation of the patterned associations for longitudinal Bernoulli data: Weight matrix versus true correlation structure. Department of Biostatistics Working Paper 26, University of Pennsylvania.

Kim, H., J. Shults and J.M. Hilbe (2009). "Quasi-Least Squares for Modeling Longitudinal Binary Data" (working paper, Department of Biostatistics, University of Pennsylvania.)

Kim, H. and J. Shults (2008). %QLS SAS Macro: A SAS macro for analysis of longidudinal data using quasi-least squares. *Journal of Statistical Software* (under review).

King, G. and L. Zeng (2002). Estimating risk and rate levels, ratios and differences in case-control studies. *Statistics in Medicine* 21: 1409–1427.

Kirkwood, B.R. and Sterne, J.A.C. (2005). *Essential Medical Statistics*, 2nd edition. Malden, MA: Blackwell Science.

Kleinbaum, D.G. (1994). *Logistic Regression: A Self-Teaching Guide*. New York: Springer-Verlag.

Kropko, J. (2008). Choosing between multinomial logit and multinomial probit models for analysis of unordered choice data. *Presented at the MPSA Annual National Conference*, April 3, 2008. http://www.allacademic.com/meta/p265696_index.html.

Lachin, J.M. (2000). *Biostatistical Methods: The Assessment of Relative Risks*. New York: John Wiley & Sons.

Lee, Y. and J. Nelder (2001). Hierarchical generalized linear models: A synthesis of generalized linear models, random effect models and structured dispersions. *Biometrika* 88: 957–1006.

Lee, Y., J. Nelder, and Y. Pawitan (2006). *Generalized Linear Models with Random Effects: Unified Analysis via H-Likelihood*. London: Chapman & Hall/CRC Press.

Lian, Ie-Bin (2003). Reducing over-dispersion by generalized degree of freedom and propensity score. *Computational Statistics & Data Analysis* 43: 197–214.

Liang, K.Y. and S.L. Zeger (1986). Longitudinal data analysis using generalized linear models. *Biometrika* 73: 13–22.

Liao, J.G. and D. McGee (2003). Adjusted coefficients of determination of logistic regression. *The American Statistician* 57 (3): 161–165.

Liberman, M. (2007). Thou shalt not report odds ratios. http://itre.cis.upenn.edu/~myl/languagelog/archives/004767.html. Accessed July 30, 2007.

Lindsey, J.K. (1995). *Modelling Frequency and Count Data*. New York: Oxford University Press.

Lindsey, J.K. (1997). *Applying Generalized Linear Models*. New York: Springer-Verlag.

Lindsey, J.K. and P. Lambert (1998). On the appropriateness of marginal models for repeated measurements in clinical trials. *Statistics in Medicine* 17: 447–469.

Littell, R.C., G.A. Milliken, W.W. Stroup, R.D. Wolfinger, and O. Schabenberger (2006). *SAS for Mixed Models*, 2nd edition. Cary, NC: SAS Press.

Little, R.J. and D.B. Rubin (1987). *Statistical Analysis with Missing Data*. New York: John Wiley & Sons.

Long, J.S. (1997). *Regression Models for Categorical Dependent Variables*. Thousand Oak, CA: Sage.

Long, J.S. and J. Freese (2006). *Regression Models for Categorical Dependent Variables Using Stata*, 2nd edition. College Station, TX: Stata Press.

Lumley, T., R. Kronmal, and S. Ma (2006). Relative risk regression in medical research: Models, contrasts, estimators, and algorithms. Department of Biostatistics Working Paper 293, University of Washington.

Magder, L.S. and J.P. Hughes (1997). Logistic regression when the outcome is measured with uncertainty. *American Journal of Epidemiology* 146: 195–203.

Mahalanobis, P.C. (1936). On the generalized distance in statistics. *Proceedings of the National Institute of Science of India* 12: 45–55.

Mann, C.J. (2003). Observational research methods. Research design II: Cohort, cross sectional, and case control studies. *Emergency Medicine Journal* 20: 54–60.

Manor, O., S. Matthews, and C. Power (2000). Dichotomous or categorical response? Analyzing self-rated health and lifetime social class. *International Journal of Epidemiology* 29 (1): 149–157.

McCullagh, P. (1980). Regression models for ordinal data (with discussion). *Journal of the Royal Statistical Society B* 42: 109–142.

McCullagh, P. and J. Nelder (1989). *Generalized Linear Models*, 2nd edition. London: Chapman & Hall.

McFadden, D. (1974). Conditional logit analysis of qualitative choice behavior. In *Frontiers in Econometrics*. New York: Academic Press.

McFadden, D. (1974). Analysis of qualitative choice behavior. In Zaremba, P. (ed.), *Frontiers of Econometrics*. New York: Academic Press.

McKelvey, R.D. and W. Zavoina (1975). A statistical model for the analysis of ordinal level dependent variables. *Journal of Mathematical Sociology*. Vol. 4, 1: 103–120.

McNutt, L.-A., C. Wu, X. Xue, and J.P. Hafner (2003). Estimating the relative risk in cohort studies and clinical trials of common outcomes. *American Journal of Epidemiology* 157: 940–943.

Mehta, C.R. and N.R. Patel (1995). Exact logistic regression: Theory and examples. *Statistics in Medicine* 14: 2143–2160.

Mehta, C.R., N.R. Patel, and P. Senshaudhuri (2000). Efficient Monte Carlo methods for conditional logistic regression. *Journal of the American Statistical Association* 95 (449): 99–108.

Menard, S. (2004). Six approaches to calculating standardized logistic regression coefficients. *The American Statistician* 58: 218–223.

Mitchell, M.N. (2008). *A Visual Guide to Stata Graphics*. College Station, TX: Stata Press.

Mramor, D. and A. Valentincic (2003). Forecasting the liquidity of very small private companies. *Journal of Business Venturing* 18: 745–771.

Mundlak, Y. (1978). On the pooling of time series and cross section data. *Econometrica* 46: 69–85.

Murad, H., A. Fleischman, S. Sadetzki, O. Geyer, L.S. Freedman (2003). Small samples and ordered logistic regression. *The American Statistician* 573: 155–160.

Murrell, P. (2006). R Graphics. Boca Raton, FL: Chapman & Hall/CRC.

Nagler, J. (1994). Scobit: An alternative estimator to logit and probit. *American Journal of Political Science* 38: 230–255.

Natarajan, S., S.R. Lipsitz, and R. Gonin (2008). Variance estimation in complex survey sampling for generalized linear models. *Journal of the Royal Statistical Society C* 57 (1): 75–87.

Nelder, J.A. and R.W.M. Wedderburn (1972). Generalized linear models. *Journal of the Royal Statistical Society A* 135: 370–384.

Norton, E.C., H. Wang and C. Ai (2004). Computing interaction effects and standard errors in logit and probit models. *Stata Journal* 4 (2): 154–167.

Pan, W. (2001). Akaike's information criterion in generalized estimating equations. *Biometrics* 57: 120–125.

Peterson, M.R. and J.A. Deddens (2008). A comparison of two methods for estimating prevalence ratios. *BMC Medical Research Methodology 2008* 8:9.

Peterson, B.L. and F.E. Harrell (1990). Partial proportional odds models for ordinal response variables. *Applied Statistics* 39: 205–17.

Pregibon, D. (1980). Goodness of link tests for generalized linear models. *Applied Statistics* 29: 15–24.

Pregibon, D. (1981). Logistic regression diagnostics. *Annals of Statistics* 9: 705–724.

Prentice, R.L. (1988). Correlated binary regression with covariates specific to each binary observation. *Biometrics* 44: 1033–1048.

Punj, B. and R. Staelin (1978). The choice process for business graduate schools. *Journal of Marketing Research* 15: 588–598.

Qu, A., B. Lindsay, and B. Li (2000). Improving generalized estimating equations using quadratic inference function. *Biometrika* 87: 823–836.

Qu, A. and P. Song (2004). Assessing robustness of generalized estimating equations and quadratic inference functions. *Biometrika* 91: 447–459.

Rabe-Hesketh, S. and B. Everitt (2007). *A Handbook of Statistical Analysis Using Stata*, 4th edition. Boca Raton, FL: Chapman & Hall/CRC Press.

Rabe-Hesketh, S. and A. Skrondal (2008). *Multilevel and Longitudinal Modeling Using Stata*, 2nd edition. College Station, TX: Stata Press.

Raftery, A. (1986). Bayesian model selection in social research. In Harden, P.V. (ed.), *Sociological Methodology* 25: 111–163, Oxford, U.K.: Basil Blackwell.

Ratcliffe, S. and J. Shults (2008). GEEQBOX: A MATLAB toolbox for implementation of quasi-least squares and generalized estimating equations. *Journal of Statistical Software* 25 (14): 1–14.

Rochon, J. (1998). Application of GEE procedures for sample size calculations in repeated measures experiments. *Statistics in Medicine* 17: 1643–1658.

Rotnitzky, A. and N.P. Jewell (1990). Hypothesis testing of regression parameters in semiparametric generalized linear models for cluster correlated data. *Biometrika* 77: 485–497.

Royston, P. and D.G. Altman (1994). Regression using fractional polynomials of continuous covariates: parsimonious parametric modeling. *Applied Statistics* 43: 429–467.

Royston, P. (2004). Multiple imputation of missing values. *Stata Journal* 4 (3): 227–241.

Shults, J. and N.R. Chaganty (1998). Analysis of serially correlated data using quasi-least squares. *Biometrics* 54 (4): 1622–1630.

Shults, J., S. Ratcliffe, M. Leonard (2007). Improved generalized estimating equation analysis via xtqls for quasi-least squares in Stata. *Stata Journal* 7 (2): 147–166. (http://www.cceb.upenn.edu/~sratclif/QLSproject.html)

Shults, J., W. Sun, X. Tu, and J. Amsterdam (2006). On the violation of bounds for the correlation in generalized estimating equation analysis of binary data from longitudinal trials. Division of Biostatistics Working Paper 8, University of Pennsylvania (http://biostats.bepress.com/upennbiostat/papers/art8/).

Shults, J., W. Sun, X. Tu, H. Kim, J. Amsterdam, J.M. Hilbe, and T. Ten-Have. A comparison of several approaches for choosing between working correlation structures in generalized estimating equation analysis of longitudinal binary data. *Statistics in Medicine*. Forthcoming.

Small, K.A. and C. Hsiao (1985). Multinomial logit specification tests. *International Economic Review* 26 (3): 619–627.

Song, P. and Z. Jiang (2008). SAS macro QIF manual 2007: Version 0.2. http: ==www:math:uwaterloo:ca= song=QIFmanual:pdf.

Spitznagel, E.L., Jr. (2008). "Logistic regression." In Rao, C.R., J.P. Miller, and D.C. Rao (eds.), *Handbook of Statistics 27: Epidemiology and Medical Statistics*. Amsterdam: Elsevier.

Stone, C. (1985). Additive regression and other nonparametric models. *Annals of Statistics* 13: 689–705.

Swartz, J. (1978). Estimating the dimension of a model. *Annals of Statistics* 6: 461–464.

Thara, R., M. Henrietta, A. Joseph, S. Rajkumar, and W. Eaton (1994). Ten year course of schizophrenia—The Madras Longitudinal study. *Acta Psychiatrica Scandinavica* 90: 329–336.

Theil, H. (1969). A multinomial extension of the linear logit model. *International Economic Review* 10: 251–259.

Truett, J., J. Cornfield, and W. Kannel (1967). A multivariate analysis of the risk of coronary heart disease in Framingham. *Journal of Chronic Diseases* 20: 511–524.

Tukey, J.W. (1949). One degree of freedom for nonadditivity. *Biometrics* 5: 232–242.

Twist, J.W.R. (2003). *Applied Longitudinal Data Analysis for Epidemiology.* Cambridge, U.K.: Cambridge University Press.

Twist, J.W.R. (2008). Causal inference in longitudinal experimental research. In Menard, S. (ed.). *Handbook of Longitudinal Research: Design, Measurement, and Analysis.* Burlington, MA: Academic Press/Elsevier.

Vaida, F. and S. Blanchard (2005). Conditional Akaike information for mixed-effects models. *Biometrika* 92 (2): 351–370.

van Buren, S., H.C. Boshuizen, and D.L. Knook (1999). Multiple imputation of missing blood pressure covariates in survival analysis. *Statistics in Medicine* 18: 681–694. Also see http://www.multiple-imputation.com.

Van der Laan, M.J. (2008). Estimation based on case-control designs with known incidence probability. Division of Biostatistics Working Paper 234, University of California at Berkeley.

Vittinghoff, E., D.V. Glidden, S.C. Shiboski, and C.E. McCulloch (2005). *Regression Methods in Biostatistics: Linear, Logistic, Survival, and Repeated Measures Models.* New York: Springer-Verlag.

Vuong, Q.H. (1989). Likelihood ratio tests for model selection and non-nested hypotheses. *Econometrica* 57: 307–333.

Walker, S.H. and D.B. Duncan (1967). Estimation of the probability of an event as a function of several independent variables. *Biometrika* 54: 167–179.

Wang, Z. (2000a). Sequential and drop one term likelihood-ratio tests. *Stata Technical Bulletin* 54: 46–47.

Wang, Z. (2000b). Model selection using the Akaike information criterion. *Stata Technical Bulletin* 54: 47–49.

Ware, J.H., D.W. Dockery, F.E. Speizer, and B.J. Ferris (1984). Passive smoking, gas cooking and respiratory health of children living in six cities. *American Review of Respiratory Disease* 129: 366–374.

Wedderburn, R.W.M. (1974). Quasi-likelihood functions, generalized linear models and the Gauss–Newton method. *Biometrika* 61: 439–447.

Weinberg, C.R. and S. Wacholder (1993). Prospective analysis of case-control data under general multiplicative-intercept risk models. *Biometrika* 80 (2): 461–465.

Williams, D.A. (1982). Extra-binomial variation in logistic linear models. *Applied Statistics* 31: 144–148.

Williams, R.L. (2000). A note on robust variance estimation for cluster-correlated data. *Biometrics* 56: 645–646.

Williams, R.W. (2006). gologit2: Generalized ordered logit/partial proportional odds models for ordinal dependent variables. *Stata Journal* 6: 58–82.

Williams, R. (2008). Estimating heterogeneous choice models with Stata. Department of Sociology working paper, University of Notre Name (July 18, 2008).

Winter, Nick (1999). "SMHSIAO: Small–Hsiao Test for IIA in Multinomial Logit," Boston College Department of Economics, http://ideas.repec.org/c/boc/bocode/s410701.html.

Wolfe, R. (1998). sg86: Continuation-ratio models for ordinal response data. *Stata Technical Bulletin* 44: 18–21. (In *Stata Technical Bulletin Reprints*, Vol. 7, 199–204. College Station, TX: Stata Press).

Wolfe, R. and W. Gould (1998). An approximate likelihood ratio test for ordinal response models. *Stata Technical Bulletin* 42: 24–27.

Wolkewitz, M., T. Bruckner, and M. Schumacher (2007). Accurate variance estimation for prevalence ratios. In *Methods of Information in Medicine* 46: 567–571.

Wood, S. (2000). Modeling and smoothing parameter estimation with multiple quadratic penalties. *Journal of the Royal Statistical Society B* 62: 413–428.

Wood, S. (2006). *Generalized Additive Models: An Introduction with R*. Boca Raton, FL: Chapman & Hall/CRC Press.

Woodward, M. (2005). *Epidemiology: Study Design and Data Analysis*. Boca Raton, FL: Chapman & Hall/CRC Press.

Woolf, B. (1955). On estimating the relation between blood group and disease. *Annuals Human Genetics* 19: 251–253.

Xie, J. and J. Shults (2008). Implementation of quasi-least squares with the R package qlspack. *Journal of Statistical Software*.

Yu, B. and Z. Wang (2008). Estimating relative risks for common outcome using PROC NLP. *Computer Methods and Programs in Biomedicine* 90: 179–186.

Zhang, J. and K. Yu (1998). What's the relative risk? *Journal of the American Medical Association* 280 (19): 1690–1691.

Zou, G. (2004). A modified Poisson regression approach to prospective studies with binary data. *American Journal of Epidemiology* 159: 702–706.

Author Index

Subject Index

A

abic, 142, 264–265, 415
Adaptive Gauss–Hermite quadrature
       method, likelihood function
       of, 599
Adjacent category logistic model, 427–429
aic, 415
Akaike information criterion (AIC)
       statistics, 248, 259, 266, 323–324,
       327, 460
  alternate, 260–262
  and BIC statistics, 311
  finite sample, 262–263
  LIMDEP, 8, 263
  nested-model, 262
  for panel data, 504–505
  and Pearson dispersion statistics, 308
  Swartz, 263
Alternating logistic regression (ALR),
       466–470, 516
Anscombe residuals, 279–280
Apparent overdispersion, 321–334
  missing predictor, 323–324
  simulated model setup, 322–323
asprvalue, 495, 602
Attributable risk
  confidence intervals of, 130
  formula for, 129
  lung cancer example, 129–131
Autoregressive correlation structure,
       451–453

B

Bartlett's test, 535
Bayesian information criterion (BIC)
       statistics, 248, 259, 263–266

and AIC statistics, 311
  LIMDEP, 312
Ben-Akiva and Lerman adjusted
       likelihood-ratio index, 245
Bernoulli algorithm, 297, 300
Bernoulli distributions, 2, 297
Bernoulli links, 70
Bernoulli (logistic) deviance function,
       246
Bernoulli models, 68–70
  IRLS algorithm for, 58
  log-likelihood function of, 498
Bernoulli PDF, 63, 245
Beta-binomial, derivation of, 597
BIC statistics, *see* Bayesian information
       criterion statistics
Binary logistic model
  IRLS algorithm for, 58
  modeling an uncertain response,
       158–161
    R code, 185–186
    sensitivity, 158, 160
    specificity, 158, 160
  overdispersion, 338–341
  QIC/QICu, 464–466
  robust score equations for, 137
Binary overdispersion, 338–340
Binary response model, 371
Binary × Binary interactions, 191–201, 230
  graphical representation, 198
  interpretations for, 194
  odds ratios, 194
  R code, 236–237
  standard errors and confidence
       intervals, 197
Binary × Categorical interactions, 201,
       237–238

Printed in the United States
by Baker & Taylor Publisher Services